COMPUTER-ASSISTED STRUCTURE ELUCIDATION

COMPUTER-ASSISTED STRUCTURE ELUCIDATION

NEIL A. B. GRAY
Department of Computing Science
University of Wollongong
Wollongong, New South Wales, Australia

A WILEY-INTERSCIENCE PUBLICATION
JOHN WILEY & SONS
New York / Chichester / Brisbane / Toronto / Singapore

Library of Congress Cataloguing in Publication Data:

Gray, Neil A. B.
Computer-assisted structure elucidation.

"A Wiley-Interscience publication."
1. Chemical structure—Data processing. I. Title.
QD471.G755 1986 541.2′2′0285 85-29562

ISBN 0-471-89824-4

Printed in the United States of America

10 9 8 7 6 5 4 3 2 1

PREFACE

Twenty years have elapsed since the publication of the first reports of computer programs for representation, manipulation, and interpretation of chemical structural and spectral data. The first researchers in this field had quite diverse objectives. Some saw the computer as an automated compendium of reference data. Their research efforts have lead to (1) file-search methods for the retrieval of reference compounds that incorporate specified substructures and (2) systems that can be used to identify references whose spectra match with observed data. Problems associated with the interpretation of chemical spectral data have served as test cases for those interested in pattern recognition algorithms and those who have sought to develop knowledge-based artificial intelligence systems. A few far-sighted synthetic chemists perceived that computers might eventually be of value in the development of synthesis plans and thus have proceeded to develop computer-manipulable representations of chemical structures and reactions. Major programs for computer-assisted structure elucidation have been developed by different groups whose original individual interests were (1) spectrum interpretation, (2) practical structural chemistry, and (3) the problems of defining the scope of chemical isomerism and stereoisomerism.

The diverse origins of work on computer aids for structural chemists have had some adverse consequences. Research publications are scattered in everything from journals on discrete mathematics and computer science through to journals that would more typically contain reports of newly identified natural compounds. Many basic algorithms have been rediscovered and republished on numerous occasions. The style of papers expected in chemical journals is not well suited to the presentation of algorithms. Researchers have made various compromises when presenting their work and, sometimes, essential algorithmic details have been omitted from the published reports. The papers that present the most reliable algorithms require considerable mathematical sophistication on the part of the reader. The majority of papers lack the examples and discussions that are needed to explain the real purpose of the reported algorithms. Inevitably, this results in previous work being overlooked and, in consequence, repeated. For example, the relevance of paper on "graph labeling" would not be apparent to a student commencing a project on, say, the use of symmetry in the planning of reaction sequences; instead of exploiting existing proven algorithms, it is probable that such a student would invent another *ad hoc* method for recognition, representation, and exploitation of symmetries.

This book was motivated by the perceived need for some comprehensive review of all related work in computer-assisted structural chemistry. The initial outline for this text was composed by several members of the DENDRAL project at Stanford University. The DENDRAL project applied methods of artificial intelligence to problems of structure elucidation in organic chemistry. Our interests in structure elucidation provided a focus for the book. However, we intended for it to be useful to a wider audience and to illustrate how algorithms developed for problems in structure elucidation might also be applied to other areas of computer-assisted chemistry.

In developing this book, I have attempted to review and interrelate the various aspects of work on computer-assisted structural chemistry. My treatment of algorithms is informal. I use examples to introduce particular problems related to the representation and manipulation of chemical data. I have illustrated various approaches to the solution of such problems. My aim has been to provide an introduction to the graph-theoretic and other algorithms that are used. The introductory explanations given in this book should provide the reader with the background necessary for appreciation of those formal mathematical presentations that exist in the literature for the more important algorithms. Hopefully, future work will be based on the best of the existing algorithms and will focus on new developments.

The book is organized into 12 chapters; the first is simply a short introduction. Chapter II consists of a long worked example that illustrates the use of some of the programs that have been devised to aid the structural chemist. Chapter III provides an historical overview of the development of these and related programs. The remainder of the book is divided into three sections. Chapters IV, V, and VI present various approaches for computer-aided recognition and interpretation of spectral data. Chapters VII–X are concerned with methods for representing and manipulating structural information. The final part, Chapters XI and XII, reviews computer-aided methods for the prediction of spectral properties and the analysis of chemical transformations. An appendix lists some of the key papers published in each of the main areas of work in computer-assisted structural chemistry.

Chapter IV reviews file-search systems for spectrum matching. Programs for spectral identification based on various file-searching schemes are the most commonplace of the computer-assisted methods for structure elucidation. Effective systems have been in routine use for some 15 years. Yet, even here there remain some opportunities for further developments.

The review of pattern recognition approaches, given in Chapter V, is brief. These approaches are already covered in several other texts. Moreover, for reasons that are discussed in Chapter V, the pattern recognition methods have not proved to be of great utility in systems for automated structure elucidation.

Artificial intelligence methods for interpretation of spectra are covered in Chapter VI. These methods are based on approaches similar to those that a chemist might use for the interpretation of data. Several of these methods depend on substructure–subspectrum correlation tables. Frequently, the spectrum interpretation process requires comparisons of alternative substructures and the assembly of combinations of substructures. Thus a review of these interpretation methods leads to the problems of representing and manipulating structural information.

Methods for representing structure, including configurational stereochemical details, are discussed in Chapter VII. The existence of computer-manipulable structure representations makes it possible to create structure-oriented data retrieval systems. There are now search systems for retrieval of reference structures that incorporate specified substructures. Such substructure search methods are described in Chapter VIII along with details of the graph-matching algorithms that are used to establish whether a given structure incorporates a particular substructure. This chapter also introduces "constructive substructure search," a novel extension to standard substructure search that serves as the basis for an approach to structure generation.

Chapter IX reviews the problems involved in (1) the derivation of a standard, canonical form for a structure, (2) the perception and use of symmetries present in a structure, and (3) the identification of ring systems within a structure. Such problems arise in work on data bases of chemical structures, synthesis planning, and structure elucidation. The same algorithmic solutions and the same data structures are appropriate in all these domains.

A complete computerized structure elucidation system consists of (1) some interpretive module that identifies substructural fragments contained in some unknown together with constraints on their bonding, (2) a structure generator that creates all candidate structures consistent with these constraints, and (3) some evaluation module for the further analysis of these candidates. The algorithms for candidate structure generation form the core of such a computerized system. These algorithms must be capable of giving the structural chemist a guarantee that every possible structure compatible with available chemical and spectral evidence has been considered. These structure generation algorithms are discussed in Chapter X. A number of apparently quite different approaches have been presented in the literature. Most are fairly closely related and involve similar graph-matching, symmetry analysis, and canonicalization steps. The most marked differences among the various systems relate to the use of symmetry. Different algorithms are optimal for different classes of structural problems.

Chapter XI deals with computer-assisted "evaluation" of structural candidates. Most of this chapter is devoted to a discussion of methods for the prediction of spectral properties and the use of such predictions in the determination of relative plausibilities of different candidates for some unknown structure. Other evaluation techniques considered include methods for analyzing sets of candidate structures to identify structural features that could be used to group candidates into different classes.

Computerized structure elucidation systems have been extended to incorporate programs that allow the modeling of chemical degradation reactions. Such extensions are discussed in the first part of Chapter XII, where the DENDRAL-REACT program is used to (1) illustrate methods for the computerized analysis of reactions and (2) introduce some of the simpler problems faced by a synthesis planning system. The second part of Chapter XII deals with (1) the problem of giving a program a sense of chemical strategy, (2) other major problems for synthesis planning programs, and (3) some approaches used in current programs.

Structural chemists have always been receptive to new instrumental aids. The various spectrometers used in their laboratories are extensively computerized. Originally, these computers served only to reduce the raw data; increasingly, these machines are used in file-searches for compound recognition and, sometimes, for running spectral interpretation and prediction programs. The microcomputers used in laboratory instruments are becoming increasingly powerful; many current microcomputers have the capacity to run programs such as GENOA, the structure elucidation program developed for the DENDRAL project. It is inevitable that structural chemists will seek to extend the applications of their computer systems to include more elaborate structure manipulation procedures. This book should be of value to all who wish to explore new opportunities for computer-assisted structure elucidation.

NEIL A. B. GRAY

Wollongong, New South Wales, Australia
April 1986

ACKNOWLEDGMENTS

The film "Romeo and Juliet" commences with a brief acknowledgment of the work of William Shakespeare, "without whom they would have been at a loss for words." I too must acknowledge the essential contributions of many others.

This book is a final product of the entire Dendral project group as it existed at Stanford in the 1970s and early 1980s. R. E. Carhart implemented most of the algorithms in the final versions of the Dendral programs. Carhart's algorithms for "constructive substructure search" are of particular interest. These algorithms form the basis of the simplest and most elegant scheme for generating those structures that satisfy given constraints. Carhart's development of these novel algorithms was made possible by his intimate knowledge of previously existing algorithms for graph-matching and the analysis of symmetry. Hopefully, by collecting together such algorithms, this book will provide the basic material from which others will be able to effect further developments comparable to those of Carhart. J. G. Nourse gave us the ability to handle stereochemical information; stereochemical details are essential, but surprisingly such details are not yet included routinely in all computer systems intended to aid structural chemists. C. W. Crandell, H. Egli, and M. Lindley helped develop programs for exploiting NMR data. A. Buchs, A. Lavanchy, and T. Varkony contributed to the systems for analyzing mass spectral data. T. Varkony contributed also to the programs that help solve structural problems through the use of results obtained in chemical reactions. C. Djerassi intervened occasionally to bring us back to chemical realities. B. G. Buchanan made intelligent contributions (both human and artificial).

Many others contributed to the Dendral project, beginning with its inception in the 1960s through 1977, the year I joined the group. These contributors include J. Lederberg, E. A. Feigenbaum, H. Brown, and L. Masinter. Their contributions to this book are indirect. They and their coworkers established the project, set directions, and provided a core of algorithms subsequently developed by Carhart and Nourse.

D. H. Smith deserves special acknowledgment for his contributions. Dr. Smith was an active member of the Dendral project from the time when work on Heuristic Dendral was just being completed, right through to the final work on configurational and conformational stereochemistry. Dr. Smith provided much of the continuity for the project through the many changes of research personnel. Working with Carhart and Nourse, he helped develop the first practical applications of all the exotic algorithms. Originally, this book was a joint effort (hence the use of "we" in some commentaries and discussions). Dr. Smith drafted the first versions of the chapters on historical perspectives, structure representations, and substructure search; he contributed to early drafts of many other sections. Unfortunately, geographical distance prevented effective cooperation after we had both departed from Stanford. I have completed the book along the lines that we had jointly planned.

N.A.B.G.

CONTENTS

COMPUTER-ASSISTED STRUCTURE ELUCIDATION

I

INTRODUCTION

In this book we discuss several areas of research in structural chemistry where computational techniques already have made important contributions. We emphasize from the beginning that in essentially every area the influence of the computer has been as an *aid* to the chemist rather than a replacement for sound chemical reasoning. For this reason we stress the word *"assisted"* when we speak of computer-assisted structural analysis. Throughout the history of chemistry chemists have been able to solve complex research problems with the use of manual techniques, and the next few years will see little change in the requirement for human intervention at all phases of computer-assisted analysis.

Why, then, should chemists wish to use the computer at all? The answer is that when onerous tasks for which computers are ideally suited are given over to a computer, the chemist is free to pursue more creative tasks. This saves both time and money and makes the conduct of research a more pleasant and rewarding occupation. In the future, ways will be found to make computers perform more complex problem-solving tasks, and some approaches are discussed in subsequent chapters. In our opinion, however, the requirement for the creative thinking of chemists who are actually analyzing their own problems will remain. The most successful programs will be those able to take full advantage of knowledge possessed by the chemist about a given structural problem.

We approach the topics of this book with a definite paradigm in mind for problem-solving in structural chemistry. This paradigm includes the following steps:

1. Determining some of the structural characteristics of a newly isolated, unknown compound through analysis of its spectral and chemical properties.
2. Deriving all structural candidates consistent with the initially available data.
3. Analyzing further these candidates, possibly in the context of additional chemical and spectral data, to determine which candidate represents the correct structure of the unknown.
4. Planning a synthesis to confirm the structure and derive related compounds.
5. Investigating a series of structure–activity relationships for the set of related compounds.

Now, obviously not any one chemist nor any one computer program is capable of carrying out more than one or two of these steps. Most chemists, however, are well aware of these basic steps and perform research in one area with complete cognizance of the relation their own results have to those of their colleagues who are specialists in one of the other areas. The preceding sequence of steps also represents a convenient framework around which to organize this book because it does represent a logical progression in the study of many molecular structures, particularly those that display some unique or important properties, such as biological activity.

Topics covered in subsequent chapters include computer aids for helping the chemist to infer structural information from spectral data, algorithms for assembling inferred substructures into possible candidate structures, and methods for evaluating how compatible the various candidate structures are with observed spectral data. The evaluation of candidate structures further includes consideration of the effects of chemical transformations. This consideration of computer modeling of chemical transforms leads to the related but more general field of computer-aided planning of syntheses. The treatment of computer aids for planning syntheses (and of programs for correlating structures with biological activity) is brief, for such work really lies outside the scope of this book.

Before beginning the presentation of computer-assisted techniques, we provide an example of the kinds of structural problems that can be solved with the aid of some *existing* programs. This problem is solved retrospectively in that there already exist in the literature several descriptions of the elucidation of structure, synthetic methods, and structure–activity studies. Thus, this example serves primarily to introduce the reader to the kinds of programs available and adheres to the theme of illustrating how computers will be used solve related problems in the future.

II

AN EXAMPLE: STRUCTURAL ANALYSIS OF WARBURGANAL

The following example illustrates how some current computer programs can be applied to aid in structure elucidation, synthesis planning, and studies on structure-property relationships. The example is the compound *warburganal*, a potent insect antifeedant originally identified by Kubo et al.[1] The structure of warburganal was established by conventional analysis of its chemical and spectral properties. Several syntheses have subsequently been devised,[2–8] and the biological activities of warburganal and related compounds have been investigated.[9]

II.A. COMPUTER-ASSISTED STRUCTURE ELUCIDATION OF WARBURGANAL

There are several computer programs in existence that can aid chemists in determining the structure of an unknown compound, using data obtained from chemical and spectroscopic studies. These programs are discussed in detail in Chapter X. For this example we focus on just one of those programs, **GENOA**, developed recently by our group.[10] Typically, the process of structure elucidation involves three distinct phases:

1. *Data Interpretation*. The presence and absence of particular substructural fragments are inferred through analysis of spectral and chemical data that characterize the unknown.
2. *Structure Generation*. The substructures thus derived are pieced together to yield one or more complete structures that represent possible candidates for the unknown.
3. *Structure Evaluation*. The candidate structures are evaluated against existing data to eliminate some possibilities, and new experiments are planned to discriminate among remaining candidates.

These phases are discussed in detail in subsequent sections: phase 1 in Chapters IV, V and VI; phase 2 in Chapters VIII and X; and phase 3 in Chapter XI. Much of phases 1–3, particularly the spectral interpretation in phase 1 and structure evaluation in phase 3, is intuitive. A chemist exploits analogies to other, previously identified compounds, intelligent guesswork, and general chemical experience. In contrast, the process of assembling the fragments into valid structures (phase 2) is essentially combinatorial. Computer programs can assist in all three phases of the structure elucidation process. The area in which programs currently excel is the process of combinatorial structure assembly, or structure generation (phase 2). This area is least intuitive and most difficult for the chemist because there are no simple techniques or procedures to follow to guarantee that *all* possible candidates have been written down. Structure generating programs such as **GENOA** and its predecessor **CONGEN**[11] provide this guarantee.

Such programs are more correctly called *isomer* generating programs because they generate, in the computer, *structural isomers* of a given molecular formula. Structure generators are often interfaced to other programs that perform automated spectral interpretation and structure evaluation. Such combined programs are discussed in subsequent sections. For the purposes of this example, however, we illustrate only computer programs that assist in phase 2, structure generation (Section II.A.1) and phase 3, structure evaluation (Section II.A.2).

II.A.1. Structure Generation: Obtaining Structural Candidates

In this section we use **GENOA** as a prototypical structure generator, a tool that assists a chemist in establishing the scope of the structural problem and can explore the implications of various structural hypotheses. Programs like **GENOA** are typically interactive. A chemist, working with the program, defines first the molecular formula of an unknown compound, and then, as they are identified, the substructural constraints that can be inferred from available data. We illustrate the processes of defining and using substructural constraints and obtaining a *complete* set of structural candidates that obey all constraints, for the structure of warburganal. For your amusement, and to put

TABLE II-1. IR Spectral Data and Related Substructural Constraints for Warburganal

Absorption[a]	IR Spectrum Substructural Constraint
3460	OH (intramolecular hydrogen bonded)
2850, 1722	CHO
1687, 1650	Enal

[a] In cm^{-1}.

you into the mode of thinking systematically about the structural inferences, as **GENOA** must do, you should get a pencil and piece of paper to see if you can verify the results of the program at each step of the analysis.

Warburganal was one of four components isolated, using several chromatographic techniques, from the antifeedant fraction obtained by extraction of the ground bark of *Warburgia ugandensis*.[1] The molecular formula, determined by mass spectrometry, is $C_{15}H_{22}O_3$. Key infrared (IR) spectral data and substructural constraints derived therefrom[1] are summarized in Table II-1.

The ultraviolet (UV) spectrum (λ_{max} 224, ϵ, 6300) and circular dichroism (CD) spectrum indicated the presence of an enal and overlapping n, π^* extrema of two aldehydes.

Additional spectral data for warburganal are given in:†

Figure II-1, the ^1H nuclear magnetic resonance (NMR) spectrum
Figure II-2, the ^{13}C NMR spectrum
Figure II-3, the low-resolution mass spectrum.‡

To arrive efficiently at a set of candidate structures, the **GENOA** program must be supplied with all known information about a structure. In principle, any piece of data that can be expressed as a substructural constraint can be used within the program as a substructure. The first and most important constraint is the molecular formula, and this is required before **GENOA** will begin to construct structures. Other constraints consist of a substructural feature together with some statement about the number of occurrences of that feature in the unknown molecule. The program possesses a library of common substructural fragments that can be utilized directly without requiring their continued redefinition. However, most structural problems involve specialized substructures that could not reasonably be provided in a general-purpose

†We express our thanks to Professor Koji Nakanishi for making available to us a sample of warburganal that was used to obtain these complete spectra.

‡The low-resolution spectrum was derived by plotting a high-resolution mass spectrum, obtained in our laboratory, at nominal mass resolution.

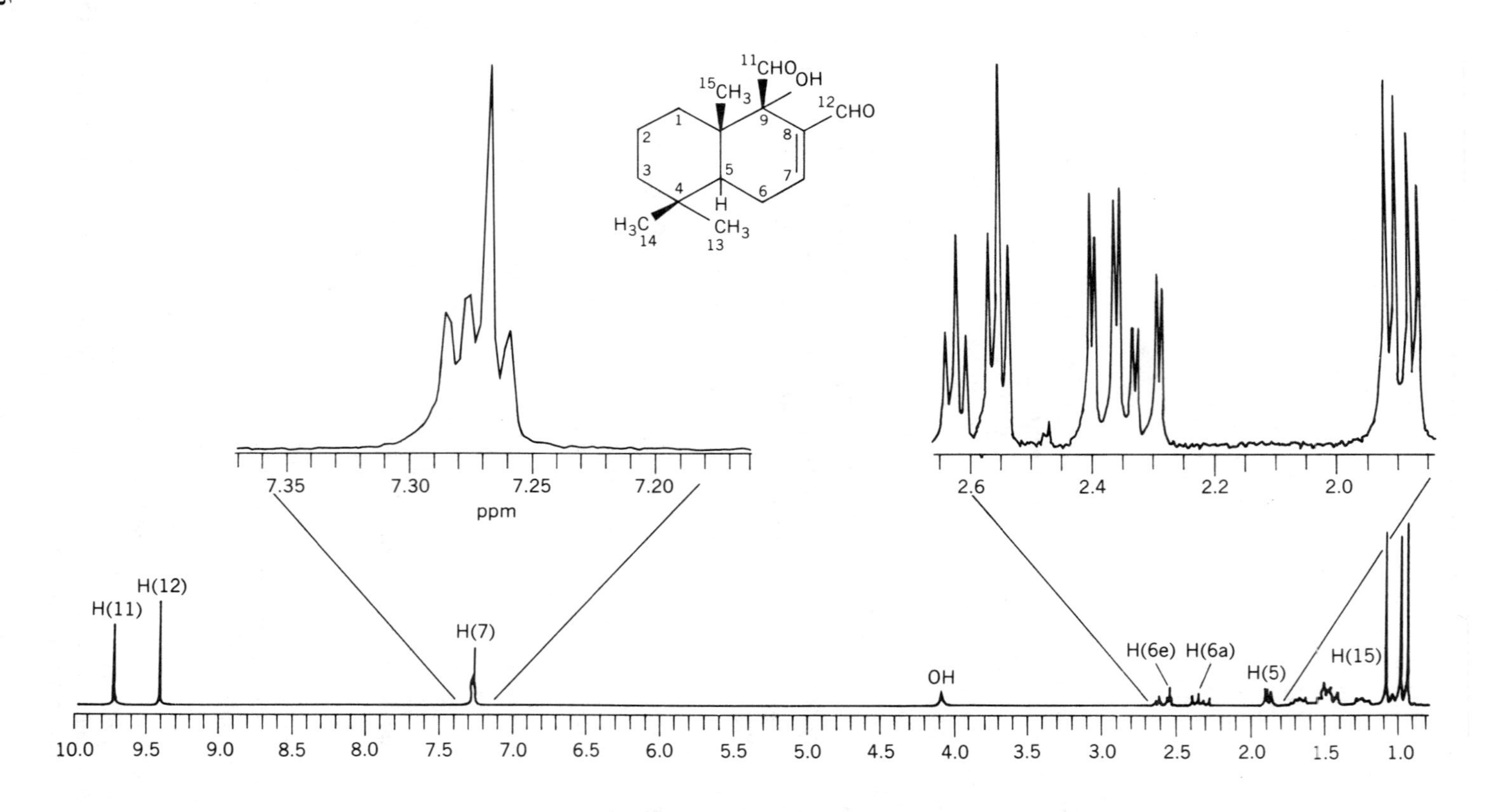

FIGURE II-1. The 360 MHz ^{1}H NMR spectrum of warburganal using TMS as internal standard.

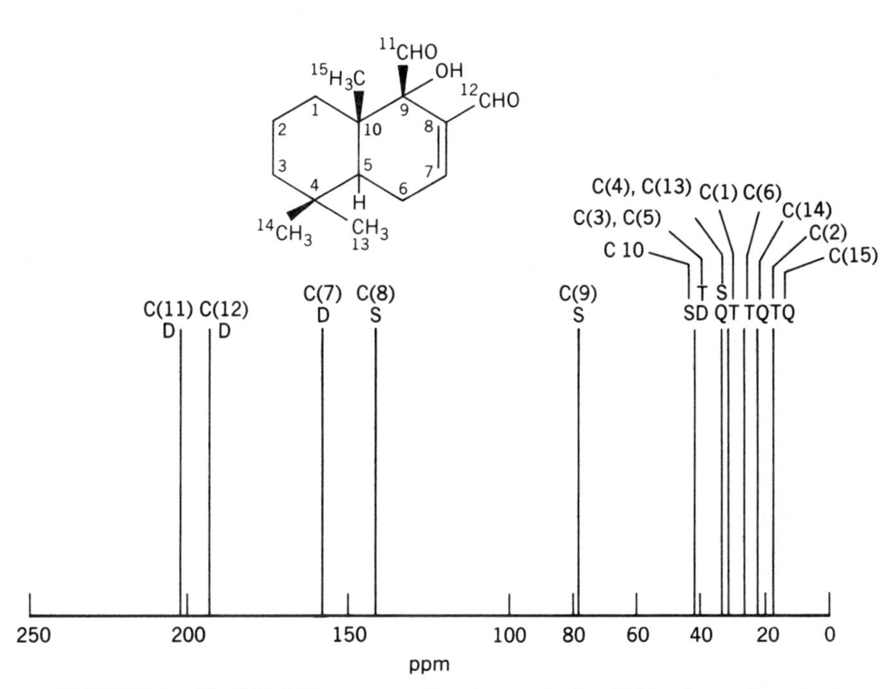

FIGURE II-2. The ^{13}C NMR spectrum of warburganal using TMS as internal standard.

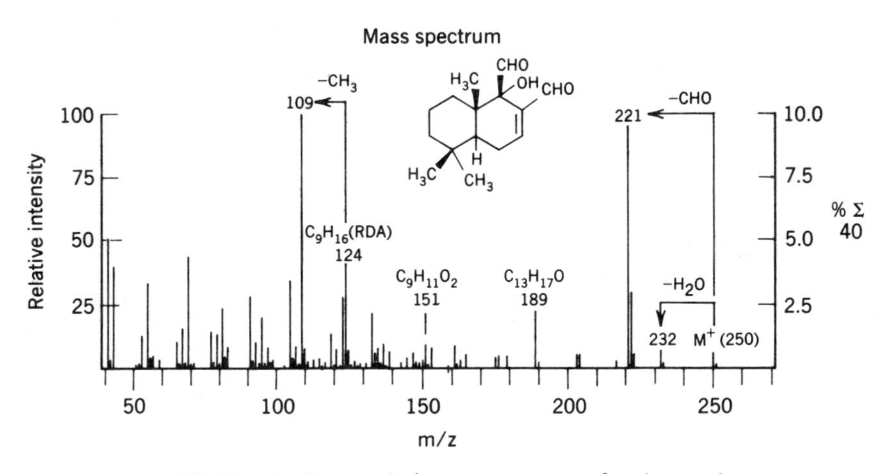

FIGURE II-3. Low resolution mass spectrum of warburganal.

library, for example, a specific chain of methylene and methine carbons derived through a proton decoupling experiment. Therefore, the GENOA program provides a simple mechanism for substructure definition that allows the chemist to build up any desired substructure through an appropriate sequence of commands for creating rings and chains of atoms of specified types.

In the subsequent discussion we define and use several substructural constraints with the use of **GENOA**. The order in which these constraints are used is at the discretion of the chemist who interacts with the program. Some ways may be more computationally efficient than others, but generally the differences are of little consequence. After each new constraint is applied, the results are immediately available to the chemist for examination. In this way the gradual growth of complete structures can be monitored. This ability to monitor progress sometimes makes it possible for additional constraints to be conceived; these additional constraints can be used to limit the problem further. Examples of this step-by-step examination of results, stimulation of new ideas, and use of new constraints are given below.

```
(1)      #DEFINE MOLFORM
         MOLECULAR FORMULA:C 15 H 22 O 3
         MOLECULAR FORMULA DEFINED

(2)      #DEFINE SUBSTRUCTURE C-OH
         (NEW SUBSTRUCTURE)
  (2a)   >CHAIN 2
  (2b)   >ATNAME 2 O
  (2c)   >HRANGE
         ATOM:1
         MINIMUM NUMBER OF H'S:0
         MAXIMUM NUMBER OF H'S:0
  (2d)   >HRANGE 2 1 1
  (2e)   >HYBRIDIZATION
         ATOM:1
         TYPE:SP3
  (2f)   >DONE
         C-OH DEFINED

(3)      #CONSTRAINT
         SUBSTRUCTURE NAME:C-OH
         RANGE OF OCCURRENCES:AT LEAST 1
         .
         1 CASE WAS OBTAINED
```

FIGURE II-4. Initial interaction with the **GENOA** program to define the molecular formula of warburganal and to define and use as a constraint a substructure representing a tertiary hydroxyl group.

We illustrate the initial steps of interaction with **GENOA** in Figure II-4, which is an annotated transcript of actual use of the program for this example.† At step 1 in Figure II-4 the molecular formula of warburganal, $C_{15}H_{22}O_3$,[1] verified by our measurements of the mass spectrum at high resolving power, is defined.

The presence of some form of hydroxyl group was obvious from the IR spectrum (Table II-1). Analysis of the ^{1}H NMR spectrum (Figure II-1) and ^{13}C NMR (Figure II-2) suggested that the compound must be a tertiary alcohol. Therefore, a named substructure representing a carbon bearing a hydroxyl group and no hydrogens must be defined to **GENOA**. Although this simple substructure is in fact in a substructural library, we define it in this example to provide a brief illustration of how all substructures, of arbitrary complexity, can be defined. The name assigned to the substructure is important only as a "tag" that can be used to refer to the substructure at other points in the program (e.g., in the ***CONSTRAINT*** command Figure II-4). Subsequently in this example we assume that all other substructures are named and defined to **GENOA** in like manner.

The substructure, arbitrarily named **C-OH**, is created by specifying first a chain of two atoms, by default both carbons, (step 2a, Figure II-4). Next we specify the name of atom 2 to be an oxygen (step 2b, Figure II-4). Persons experienced with **GENOA** and related interactive programs soon learn how to abbreviate and combine commands. Persons using the program only occasionally are assisted by the program prompting for missing information. This is illustrated at steps 2c and 2d in Figure II-4, where at 2c we step through the ***HRANGE*** command to fix the number of hydrogens on the carbon as zero, whereas at 2d we merely type all information about the single hydrogen atom on the oxygen atom on one line. Additional information on the properties of the carbon atom is required because we know that this carbon is sp^3 hybridized. The ***HYBRIDIZATION*** command, at step 2e, specifies this information, thereby avoiding the potential ambiguity of the requirement for a tertiary alcohol being accidentally satisfied by the presence of an enol system. Finally, the substructure definition is completed by issuing the ***DONE*** command at step 2f in Figure II-4.

The sequence of commands in step 2 in Figure II-4 merely defines to **GENOA** the substructural details of something called "**C-OH**"; we have not yet specified how many such substructures are required for this structural problem. The ***CONSTRAINT*** command is used to tell GENOA the number of occurrences of any defined substructure; a substructure may occur no times (i.e., it is forbidden) or it may a given number or range of times under the complete control of the chemist.[10] Each constraint causes **GENOA** to construct, in the computer, the given number of occurrences of the specified substructure. The starting point in each instance is what was obtained from the previous constraint. The construction procedure has been described previously[10] and is briefly reviewed in Chapter VIII. For the first constraint

†Throughout this book, whenever transcripts of program interaction are presented, we adopt the convention that input to the program, typed in by a chemist, is in boldface type.

(Figure II-4) the starting point is the molecular formula and the constructive procedure is trivial in that there is only one unique way to construct the ***C-OH*** group from the atoms in the molecular formula.

The results of incorporating a constraint are *cases* (Figure II-4). Cases are partial structures in which all previous constraints are incorporated. Complete structural isomers for a problem are never constructed until later in a problem when considerably more substructural constraints have been specified. In examples given below we illustrate cases obtained for warburganal after additional constraints have been supplied.

Before inferring and using new constraints we summarize the results obtained and refer back to these results in subsequent discussion. In Table II-2 we show the number of cases that result from incorporation of each named substructural constraint. The substructures themselves are given in Figure II-5.

The spectral data characterizing warburganal imply the presence of two aldehyde groups, one forming part of an enal system (Table II-1). These aldehyde groups are also evidenced by the resonances at 9.72 (aldehyde) and 9.41 ppm (enal) in the ^{1}H NMR spectrum (Figure II-1) and at about 201 and 192 ppm in the ^{13}C NMR spectrum (Figure II-2). The aldehyde can be assumed to be attached to a quaternary alkyl carbon because its only coupling in the proton spectrum is a long-range 1-Hz coupling to the hydroxyl proton. The first aldehyde functionality is expressed by defining **C-ALDEHYDE** (Figure II-5). Use of this substructure as a constraint yields two new cases (Table II-2) because there are two ways to incorporate this new fragment. The quaternary alkyl carbon attached to the aldehyde may or may not be the same as the carbon to which the hydroxyl group in substructure **C-OH** is bonded, resulting in partial structures **II-1** and **II-2**.

TABLE II-2. Number of Cases Obtained by GENOA on Incorporation of Each Substructural Constraint, and Final Number of Structures[a]

Substructure Name	Range of Occurrences	Resulting "Number of Cases"[b]
C-OH	At least 1	1
C-ALDEHYDE	At least 1	2
ENAL	At least 1	2
C-SP2	Exactly 4	2
C-CH3	At least 3	17
T-BUTYL	None	15
CH3	Exactly 3	14
CH3-C-OH	None	6
DECOUPLE	At least 1	9
CH2	Exactly 4	9
CHAIN3	At least 1	9
CYCP	None	9

[a]See Figure II-5 for the substructure corresponding to each name.
[b]Final number of structures = 42.

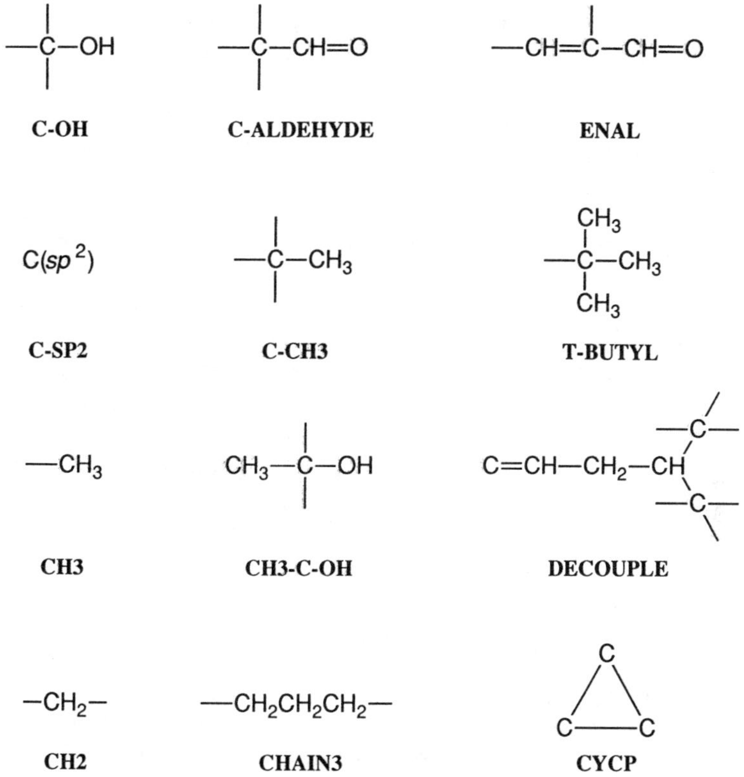

$$-\overset{|}{\underset{|}{C}}-OH$$ **C-OH**

$$-\overset{|}{\underset{|}{C}}-CH{=}O$$ **C-ALDEHYDE**

$$-CH{=}\overset{|}{C}-CH{=}O$$ **ENAL**

$C(sp^2)$ **C-SP2**

$$-\overset{|}{\underset{|}{C}}-CH_3$$ **C-CH3**

$$-\overset{CH_3}{\underset{CH_3}{C}}-CH_3$$ **T-BUTYL**

$-CH_3$ **CH3**

$$CH_3-\overset{|}{\underset{|}{C}}-OH$$ **CH3-C-OH**

$C{=}CH-CH_2-CH(-C\lt)_2$ **DECOUPLE**

$-CH_2-$ **CH2**

$-CH_2CH_2CH_2-$ **CHAIN3**

C–C–C (cyclopropane ring) **CYCP**

FIGURE II-5. Substructural constraints used in structure elucidation of warburganal (see Table II-2 for range of occurrence of each substructure).

The next few substructures defined and used as constraints included (see Table II-2 and Figure II-5):

1. The required enal system, **ENAL**.
2. A limit of exactly four sp^2 hybridized carbons, **C-SP2**, as established from the number of resonances in the 100–240-ppm region of the ^{13}C NMR spectrum (Figure II-2).
3. A requirement for at least three methyl groups bonded to quaternary alkyl carbons, **C-CH3**. These methyls correspond to the 3H singlets at 1.10, 1.00, and 0.96 ppm in the ^{1}H NMR spectrum (Figure II-1).

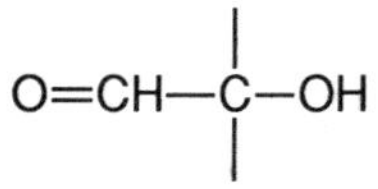

$$O{=}CH-\overset{|}{\underset{|}{C}}-OH$$

II-1

$$\gt C-CH{=}O \qquad \gt C-OH$$

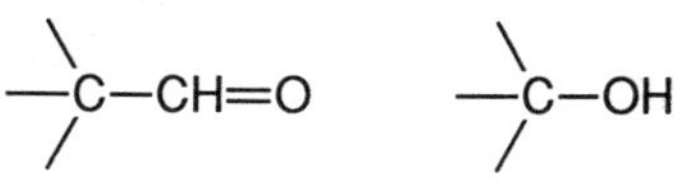

II-2

These constraints result in 17 possible cases. Although each case is still comprised of only several small, disconnected substructural fragments, it is instructive to examine several of them to illustrate the point mentioned earlier that such examination often reveals that additional constraints can be implemented. Four of the 17 cases are shown as structures **II-3-II-6**.

In this example, some of the partial structures proved to contain *t*-butyl groups (e.g., **II-5**). Such substructures do not seem compatible with the proton resonance data. The constraint **T-BUTYL** with range of occurrence "none" removes two cases containing a *t*-butyl group. Case **II-6** possesses six methyl groups. In **II-6**, three new quaternary centers were created to satisfy the requirement for three **C-CH3** substructures. The **GENOA** program has perceived that, for **II-6**, given the other constraints used, there are so many remaining hydrogens that *three additional methyl groups would have to be formed*. Constraining the problem so that exactly three methyls, **CH3**, are allowed eliminates **II-6**, the only case that possesses more than three, leaving 14 cases to this point (Table II-2).

A partially completed problem can be saved at any stage in the analysis in order to wait for additional data. Meanwhile, the analysis can proceed on the basis of plausible, but unproven, structural inferences. In this example the methyl resonance shifts in the proton spectrum suggest that it is unlikely that any methyl be beta to the hydroxy group. Methyl groups in environments such as $CH_3{-}C{-}OH$ typically have shifts of 1.3 ppm or greater. Also, addition of lanthanide shift reagent was not reported to result in shift of methyl resonances.[1] A constraint expressing these observations, **CH3–C–OH**, was

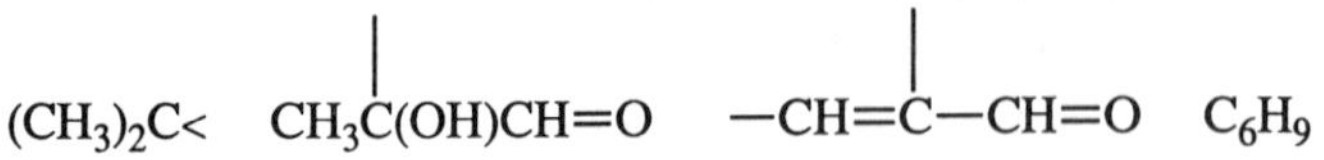

$$(CH_3)_2C< \quad CH_3\overset{|}{C}(OH)CH{=}O \quad -CH{=}\overset{|}{C}-CH{=}O \quad C_6H_9$$

II-3

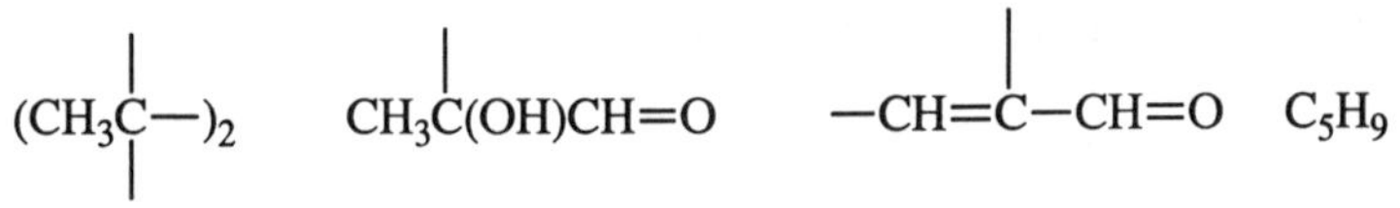

$$(CH_3\underset{|}{\overset{|}{C}}-)_2 \quad CH_3\overset{|}{C}(OH)CH{=}O \quad -CH{=}\overset{|}{C}-CH{=}O \quad C_5H_9$$

II-4

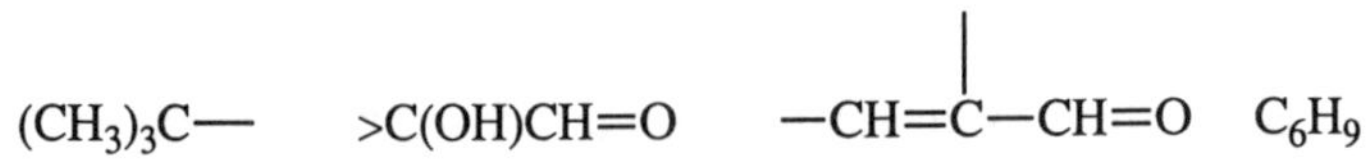

$$(CH_3)_3C- \quad >C(OH)CH{=}O \quad -CH{=}\overset{|}{C}-CH{=}O \quad C_6H_9$$

II-5

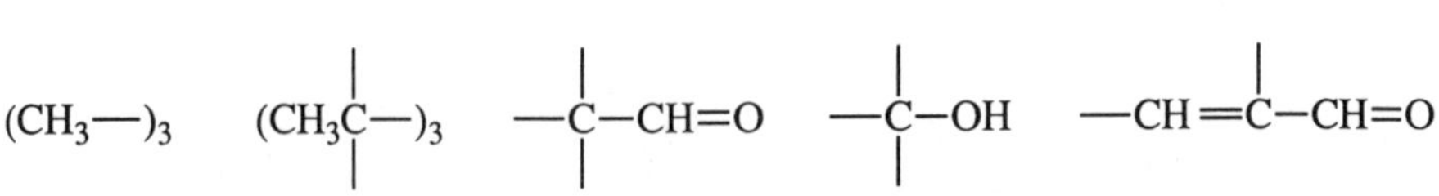

$$(CH_3-)_3 \quad (CH_3\underset{|}{\overset{|}{C}}-)_3 \quad -\underset{|}{\overset{|}{C}}-CH{=}O \quad -\underset{|}{\overset{|}{C}}-OH \quad -CH{=}\overset{|}{C}-CH{=}O$$

II-6

defined and used with a range of occurrence “none,” leaving only six cases (Table II-2).

The structure of warburganal was proposed on the basis of the few observations mentioned above, together with comparison of spectral properties with analogs whose structures were determined previously.[1] For the purposes of our example, we proceed with further examination of the spectral data for warburganal to illustrate the circumstance that analogs with proven structures are not available. Decoupling experiments in the ^{1}H NMR spectrum provide some of the most definitive substructural information about a molecule. Unlike constraints based on correlations of substructure and chemical shift–absorption frequency, constraints based on positive results from proton decouplings are unquestionable. Such experiments frequently allow substantial fragments of a molecule to be mapped out clearly. We have performed a series of decoupling experiments in our laboratory that resulted in the specification of an additional substructure that could be used as a constraint.

The vinylic proton at 7.26 ppm (Figure II-1) is coupled (J = 2.6, 5.0 Hz) to the two protons comprising a non equivalent methylene that resonate at 2.58 and 2.33 ppm. These protons display a geminal coupling of 21 Hz and are further coupled to a methine proton at 1.88 ppm (J = 5.0 and 11.8 Hz, respectively). The methine exhibits no further coupling. Substructure **DECOUPLE** is implied by these data. The assumption that the methine is in fact bonded to two quaternary carbons is reasonable but not proven. Substructure **DECOUPLE** was used as a constraint, bearing in mind that the data may have been over interpreted, leading to nine new cases (Table II-2). Four of the nine are shown as **II-7-II-10**. With the addition of the new constraint the structures are beginning to take definite shape.

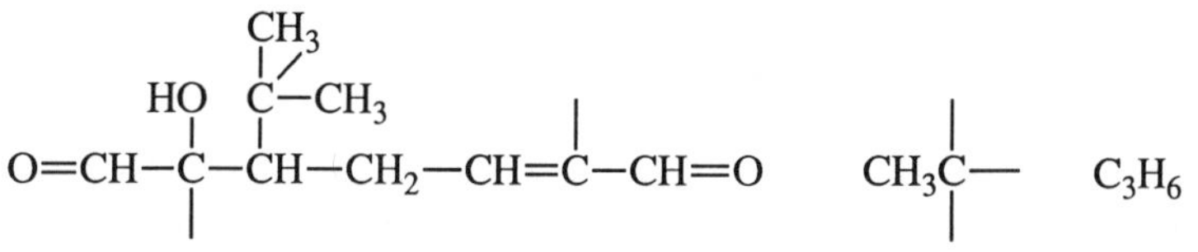

CH_3
HO C—CH_3
O=CH—C—CH—CH_2—CH=C—CH=O CH_3C— C_3H_6

II-7

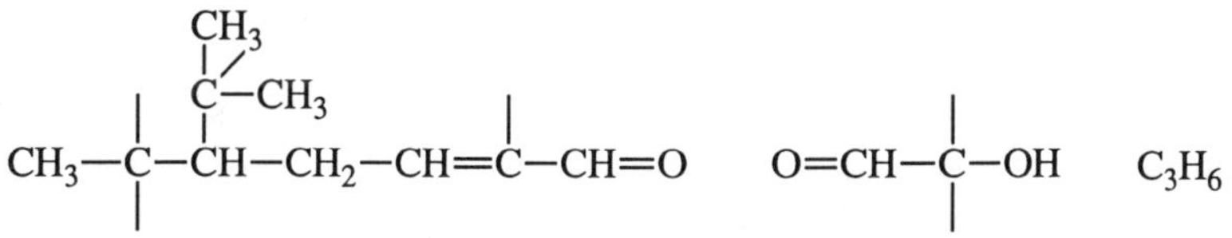

CH_3
C—CH_3
CH_3—C—CH—CH_2—CH=C—CH=O O=CH—C—OH C_3H_6

II-8

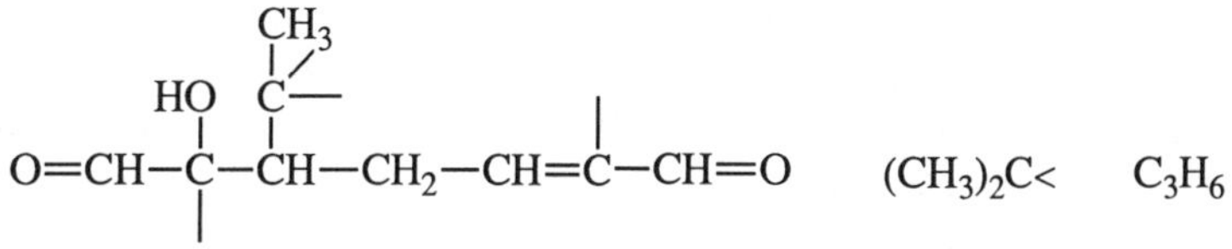

CH_3
HO C—
O=CH—C—CH—CH_2—CH=C—CH=O $(CH_3)_2$C< C_3H_6

II-9

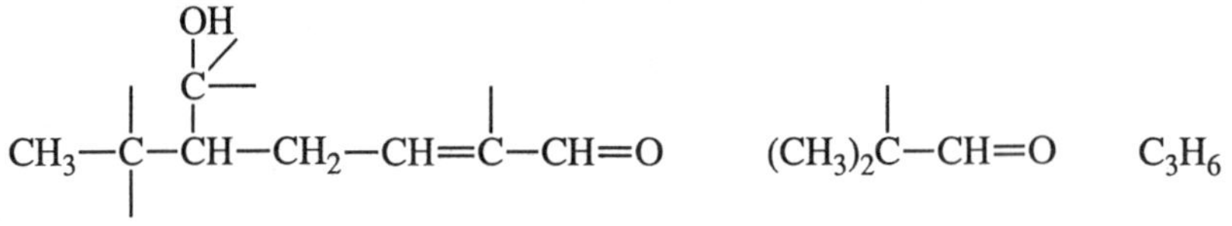

OH

$CH_3{-}C{-}CH{-}CH_2{-}CH{=}C{-}CH{=}O$ $(CH_3)_2C{-}CH{=}O$ C_3H_6

II-10

The ^{13}C spectrum indicates the presence of exactly four methylenes (Figure II-2). The requirement for the presence of exactly four of substructure **CH2** merely fixes the degree of the three remaining carbons (e.g., **II-7–II-10**). The proton spectrum reveals no sharp methylene 2-H singlets, indicating the absence of isolated methylene groups (Figure II-1). Given the structural inferences already made, the requirement for no isolated methylenes implies that three must form a chain, substructure **CHAIN3**. This constraint results in connecting the three methylenes together without changing the number of cases (Table II-2).

The constraints applied so far represent all that one can obtain easily from the available spectral data. At this point it is reasonable to generate final structures from the existing cases, thereby obtaining the complete set of candidate structures. Prior to structure generation, it is useful to specify additional constraints to prevent the construction of undesirable structural features, thereby saving the time required to find and remove them from the set after all are built. In this example there is no evidence for a cyclopropyl group, so the constraint **CYCP** is defined and applied with range of occurrence "none." This constraint is justified by the absence of high-field resonances characteristic of cyclopropyl hydrogens in the ^1H NMR spectrum. This constraint has no effect on the existing cases, none of which contain a cyclopropyl group, but it will discard automatically any structure that possesses such a functionality during structure generation. A total of 42 final structures are obtained; these are shown in Figure II-6. As with many other structural problems, the number of possible solutions consistent with numerous and detailed structural constraints is often surprising.

A distinct advantage of carrying out structural studies on the computer is that no sample and little time is consumed by exploring alternative interpretations of data. Thus we can examine the results, in terms of numbers of structural possibilities, using different assumptions. For example, if we assume that the ^1H NMR data do not exclude CH_3–C–OH groups, the only limit placed on methylene groups is to exclude–CH_2–groups between two quaternary carbons. Further, if no assumptions are made about the bonding of the alkyl methine, then some 2000 structures are compatible with these unambiguous spectral inferences. However, if the origin of warburganal is assumed to imply that it incorporate a standard decalin system or perhydroazulene skeleton (as are found in many sesquiterpenes), then only 80 of the 2000 structures are compatible with these skeletons and unambiguous spectral interpretations.

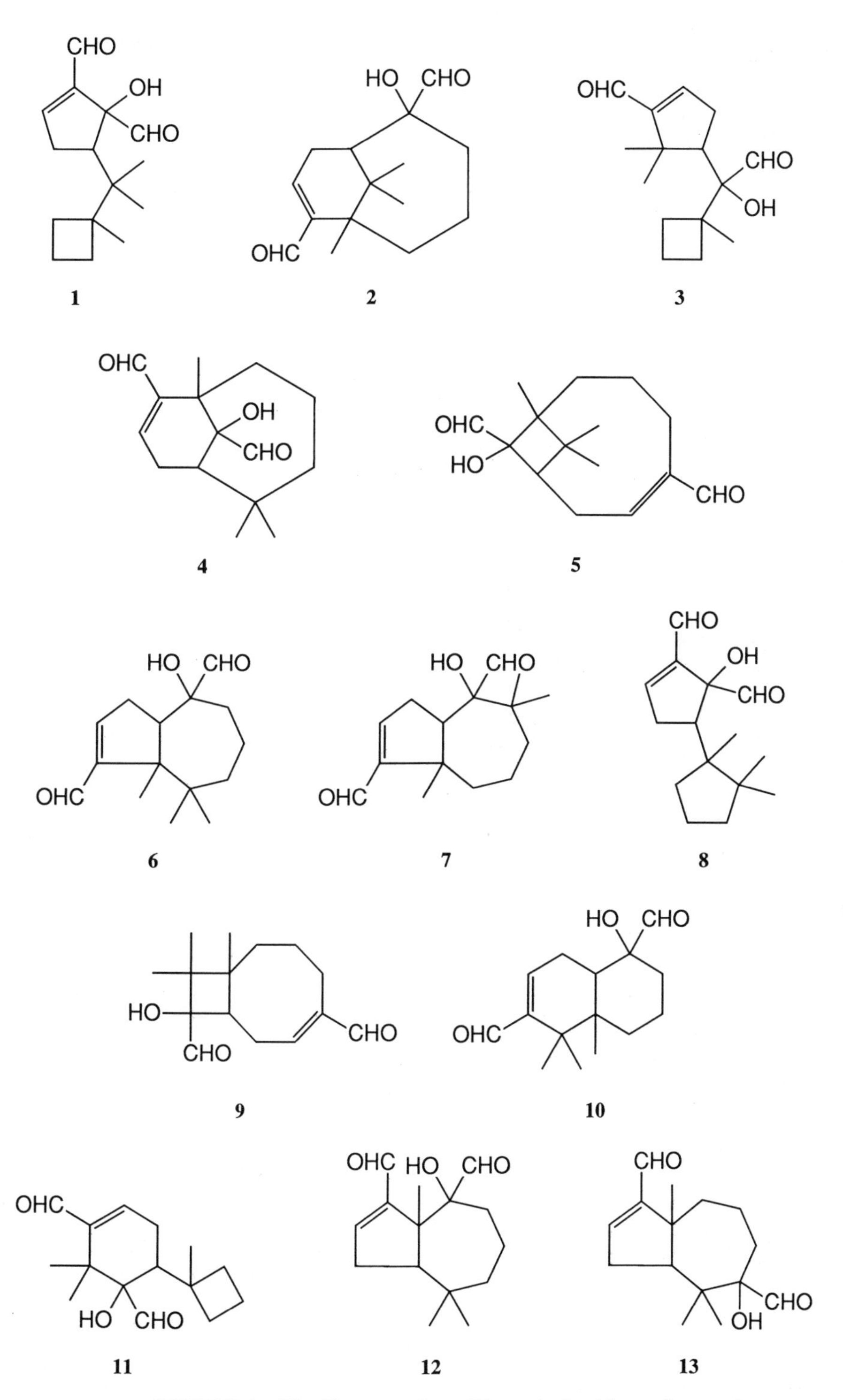

FIGURE II-6. The 42 structural candidates obtained for warburganal.

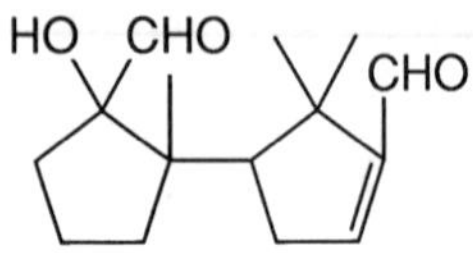

14 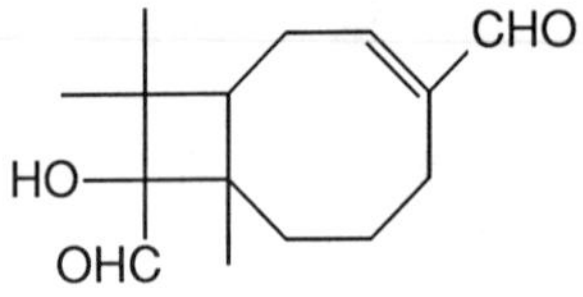 15 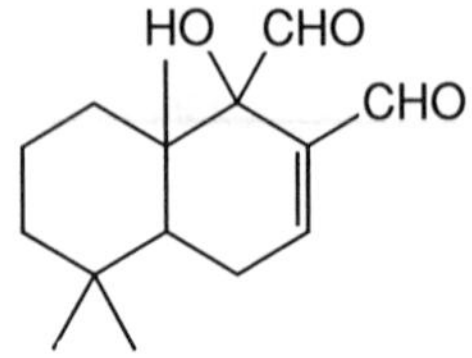 16

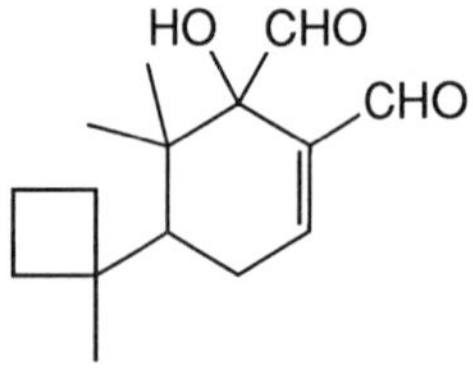

17 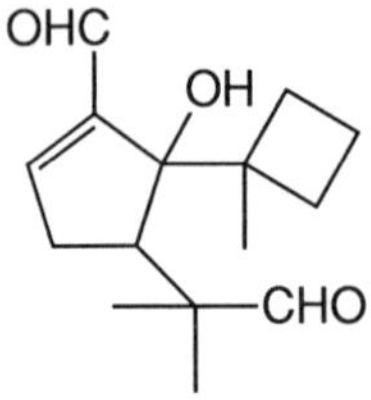 18 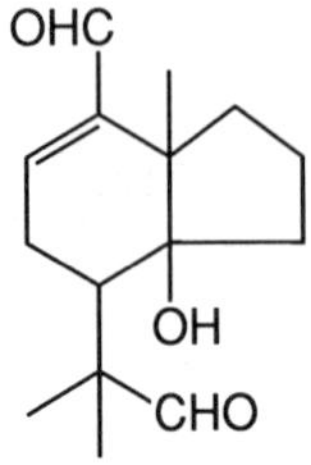 19

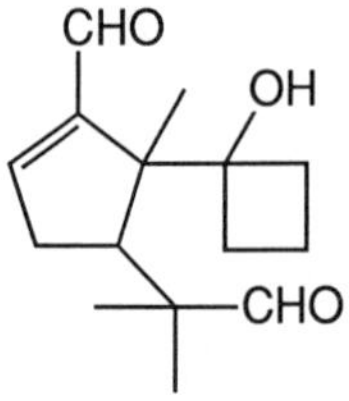

20 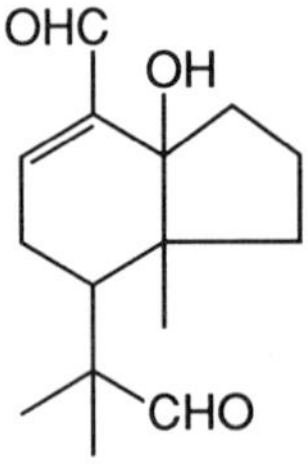 21 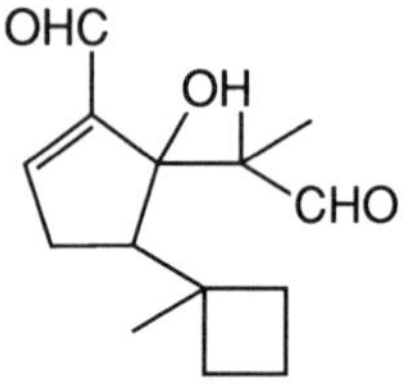 22

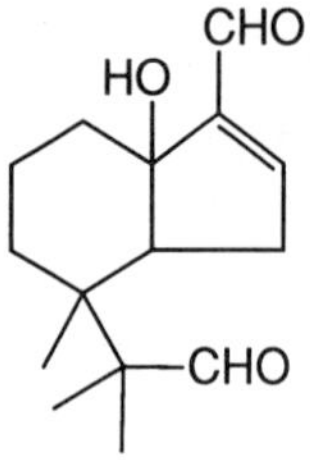

23 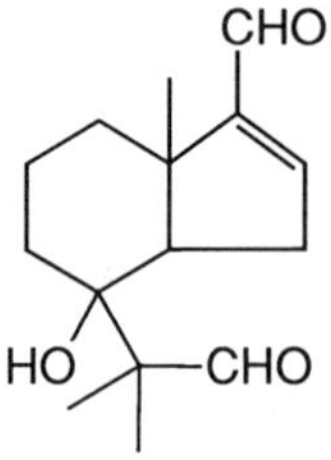 24 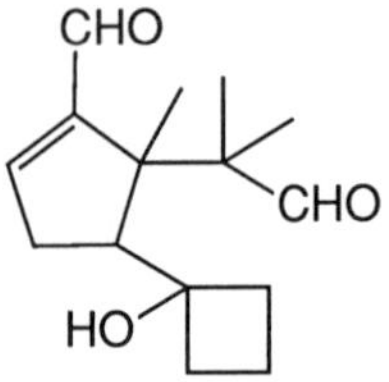 25

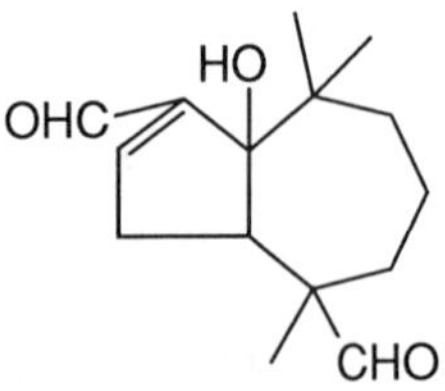

26 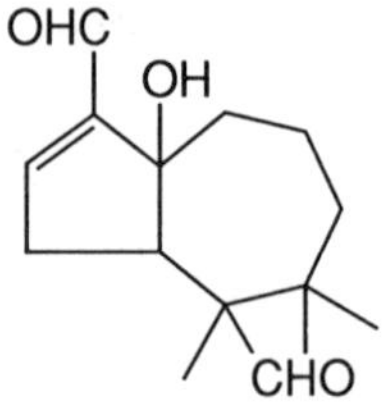 27 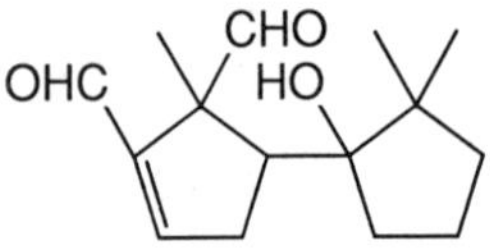 28

FIGURE 11-6. (*Continued*).

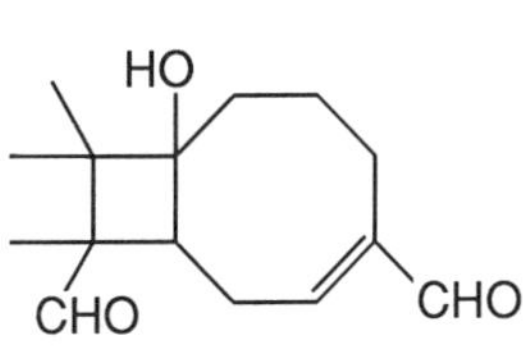

29 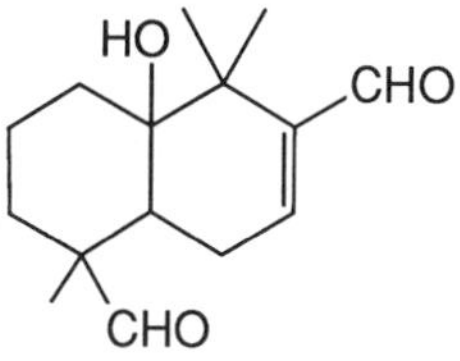 30 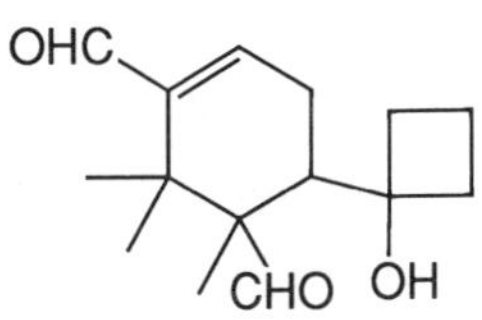 31

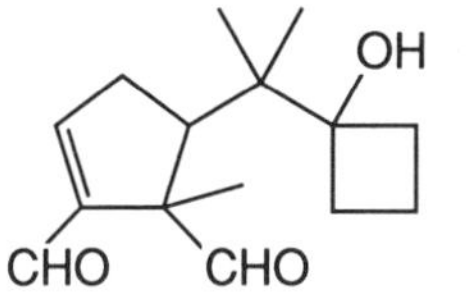

32 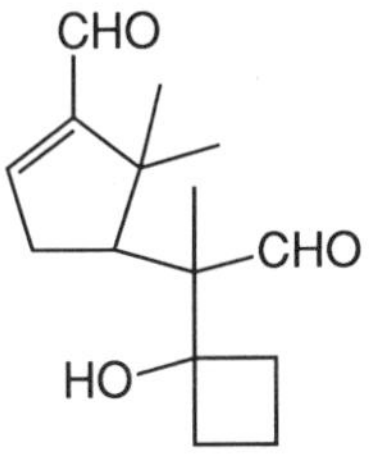 33 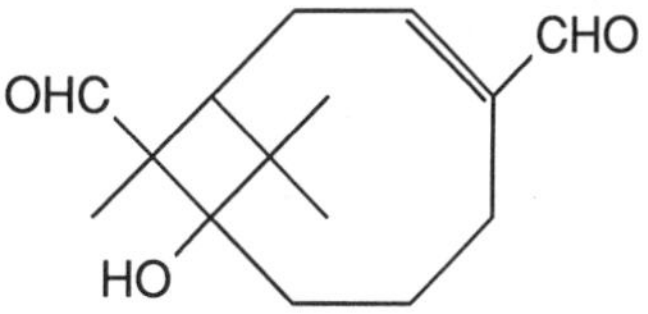 34

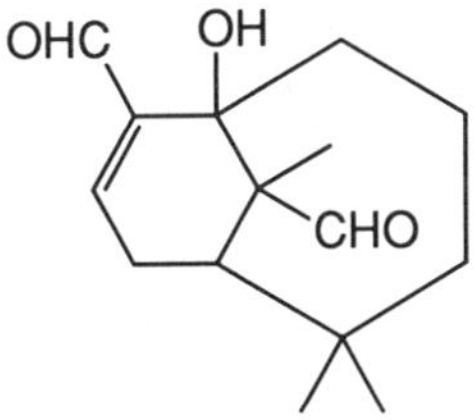

35 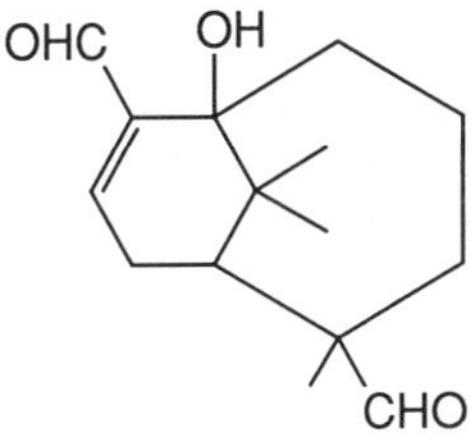 36 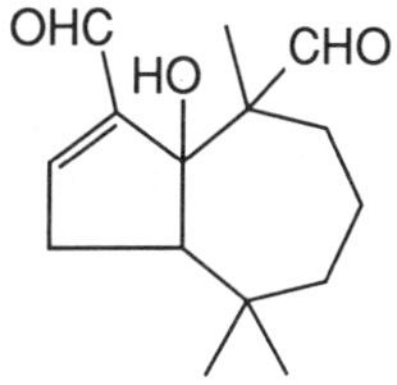 37

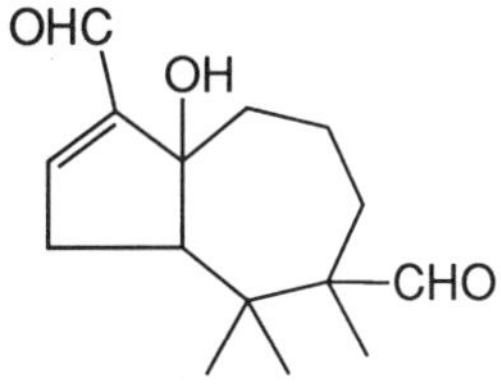

38 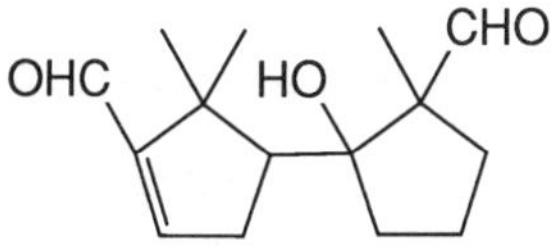 39

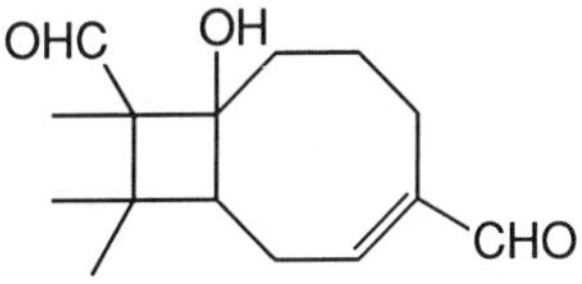

40 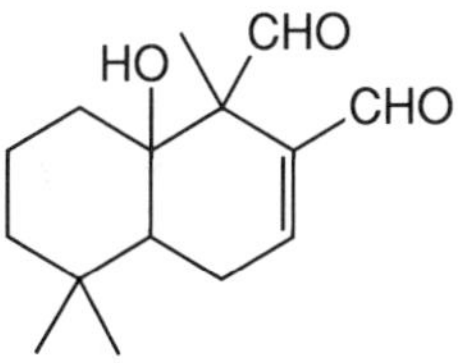 41 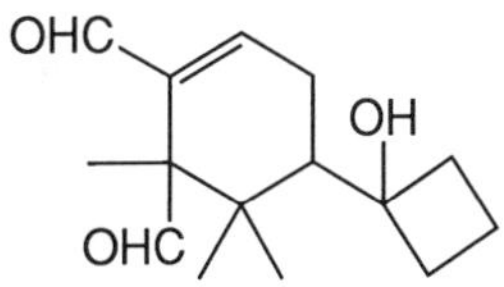 42

FIGURE 11-6. *(Continued)*.

II.A.2. Structure Evaluation: Focusing on the Correct Structure

Once structure generation is complete, transfer is made to the **STRCHK** (**STR**ucture **CH**ec**K**ing) program, which provides the chemist with assistance in carrying out a number of procedures to evaluate the structural candidates.[12,13] It includes program modules to:

1. Survey the generated structures to identify any containing standard skeletal systems, or any other specified combination of substructural features.
2. Evaluate the candidates through automated spectral analysis.
3. Explore stereochemical aspects of the structural problem.

II.A.2.a. Surveying the Structural Candidates

The first step in the evaluation of the 42 candidates found for warburganal would most appropriately be surveying them for the presence of standard sesquiterpene skeletons using the ***SURVEY*** module. The computer interaction with the **STRCHK** program to perform this analysis is shown in Figure II-7. The library of standard skeletons, ***BSTERP***, bicyclic sesquiterpenes, used in ***SURVEY***[14] was taken from an earlier compilation,[15] which is now somewhat out of date. This compilation includes some of the more common sesquiterpenes (e.g., chamigrane, valerane, tutin, trichothecane) and some skeletal fragments (identified by code names such as S005 and S076).

Eight candidates, whose structure numbers are shown in Figure II-7, incorporate a perhydroazulene skeleton, for example, structure 13 in Figure II-6 (**II-11**).

Only one structure (Figure II-7) incorporates a complete, standard skeleton, that of drimane (**II-12**), structure 16 in Figure II-6 (**II-13**).

II.A.2.b. Prediction of Spectral Properties

Although many spectral data are exploited in the initial phase of interpretation, it is often possible to obtain considerably more structural information from the

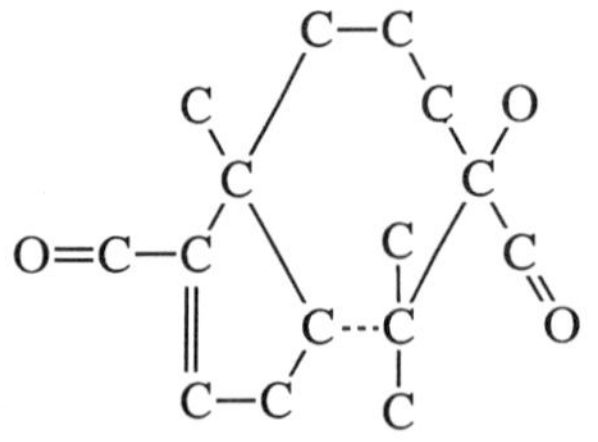

II-11 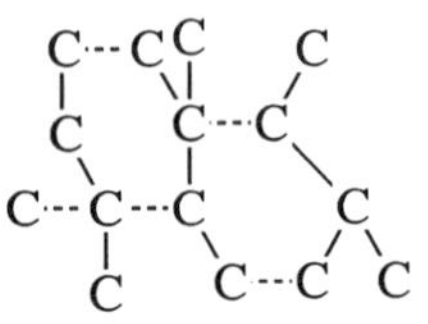 **II-12** 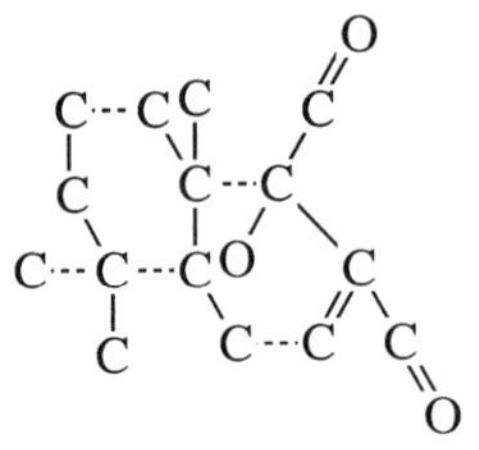 **II-13**

```
#SURVEY
Do you want to use a library of substructures? Y
Which library file? BSTERP
Which substructures: ALL

READING ENTRIES FROM LIBRARY FILE.
...............................................................

SCANNING THROUGH STRUCTURES.
...............................................................

THE FOLLOWING LIBRARY FEATURES WERE NOT FOUND IN ANY STRUCTURE.

CHAMIGRANE          VALERANE        TUTIN           CARYOPHYLLANE
CUPARANE            ACORANE         CAROTANE        PSEUDO-GUAIANE
    .                   .               .                .
    .                   .               .                .
TRICHOTHECANE       S023            S016            S076
STRUCTURES WITH DISCRIMINATING FEATURES:
  1  DRIMANE
  2  S027
  8  S005
  8  S007
Do you want to select structures with combinations of features? Y
→SELECT
Desired features>S005
8 structures.

→INDEX
8 structures currently selected.
Index numbers are:
  6  7  12  13  26  27  37  38
→RESET
42 structures.
→SELECT DRIMANE
1 structures

→INDEX
1 structures currently selected.
Index numbers are:
  16

→DONE
```

FIGURE II-7. Results of *SURVEY*'s comparison of the 42 candidate structures for warburganal against a library of bicyclic sesquiterpane skeletons.

data once candidate structures have been produced. During spectral interpretation, an analyst must rely on spectral-feature/substructure correlations that generally yield only small fragments of a much larger structure. Once candidate structures are available, however, more extensive investigations can be madc by exploiting spectral correlations that depend on larger substructures and in some cases on complete structures. An example of the latter is configurational stereochemistry. Given complete structures, we can predict the spectral properties of each candidate structure on the basis of a simple algorithmic model and rank the candidates on the basis of comparison between predicted and observed spectra.

Mass Spectral Prediction and Ranking. There are a number of advantages in using mass spectral prediction and structure ranking procedures as an aid to structure evaluation. Although configurational stereochemical differences can induce subtle changes in the mass spectra of stereoisomers, most of the features in a compound's mass spectrum can be adequately described in terms of cleavages of the bonds in a constitutional (topological) representation of that structure. In contrast, prediction and ranking using ^{13}C NMR spectra, discussed later in this chapter, require consideration of configurational stereochemistry. A representation that included molecular conformation would be required for other techniques such as ^{1}H NMR. A further advantage of mass spectral analysis is that we can exploit some fairly general models describing how molecules fragment within a mass spectrometer. Empirical methods for predicting other spectral properties usually depend on data bases of specific rules derived from standard reference compounds.

The 42 candidate structures for warburganal were rated according to their compatibility with the recorded high-resolution mass spectrum. The "half-order" model was used for mass spectral predictions.[16] Given a set of constraints on the complexity of fragmentation processes allowed, the half-order model finds the most plausible fragmentation process that, applied to a given structure, could lead to each specific observed ion. A score is derived that describes how readily the observed mass spectrum can be rationalized in terms of simple fragmentations of a structure. These scores are used to rank the various candidates.[16]

In Figure II-8 we present the interaction with our program for mass spectral prediction and ranking. This figure summarizes how one can control maximum complexity of the fragmentation processes considered by the program. In this example the program was restricted to simple single-step fragmentations involving at most a ring cleavage and the transfer of one hydrogen atom into or out from the ion.

In Figure II-9 we summarize the results of prediction and ranking, using elemental compositions and relative ion abundances as determined in the high-resolution mass spectrum. Our goal in this analysis is not to identify the correct structure unambiguously, but to obtain a subset of candidates, each of which yields a significantly better explanation of the mass spectrum than any

```
Are you using low-resolution mass spectra? N
Processing option: ?
(Rank structures, or just predict their spectra.)
Processing option: RANK

--FUNCTIONS FOR MASS SPECTRAL ANALYSIS--

Prediction method: ?
Either RULES for rule based approach or HALF for ''half-order''
theory.
Prediction method: HALF

Parameters.
________________________________________
Single bond breaks, plausibility: 1
Aromatic bond breaks, plausibility: 0
Multiple bond breaks, plausibility: 0
Adjacent breaks, plausbility: 0.2
Do you want to define bond break plausibilities in
substructures? N

________________________________________
Molecular ion, plausbility: 1
Minimum plausibility for ions to be accepted: 0.1
________________________________________

Define complexity of fragmentation process.
Allow fragmentation of fused/bridged rings? N
Allow simple ring cleavages? Y
Relative plausbility of a simple-ring cleavage step: 0.9

Maximum steps per process: 1
Maximum bonds to break in a process: 2

________________________________________
Any neutral losses? N

Any hydrogen transfers? Y
How many different hydrogen transfers: 3
#Hs: -1
Plausibility: 0.7
#Hs: 0
Plausibility: 1
#Hs: 1
Plausibility: 0.7
```

FIGURE II-8. Specification of fragmentation processes considered in mass spectral prediction and ranking program as applied to the 42-candidate structures for warburganal.

Is your spectrum in a file? **N**

Ion: **?**

An integral mass, semicolon integer intensity or a list of atom names and numbers, a semicolon, and an intensity value.

Ion: **C_3H_5; 530**

Ion: **C_23_3O; 305**

.

. (Remainder of spectrum)

.

Ion: **$C_{15}H_{22}O_3$; 73**

Ion:

Distribution of scores.

Maximum:	24
Mean:	13
Median:	12
Mode:	12

Score:	24	23	22	21	20	19	18	17	16	15	14	13	12
Frequency:	2	0	0	0	0	0	0	4	3	2	7	2	9

Top ranked structures and their scores:

#16	24
#17	24
#8	17
#26	17
#27	17

.

.

.

Do you want to prune the structure list by deleting structures with low scores? (Y or N) : **N**

FIGURE II-9. Results of mass spectral prediction and ranking for warburganal.

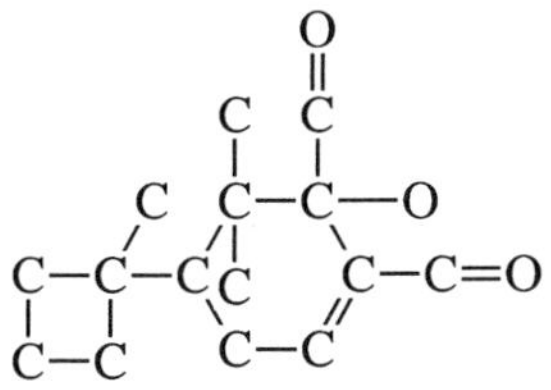

```
               O
               ||
            C  C
            |  |
       C    C—C—O
      /    /|   \
  C—C—C C       C—C=O
  |  |    \    //
  C—C     C—C
```

II-14

candidate in the remainder of the set. In this example two candidates, 16 (**II-13**) and 17 (**II-14**) in Figure II-6, rationalize the observed mass spectra better than do any of the other 40 candidates. Structure **II-13** is the drimane-type sesquiterpene identified previously in the search against the file of standard sesquiterpenes.

Generation of Configurational Stereoisomers. We might apply several other techniques to further evaluate a set of structural candidates. For example, we might wish to perform spectrum prediction and ranking using NMR data or construct geometric representations of the structures for viewing on a computer graphics system. These techniques require stereochemical representations of the structures. Thus a logical next step is to obtain the set of *configurational stereoisomers* for the candidates. This, too, can now be done with the aid of a computer program. This program, called **STEREO**, is itself a constrained structure generator of a type, devoted strictly to obtaining a set of *candidate stereoisomers under stereochemical constraints*.[17,18]

Few data given in our example of warburganal allow formulation of a large number of stereochemical constraints. The molecule is clearly chiral from the observed CD spectrum.[1] The results of experiments using lanthanide shift reagents revealed no shifts of the methyl singlets, indicating that they are remote from the hydroxyl or disposed of on opposite sides of a ring if on adjacent carbons. We can presume that the double bond, if in a ring, is in the *cis* configuration.

These constraints can be applied to the list of possible stereoisomers of the 42 candidate structures. The results obtained are as shown in Table II-3. There are a total of 576 stereoisomers for the candidates; all these stereoisomers are chiral, so no reduction is obtained from the chirality constraint. The constraint on the C=C bond, (requiring that it be in the *cis* configuration if in a ring of size less than 8) does reduce the total number of stereoisomers but does not eliminate any of the 42 candidates.

A prohibition on substructures with a *cis* relationship between methyl and hydroxyl groups on any ring of size less than 8 will restrict the stereoisomers to those consistent with the results of the lanthanide shift reagent experiment. This constraint reduces the number of possible stereoisomers to 152. Furthermore, this constraint eliminates one-third of the candidate constitutional

TABLE II-3. Summary of Cumulative Effects of Applying Stereochemical Constraints to the 576 Stereoisomers of the 42 Candidate Structures Generated for Warburganal

Constraint	Number of Structures	Total Number of Stereoisomers
None	42	576
Chiral	42	576
cis C=C	42	336
No *cis* CH_3,OH	28	152

isomers originally generated for warburganal. These are those structures where every stereoisomer must have such a *cis* relationship between methyl and hydroxy groups (e.g., those with both a *gem*-dimethyl in and a hydroxy substituent group on the same small ring).

Prediction and Ranking Using ^{13}C NMR Data. In most structural problems, the initial set of candidates resulting from the generation step can be reduced by pruning away those that do not satisfactorily explain observed magnetic resonance spectra. Quite often, NMR measurements needed to explore couplings are best deferred until candidate structures have been generated. Once candidates have been created, ***SURVEY*** and similar programs can be used to identify different chains and groupings of atoms in the various structures; these data can facilitate identification of the decoupling experiments that would most effectively discriminate among the candidates.

Results of decoupling experiments can be expressed in terms of substructures, like **DECOUPLE** as in Figure II-5, and used to prune the set of generated candidates. If, as with warburganal, the substructural constraints implied by the decoupling data are available from the outset, they should be used to constrain candidate generation. However, any further constraints derived through additional decoupling experiments can be used unequivocally to prune the list of candidates.

Chemical shifts can also serve as the basis for a scheme for evaluating candidates. Shifts, particularly carbon NMR shifts, can be correlated quite well with the (stereo)chemical environment of resonating nuclei. Details of known shift-substructure correlations can be used to predict chemical shifts for each atom in any given structure. Different candidate structures can be ranked according to how well their predicted spectra match with recorded data.

We have developed methods for predicting ^{13}C and ^{1}H NMR shifts for given structures.[19–21] These prediction methods depend on empirical correlations of substructures and shifts. The substructural environment of each resonating atom in each given candidate structure is analyzed, and a data base is then searched to find the most closely analogous known reference substructure. In the data base, each reference substructure is characterized by a *chemical shift range*. This shift range encompasses all the shifts that, in previously analyzed reference compounds, have been assigned to atoms in that specified substructural environment. The result of the prediction process is a "fuzzy" spectrum in which each atom in a given structure is associated with a shift range.

These chemical shift prediction programs can process either constitutional isomers or configurational stereoisomers. Predictions made for constitutional isomers yield shift ranges that span the ranges associated with all possible configurational forms. Consequently, such predictions result in "fuzzy" spectra with still wider shift ranges than those that would be obtained if the predictions were being made for specific stereoisomers.

The programs attempt to find the largest analogous substructural environment. For example, the generated candidate 41 (Figure II-6) for warburganal includes a methylene in the four-bond environment represented as **II-15**. When attempting to predict the ^{13}C chemical shift of this methylene carbon, the program would seek an identical reference substructure (using configurational stereochemical data if these are defined in the structure for which chemical shift predictions are being made). If no matching four-bond environment were found, the program would try in turn the three-, two-, and one-bond environments represented as **II-16–II-18**.

The size of the reference substructure found for any given atom environment depends on the comprehensiveness of the data base. Neither substructure **II-15** nor **II-16** is in the current ^{13}C data base; consequently, the best reference model for this substructural environment is **II-17**. This represents the atom environment $\rangle C\text{–}CH_2\text{–}C^*H_2\text{–}CH_2\text{–}C\langle$ (with the asterisk marking the central resonating atom). The shift range associated with such a general substructure is naturally quite broad, in this case the range of ^{13}C chemical shifts is 16.7–31.3 ppm. If the data base had not included substructure **II-17**, predictions would have had to been made on the basis of the still more generalized model substructure **II-18** (with shift range 15.9–34.6 ppm).

The candidates for warburganal include some unusual structural forms, and this, together with the rather high degree of functionality, results in many substructural environments for which the best reference substructures match only to a one-bond radius. (The ^{13}C data base used by the program contains reference information derived from about 2000 naturally occurring steroidal,

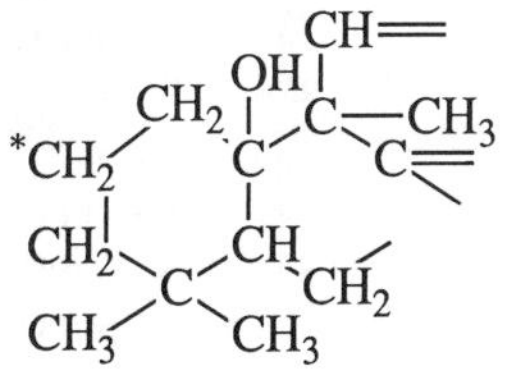

II-15 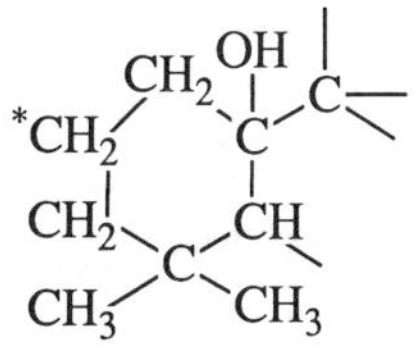 **II-16** 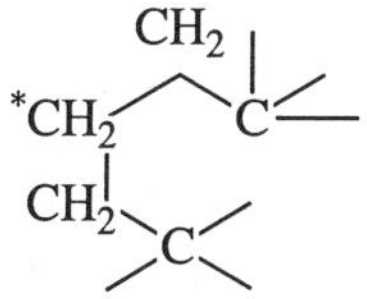 **II-17** 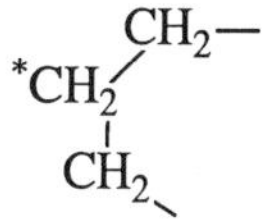 **II-18**

terpendoidal and alkaloidal compounds; it lacks any detailed models appropriate for the candidates with four to eight ring fused systems or for those with two disjoint five-membered rings, etc.) For these structures, there is little purpose in predicting spectra of individual stereoisomers as a two- or three-bond environment is usually required before different configurational forms can be distinguished. Thus ^{13}C spectral predictions for the 42 candidates were based on their constitutional forms.

Some example "fuzzy" spectra are shown in Table II-4; the data give the shift range predicted to encompass the resonance for each atom along with details of the size of the reference substructure used as a model when making the predictions. For most atoms, the predictions are based on very crude one-bond (alpha) environments. For one >C< atom, no appropriate reference substructure could be found and the "prediction" is simply for a singlet resonance in the range 16–115 ppm. Usually, predictions based on larger model substructures are associated with narrower shift ranges; there are exceptions, such as the *gem*-dimethyl methyl groups in structure 41 in Figure II-6, where predictions for the diastereotopic methyls in this constitutional isomer have ranges spanning the quite different shifts associated with their differing stereochemical environments.

Once generated, the fuzzy spectra derived for each candidate structure must be put in correspondence with and scored against the observed spectrum.[22] Of course, any scoring scheme must take cognizance of the quality (degree of detail) of the model substructures used in making predictions. Predictions based on detailed substructural models should provide a much more accurate

TABLE II-4. Examples of the "Fuzzy" ^{13}C Spectra Predicted for Some of the Structural Isomers Generated as Candidates for Warburganal

	8	*17*	*41*
$-CH_3$	11.4– 26.3 (2)	14.9– 28.8 (2)	22.6– 22.8 (2)
$-CH_3$	20.6– 28.1 (2)	18.0– 26.6 (2)	19.9– 34.5 (3)
$-CH_3$	20.6– 28.1 (2)	18.0– 26.6 (2)	19.9– 34.5 (3)
$-CH_2-$	15.9– 34.6 (1)	15.9– 34.6 (1)	16.7– 31.3 (2)
$-CH_2-$	17.7– 47.3 (1)	17.7– 47.3 (1)	29.7– 29.7 (2)
$-CH_2-$	17.7– 47.3 (1)	17.7– 47.3 (1)	33.6– 42.9 (2)
$-CH_2-$	18.9– 51.1 (1)	21.3– 25.8 (2)	21.3– 25.8 (2)
>CH–	24.8– 67.0 (1)	24.8– 67.0 (1)	24.8– 67.0 (1)
>C<	32.8– 37.3 (1)	32.1– 49.1 (1)	32.4– 35.6 (2)
>C<	34.6– 51.5 (1)	46.4– 53.7 (1)	48.0– 48.3 (1)
>C<	16.7–114.7 (0)	77.4– 77.4 (1)	74.1– 86.1 (1)
–CH=	135.4–157.6 (1)	157.6–157.6 (2)	135.4–157.6 (1)
=C<	138.8–139.6 (1)	139.6–139.6 (2)	138.8–139.6 (1)
–CHO	192.7–195.2 (2)	192.7–192.7 (3)	192.7–195.2 (2)
–CHO	195.7–211.6 (1)	201.3–201.3 (2)	195.7–211.6 (1)

rendition of the actual spectrum of any given compound than those predictions that are based on poor models. When matching the observed data of some compound against predictions made for different candidate structures, a greater tolerance must be allowed for any candidate structure for which predictions have had to be based on poor model substructures.

Various empirical schemes have been evaluated and a number of stable scoring functions developed. These functions provide a measure of dissimilarity between predicted ranges and observed chemical shift patterns. The computed dissimilarity score makes allowance for the average sizes of the substructures used for predicting shifts for each atom in a structure. This average size is measured in terms of "shell level" (i.e., number of bonds radius from the resonating atom to which the environment is matched); this average shell level provides a direct measure of the quality of the predictions.

Results from the ^{13}C prediction programs are summarized in Table II-5. (We did not attempt similar prediction and ranking of ^{1}H NMR spectra because the corresponding data base of substructures and ^{1}H chemical shifts is too small and too specialized to be of value in this context.) The data in Table II-5 give the dissimilarity scores and average prediction shell levels for the 20 top-ranked structures.

Structure 16 in Figure II-6 is top ranked by this procedure; the chemical shift predictions for many of the atoms in this structure are based on two- or three-shell models. The data base used contains reference spectral data for drimenol (**II-19**) and eperua-7–13-dien-15-oic acid (**II-20**). These, and other similar compounds, incorporate many substructural environments similar to those present in structures 16, 41 and so on (Figure II-6) and thus allow for reasonably accurate spectral predictions.

TABLE II-5. Partial Results of Ranking of Predicted ^{13}C NMR Spectra of the 42 Candidate Structures[a]

Rank	Score	Shell	Structure[b]	Rank	Score	Shell	Structure[b]
1	46.2	2.3	**16**	11	66.3	1.4	**7**
2	53.5	1.2	**9**	12	66.3	1.5	**35**
3	53.6	1.2	**29**	13	67.2	1.6	**4**
4	60.8	1.3	**34**	14	69.7	1.8	**41**
5	62.2	1.2	**8**	15	69.9	1.1	**21**
6	62.9	1.2	**23**	16	71.8	1.1	**19**
7	63.0	1.2	**27**	17	73.1	1.1	**39**
8	64.0	1.2	**24**	18	73.5	1.2	**38**
9	64.0	1.2	**28**	19	74.1	1.1	**33**
10	64.0	1.4	**40**	20	75.6	1.4	**37**

[a]The "scores" give a measure of the incompatibility between the predicted shifts and the resonces observed in the spectrum recorded for warburganal. The "shell" value is a measure of the assessed reliability of prediction.
[b]From Figure II-6.

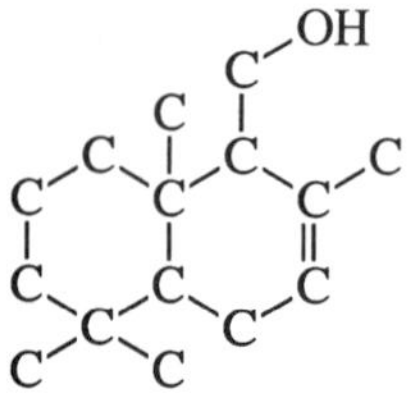

II-19 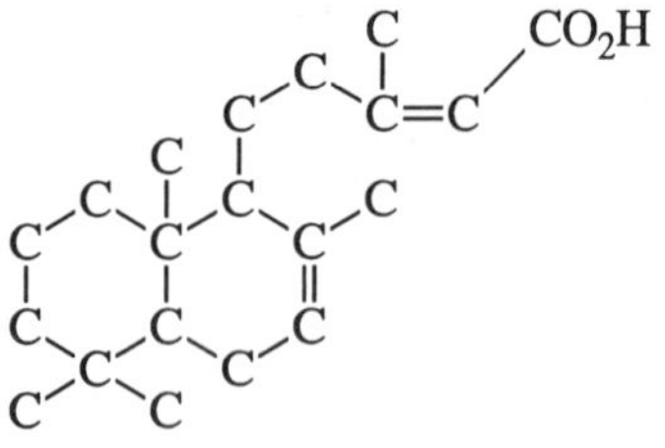 **II-20**

Many of the other "highly ranked" structures are characterized by very low shell levels. For example, structures 21, 33 and 39 in Figure II-6 have an average shell level of only 1.1. Because the ^{13}C prediction program finds that it has no good models for the atoms of these structures, it scales down computed mismatches between predicted and observed data and assigns low dissimilarity scores to these structures. In this way, the program avoids bias against structures incorporating novel features.

Again, the objective of this spectrum prediction and ranking process is not the "identification of the correct structure." Since both the spectrum prediction process and the scoring function are empirical, it would not really be appropriate to discard candidate structures solely on the basis of their scores or rankings. The rankings are intended merely to help in focusing attention on the most plausible of candidates. Here, structure 16 in Figure II-6 is again found to be reasonably consistent with the observed data, whereas structure 17 (Figure II-6), which provided an equally good rationalization of the MS data, does not appear among the 20 most plausible. Of the other structures containing commonly found sesquiterpene fragments S007 or S005, only structure 27 (Figure II-6) is identified as reasonably consistent with both the observed MS and ^{13}C data. Structure 8 (Figure II-6), although not incorporating one of the standard substructures, is compatible with the spectral observations.

Once NMR, MS, survey, and similar analyses have identified structures such as 8, 16 and 27 (Figure II-6) as being plausible, the chemist has a starting point for further investigation. The literature can be searched for references to these specific structures or to analogous compounds isolated from similar sources; further experiments can be planned that would discriminate among these more plausible candidates. Thus, although empirical, these various analysis programs can provide much assistance to the structural chemist.

II.A.3. Conclusions

The constitution of warburganal is indeed represented by structure 16 (**II-13**) of the 42 candidates shown in Figure II-6. The results of our more detailed evaluation of the spectral data add significant weight to the structural assignment because this structure is found to possess a common sesquiterpene skeleton and to offer among the best explanations of both the mass and ^{13}C NMR spectra. However, it is not possible on these data alone to assign the structure unambiguously. One method for structure verification is unambiguous synthesis, and in the next section we illustrate briefly how computer programs can be used to help development of such syntheses.

II.B. COMPUTER-ASSISTED SYNTHETIC ANALYSIS OF WARBURGANAL

Structures inferred by interpretation of spectral and chemical data must be confirmed by X-ray crystallography, unambiguous synthesis, or chemical conversion to a known structure. For the example of warburganal, an unambiguous synthesis would obviously be desirable, for—in addition to confirming the structure—synthesis would provide routes to other, related compounds whose biological activity could then be evaluated.

Chemists exploit a variety of problem-solving methods in their analyses of synthetic plans.[23] In a few cases, required starting materials and synthetic steps can be recognized immediately from the form of the desired product by direct associative techniques. Much more commonly, compounds similar to the desired structure may already be available from previous related studies, and the planning of the synthesis involves an analysis of the differences between available and desired structures and choice of a sequence of reaction steps to effect their interconversion. A final approach to synthesis planning is the *logic-centered method*.[23] The logic-centered method uses a detailed analysis of the structure of a desired target compound in order to identify features, *synthons*, that suggest chemical reactions, or *transforms*, that might be used in the final step of a synthesis. Thus precursors that require a single transform to obtain the synthetic target are identified. Subsequently, these precursor structures are analyzed similarly in turn to derive new precursors that may be converted to the target compound through two synthetic transformations. This analysis, carried out repeatedly, results in a synthetic "tree" of intermediates leading from simple starting materials to the desired structure.

Synthetic "trees" obtained in the logic-centered method can become quite cumbersome to maintain and manipulate when more than a few synthetic procedures are considered. However, the storage and manipulation of such information is a problem amenable to solution with computer-assisted techniques. There now exist several computer programs that can be used to suggest plausible synthetic routes to a desired structure (see Chapter XII). In this section we illustrate the use of an existing program for computer-assisted synthetic analysis in planning syntheses for warburganal and contrast the results with the several published routes.

Investigation of Synthetic Approaches to Warburganal Using the SECS Program. The program used in the study of synthetic routes to warburganal is an early version of the Simulation and Evaluation of Chemical Synthesis, or

SECS, program developed by Wipke and co-workers.[24,25] Like the **GENOA** and **STRCHK** programs described previously in the structure elucidation of warburganal, **SECS** is an interactive computer program intended to assist the chemist in exploring alternative solutions.

The initial **SECS** analysis of a target structure identifies functional groups, ring systems, and *strategic* bonds whose cleavage (or preservation) should be an important consideration in the determination of the next transform and possible precursor structures. The **SECS** program builds a geometric model of the target structure to determine factors such as steric congestion that may inhibit reactions or allow selective modification at only one of several examples of a particular functional group. **SECS** incorporates a data base of transforms that can be retrieved according to the match between their required synthons and the features characterizing the structure being analyzed. These transforms detail (1) reaction conditions, (2) the structural differences between the target compound and precursor compound(s), and (3) rules for estimating the merit of a proposed synthetic step.

The program applies the transforms to the target structure thereby deriving the structural forms of possible precursors. Precursors are assigned priority ratings expressing both the intrinsic merit of the reaction step by which they may be converted to the target molecule and the degree to which other goals, such as cleavage of strategic bonds, has been accomplished. The program performs all the bookkeeping entailed in developing the synthetic tree. The chemist exerts final control over the choice of synthetic route. The chemist can to a limited degree modify the rules the program uses for identifying strategic bonds and change the evaluation functions that **SECS** uses to eliminate disfavored generated precursor structures. Control is, however, expressed mainly through the chemist's selection of the generated precursor structures that should themselves be further analyzed. This selection dictates the form of the synthetic tree and the types of synthetic route that will be determined.

During initial analysis of the structure of warburganal (**II-13**), **SECS** perceives several bonds as strategic. These bonds are those attaching functionalized appendages to the basic cyclic skeleton and those whose cleavage could yield a simpler ring system. The strategic bonds are identified by **SECS** in Figure II-10. Transforms involving cleavage of strategic bonds (i.e., formation of these bonds in the real synthetic process) are particularly favored because such transforms will relate substantially simpler precursor structures to the desired warburganal structure.

In selecting transforms, the program was instructed to consider Functional Group Interchange and Functional Group Introduction transforms in addition to more substantial changes. Generated precursor structures were evaluated by using the **SECS** default criteria, summarized in Figure II-11. Thus any precursor structure that proved to incorporate a cumulene system or an sp^2 carbon at a bridgehead position would automatically be eliminated. Similarly, possible precursor structures would be eliminated if the rating for the transform relating them to warburganal should, as a result of modifying factors, fall below −50.

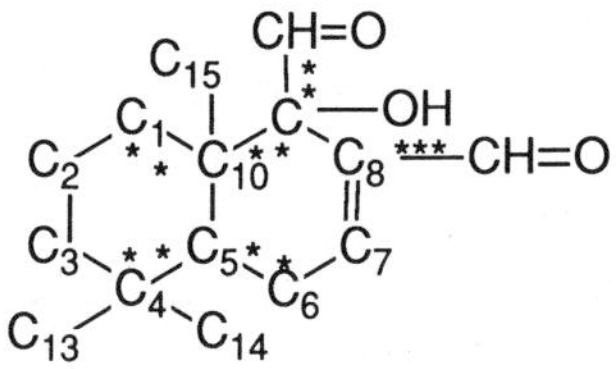

```
THE MACHINE GENERATED STRATEGY IS:
SUBLIST# 2

LOGIC:
IF ANY OF THE FOLLOWING ARE ACHIEVED ADD 100

    BREAK 4  5
    BREAK 5  6
    BREAK 9  10
    BREAK 1  10
    BREAK 9  11
    BREAK 8  12
```

FIGURE II-10. Strategic bonds of warburganal (**II-13**) identified by **SECS** during the perception phase of synthetic analysis.

Precursor elimination control parameters:

```
BRIDGEHEAD SP2 ATOM IN RING SMALLER THAN SIZE 7
ANTIAROMATIC RING
CUMULENE
TRIPLE BOND IN RING SMALLER THAN SIZE 6
TRANS DOUBLE BOND IN RING SMALLER THAN SIZE 7
TRANS-BRIDGED RING SYSTEM OR THREE MEMBERED RING FUSED
TO RING OF SIZE LESS THAN 8
KETONIZE ANY ENOLS
VALENCE VIOLATIONS FOUND
DIANION,DICATION,DIRADICAL
```

Strategy options:

```
TRANSFORM PRIORITY CUTOFF VALUE =  -50
USER CANNOT MODIFY STRATEGY
FUNCTIONAL GROUP INTERCHANGE ALLOWED
CORRECTIONS OF BAD TRANSFORMS NOT ALLOWED
DO HUECKEL CALCULATIONS
```

FIGURE II-11. Default controls for evaluating precursors in the **SECS** program.

We present in Figure II-12 an example of a transform, taken from the **SECS** transform library, applicable to warburganal. This transform contains commentary on the literature source for the reaction, a description of the functionalities involved and their relative locations, and a variety of constraints that modify the initial priority or eliminate the reaction from further consideration. The synthon required by this transform consists of a hydroxyl group and a ketone or aldehyde group separated by two bonds and thus matches the warburganal skeleton at C(9)-C(11) (see Figure II-10).

The default priority, 50, for this transform is reduced for warburganal to reflect the facts that the oxo group is an aldehyde and the alcohol is on a quaternary center. The presence of the C(7)-C(8) double bond alpha to C(9) enhances reactivity and improves the priority score. Also, because the transform involves cleavage of the strategic C(9)-C(11) bond, the priority is further

```
;;;;;
;   HO-C-C=O => C=O + DITHIANE
; ADDITION OF DITHIANE ANION TO A KETONE OR ALDEHYDE
; REF: COREY ANG. CHEM. 4, 1075 (1965)
; REAGENT: (1) DITHIANE ANION; (2) HG++/H3O+
DITOXOGROUP
ALCOHOL OXO PATH 2 PRIORITY 50
CHARACTER BREAKS CHAIN
        IF GROUP 2 IS ACID THEN KILL
        IF GROUP 2 IS ESTER THEN KILL
        IF GROUP 2 IS AMIDE THEN KILL
        IF GROUP 2 IS ACIDX THEN KILL
        IF GROUP 2 IS ALDEHYDE THEN SUBT 20
        IF BOND 1 IS RING BOND THEN KILL
        IF ATOM 1 IS QUATERNARY THEN
        BEGIN IF ATOM 1 IS ATOM (1)
        IF ATOM IS ALPHA TO ATOM 1 OFFPATH (2)
        PUT (1) OR (2) INTO (1)
        IF ATOM 2 IS ON MOST HINDERED SIDE OF (1) THEN SUBT 50
        DONE
        CONDITIONS ELECTROPHILIC
        CONDITIONS SLIGHTLY OXIDIZING AND ACIDIC
        IF ATOM 1 IS QUATERNARY THEN SUBT 10
        IF ATOM ALPHA TO ATOM 1 OFFPATH IS OLEFIN THEN ADD 20
        BREAK BOND 1
        MAKE BOND FROM ATOM 1 TO ATOM 1 IN GROUP 1
        END
;       NAME:  DITOXOGROUP
;
```

FIGURE II-12. Description of the DITOXOGROUP transform. (Reprinted, in part, with permission from W. T. Wipke, University of California at Santa Cruz.)

```
(LEFT FRAGMENT)                    (RIGHT FRAGMENT)
TRANSFORM = DITOXOGROUP            TRANSFORM = DITOXOGROUP
PRIORITY = 140                     PRIORITY = 140

          O
      C   ||
   C  |   C      C=O
C    C      C
|    |      ||
C    C      C                          O=C
   C    C
C    C

     II-21                              II-22
```

FIGURE II-13. Precursors generated by application of the DITOXOGROUP transform to warburganal.

increased by 100. Thus the precursor structure (**II-21**), left fragment in Figure II-13, is identified and associated with an overall priority of 140. The other precursor needed for this transform is, of course, simply formaldehyde **II-22**, Figure II-13.

```
;;;;;
;      W-C=C-C  =>  W-C-C=C
; CONJUGATION OF BETA-GAMMA UNSATURATED WGROUP
; REF:
; REAGENT:  ACID OR BASE
CONJUGATION
WGROUP  WGROUP  PATH 4  PRIORITY 0
CHARACTER MIGRATES BOND
        IF BOND 2 IS NOT DOUBLE BOND THEN KILL
        IF BOND 2 IS AROMATIC THEN KILL
        IF ATOM 4 IS QUATERNARY THEN KILL
        IF ATOM 4 IS UNSATURATED THEN KILL
        IF ATOM 4 IS BRIDGEHEAD ATOM THEN KILL
        IF GROUP 1 IS AMIDE THEN SUBT 10
        IF ATOM 4 IS DGROUP THEN KILL
        IF ATOM 4 IS AMINE THEN KILL
        IF ATOM 4 IS ETHER THEN KILL
        IF UNSATURATED ATOM IS ALPHA TO ATOM 4 OFFPATH THEN SUBT 20
        BREAK BOND 2
        MAKE BOND 3
        END
;       NAME:    CONJUGATION
;
;
```

FIGURE II-14. Description of the CONJUGATION transform. (Reprinted, in part, with permission from W. T. Wipke, University of California at Santa Cruz.)

The CONJUGATION transform, defined in Figure II-14, applied to warburganal results in **II-23** being identified as another precursor.

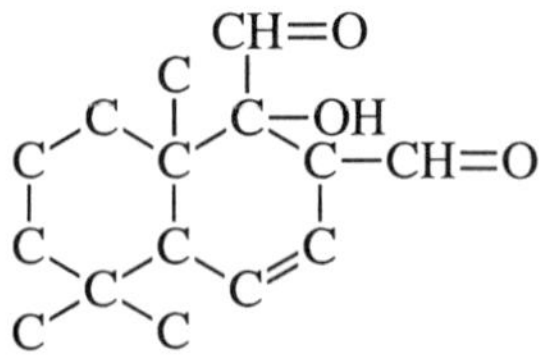

II-23

Some of the other precursors created are shown as **II-24-II-28**, together with the shorthand description of the transform. Additional precursors were derived through consideration of Grignard transforms and a large variety of functional group interchange–introduction transforms.

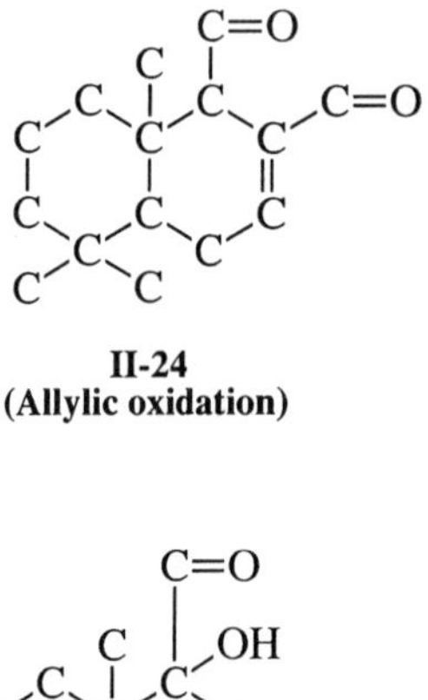

II-24
(Allylic oxidation)

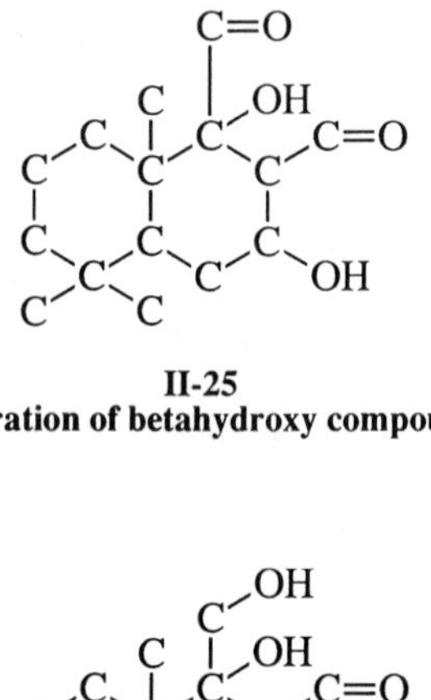

II-25
(Dehydration of betahydroxy compound)

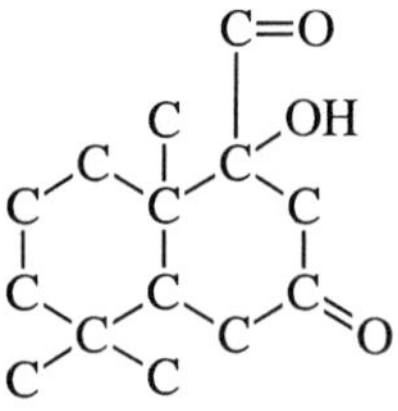

II-26
Aformylreduction

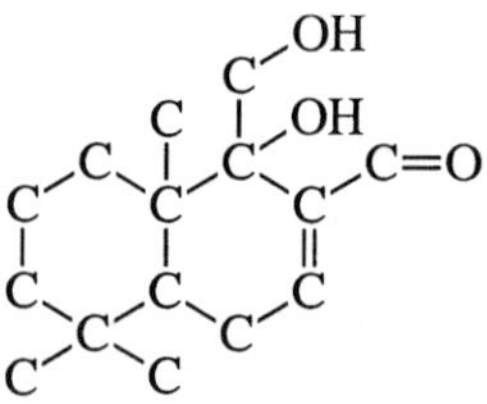

II-27
(CH=O → CH–OH)

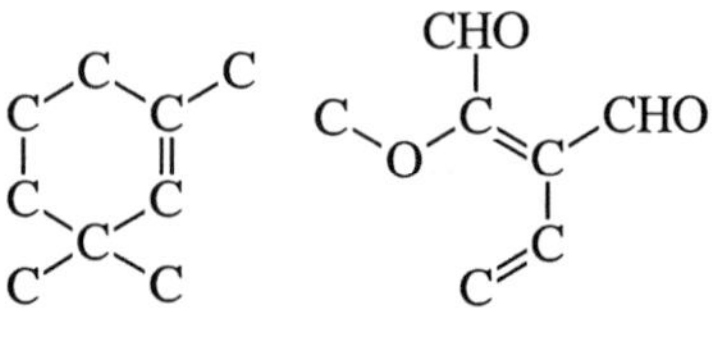

II-28
(Diels Alder)

Selected precursors were further analyzed. Structure **II-21** was one of the precursors chosen; some 14 possible precursor structures were identified, including **II-29**.

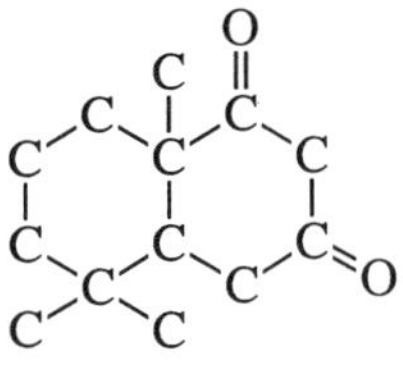

II-29 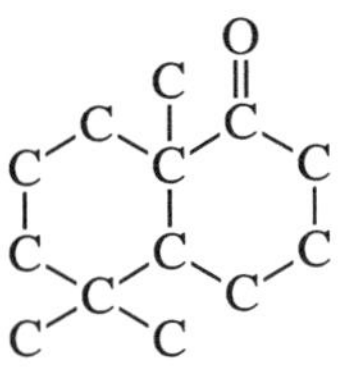 **II-30**

The route proposed by SECS through **II-21** and **II-29** is somewhat analogous to the model synthesis of warburganal and muzigadial proposed by Peterse et al.[3] The Peterse model synthesis would have involved **II-30** rather than **II-29** with the conversion of **II-30** to **II-21** achieved through a variant of the formylation reaction proposed by **SECS**. Some other complete syntheses have passed through structure **II-21** with varying methods used to complete the synthesis from this somewhat unreactive keto aldehyde.[6,7]

The **SECS** program did not reproduce any of the literature syntheses in their entirety. This is not surprising, for most of the literature synthetic routes were dictated by the availability of specific starting materials, and, in addition, the version of **SECS** employed utilized only a very small library of transforms. The problem of a small transform library was identified previously as being among the most critical to be resolved before such synthesis programs can be effectively exploited in an industrial setting.[26]

II.C. STRUCTURE–ACTIVITY CORRELATIONS IN WARBURGANAL AND RELATED COMPOUNDS

Warburganal and a number of related compounds have been proved to possess potent antifeedant activity against African army worms. In addition, these same molecules are powerful heliocides. It is possible that such compounds could be exploited for controlling pest insects in the field or in crop storage. The heliocidal properties are also of considerable interest for a number of diseases, such as schistosomiasis, are spread by parasitic nematodes transmitted by snails. It is the bioactivity of these compounds, in addition to the challenge presented by their unusual structures with enal and hydroxy aldehyde units in the same ring, that has helped to inspire the numerous synthetic studies that have yielded a number of analogs. Several of these compounds have been tested for their biological activity, and correlations between specific structural features and activity have been developed.[9] These structure activity

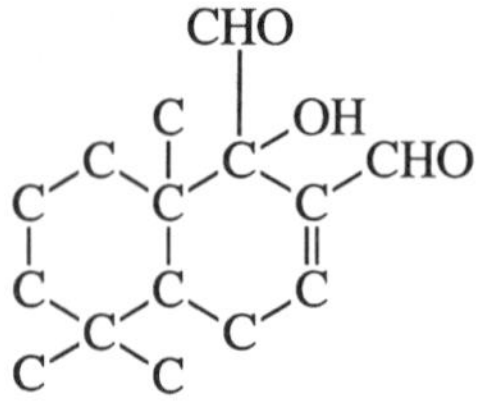

Warburganal

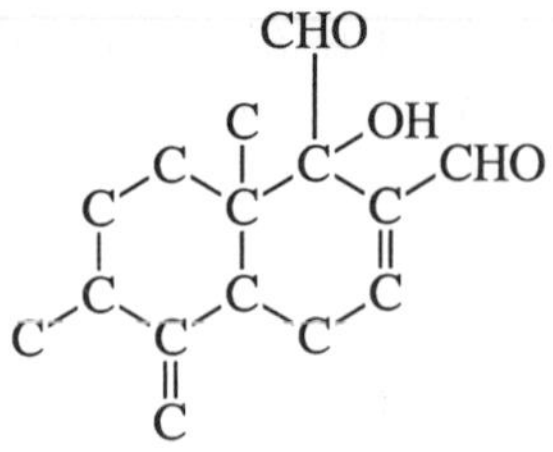

Muzigadial

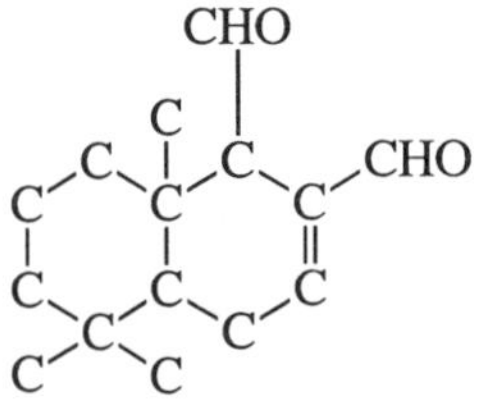

Polygodial

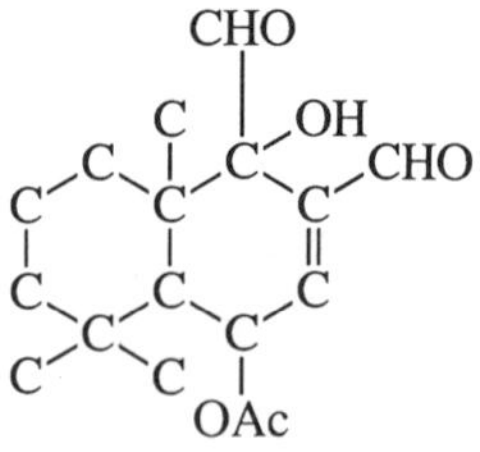

Ugandensidial

studies have involved the compounds *warburganal*, *muzigadial*, *polygodial*, and *ugandensidial.*

Biological activities of chemicals depend to a large degree on their three-dimensional shape. Compounds must "fit" receptor sites; in some cases the degree of fit can be more important than the actual chemical nature of the compound. Structures with, say, $-NH_2$ and -OH groups may show similar activity because these different chemical groups happen to fit a particular receptor site in some enzyme. Because three-dimensional shape is important, a search for a pharmacophore in a set of related compounds cannot depend on some scheme for finding common substructures defined as sets of bonded atoms.[27] Instead, a search must be made for common *three-dimensional substructures* defined in terms of sets of atoms at equivalent separations in three-dimensional space.

A prerequisite for any such search is an *X,Y,Z* coordinate model for each structure. For some of the structures involved in any particular study, coordinate data might be available from crystal X-ray studies. Apart from muzigadial,[28] X-ray coordinates are not available for the insect antifeedant structures. In cases such as this, model three-dimensional structures must be generated. One extension to the **GENOA-STRCHK** system is a generator of molecular conformations that provides coordinate models of structures of defined constitution and configuration.

The first phase of this system, **BUILD3D**, is a generator of *arbitrary* conformations of structures.[29] This generator is constrained so that the only conformations built are those that preserve defined configurations at stereocenters, maintain standard bond lengths and bond angles for bonded atoms, and appropriately restrict separations of nonbonded atoms. The generated conformations are defined in terms of the X, Y, Z coordinates of the constituent atoms of the compound. The coordinates resulting from this first step are subsequently refined. The energy of a given conformation of a molecule can be estimated from empirical force field calculations, and then the coordinates of the atoms can be systematically varied so as to minimize this energy.

In this example study, BUILD3D was used to generate initial coordinates for each of the four insect antifeedant compounds. The coordinates for each structure were then further refined by using Allinger's MM2 programs for minimizing strain energies.[30] Definitions of the structures, comprising both coordinate data and a specification of atom types, configurations, and connectivity were passed to the programs for finding common three-dimensional substructures.[31] There, the coordinate data were used to create a distance matrix with entries specifying the distance separating each pair of atoms.

The program for finding three-dimensional substructures works in terms of structure descriptors of the form (atoms of type x and y at a separation value in class z).[31] The first step in the processing entails the development of appropriate *atom-type* and *separation-class* descriptors. The atom types could involve chemical nature, number of substituent hydrogens, hybridization, aromaticity, or any number of other computed characteristics. The data that are appropriately included in the atom type depend on the specific application and are determined by the chemist using the program.

The substructure finding program has an interactive interface that allows the chemist to (1) enumerate appropriate atom properties, (2) define what *distance ranges* should be treated as equivalent (and so constitute separation class values), and (3) specify controls on maximum size of substructure sought[31]. An additional option allows the user to define a known common substructure. If, for example, all the compounds being studied incorporated an aromatic ring, it would be pointless to let the program expend central processing unit (CPU) cycles on rediscovering this feature. The program takes definitions of common substructures, collapses them into single pseudoatoms, and so reduces the effective size of the structures and the search space to be explored for common substructures. Once these user constraints have been provided, the first phase of the program is completed by generating the structure descriptions in terms of appropriately defined atom types and separations.[31]

The next phase of the program involves the "growth" of common substructures. First, substructures comprising two atoms of defined "type" and "separation" are identified in each structure; those substructures common to all the compounds are retained for further analysis. The major tasks of a common substructure finding program are really "bookkeeping" tasks. The program must maintain lists of each distinct three-dimensional substructure found along with details of where that substructure was found in each compound. For example, in three of these four compounds there are two instances of the substructure ($-CH_2-$,$-CH_2-$,"@-one-bond-length") but muzigadial has only one instance of this substructure.

Successively larger three dimensional substructures are found by adding, in all ways, a single atom to the substructures resulting from the previous steps. Unique descriptors are then generated for each of these enlarged substructures. (Descriptions of the larger substructures must incorporate stereochemical information; methods for combining stereochemical detail into the substructures have been briefly described[31]). A list is first formed that contains all substructures of size n created by expansion of size $n - 1$ substructures occurring in the first of the set of compounds being analyzed. All other compounds are then analyzed in turn and their size n substructures created. Any generated substructure that is not in the original formed list can be discarded. Substructures can also be eliminated from this list if absent from any of the subsequently analyzed compounds. The pruned list, subsequent to processing of all compounds, includes all common substructures of size n and forms part of the input to the generation procedures used to create substructures of size $n + 1$.

If a multiatom substructure is common to all the compounds, there may be very large numbers of substructures being manipulated at intermediate stages. The problem is a simple combinatorial one. If, for example, the actual common three-dimensional substructure involved eight atoms, it could result in as many as ${}^{8}C_4$ different common four-atom substructures being found. For the example compounds, there were in fact some 500 different types of five-atom common substructure.[32] Since some of these substructures occurred more than once in particular compounds, more than 2000 different sets of atoms had to be maintained by the program. The largest common substructures found by the programs incorporated 10 atoms; five distinct 10-atom substructures were found.

Figure II-15 illustrates the atoms that, in warburganal, correspond to each of the five common substructures of size 10 found for these compounds.[32] These are *geometrical* substructures (even though represented as simply atoms and edges). Their discovery by the program confirms that these atoms are similarly spatially disposed in each of the active compounds.

Development of computer techniques for representing three-dimensional substructures and searching for potential pharmacophores has only recently begun. Existing programs have many limitations. Thus the three-dimensional substructure search system required manual intervention and modification of controls before it could extend its development beyond the 10-atom substructures shown in Figure II-15.[32] The current programs for generating conformations create arbitrary forms only and do not allow for the systematic production of all conformations satisfying given constraints.[29] The decalin-based compounds considered in this example study are relatively rigid and possess few low-energy conformations, but other studies of pharmacophoric activity are likely to involve conformationally mobile structures. Such studies might well have to involve more elaborate searches with several alternative initial

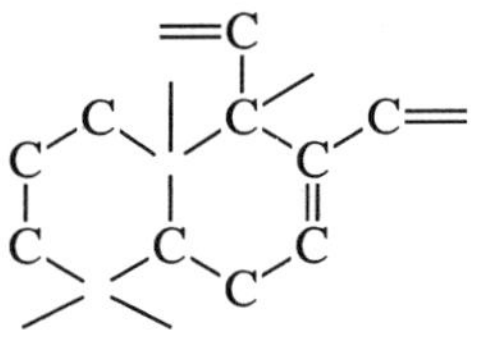

FIGURE II-15. The five common substructures of size 10 illustrated for warburganal. Bonds connecting common atoms have been marked for clarity, but are not formally considered part of the substructure because connectivity is not a consideration in the search for three-dimensional common substructures.

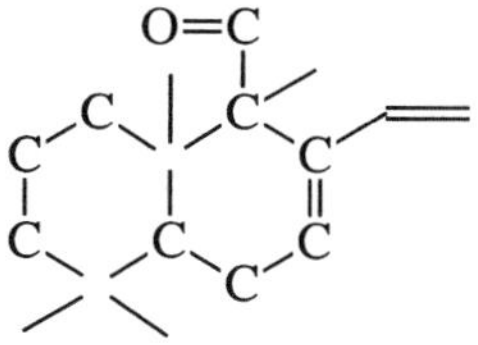

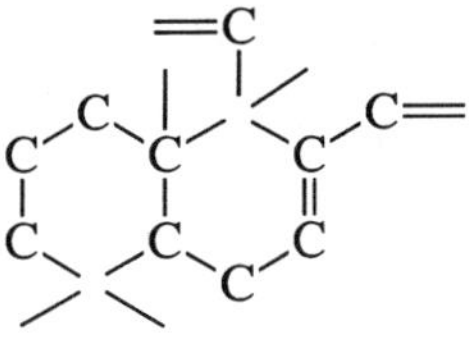

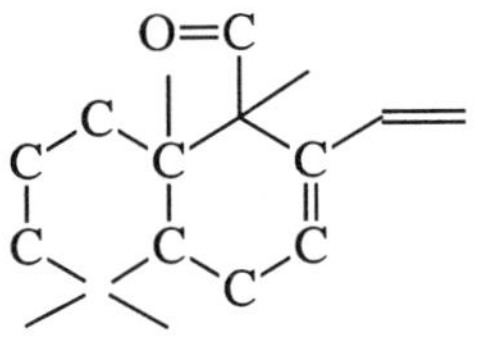

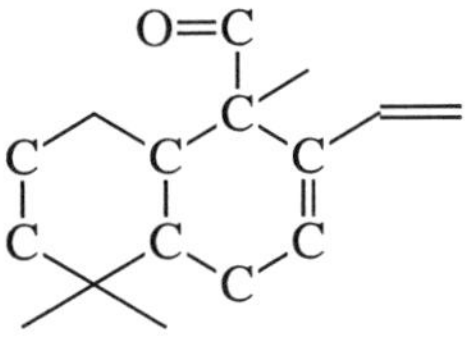

conformations being considered for each compound. Much current research and development work is devoted to these problems. Eventually, general and effective computer based methods will be available to assist in such structure-activity studies.

REFERENCES

1. I. Kubo, Y. W. Lee, M. Pettei, F. Pilkiewicz, and K. Nakanishi, "Potent Army Worm Antifeedants from the East African Warburgia Plants," *J. Chem. Soc., Chem. Commun.*, **1976**, 1013.
2. A. Ohsuka and A. Matsukawa, "Syntheses of (+/-)-Warburganal and (+/-)-Isotadeonal," *Chem. Lett.*, **1979**, 635.
3. A. J. G. M. Peterse, J. H. Roskam, and A. de Groot, "A Model Synthesis of Ring B of Warburganal and Muzigadial," *Recl. Trav. Chim. Pays-Bas*, **97**, (1978), 249.
4. S. P. Tanis and K. Nakanishi, "Stereospecific Total Synthesis of (+/-)-Warburganal and Related Compounds," *J. Am. Chem. Soc.*, **101**, (1979), 4398.
5. T. Nakata, H. Akita, T. Naito, and T. Oishi, "A Total Synthesis of (+/-)-Warburganal," *Chem. Pharm. Bull.*, **28**, (1980), 2172.
6. A. S. Kende and T. J. Blacklock, "Stereoselective Total Syntheses of (+/-)-Warburganal and (+/-)-Isotadeonal," *Tetrahedron Lett.*, **21**, (1980), 3119.
7. D. J. Goldsmith and H. S. Kezar, III, "A Stereospecific Total Synthesis of Warburganal," *Tetrahedron Lett.*, **21**, (1980), 3543.
8. S. V. Ley and M. Mahon, "Synthesis of (+/-)-Warbugranal," *Tetrahedron Lett.*, **22**, (1981), 3909.

9. K. Nakanishi and I. Kubo, "Studies on Warburganal, Muzigadial, and Related Compounds," *Israel J. Chem.*, **16**, (1977), 28.
10. R. E. Carhart, D. H. Smith, N. A. B. Gray, J. G. Nourse, and C. Djerassi, "GENOA: A Computer Program for Structure Elucidation Utilizing Overlapping and Alternative Substructures," *J. Org. Chem.*, **46**, (1981), 1708.
11. R. E. Carhart, D. H. Smith, H. Brown, and C. Djerassi, "Applications of Artificial Intelligence for Chemical Structure," *J. Am. Chem. Soc.* **97**, (1975), 5755.
12. C. Djerassi, D. H. Smith, and T. H. Varkony, "A Novel Role of Computers in the Natural Products Field," *Naturwissenschaften*, **66**, (1979), 9.
13. D. H. Smith, N. A. B. Gray, J. G. Nourse, and C. W. Crandell, "The DENDRAL Project: Recent Advances in Computer Assisted Structure Elucidation," *Anal. Chim. Acta*, **133**, (1981), 471.
14. D. H. Smith and R. E. Carhart, "Structural Isomerism of Mono- and Sesquiterpenoid Skeletons," *Tetrahedron*, **32**, (1976), 2513.
15. T. K. Devon and A. I. Scott, *Handbook of Naturally Occurring Compounds,* Vol. II, *Terpenes*, Academic Press, New York, 1972.
16. N. A. B. Gray, R. E. Carhart, A. Lavanchy, D. H. Smith, T. Varkony, B. G. Buchanan, W. C. White, and L. Creary, "Computerized Mass Spectrum Prediction and Ranking," *Anal. Chem.*, **52**, (1980), 1095.
17. J. G. Nourse, "The Configuration Symmetry Group and Its Application to Stereoisomer Generation, Specification, and Enumeration," *J. Am. Chem. Soc.*, **101**, (1979), 1210.
18. J. G. Nourse, D. H. Smith, and C. Djerassi, "Computer-Assisted Elucidation of Molecular Structure with Stereochemistry," *J. Am. Chem. Soc*, **102**, (1980), 6289.
19. N. A. B. Gray, C. W. Crandell, J. G. Nourse, D. H. Smith, M. L. Dageforde, and C. Djerassi, "Computer Assisted Structural Interpretation of Carbon-13 Spectral Data," *J. Org. Chem.*, **46**, (1981), 703.
20. N. A. B. Gray, J. G. Nourse, C. W. Crandell, D. H. Smith, and C. Djerassi, "Stereochemical Substructure Codes for ^{13}C Spectral Analysis," *Org. Magn. Reson.*, **15**, (1981), 375.
21. H. Egli, D. H. Smith, and C. Djerassi, "Computer-Assisted Structural Interpretation of ^{1}H NMR Spectral Data," *Helv. Chim. Acta*, **65**, (1982), 1898.
22. C. W. Crandell, N. A. B. Gray, and D. H. Smith, "Structure Evaluation Using Predicted ^{13}C Spectra," *J. Chem. Inf. Comput. Sci.*, **22**, (1982), 48.
23. E. J. Corey and W. T. Wipke, "Computer-Assisted Design of Complex Organic Syntheses," *Science*, **166**, (1969), 178.
24. W. T. Wipke, "Computer-Assisted Three-Dimensional Synthetic Analysis," in *Computer Representation and Manipulation of Chemical Information*, W. T. Wipke, S. Heller, R. Feldmann, and E. Hyde, eds., Wiley, New York, 1974, 147, Chapter 7.
25. W. T. Wipke, H. Braun, G. Smith, F. Choplin, and W. Sieber, "SECS - Simulation and Evaluation of Chemical Synthesis: Strategy and Planning," in *Computer-Assisted Organic Synthesis*, W. T. Wipke and J. Howe, eds., American Chemical Society, Washington, DC, 1977, 97, Chapter 5.
26. P. Gund, E. J. J. Grabowski, G. M. Smith, J. D. Andose, J. B. Rhodes, and W. T. Wipke, "Computer-Assisted Synthetic Analysis: the Merk Experience," in *Computer-Assisted Drug Design*, E. C. Olson and R. E. Christoffersen, eds., American Chemical Society, Washington, DC, 1979, 527, Chapter 24.
27. T. H. Varkony, Y. Shiloach, and D. H. Smith, "Computer-assisted Examination of Chemical Compounds for Structural Similarities," *J. Chem. Inf. Comput. Sci.*, **19**, (1979), 104.
28. F. S. El-Feraly, A. T. McPhail, and K. D. Onan, "X-ray Structure of *Canellal*, a Novel Antimicrobial Sesquiterpen from *Canella winterara*," *J. Chem. Soc. Chem. Commun.*, **1978**, 75.

29. J. C. Wenger and D. H. Smith, "Deriving Three-Dimensional Representations of Molecular Structure from Connection Tables Augmented with Configuration Designations Using Distance Geometry," *J. Chem. Inf. Comput. Sci.*, **22**, (1982), 29.
30. N. L. Allinger, "Conformational Analysis. 130. MM2. A Hydrocarbon Force Field Utilizing V_1 and V_2 Torsional Terms," *J. Am. Chem. Soc.*, **99**, (1977), 8127.
31. D. H. Smith, J. G. Nourse, and C. W. Crandell, "Computer Techniques for Representation of Three-Dimensional Substructures and Exploration of Potential Pharmacophores," in *Structure Activity Correlation as a Predictive Tool in Toxicology*, L. Goldberg, ed., Hemisphere Publishing Corp., New York, (1982), 171, Chapter 11.
32. C. Djerassi, D. H. Smith, C. W. Crandell, N. A. B. Gray, J. G. Nourse and M. R. Lindley, "The DENDRAL Project: Computational Aids to Natural Products Structure Elucidation," *Pure Appl. Chem.*, **54**, (1982), 2425.

HISTORICAL PERSPECTIVES

In this chapter we discuss the historical development of computer programs similar to those discussed in Chapter II. This discussion is somewhat expanded to include background to the development of other computer techniques for manipulating structural information, techniques that are significant to the development of later chapters. We do not pretend that this chapter represents an exhaustive coverage of the extensive literature in these areas. Rather, we try to focus on key program developments, deferring to later chapters the more thorough coverage required in tracing development of concepts, algorithms, and their eventual expression in computer programs.

III.A. SEARCHING COMPUTER-BASED COLLECTIONS OF SPECTRAL DATA

The first steps in analysis of an unknown molecular structure involve several spectroscopic analyses, resulting in a collection of spectral data that generally includes the IR, UV, mass, and NMR spectra. The most logical next step is to compare these data with data obtained from other compounds. Computer technology could aid this comparison, given a computer program capable of searching files of spectral data accumulated on known compounds. Indeed, there exist substantial collections of IR, mass, and ^{1}H and ^{13}C NMR spectral data in book form. Over the past 20 years considerable effort has been directed toward transforming these printed spectroscopic data bases to forms that are searchable by computer techniques.

III.A.1. Searching Data Bases of Infrared Spectra

Punched card systems for storing IR spectra and retrieving spectral matches using card sorters or collators were reported as early as 1951.[1] As the ASTM file of IR spectra grew to contain tens of thousands of spectra, several groups recognized the potential utility of computer-based systems. In 1964 there appeared several reports on the use of computer techniques to search the ASTM data base of IR spectra.[2-4] By 1968, systems using rapid search of data bases on magnetic tape[5] and magnetic disks[6] were in use. These systems were already taking advantage of computer techniques of data compression, logical operators in search terms, and rank ordering of successful matches. Subsequently, several systems have been developed to analyze mixtures[7,8], and to include additional information on peak width and intensity.[9,10] A system has been reported that allows searches to be performed using both spectral and structural search terms.[11] More recently, Sadtler Research Laboratories has entered into agreements with IR instrument manufacturers for collaborative development of search programs and an associated data base to be used in conjunction with the computers that control the instruments.[12]

There are several publicly available search systems that allow access to and search of the ASTM file of IR spectra. These systems include **Iris** developed by Sadtler,[12], **IRGO**,[13] **Sirch-360**,[14] and **First**[15]. Another search system, including only gas phase IR spectra, is available as part of the NIH–EPA Chemical Information System.[16]

One interesting aspect of the ASTM spectral data base is that, from the beginning, structural information in the form of compound class and functionality was encoded with the spectra. This is not surprising, considering the strong correlation of IR absorptions with functionality. Although earlier search systems did not take advantage of this information, more recent systems have,[11,16] thereby making it possible for chemists to compare the results of searches in structural terms rather than by code number or compound name. The new Sadtler system[12] provides structural information on spectra closely matching that of the unknown by means of a microfiche system that allows retrieval of the actual spectrum and a structural diagram.

III.A.2. Searching Data Bases of Mass Spectra

As in the case of IR search systems, the earliest method in mass spectral search systems used punched cards for encoding spectral data.[17] Since the early 1960s, literally dozens of publications have appeared concerning computer-based mass spectral search systems. Key early papers include the pioneering efforts of Abrahamsson et al.[18] and Pettersson and Ryhage.[19] Subsequent work focused on alternative methods for spectrum encoding and matching.[20-24] Detailed descriptions of these and related methods can be found in the excellent review by Pesyna and McLafferty.[25]

Currently, two computer systems are publicly available and that provide access to large (35,000 - 40,000) collections of mass spectra. The "Probability Based Matching" (**PBM**) system[24] is available through the computer center at Cornell University and, more recently, as part of the Hewlett-Packard mass spectrometry systems. **PBM** provides not only spectrum matching against a data base, but also has some capabilities for analyzing spectra of mixtures. The second system is the "Mass Spectral Search System" (**MSSS**)[26,27] component of the NIH-EPA **CIS**.[16,28] **MSSS** is particularly interesting because it provides a method for interactive search of the data base[26,27] in addition to more classical search techniques.[23] The interactive search can be termed a *feature selection* search, because the chemist selects key ions, the features, from an observed mass spectrum and queries the data base for each feature in turn, stopping when the search fails or when a manageable number of compounds possessing the given ions is retrieved.

The **MSSS** has another important feature, in that spectral matches that are retrieved are indexed according to molecular structure. The structures themselves can be drawn or used as search keys in other components of **CIS**. Thus this system provides the chemist with actual structural information rather than compound names. Another important system, the **STIRS** system developed by McLafferty and co-workers,[29,30] is designed to return structural information, from mass spectral searches, in the form of substructures.

Developers of mass spectral data bases, with the exceptions of those involved in developing **MSSS** and **PBM–STIRS**, have paid little attention to encoding structural information with spectra, a situation different from development of the IR data bases. Possibly this is because mass spectra generally include few signatures for specific functional groups. Therefore, the emphasis in mass spectrometry has been on the relationship between complete spectra and complete structures. Because early mass spectral search systems did not attempt to interpret the search results, structural characterization in the form of compound names was deemed sufficient.

III.A.3. Searching Data Bases of ^{13}C NMR Spectra

The relative newness of the technique of ^{13}C NMR spectroscopy compared to IR and mass spectrometry led very quickly to the development of spectral data bases on computers, primarily because many of the techniques developed for the older spectral methods could be adapted to ^{13}C NMR data. There have been several recent studies on the development of computer systems for searching such data bases. The systems can be logically grouped into two categories, complete spectrum matching or feature selection. These systems have been reviewed recently,[31] and additional information on comparison of techniques can be found there.

Voelter et al.[32] developed one of the earliest of the spectrum matching systems. In this system as in the later systems of Bremser et al.,[33] and Mlynarik et al.,[34], the retrieval procedure relies on the correspondence, within limited tolerances, of resonances in the unknown and known spectra. Bremser has published the results of further studies, including various techniques to expedite the search procedure.[35,36] Zupan et al.[37] also developed a search system for ^{13}C NMR data that is part of a larger system for comparison of multi source data noted in Section III.A.4. Schwarzenbach et al.[38] have presented a method for matching spectra based on certain spectral characteristics rather than individual resonances. This approach has some similarities to methods for mass spectral file searching that represent chemically meaningful regions of spectra rather than complete spectra.[23] Schwarzenbach's method is also part of a larger system for analysis of multi source data.[39] It is also incorporated in the NIH–EPA **CIS** as the ***CLERC*** option for searching ^{13}C NMR data.

The NIH–EPA **CIS** provides an interactive, feature selection method of searching its ^{13}C NMR data base in addition to the ***CLERC*** option.[40] This type of search is very much like the interactive search used for MSSS in that individual resonance lines are selected and used in turn to determine a subset of spectra that possess the given resonances. This approach can lead to some difficulties in retrieval of closely related compounds due to slight shifts in resonance positions.[31] An alternative approach by Bremser uses sets of resonance lines, thereby avoiding some of these difficulties.[36] The system due to Jezl and Dalrymple uses a data base of substructures and chemical shifts that can be searched interactively.[41]

The systems mentioned in the previous paragraph are all based on collections of spectral data linked to some computer representation of structure or substructure. Thus, like related search systems for IR and mass spectra, these approaches can return to the chemist detailed structural information about close matches.

III.A.4. Searching Combined Spectral Data Bases

Given extensive collections of spectral data, it would seem logical to provide in a single search system the capability for searching several data bases for close matches to the spectra of an unknown compound. There are three systems that provide this facility. The first system, reported by Erni and Clerc in 1972,[39] provided for searching small data bases of mass, IR, and ^{1}H NMR spectra. Subsequently, a ^{13}C NMR data base search system was included.[38] A system reported by Zupan et al.[37] provided for computer search of mass, IR and ^{13}C n.m.r spectra, again using relatively small data bases (ca. 1000 compounds). The latter system has now been implemented on a minicomputer.[42] The NIH-EPA **CIS** has the capability for feature-based searching of collections of mass, ^{13}C NMR, and IR spectra;[28] environmental applications of this system have been reported recently.[16]

All these systems include structural representations in the spectral collections, Clerc's system using Wiswesser line notation, the **CIS** using connection tables, and the Zupan system using both.

III.A.5. Other Data Bases

There have only been a few reported attempts to computerize UV or ^{1}H NMR spectral library searches. Most available ^{1}H NMR spectra were acquired at 60 MHz and consequently are typically complicated by strong couplings. Broad "resonance" peaks, representing the envelope of very many individual resonance signals, render these data less suitable than simple line spectra (such as MS or decoupled ^{13}C spectra) for computer-based searches. Whereas UV-visible data can be quite valuable for specific classes of compounds, for most structures the UV spectrum comprises simply an "end-absorption" region or maybe a single broad absorption band.

The largest computerized UV and ^{1}H NMR spectral libraries appear to have been developed at the Molecular Spectroscopy Scientific Information Center at the Novosibirsk Institute.[43] In the 1970s the Novosibirsk group had more than 17000 ^{1}H NMR spectra in a computer-searchable library along with 4000 UV spectra (as well as much larger collections of mass and IR spectral data). The published examples illustrate use of the ^{1}H NMR library from an interactive program with user input of selected chemical shifts and intensities.[43]

A brief account has been published concerning another system using some 8000 spectra from the Sadtler ^{1}H NMR collection.[44] The search system encodes query spectra as boolean vectors with bits indicating where resonances occur; spectral matching is based on boolean operations combining the search bit patterns with those data characterizing reference compounds.

III.A.6. Summary

In spite of these efforts, few structural problems are solved with the aid of these techniques for a variety of reasons, including at least the following:

1. Few of the systems described are accessible to other chemists because they were originally special purpose and no longer exist, are proprietary, or are not transportable to computers in other laboratories.
2. The number of spectra in available collections is severely limited in comparison to the variety of structures encountered in daily research. The rate at which new compounds are reported in the literature greatly exceeds the rate at which new spectra are added to collections.
3. Maintenance and certification of large data bases is an extremely time consuming and expensive job. Funding for such work is not available from research grants, so it requires the involvement of some larger organization, such as a government agency. The NIH–EPA **CIS** is one example of government involvement, but the coverage of spectral data bases in terms of number and variety of classes of compounds is still severely limited.

Reason 1 (above) may be resolved by new technological developments, including more transportable software systems and development of extensive nationwide and international computer networks to allow persons with access to a computer terminal access to a centrally operated system. Reason 2 can be resolved by developing systems that provide mechanisms for building local or personal data bases of compounds relevant to a small group of researchers or a company. Examples of such systems already exist.[42] A partial solution might be to commit additional resources to maintenance of large, central collections, such as the NIH–EPA **CIS**. Reason 3 will not be circumvented until scientific organizations realize that spectroscopic data collections are an extremely valuable community asset that can now be shared among the scientific community much more readily with computer techniques than in the past. The Cambridge Crystal File of lattice parameters of three-dimensional structures represents one data base that is maintained and updated continually, under partial support from user fees.

Perhaps the best way to view the utility of search systems is to acknowledge the fact that few compounds will be represented in data bases. It is likely, however, that data bases *will* contain spectra of related compounds, or at least spectral signatures relating to important substructures in an unknown compound. A good strategy is then to use the search systems to provide clues to the identity of the structure or portions thereof.

III.B. COMPUTER AIDS TO STRUCTURE ELUCIDATION

As a means for placing the subsequent sections into a coherent framework, we distinguish three distinct phases in the structure elucidation process—a three-phase approach that epitomizes one method of problem solving—the *Plan-generate-test* approach: (1) in the *plan* phase, spectral and chemical data are interpreted to yield constraints on the forms permitted for the unknown molecule undergoing analysis; (2) in the *generate* phase, some combinatorial process is used to produce all structures compatible with these constraints; and (3) in the *test* phase, the generated structures are evaluated in the context of any additional data that might become available and further discriminating experiments are planned. These different phases involve quite different kinds of analysis and vary in their suitability for computerization. In the main body of the text we consider these three phases in their natural sequence, that is plan first, generate second, and test last. In this introductory overview section it is more convenient to consider them in another order—an order that reflects the degree to which these various phases of the overall analysis can be realized through effective computer programs. Other recent reviews of these various topics include the major paper by Gribov and Elyashberg[45] and the shorter paper by Abe et al.[46]

III.B.1. The Generation of Structures

Requirements for a method for generating structural isomers arose in a number of distinct contexts during the 1960s. Lederberg's interest in computer assisted structure elucidation originated in studies on the scope of structural isomerism and of methods for representing chemical structures in terms of mathematical models.[47] Sasaki et al.[48], and Gribov[49] separately identified the need for structure generation procedures that could assemble structures from *standardized substructural components* whose presence had been inferred automatically from features of a compound's magnetic resonance or IR spectra. In the course of some conventional structure elucidation studies, Munk perceived the need for some more automated and uniform approach for assembling a complete structure from those *arbitrary structural fragments* that had been established on the basis of an analyst's interpretation of chemical and spectral data.[50,51] Of the three phases of the overall structure elucidation process,it is to this *generation* phase that computers are most ideally suited.

A structure generation problem will typically be defined in terms of a set of component parts, from which the structure must be assembled, together with some statement of known restrictions on the allowed combinations of these parts. The component parts may be simple atoms, of defined type and valence, or may be larger, user-defined assemblies of bonded atoms (*superatoms*). The bonding among the constituent atoms of a superatom is defined; a superatom can be attached to other atoms or superatoms by using any remaining unspecified *free valences* of its constituent atoms. The restrictions that are applied during the assembly process can take the form of descriptions of excluded and required arrangements of bonds between the component parts and, possibly, also more general specifications such as details of known ring systems. Generation processes essentially involve the exhaustive application of some combinatorial function for finding all possible ways of coalescing a given set of atoms and superatoms into a complete structure subject to available constraints on their bonding.

A valid structure generation process must be both *exhaustive*, guaranteeing that all possible structures are indeed considered, and *irredundant*, thus guaranteeing that no duplicate structures are produced. Exhaustiveness is obviously an essential requirement. If a generation procedure were in some sense incomplete, for example, unable to represent or manipulate some forms of fused ring systems, it could omit certain possible structures, maybe even the correct structure, from the lists of candidates that it produced for a given structural problem. This would undermine the entire basis of the normal structure elucidation process, which depends on the ability to consider all solutions and then find additional evidence that can be used to eliminate incorrect structural hypotheses. In principle, an exhaustive structure generator is easily realized through a recursive algorithm that simply explores all possible permutations of bonds among the available atoms and substructural fragments.[52] Practical algorithms are, perforce, more complex.

A computer program for structure generation should not permit redundant output; that is, it should not generate numerous alternative representations of the same solution. As a trivial example of the problem of redundancy, consider a program generating structures, with the composition $C_6H_{12}O$, at the stage when it has assembled a six carbon atom ring onto which it must substitute a hydroxy group and 11 additional hydrogens. It could at this point generate internal representations of structures that one might conceive of as cyclohexan-1-ol, cyclohexan-2-ol, ... and cyclohexan-6-ol. At some point, the program must determine that all these structures are in fact identical and, rather than produce six equivalent but differently numbered forms, must select just one single prototypical or *canonical* example.

Such problems of possible duplication are most commonly associated with symmetries but may also arise through the combination of constraints. For example, the constraints given to a structure generator might define a particular structural problem in terms of a molecular skeleton to which various substituents are to be attached, two of these might be required to be ester linkages with one further specified as an acetate ester. Several different substitution patterns of the acetate and the other ester group would, in general, be possible. Many of these different substitution patterns might later prove redundant if it were established, later, that the second ester grop might also take the form of an acetate.

With virtually all structure generation algorithms, there are circumstances wherein additional symmetries are either induced by, or only revealed through the final construction step. Consequently, *canonicalization* procedures are required in all structure generation programs. These canonicalization procedures are applied to the final structures, converting each to its corresponding standardized form and numbering. These standardized forms can then be checked against a list of structures already created and filed by the generation program, thus allowing duplicates to be recognized and discarded.

It is, however, impractical to proceed with just a simple combinatorial generation algorithm, such as one that merely permuted bonds between atoms in all possible ways, and rely on the final canonicalization procedure to eliminate redundant structures. Far too many duplicate structures would be created, tested, and eliminated. Rather than generating six isomers and choosing the canonical one as in the preceding example, it is far better to perceive the essential equivalence of the six possible bonding sites for the hydroxy group and choose to generate only a single structure. Generation algorithms must be capable of perceiving and exploiting symmetries of partial structures, substructures, and collections of atoms at all intermediate stages of the construction process. The perception of symmetry in, and the selection of a canonical form or numbering for a structure are, in fact, closely related algorithmic processes.

The structures of *acyclic* compounds can be irredundantly generated by a relatively simple recursive algorithm that avoids the generation of duplicate forms. A first solution, for acyclic alkanes, was developed by Henze and Blair in 1931.[53] A general procedure for acyclic compounds, given by Lederberg et al,[54] involves identification of a unique centroid for a chemical structure. This centroid, either a specific atom or bond, is then used as the starting point for mapping out the structure, as a tree graph, by arranging its constituent radicals in a systematic sequence. The **DENDRAL** program, implementing this procedure, allowed for constraints in the form of excluded and required substructures. These constraints could, at least to some extent, be used prospectively to limit the number of structures created. Applications of the algorithm included studies of the scope of isomerism in acyclic structures as well as its use in the *Heuristic **DENDRAL*** program for exploring methods of spectrum prediction and interpretation.[55,56] The basic algorithms for acyclic structure generation have subsequently been recreated in a number of other implementations devised for special applications.[57–60]

The general problem of structure generation is, however, considerably more complex. The complexity is in large part a consequence of the varied kinds of symmetry that are possible in cyclic systems. Analysis of such symmetries requires exact, mathematical models of molecular structures. Such mathematical descriptions of molecular topology are expressed in terms of *graphs*; a graph consists of nodes (corresponding to chemical atoms) and edges (which correspond to bonds). Tree-structured graphs, which correspond to acyclic chemical structures, are those that can be cut into two components by the scission of a single edge. Cyclic graphs are multiply connected so that cutting any one edge will still leave a path between any pair of nodes through the remaining edges. The fundamental structure of any cyclic graph can be more clearly perceived by suppressing all nodes of degree less than 3. The resulting graph is known as a *vertex graph*. Lederberg has shown that the topology of all molecules can be defined in terms of trees and vertex graphs[47] and that it is possible to enumerate and specify all vertex graphs that can be constructed from nodes of given valences.[61,62]

The first complete solution to the problem of exhaustive and irredundant structure generation was based on Lederberg's vertex graph formulation.[61] Given a set of specified numbers of atoms of known type and valence, together with the overall molecular unsaturation, it is possible to determine the appropriate set of vertex graphs and extract representations of these graphs from a catalog. Structure generation is then completed by first replacing nodes of degree 2 back on the edges of the graph and then by labeling all nodes with appropriate atom names. The labeling process,[63] used first to find all possible placings of bivalent nodes and then to find all namings of nodes with atom symbols, took full advantage of the complex symmetries present. Typically, a graph may have several symmetrically equivalent nodes to be labeled, and there may be several sets of identical labels to be used. There is a subtle interplay between the various symmetries of graphs and permutations of identical labels, and complex mathematical analyses are required if duplicate labelings—and hence duplicate structures—are to be avoided. The algorithms for identifying appropriate vertex graphs, elaborating these to give complete graphs and for graph labeling have all been proved mathematically.[64–66] The program ***STRGEN*** constitutes an implementation of these algorithms.[61]

The ***STRGEN*** program, based on this vertex graph analysis, was used in studies of the scope of isomerism.[67] The labeling procedures[63] are directly applicable also to problems of determining the numbers of substitutional isomers[68]—a task for which various special-purpose algorithms have sometimes been devised.[69] The ***STRGEN*** program also formed the computational core of the original **CONGEN** system for structure elucidation.[70]

The **CONGEN** (***CONSTRAINED GENERATOR***) program provides an environment in which the chemist can define a structural problem and invoke the structure generation algorithms of ***STRGEN*** within a constrained context.[70] The **CONGEN** program allows the user to define arbitrarily complex superatom components for a molecule and provides sophisticated mechanisms for defining both constraints on the bonding of the various components and general constraints on the form of required structures.

Generation of structures using **CONGEN** involves a two-step approach. First, the ***STRGEN*** program is used to generate *intermediate structures* from the sets of atoms and superatom parts identified by the user. Then, in a second stage (*embedding*), superatom parts are expanded to yield complete structures; this embedding procedure is another example of graph labeling.[63] A somewhat oversimplified view of the different processes involved in **CONGEN**'s two generation steps can be obtained by considering a structural problem where the chemist has defined some superatom parts, including, for example, a decalin ring bearing free valences, and has specified the residual atoms and unsaturations. Then, in the main generation procedure, the ***STRGEN*** program will ignore the internal structure of the superatoms, such as that representing the decalin ring, and will concentrate on building up all possible arrangements of bonded atoms and superatoms. The intermediate structures resulting from this process will specify different radicals bonded to the decalin and other superatoms. Subsequently, in the embedding stage, the **CONGEN** program takes into account the internal structure of the superatom parts and elaborates each intermediate structure into, usually, several final structures. Thus each individual intermediate structure, specifying a particular set of substituent radicals on the decalin system, will in general yield many final structures when the program considers the different permutations possible for placing such radicals about a ring system.

The ***STRGEN*** structure generation algorithm represents the most effective way of assembling structures comprised of several component sets, each set consisting of large numbers of identical atoms or substructural parts. However, with the exception of exotic applications such as isomer enumeration, few chemical problems really require such an exacting approach. Although their empirical formulas may have dozens of non-hydrogen atoms, in most structure elucidation problems, as handled by **CONGEN**, there are typically only 5–10 superatoms or atoms of several different types that must be manipulated, usually without an overabundance of atoms of any particular type. This kind of structural problem can be handled by simpler and faster algorithms.[71]

One such algorithm has been described by Kudo and Sasaki.[72] This algorithm forms the basis for the **CHEMICS** system for automated structure elucidation.[73] This system analyses structures in terms of a predefined set of multiatom substructural fragments.[74] For molecules comprised solely of carbon, hydrogen, and oxygen there are some 189 standard fragments. Typically, the presence or absence of each of these specific fragments is determined through an automated spectral interpretation system; a chemist, working with the **CHEMICS** program, may suggest larger macrofragments, but the analysis still proceeds largely in terms of the standard fragments.[75,76] Through an analysis of the empirical composition of the unknown and the fragments identified as consistent with the spectral data, the **CHEMICS** program identifies various self-consistent sets of standard fragments. Each such set of fragments represents a specific variant of the overall structural problem; each variant is analyzed individually by the structure generation algorithm.

The Kudo-Sasaki algorithm creates structures by filling in a *connectivity* or *adjacency* matrix of the constituent molecular fragments (the program does not actually use matrices directly but instead works in terms of an equivalent but computationally more convenient *connectivity stack* representation).[77] The matrix generation procedure finds all unique ways of successively filling out larger square subblocks of the matrix in such a way that connectedness can be guaranteed at each stage (so that the procedure does not result in the generation of disjoint final "structures"). If all the structural fragments being combined are distinct, the problem of expanding submatrices to give unique structures is relatively simple. More often, there are multiple instances of some fragments; in such cases distinct completed connectivity matrices will represent the same structure just with the atoms numbered in some different order. Such duplicates have to be eliminated by some final canonicalization procedure. A canonical matrix, from a set of equivalent matrices, can be taken as that having the greatest (or least) value as defined through some specified order for comparing matrix elements. Kudo and Sasaki developed a method whereby tests for identifying canonical representative connectivity matrices are performed during the construction procedure. As the matrix, or connectivity stack, is filled in, possible permutations that interchange atoms of the same type and which also affect only the filled locations of the matrix can be considered. If such a permutation applied to a particular submatrix yields a submatrix of greater value, it has already been revealed that the given submatrix can never result in a canonical structure and can be abandoned. Through such procedures, the Kudo-Sasaki algorithm can exclude most duplicate structures and help to minimize the computational effort needed to generate final canonical structures.

The method used by Kudo and Sasaki only works with components having bonding sites of a single nature. Frequently, the substructures that can be identified by chemical and spectral analysis have bonding of a plural nature. For example, an ester group has free valences on both a carbonyl carbon and an acyl oxygen. Substructures such as ester groups must be broken down into simpler components that have just one type of valence. Thus an ester must be considered as a carbonyl group and an oxy link, both of which bear two identical free valences. These smaller substructures must then be tagged with restrictions on their allowed bonding to recapture the original structural data.

Carhart[71] extended the Kudo-Sasaki approach with a more complete analysis of the problem of duplicate removal. Carhart developed a mechanism, related to the Morgan algorithm for canonicalizing structures, that defined rules for numbering the atoms of a molecule.[78] These rules could be applied during a structure generation process to limit the range of possible expansions of each submatrix. These rules were combined, in the program ***AMGEN***, with an exhaustive connectivity matrix generating algorithm. The ***AMGEN*** connectivity matrix generating program replaced the vertex graph analysis in the generation phase of the final production version of the **CONGEN** program. This version of **CONGEN** was still based on a two-stage structure generation process. Assembly of intermediate structures from atoms and superatoms was accomplished using ***AMGEN***. Expansion of the superatoms to express their full internal structure was still performed in a second embedding step. This two-step procedure does, of course, avoid the problems associated with components having bonding of plural nature that arose in the Kudo-Sasaki implementation.

Several other general purpose structure generators have been described in the literature. These include ***ASSEMBLE***,[79] the generator algorithm for Munk's "*Computer **A**ssisted **S**tructure **E**lucidation*" (CASE) system,[80,81] the ***MASS***,[82] program (***M**athematical **A**nalysis and **S**ynthesis of **S**tructures*) that is the generator for Gribov's structure recognition system **STREC**,[83] and the generator functions for **SEAC**.[84]

Like **CONGEN**, the **CASE** system is intended to handle arbitrary user-defined substructural fragments along with multiatom fragments identified by its various automated spectral interpretation functions. The **CASE** program allows each atom in such substructures to be *tagged*. These tags indicate particular environmental constraints that apply to that specific atom. The constraints that can be expressed include restrictions on the atom type or hybridization of possible neighbors for an atom and requirements that the atom be in a ring of specified size. The ***ASSEMBLE*** function creates complete structures from these fragments by use of an algorithm that defines a form of depth first recursive search. Each step in the recursive algorithm involves creation of a new bond from an atom in a partially assembled structure; a bond to an atom not yet incorporated results in the expansion of the partial structure, whereas a bond to another already incorporated atom yields a new ring. Each such bond-building step is verified for consistency with the constraints defined for the participating atoms. The ***ASSEMBLE*** algorithm analyzes the symmetry of both the partially assembled structure and the sets of as yet unincorporated atoms and fragments, in order to reduce the potential for duplicate structure generation.[85,86] A final canonicalization procedure eliminates such duplicates as are created.[87]

Like **CHEMICS**, *MASS*[82] works in terms of fragment sets selected from a library of standard fragments. The fragment sets appropriate for a particular unknown are identified by automated spectral interpretation systems. The *MASS* system generates all connectivity ("incidence") matrices which satisfy both the empirical formula and the predetermined distribution of valences for the selected structural fragments. The algorithm used can generate disjoint structures as well as duplicates, and consequently structures produced through this process must be checked for connectivity as well as having to be put into canonical form and checked for uniqueness. As in **CONGEN**, the elaboration of final structures involves two steps, with the *MASS* algorithm used to construct intermediate structures defined in terms of "structurally discrete units" (i.e., the superatoms inferred by the automated spectral interpretation procedures) and with a second expansion step used to define final structure representations. The structure generation procedure used in **SEAC** has not been described in detail but seems to be another variant of the basic connectivity matrix approaches of **CHEMICS** and *MASS*.[84]

Structure generators such as **CONGEN** and **CASE** require that the chemist be able to define a structural problem in terms of a number of discrete superatom parts and some residual atoms. The bonding allowed to specific free valences of these superatom parts can usually be constrained in various ways; thus the larger environment of any given superatom component can be partially specified. Arbitrary overlaps of structural fragments cannot readily be described. Unfortunately, structural information obtained through an anaylsis of chemical and spectral data is typically both redundant and ambiguous. The information is redundant in the sense that substructures inferred from different data can overlap one another to arbitrary degrees. Conventional structure generators require careful analysis by the chemist who must derive separate and distinct substructural components or superatoms from the inferred substructures. Additional constraints, expressed either in terms of restrictions on the bonding of superatoms or as atom tags, must then be provided to reexpress that structural information that was effectively lost when abstracting the discrete components from the original inferred substructures. Quite apart from the inconvenience of this general procedure, the enforced use of many small structural components in the generator, followed by some form of filtering procedure to select the desired structures, adds considerably to the computational costs. Moreover, spectral and chemical data are often ambiguous in the sense of admitting more than one possible interpretation. Most programs require a unique and rigorously specified set of substructures; if alternative interpretations must be considered, each such possible interpretation leads to a distinct structural problem that must be separately analyzed. These limitations of standard generation mechanisms have been addressed in the **GENOA** program,[88] the most recent in the series of structure generating programs developed through the Stanford **DENDRAL** project.

The **GENOA** program adopts an approach to structure generation that is quite different from that of more standard procedures. In a sense, **GENOA** does not generate structures at all; instead, it searches for ways of combining each new piece of structural information, as provided by the chemist, with the results of previously specified structural constraints. The chemist can define substructures of arbitrary complexity and can specify how these substructures are to be used by either indicating their number of occurrences or identifying a set of substructures postulated as being alternative interpretations of some chemical or spectral data.

The **GENOA** construction procedure is closely related to *graph-matching* algorithms. Graph-matching algorithms define methods for identifying the equivalence of two structures or finding a specified component within some larger structure. Such algorithms have many applications. Graph matchers are used, for example, at several stages in conventional structure generators as when checking generated (partial) structures for the presence of particular required or prohibited combinations of atoms and superatoms. Such algorithms are also employed in structure evaluation programs for such purposes as identifying those generated structures that incorporate some standard skeleton or when predicting spectral characteristics on the basis or some substructure–subspectrum model.

Carhart extended the standard substructure search procedures of conventional graph matchers and developed the concept of *constructive substructure search*. In a constructive substructure search, if the number of required instances of some specified substructure is not already satisfied in some partially assembled structure, the algorithm identifies all alternative ways of expanding that partial structure until the constraint is satisfied. Unless otherwise specified, substructural constraints given to **GENOA** are presumed to be potentially overlapping; substructural components that are known to be distinct can be defined, of course, and mechanisms exist for expressing limitations on potential overlaps.

The **GENOA** constructive substructure search procedure builds up a set of partially assembled structures, or *cases*, each of which represents one distinct way of realizing all the structural constraints provided. When all the available structural constraints have been utilized, the final cases frequently correspond to unique structures. A simple connectivity matrix generation algorithm is used to create final structures from those cases that do still represent collections of disjoint structural components.

Another recent development has been the creation of algorithms for generating configurational stereoisomers.[89,90] Standard structure generators work in terms of molecular constitution, ignoring all aspects of stereochemistry. The results of such programs are, of course, in a sense incomplete as they leave stereochemistry undefined. This leads to problems when attempting to evaluate candidate structures by predicting their spectral properties. Many physical properties of molecules are quite markedly influenced by stereochemical factors. Spectral properties predicted for a constitutional structure are essentially some form of average of the spectral properties of all the possible stereoisomers of that structure. Such averaged predictions are inherently imprecise, and there are limits to the degree to which these predictions may be used as a basis for discriminating among different candidate structures. Since so many structure-activity and structure-property relationships are dependent on stereochemistry, it is essential for a complete structure elucidation system to incorporate stereochemistry.

In current implementations, stereochemistry is handled separately from and subsequent to the constitutional analysis. Constitutional isomers must first be generated through **CONGEN** or **GENOA**, and then the stereochemical analysis must be performed by the separate STEREO program. Current stereoisomer generators allow for a wide variety of stereochemical constraints to be expressed and used to limit the stereoisomer generation process. Configurational stereochemistry may be more thoroughly incorporated in future structure generation procedures because there are examples where stereochemical constraints can exclude all stereoisomers corresponding to a particular constitutional structure and, consequently, eliminate that constitutional form from the list of candidates. Further extensions, to handle conformational stereochemistry, are currently being developed.[91]

The computer's power at handling the combinatorics of structure generation perfectly complements the human reasoning needed to infer structural constraints from spectral and chemical data. A synergy of human and computer problem-solving has been exploited in a number of systems, including **CONGEN**, wherein the computer program performs only the generation process using constraints provided by the chemist through a flexible interface procedure.[70] More generally, we would like the computer to aid in the other phases of the structure elucidation process, by helping to evaluate the numerous candidate structures that it so readily produces and, possibly, even by helping to interpret data to derive the constraints that are used to guide and delimit the generation procedure itself.

III.B.2. Evaluation of Structural Candidates

In conventional analysis, the analyst usually soon realizes when the data are inadequate to uniquely define a structure and thus abandons any attempt to list all possible isomers and instead returns to the laboratory in an effort to obtain more structural information through further experiments. If computerized structure generators were used routinely, chemists would have somewhat greater flexibility as to how they might proceed. Provided the initial structural constraints are sufficient to limit the number of candidates to a few thousand at most, it is quite feasible to use a computer program to generate all candidates. Other computer programs may then be used to analyze these candidates, identify those that seem most plausible, and help in planning of additional, more discriminating, experiments.

There are at least two facets to such a computerized analysis. First, a program can search through the generated structures, grouping them according to substructural features. Such groupings can be based on logical combinations of standard substructural components as taken from a library or defined by a user.[81,92,93] Or, more elaborately, programs can group structures using computed properties such as estimated strain energy, cyclic skeletal type,[81] or even according to predictions of expected products in defined chemical reactions.[94,95] Second, a program can undertake a more detailed analysis of all available spectral data. Although some spectral and chemical data will have been exploited in the process of inferring constraints for the structure generator, this exploitation is rarely comprehensive. Spectral interpretation processes, whether manual or automatic, rely primarily on subspectral–substructural correlations. Given hypothesized candidate structures, preferably incorporating stereochemistry, it is possible to analyze spectral properties that depend on the more global molecular context, and properties, such as mass spectral fragmentation patterns, that are readily rationalized only in terms of complete structures.

Possibly because one requires a working structure generator before such evaluation procedures are of practical value, only limited work has been done in this area. Most structure generators do incorporate at least rudimentary facilities for surveying the set of generated candidate structures to determine which of these incorporate either individual standard-skeletons or substructures, or chosen combinations of substructural features.[81,92,93,96] Elaborate methods for automatically grouping structures, according to presumably discriminating features, have received little, if any, practical application.[81,93] Methods for analyzing hypothesized structures and their expected chemical transformations, and exploiting the observed results of the actual experiments to yield new structural constraints have been demonstrated and provide an interesting link to the studies of computer-assisted chemical synthesis planning.[94,95] However, such methods have again received relatively little practical application.

An approach that has been somewhat more widely investigated involves prediction of spectral properties for each candidate and ranking the candidates according to some measure of the consistency of the observed data and the predictions made for each candidate. Many structural properties could in principle be predicted through either *ab initio* or semi empirical quantum-mechanical calculations. Few attempts have been made to adapt such quantum-chemical techniques for use with structure elucidation systems.[97] There are two reasons why such quantum-chemical calculations are of little immediate applicability: (1) the "structures" produced by generation procedures are usually merely definitions of molecular constitution—substantial computations are necessary for generation of even approximate three-dimensional models necessary for quantum-chemical programs;[98] and (2) the quantum-chemical calculations are themselves costly and are rarely of such precision for the results to provide a reliable basis for discrimination among very similar isomeric structures.

Instead, spectral predictions have, for the most part, been based on empirical schemes. Such schemes employ simplified models for spectral prediction such as, for example, additivity rules for predicting chemical shifts. Predictions based on such models are computationally inexpensive, and it is quite feasible to process hundreds of structures. Such predictions are, of course, approximate, and limitations inherent in the theory or model used may bias the results in favor of particular structural types. To some extent, such biases and other limitations can be taken into account through the use of a well-chosen scheme for "matching" predicted and observed data. In any case, the actual rank ordering of hypothesized structures is not of particular significance; instead, the valuable result from such spectrum prediction, matching, and ranking processes is a division of the set of possible candidates into those that seem relatively plausible and those that are, in one way or another, inconsistent with more global spectral characteristics. The chemist retains responsibility for the evaluation of the candidate structures; the spectral analysis programs simply help to identify those candidates that seem most likely and, consequently, should be considered first.

Sasaki et al[73] have described a system in which spectral properties of generated candidate structures are retrieved from a reference data base and compared with those observed for the unknown; such comparisons can be used only to exclude structures for which data are available in the data base. Such methods are of limited application, for few candidate structures would be in a reference data collection and, in general, it would be more effective simply to identify such standard molecules as do arise by conventional file search methods rather than exploit the complete structure generation–evaluation procedures.

The **STREC** system uses substructure-subspectra correlations to predict ^{1}H NMR, MS, IR and UV spectra of hypothesized candidates.[45,83] The library of substructural fragments used in this system contains about 20 UV–, 80 IR–, 150 ^{1}H NMR–, and 270 MS–substructure/subspectra combinations. The resulting predictions are checked for gross inconsistencies with the observed data, thus allowing some inappropriate structural candidates to be excluded. In addition, Gribov et al[99,100] have developed more refined methods for calculating the IR spectra of structures and using such calculated spectra to filter the list of candidates.

A number of studies on the use of ^{13}C NMR for structure evaluation have been reported and recently reviewed.[31] Munk[101] and Sasaki[102] and their coworkers have devised methods for predicting the number of distinct resonances to be expected in the ^{13}C spectra of structures. Such predictions are, however, based solely on the analysis of constitutional symmetry and provide only a relatively crude basis for eliminating inappropriate structures.[31] Additivity rules have been used in some programs for predicting approximate chemical shifts for carbon atoms.[103] More elaborate methods for predicting chemical shifts, through the use of detailed and extensive libraries of substructure–shift correlations, have been reported by a number of authors.[33,35,41] Recently, such studies have been extended to include configurational stereochemical information.[104,105] Predictions based on such substructure-shift correlations result in "fuzzy spectra," wherein a resonance for an atom is given as a shift range rather than as a specific value. Predictions made for different candidate structures vary greatly in precision depending on the detail of substructural models found in the reference library. A study of how these problems affect the matching and ranking processes has been presented.[106]

Since only constitutional structures are produced by most structure generators, properties that are markedly influenced by stereochemical factors (e.g., ^{1}H NMR spectra) cannot readily be predicted with any accuracy. Mass spectrometry is the technique that is probably the least strongly influenced by stereochemical factors. Consequently, mass spectra are in some ways the data most readily and appropriately analyzed in the evaluation of candidate constitutional structures. Simple mass spectral prediction functions were included in Heuristic **DENDRAL**.[55,107] Two basic approaches to mass spectral prediction have subsequently evolved.

Detailed, class-specific rules can be used to define the standard fragmentations of known skeletons; such rules may subsequently be applied to new structures incorporating the same skeletons. Such predictions can be used to establish the nature and position of substituent groups in the new structures. Such rule-based schemes are capable of extremely subtle and precise discriminations between closely related isomers.[108] They are, however, of limited applicability, and the rule development process[109,110] is relatively costly.

The **DENDRAL** *half-order theory of mass spectrometry* constitutes a more generally applicable, but much less precise, spectral "prediction" procedure.[109] The half-order theory provides a method for systematically exploring the possible fragmentations of a given structure under specified constraints concerning the overall complexity of allowable fragmentation processes. Fragmentations involving the cleavage of too many bonds or involving excessively complex combinations of neutral losses and hydrogen transfers can be excluded from consideration. Through this exhaustive procedure, all possible fragmentations of a given hypothesized structure that could lead to an observed ion may be systematically explored. If some observed fragment ion in the spectrum of an unknown is readily rationalized in terms of fragmentations of one hypothesized structure but can result only through relatively complex fragmentations of another candidate, one may reasonably infer that the first structure is the more plausible.[111] Candidate structures can be ranked according to a measure of how readily they serve to rationalize all the ions in an observed spectrum.[112] The half-order theory is known to have biases to certain structural types[112] but still provides a convenient basis for a crude division of structural candidates into more and less probable classes.

III.B.3. Structural Interpretation of Spectra

Computer programs are least readily applied to the problem of "planning"—that is, helping the chemist to infer structural constraints from spectral or chemical data. Chemists exploit a great deal of background information when interpreting spectral data. The spectral interpretation process itself is to some degree an art, a skill that must be acquired through practice. Although some aspects of spectrum interpretation are based on strictly factual rules (e.g., *carbonyl groups are characterized by strong IR absorption in the 1850–1650cm^{-1} region*) many other more heuristic rules are employed, such as in determining the relevance of special cases or when making more subtle, possibly subjective, judgmental discriminations. It is extremely difficult merely to encode, in a computer-compatible form, the kind of background information necessary for the more sophisticated forms of spectral interpretation. It is even more difficult to devise an effective program for exploiting such information. These difficulties have constituted something of a challenge to developers of computer systems, and a number of distinct approaches have resulted from attempts to meet this challenge.

One approach has been to attempt to exploit information obtained through spectrum matching *file searches*. Most file search systems are, of course, concerned primarily with the recognition of standard compounds by matching a recorded spectrum with reference data pertaining to known compounds. Some file search systems do go further and attempt to provide specific structural information for those compounds not yet represented in the reference file. It is possible to construct files containing not spectrum–structure combinations but, instead, subspectrum–substructure combinations. Through the use of such files and appropriately modified versions of conventional file search algorithms, it is possible to identify the most plausible substructural components in a molecule.[35] More commonly, substructures are inferred from some analysis of those reference structures retrieved on the basis of a conventional spectrum matching file search method.[29] Typically, if a specific substructure is present in more than a certain percentage of the most closely matching reference structures, its presence in the unknown structure is assumed.[30,113] Such use of information derived through file searches is essentially an informal application of *k-nearest-neighbor* (k-n-n) pattern recognition methods wherein an object is assigned to the same class as includes the majority of its k nearest neighbors (with k being a parameter for the algorithm).

The k-nearest-neighbor method is just one of a number of *statistical pattern recognition* methods whose potential applications have been subject to widespread investigation by analytical chemists.[114] Although one or two earlier studies of such methods had been reported, the seminal work in this area was that of Jurs, Isenhour, Kowalski, and Reilley. In 1969 these authors illustrated how *linear learning machine* methods could be applied to the interpretation of mass, IR and multisource, spectral data.[115–117] Subsequently, these initial researches have been elaborated,[118] and several other pattern recognition procedures have also been applied.[119–124] The field of pattern recognition systems and their application to chemical problems has recently been comprehensively reviewed.[125]

One of the most characteristic features of the pattern recognition approaches is the fact that they establish direct correlations between spectral data and structural information. These direct correlations are made without resource to any of the intermediate interpretation levels or concepts that would normally be employed by a chemist when interpreting spectra. There are many advantages to the use of such direct correlations. Obviously, pattern recognition procedures are not restricted to those applications for which interpretation mechanisms have already been devised. Instead, they may be applied wherever reliable correlations can be established—even in cases where, given our current knowledge, it is not possible to identify the physical basis for the correlation. Further, because intermediate, technique-dependent conceptual stages are not involved in the process of spectral analysis, it is practical to adopt a uniform algorithm for interpreting all types of spectral data.

Such benefits of the pattern recognition approaches are not realized without significant costs. Typically, a pattern recognition procedure makes a single binary decision: based on spectral evidence a compound either is, or is not of class x. Obviously, a complete structure determination process will have to involve a very large number of such simple classification steps. Thus, although any one computational step may be simple, a relatively complex overall system may be required to group and sequence the application of many such steps. Furthermore, the only justification that a pattern recognition process can give for any decision is that "that is the way the numbers come out." Most pattern recognition methods use classification functions that involve parameters, such as weights in some weight vector, whose values are determined through some training procedure using sets of data from compounds whose classification is known. The training procedure guarantees that all these standard data are correctly classified (a class-member recognition process) and that the resulting classifier also achieves a high success rate when applied to additional examples (a class-member prediction process). It is, however, rare for a 100% success rate to be achieved in prediction. Usually, some false-positive results are found, where compounds are erroneously identified as class members; other compounds, which should be identified as class members, are omitted. Since the pattern recognition methods cannot give any real justifications for a proposed classification, the chemist has little guidance as to whether to accept or reject any specific result and has no means of checking the reasoning leading to the program's conclusion.

Instead of ignoring existing human expertise in spectrum interpretation, we can attempt to incorporate such skills in computer programs. Such efforts lead to the third general class of spectral analysis programs. These programs are commonly described as *artificial intelligence programs*, but a possibly more apt description might be *knowledge-based programs*. Such programs are characterized by a *rule base* and a *rule interpreter*; the specific form of the rules and the nature of the interpreter can vary widely. The rules are provided by the chemists developing the system and attempt to express and codify their knowledge of molecular spectroscopy. (It is also possible, in some circumstances, to generate the rules automatically through an analysis of known structures and their spectra.[126]) An interpreter is typically some form of search procedure that determines which rules are apposite to the analysis of particular spectral data.

In a number of the early studies, the rules for spectral interpretation were expressed through FORTRAN subroutines or equivalent program constructs.[52,127–134] Encoding of interpretation rules as subprograms eliminates the distinction between *rule* and *rule interpreter*, and this has a number of adverse consequences such as the following: (1) it becomes difficult to amend or extend rules; and (2) the rules are no longer in a form amenable to simple analysis by other programs, and, consequently, it may well prove difficult to generate explanations for decisions based on the use of particular combinations of rules, or to examine a set of rules to detect inconsistencies. Far greater flexibility is possible if the spectral interpretation rules are represented as data for an interpretive program.

Some interpretation schemes impose a detailed structure on the organization of "data-type" rules; the major benefit of such structuring is in the efficient use of the rules. Most typically, a tree-structured hierarchy of classes and subclasses is defined. Associated with each class in the hierarchy is a set of rules defining tests for its distinctive spectral features. Each such rule typically takes the form of a set of parameterized calls to standard spectrum checking functions built into the program. These built-in functions define, and limit, a program's capacity for analyzing spectra. The rules are essentially merely data; each rule identifies a subroutine and its appropriate parameters. Rules for identifying particular classes can be readily modified without requiring changes to the interpretation program. In one such scheme, for the analysis of mass spectra on laboratory computers,[135] the chemist used an interactive program to define the form of the classification hierarchy and specify the spectral features characteristic of each compound class. These data were encoded for subsequent use in a program that identified probable structural class of compounds as their spectra were acquired by a computerized gas chromatographic-mass spectrometric (GC-MS) system. A more recent and more elaborate example of such a hierarchical classification system is provided by Woodruff and Smith's PAIRS IR interpretation program.[136,137] The PAIRS system includes both a rule interpreter, which applies encoded rules to recorded data, and a "compiler" that generates the appropriately encoded form of rules from their original source form as expressed in the CONCISE language.[137]

The simplest and yet in many ways the most versatile rule-based schemes exploit what are in essence simply correlation charts. The interpretation procedures of Sasaki's **CHEMICS** system[74,73] and Gribov's **STREC**[83] system are both based on such correlation charts. The **CHEMICS** interpretation system uses tables that define those features in the IR and ^{1}H NMR spectra that would have to be observed if the unknown structure were to incorporate any one of the standard substructural fragments defined in the system. By successively comparing the observed IR, ^{1}H NMR, and other spectra with the requirements, as specified in these tables, increasing numbers of the standard components may be eliminated from further consideration. Gribov's **STREC** system works primarily in terms of a detailed analysis of IR spectral-substructure correlations.

Mitchell and Schwenzer[126,138] developed a system for inferring structural information from ^{13}C spectra and using this information to guide a special purpose structure generator. Their method is described in terms of *production rules* and a rule interpreter. It is essentially a rather refined correlation chart approach. Each production rule relates a specific substructure to a defined spectral pattern. These spectral patterns consist of a primary resonance for a central atom of the substructure and secondary-support predictions that define broader resonance ranges for other constituent atoms. The rule interpreter works by matching the patterns, associated with particular substructures, against the observed spectrum. The successful application of some rules provides this rule interpreter with a more detailed context in which to explore the possible application of subsequent rules.

The wealth of structural information present in a ^{13}C spectrum has inspired several other attempts at developing interpretation systems. An early, only partially automated system was described by Jezl and Dalrymple.[41] More recent developments include the ^{13}C interpreters for both the **DARC** system[139,140] and the **DENDRAL** system.[104,105] These systems all exploit very large libraries or data bases of subspectra-substructure combinations. In the **DARC** system, substructures are keyed to a resonance pair; as in the Mitchell–Schwenzer system[138], the interpreter is guided in its search for possible substructures by context information established at earlier stages in the procedure.[140] The **DENDRAL** ^{13}C interpretation programs correlate substructures with single chemical shifts (of specified multiplicity); these programs can exploit known structural constraints, as may be available from other data, to aid in their analysis of ^{13}C data.[104,141]

A major problem facing these more elaborate interpretive procedures is that of *ambiguity*. Several rules may provide alternative substructural interpretations for certain spectral features; it is in this sense that the spectral features are "ambiguous." Interpretation procedures have to resolve such ambiguities by identifying mutually consistent interpretations for different features. Simple versions of such procedures are incorporated in the systems due to Sasaki[73–75] and Gribov[45,83]. Basically, these procedures must select, from among the set of fragments identified as consistent with the spectral data, combinations of fragments necessary to make up the empirical composition. As well as the restrictions provided by the molecular composition, other constraints can be applied to this selection procedure. For example, one resonance in the ^{13}C spectrum of a compound might have admitted interpretations in the form of an aromatic carbon bonded either to an oxygen or a saturated carbon; if this is the only structural component identified as containing an aromatic carbon, it can be excluded by the component selection process (because its implied requirement for other aromatic components can not be satisfied). Szalontai et al.[142] have developed a more extensive analysis procedure of this type that can identify mutually consistent interpretations of features in ^{13}C and IR spectra.

As the size of the correlation tables used are expanded from a few hundred entries[73,142] to several tens of thousands of entries[104], the problems of ambiguity are greatly magnified. Considerably more complex mechanisms are necessary to allow for effective searches for mutually consistent combinations of interpretations of individual spectral features.[143] Research has only recently started on these problems, and considerable further refinements are necessary before practical interpretation programs can be realized.

Although it is useful to distinguish *file search*, *pattern recognition* and *knowledge-based* systems as the three prototypical methods for spectrum analysis, it is not uncommon for a particular program to straddle more than one of these categories. For example, McLafferty's **STIRS** system for mass spectral analysis is simultaneously a (1) file-search method for spectrum matching using elaborate highly discriminating computed spectral features,[29] and (2) substructure identification program using essentially a k-nearest-neighbor approach to interpreting the results of the initial file search[30,113]; further, the combination of **STIRS** and a maximum subgraph analysis program[144] has much of the character of certain of the knowledge-based systems.

III.C. SEARCHING COLLECTIONS OF STRUCTURES AND THEIR ASSOCIATED PROPERTIES

In most structural studies, structure elucidation is not an end in itself. If the previously unknown compound displays interesting properties, such as biological activity, a new set of important questions arises. Was the structure of the compound reported previously? How do the properties of this compound relate to other, similar, known structures? Are similar structures available to simplify the synthesis of this and related compounds?

The traditional way to answer these questions is to go to the chemical literature. The problems here are obvious. Literature is keyed to chemical names, not structures. A search for important properties will succeed only if those properties were previously reported *and* abstracted properly. There is no direct way to search manually for related structures. This is obviously another area in which computational techniques can be of assistance, and several systems have been developed.

These systems fall under the name *chemical information systems* (CIS). The general strategy in such systems is to build a computer file of chemical structures with associated properties. The structures will be in one of a number of different representations, but one that allows the data base to be queried at least in structural terms. The properties may include chemical names, accession numbers, or literature references, together with subfiles or cross-references to other files that contain, for example, toxicologic data, spectral data, and bioactivities. Much of the development related to construction and use of information systems has been and remains centered in the chemical industries, because they have a vested interest in rapid retrieval of information on structures and related properties. A summary of early developments of chemical information systems has been presented.[145] This summary discusses the major systems under development over the period 1962–1968. Since that time, most large chemical firms have developed their own versions of information systems. More recent efforts have been discussed in a review by Ash and Hyde[146]. The focus of most of these systems is on chemical structure, making it possible to search large data bases on the basis of structure or substructure. However, seldom is a mechanism provided for the alternative method of searching, that of directly retrieving structures based on searching for specified properties.

Tracing the developments of chemical information systems would be tangential to our presentation because we are focusing on systems that are now publicly available to all chemists. Important aspects of the concurrent development of notational systems are given in Chapter VII, and searching is presented in Chapter VIII.

In publicly available chemical information systems the situation is somewhat better than that pertaining to spectroscopic data bases. The Chemical Abstracts Service (CAS) has, of course, pioneered in the applications of computer technology to handling of chemical information. Here, too, the system is keyed to chemical structure. It is interesting, however, that only recently has it been possible to actually *search* CAS files on the basis of structure.[147] In CAS, the "properties" relating to structure are the CAS numbers, literature references, and associated abstracts. For many years, there have been commercial systems available to search on the basis of such properties, including Lockheed **DIALOG**[148], the **ORBIT** system[149] and the **BRS** system[150]. The addition of the ability to search on the basis of structure should make *structural* access to the chemical literature much simpler in the future. The **QUESTEL** system[151] is a commercially available system that is designed to allow structure searching of CAS structure files.

Another publicly available system, the NIH/EPA **CIS**, has been mentioned previously.[28] This system is intimately keyed to chemical structure as the primary basis for gaining access to several files of structural properties in addition to spectroscopic data.[28] This system would be ideal for many applications except that the coverage of compounds is limited and there is no mechanism for adding private collections of structures and properties to the file.

In summary, then, there still are significant limitations to available systems. Private systems are well-developed for obtaining information on structures and related properties, but they are not accessible publicly. Publicly available systems such as **CASONLINE** and **QUESTEL** are not designed to capture directly critical information on properties, or, for **CIS**, have limited data bases. There is now a commercially available system, the **MACCS** system from

Molecular Design, Ltd.[152], which provides the capability of building and maintaining private data bases of structures and properties. These data bases can be searched for either structures or properties. The relatively high cost of the system precludes purchase by individual investigators or small research groups, however, although this situation may well change in the future as the cost of computer hardware continues to decline.

III.D. COMPUTER AIDS TO SYNTHESIS PLANNING

The first reports of computer programs for assisting in the planning of chemical syntheses appeared in the late 1960s, simultaneous with the first reports of computerized structure generators and file-based compound recognition systems. In their 1969 "***O**rganic **C**hemical **S**imulation of **S**yntheses*" (OCSS) program,[153] Corey and Wipke realized a computer implementation of the *logic-centered* approach to synthesis previously enunciated by Corey.[154] Although other researchers have subsequently introduced new concepts, the work by Corey and Wipke constitutes a paradigm for all computerized approaches to synthesis planning. There are several reviews of the development of synthesis planning programs.[155–161]

Computerized synthesis planning programs start with a given "target" molecule. The target structure is analyzed, and synthetically important features, "synthons" or "synthemes," are identified; these features may include simple functional groups, sets of functional groups having some defined structural relationship, ring systems, and bonds whose cleavage would greatly simplify the structure.[162] These identified features then serve as keys for accessing a data base of chemical reaction data.

This data base has to be oriented to the description of the product; one knows features in the product and needs to identify those chemical changes that could yield such features. The entries in the data base are known as *transforms* and define the retrosynthetic structural changes that allow identification of the precursor(s) that could, by real synthetic reactions, be converted into the desired target. The transforms in the library are indexed by the various structural features that the perception processes are designed to identify in structures.[162] The transforms themselves include a complete description of the required substructural features in the target, details of bonding changes, and appropriate literature citations.

Once transforms relevant to the target have been retrieved, they can be applied by some interpretive procedure that effects the required changes of bonding and generates precursor structures.[163] There is normally some evaluation step in which factors potentially affecting the yield of a reaction are considered. Potential precursors are discarded if they appear to be related to the target by reactions of inadequate yield or specificity. Precursors may also be eliminated if they contain specified undesirable structural features or, possibly, if their structures are assessed as being significantly more complex than the target structure.

Usually, there will be many transforms applicable to a given target structure; in some cases a particular transform may be applied to a target structure in several different ways and so result in several alternative precursors. A single step in the development of a synthesis plan involves the identification of all relevant transforms, their application to create precursors, and the filtering of the generated precursors to remove any duplicates or any rejected by evaluation procedures. This first step yields those structures that could be converted to the desired target by means of a single synthetic reaction.

Multistep synthesis plans can be developed by taking all these precursor structures, or some selected subset of these precursors, and subjecting them to an identical analysis. This first step in the further analysis yields precursor structures that could be converted into the desired target by two synthetic steps. The same analyses can be applied repeatedly to develop longer synthetic routes. In this way, a "search tree" is built up, with different paths through this tree representing alternative syntheses. Search along particular paths can be terminated by a number of conditions. Many paths terminate with precursor structures that could not themselves be readily synthesized or for which the total sequence of synthetic reactions would result in an inadequate product yield. Other paths, representing reasonable synthetic routes, terminate at precursor structures identified as available. (Possibly, this identification proceeds automatically and involves some search of a catalog of available chemicals.)

Work on computer-assisted synthesis has been more varied in nature than that on structure elucidation or on data base oriented systems. For some researchers, the computer has essentially offered an opportunity for developing new, abstract approaches to the representation of synthetic processes. Research by Hendrickson is more concerned with representation of chemical changes.[164–167] Hendrickson's formal notations allow classification of types of chemical reaction and can facilitate identification of conversion steps that would be synthetically valuable but for which no reactions are yet known. Work on synthesis planning by Ugi and co-workers forms part of a more general approach to providing a mathematical and logical structure to chemistry.[168–170]

However, most researchers have concentrated on the development of practical tools that could actually help chemists to solve specific synthetic problems. These approaches are empirical; the data in the transform libraries summarize standard, known synthetic reactions. The evaluation of transforms and the analyses of possible precursor structures all involve empirical estimates of effects of steric congestion, or interfering functionality, on potential yields. These empirical approaches have a strong overlap of interest with methods developed for spectral interpretation and structure evaluation that form major themes of this book.

Many of the problems facing the designers of empirical synthesis planning systems are similar to those arising in structure elucidation. For example, the programs must have mechanisms for representing and manipulating structures. The issues here are essentially the same as those discussed in Chapter VII. There are some differences in representations typically chosen. It can be advantageous for a synthesis planning program to use redundant structure representation (e.g., with information on bonding represented in both atom connection tables and bond tables) if this allows for more rapid testing for requirements specified in all the various transforms that must be considered. A structure generator, with less need to test for standard predefined substructures, would gain little from such redundancy and could well omit tables that duplicate information on bonding.

Synthesis planning programs must also include functions that can test whether a structure incorporates some particular substructure. Such functions depend on methods similar to those described in Chapter VIII. Planning programs must use some *canonical* form for structures (so that it is possible to determine whether a postulated precursor is in some list of available compounds or is identical to a previously generated precursor found on some alternate synthesis route). The canonical representation of structures, the discovery of symmetry in structures, and so forth, all involve algorithms similar to those described in Chapter IX.

The other major requirement in a synthesis planning program is the transform library. The transform library can be viewed as the set of empirical rules that relate substructures to synthetic methods and thus is analogous to the sets of empirical substructure-subspectrum correlation rules used in structure elucidation systems. Unlike the authors of some of the early spectral interpretation programs, the developers of synthesis planning programs very largely avoided the temptation to incorporate their empirical rules into the actual code of the synthesis planning program. Transforms have always been data structures, defined separately from, and processed interpretively within the synthesis planning programs.

Transforms are normally defined in something similar to a restricted, special-purpose programming language. Transforms defined in these languages are converted and compiled by some auxiliary program into a form that can be used by the interpretive procedures of the main synthesis planning program. Essentially, each transform specification is a description of a substructure to be found in a target molecule and a list of bonding changes that describe the differences between target structure and precursor molecule(s).[†] Transform descriptions typically contain large numbers of tests for nonparticipating functional groups or ring systems whose presence in a structure could

[†]Some transform specifications include a kind of description of how to find the desired substructure (with "transform program code" detailing tests for particular functional groups and subsequent branches to alternative parts of the transform description). Such descriptions are not essential; general-purpose routines can always be used to match arbitrary substructures onto complete structures.

affect the reactions. Since similar tests are used in all transforms, it is advantageous for synthesis programs to complete some exhaustive process of functional group and ring identification in a structure prior to any search of the transform library. Such preliminary analysis avoids the reidentification of the same structural attributes during analysis of individual alternative transforms. It is simple to identify situations where interfering functional groups can be protected; descriptions of the proposed synthetic reaction can then be supplemented by details of necessary blocking-deblocking reactions.[171]

Practical syntheses often involve the introduction, exchange, or removal of functionality. Starting materials may need to contain additional functional groups that serve to activate intended reaction sites. When the synthetic reaction step has been completed, such groups are removed. Consequently, when a desired target is analyzed, it may not be found to exhibit the substructural features produced by any major synthetic step. Additional functionality must be introduced into the target before it matches any of the transforms in the library. It is, however, impractical to allow a synthesis planning program to arbitrarily introduce or modify functionality in a target molecule in the hope that this will result in a match with the transform library. Such arbitrary use of functional group introduction–exchange would lead to a combinatorially explosive increase in the number of possible precursors.

Transforms developed for the "*Logic and* ***H****euristics* ***A****pplied to* ***S****ynthetic* ***A****nalysis*" (LHASA) program exemplify one approach to solving such problems.[172–176] Several major classes of synthetic reactions, such as Diels-Alder ring formation, have been analyzed in detail by the synthetic chemists assisting in the development of **LHASA**. Detailed strategies have been developed that determine appropriate sequences of functional group introductions–exchanges necessary in different types of target molecule. The corresponding transform descriptions contain rather general models of necessary attributes in a target, possibly simply a requirement for a particular ring system. The remainder of these elaborate transform descriptions contain details of the sequence of structural modification steps that allow application of these strategically important transforms.

An emphasis on routine use by chemists has lead to the development of very sophisticated user-interfaces for many of these synthesis planning programs. It is, of course, possible to require that all details of transforms for the libraries, all specifications of problems to be solved, and so on, be defined in some linear notation such as WLN.[177] Multistep plans can be developed by programs run in batch mode with the user receiving output only on program termination.[159,177,178] More attractively, the programs can be run interactively with some more flexible interface that does not require linearization of structures. A command language interface, such as that of **GENOA**[88] illustrated in the examples in Chapter II, can suffice. However, such command languages are not ideal—the conventional "language" of the synthetic chemist is the structural diagram, and it is desirable for a chemist to be able to use structural diagrams when communicating with synthesis planning programs.

The programs **LHASA** and "***S****imulation and* ***E****valuation of* ***C****hemical* ***S****ynthesis*" (SECS) both incorporate graphics interfaces that allow users to draw diagrams of target structures and view realistic models of possible precursors on a video screen.[179,180] (There have even been experiments with three-dimensional input devices that allow the user to trace out thc shape of a physical model of a target structure).[180,181]

There is one major difference between structure elucidation and synthesis planning systems. *The heart of any structure elucidation system is a combinatorial algorithm for the generation of* ***all*** *candidate structures compatible with known constraints.* Whatever it is that one normally wants from an automated synthesis planner *certainly does* ***not*** *include the combinatorial generation of "****all****" possible synthesis plans*.†

The most difficult and important component of a synthesis planning program is the mechanism used to focus the search for suitable precursors. Much of this focusing effort can be accommodated in the procedures for applying and evaluating the chemical transforms included in the program's library. The constraints specified in a transform can identify interfering functionality or congestion in the structure that would invalidate a proposed synthetic step. However, these *tactical*-level evaluations are in themselves incomplete and, further, provide no real sense of direction, no focusing on the development of a practical synthesis plan.

The best known of the synthesis planners are probably **LHASA**[161] and **SECS**.[158,180] Both these programs involve the user in the processes of structure evaluation and synthesis plan development. The analysis of individual transforms and possible precursors proceeds as already outlined, but control is returned to the chemist after each transform step. The precursor structures on all the various branches of the partially developed synthesis tree can be reviewed, and the chemist can select the structure most deserving of further analysis. Thus it is possible to pursue a single route to some considerable depth, examine all reasonable alternatives at some key step, or simply browse through all precursors related to the target by one or two transforms (and so obtain some general idea of different lines of synthesis). As well as control over the development of the route, with these programs the chemist has the opportunity to exercise some control over the classes of chemical transform to be applied and over the criteria used in the automatic phase of structure and transform evaluation.

These systems should best be viewed as "consultancy services" or "expert systems".[183,184] Expertise of other synthetic chemists is encoded in the transform libraries and made available to assist in the feasibility analysis of the individual steps of some hypothesized, incomplete synthesis plan. The major

†The programs of Ugi et al[169], and related programs,[182] are more closely related to (constrained) combinatorial structure generators; such programs could be used to solve problems where "all" synthetic routes, subject to any chemically oriented constraints on the interconversion steps, were indeed desired.

impediment to their routine use is the cost of developing comprehensive reaction libraries that encompass a significant amount of expertise.

The initial implementation of programs that could represent structures, transforms, and synthesis routes,[153,177,178,180] and their subsequent extensions to incorporate stereochemistry,[185–187] steric,[188] and conformational effects[189,190], really represents the *"easy"* phase of the development of practical synthesis planners. Subsequent progress has been less rapid. Programs have been extended to new domains, such as phosphorous chemistry,[191] and additional generally important synthetic routes have been analyzed and encoded as transforms.[174–176] Research continues into the major outstanding problems. It is still necessary to provide greater sophistication at the tactical level of transform and precursor evaluation. More sense of strategy should be incorporated into the search system. (Even in the interactive programs like **SECS**, when analyzing the precursors at step n of some synthetic plan, the analyst is all too often troubled by a host of possible precursors created by transforms similar to those yielding structures previously rejected at step $n-1$). Methods for directing search toward (or from) arbitrary starting materials are also required.[161]

The various empirical synthesis planning programs have already been exploited to some extent in industry;[159,192] however, there are few accounts of their applications. (Inevitably such data are proprietary.) Some of the programs have been adapted to serve as the basis of systems for mechanistic studies.[193,194] The **LHASA** program has also been quite successfully adapted to a teaching role (something that has not been achieved for any of the structure elucidation systems).[195]

REFERENCES

1. L. E. Kuentzel, *Anal. Chem.*, **23**, (1951), 1413.
2. R. A. Sparks, *Storage and Retrieval of Wyandotte-ASTM Infrared Spectral Data Using an IBM 1401 Computer*, ASTM, Philadelphia, PA, 1964.
3. L. D. Smithson, L. B. Fall, F. D. Pitts, and F. W. Bauer, *Storage and Retrieval of Wyandotte-ASTM Infrared Spectral Data Using a 7090 Computer*, Technical Report RTD-TDR-63–4265, Research and Technology Division, Wright Patterson Air Force Base, OH, 1964.
4. T. A. Entzminger and A. E. Diephaus, *Storage and Retrieval of Wyandotte-ASTM Infrared Spectral Data Using a Honeywell-400 Computer*, U. S. Public Health Service, Robert Taft Sanitary Engineering Center, Cincinnati, OH, 1964.
5. D. H. Anderson and G. L. Covert, "Computer Search System for Retrieval of Infrared Data," *Anal. Chem.*, **39**, (1967), 1288.
6. D. S. Erley, "Fast Searching System for the ASTM Infrared Data File," *Anal. Chem.*, **40**, (1968), 894.
7. R. W. Sebesta and G. G. Johnson, Jr., "New Computerized Infrared Substance Identification System," *Anal. Chem.*, **44**, (1972), 260.
8. J. Zupan, D. Hadzi, and M. Penca, "A New Retrieval System for Infrared Spectra," *Comput. Chem.*, **1**, (1976), 71.

9. E. C. Penski, D. A. Padowski, and J. B. Bouck, "Computer Storage and Search System for Infrared Spectra Including Peak Width and Intensity," *Anal. Chem.*, **46**, (1974), 955.
10. R. C. Fox, "Computer Searching of Infrared Spectra Using Peak Location and Intensity Data," *Anal. Chem.*, **48**, (1976), 717.
11. H. B. Woodruff, S. R. Lowry, and T. L. Isenhour, "A Text Search System Using Boolean Strategies for the Identification of Infrared Spectra," *J. Chem. Inf. Comput. Sci.*, **15**, (1975), 207.
12. R. H. Shaps and J. F. Sprouse, "Fast Matching with IR Spectral Search and Display," *Ind. Res. Devel.*, **1981**, 168.
13. Chemir Laboratories, 761 West Kirkham, Glendale, MO 63211.
14. American Society for Testing and Materials, 1916 Race St., Philadelphia, PA 19103.
15. DNA Systems, Inc., 1258 S. Washington St., Saginaw, MI 48605.
16. G. W. A. Milne, C. L. Fisk, S. R. Heller, and R. Potenzone, Jr., "Environmental Uses of the NIH-EPA Chemical Information System," *Science*, **215**, (1982), 371.
17. P. D. Zemany, "Punched Card Catalog of Mass Spectra Useful in Qualitative Analysis," *Anal. Chem.*, **22**, (1950), 920.
18. S. Abrahamsson, S. Stallberg-Stenhagen, and E. Stenhagen, "Mass-Spectrometry in Organic Structure Determination - A Problem of Storage and Identification," *Biochem. J.*, **92**, (1964), 2.
19. B. Pettersson and R. Ryhage, "Mass Spectral Data Processing 1. Computer Used for Identification of Organic Compounds," *Ark. Kemi*, **26**, (1966), 293.
20. L. R. Crawford and J. D. Morrison, "Computer Methods in Analytical Mass Spectrometry," *Anal. Chem.*, **40**, (1968), 1464.
21. B. A. Knock, I. C. Smith, D. E. Wright, R. G. Ridley, and W. Kelly, "Compound Identification by Computer Matching of Low Resolution Mass Spectra," *Anal. Chem.*, **42**, (1970), 1516.
22. S. L. Grotch, "Matching of Mass Spectra When Peak Height is Encoded to One Bit," *Anal. Chem.*, **42**, (1970), 1214.
23. H. S. Hertz, R. A. Hites, and K. Biemann, "Identification of Mass Spectra by Computer-Searching a File of Known Spectra," *Anal. Chem.*, **43**, (1971), 681.
24. F. W. McLafferty, R. H. Hertel, and R. D. Villwock, "Probability Based Matching of Mass Spectra," *Org. Mass Spectrom.*, **9**, (1974), 690.
25. G. M. Pesyna and F. W. McLafferty, "Computerized Interpretation of Mass Spectra," in *Determination of Organic Structures by Physical Methods*, F. C. Nachod, J. J. Zuckerman, and E. W. Randall, eds., Academic Press, New York, 1976, 91, Chapter 2.
26. S. R. Heller, "Computer Techniques for Interpreting Mass Spectrometry Data," in *Computer Representation and Manipulation of Chemical Information*, W. T. Wipke, S. Heller, R. Feldmann, and E. Hyde, eds., Wiley, New York, 1974, 175, Chapter 8.
27. S. R. Heller, H. M. Fales, and G. W. A. Milne, "A Conversational Mass Spectral Search and Retrieval System - II. Combined Search Options," *Org. Mass Spectrom.*, 7, (1973), 107.
28. G. W. A. Milne and S. R. Heller, "NIH/EPA Chemical Information System," *J. Chem. Inf. Comput. Sci.*, **20**, (1980), 204.
29. K.-S. Kwok, R. Venkataraghavan, and F. W. McLafferty, "Computer-Aided Interpretation of Mass Spectra. III. A Self-Training Interpretive and Retrieval System," *J. Am. Chem. Soc.*, **95**, (1973), 4185.
30. H. E. Dayringer, G. M. Pesyna, R. Venkataraghavan, and F. W. McLafferty, "Computer-Aided Interpretation of Mass Spectra. Information on Substructural Probabilities from STIRS," *Org. Mass Spectrom.*, **11**, (1976), 529.

31. N. A. B. Gray, "Computer Assisted Analysis of Carbon-13 NMR Spectral Data," *Prog. Nuc. Magn. Reson. Spectrosc.*, **15**, (1982), 201.
32. W. Voelter, G. Haas, and E. Breitmaier, "Electronic Data Processing of the Parameters of Impulse Fourier Transform Carbon-13 NMR Spectroscopy for the Structure Clarification of Organic Compounds," *Chem. Ztg.*, **97**, (1973), 507.
33. W. Bremser, M. Klier, and E. Meyer, "Mutual Assignment of Subspectra and Substructures - A Way to Structure Elucidation by C-13 NMR Spectroscopy," *Org. Magn. Reson.*, **7**, (1975), 97.
34. V. Mlynarik, M. Vida, and V. Kello, "Computer-Aided NMR Spectra Interpretation. Part 2. *Anal. Chim. Acta*, **122**, (1980), 47.
35. W. Bremser, "The Importance of Multiplicities and Substructures for the Evaluation of Relevant Spectral Similarities for Computer Aided Interpretation of C-13 NMR Spectra," *Z. Anal. Chem.*, **286**, (1977), 1.
36. W. Bremser, H. Wagner, and B. Franke, "Fast Searching for Identical C-13 NMR Spectra via Inverted Files," *Org. Magn. Reson.*, **15**, (1981), 178.
37. J. Zupan, M. Penca, D. Hadzi, and J. Marsel, "Combined Retrieval System for Infrared, Mass, and Carbon-13 Nuclear Magnetic Resonance Spectra," *Anal. Chem.*, **49**, (1977), 2141.
38. R. Schwarzenbach, J. Meile, H. Konitzer, and J. T. Clerc, "A Computer System for Structural Identification of Organic Compounds from C-13 NMR Data," *Org. Magn. Reson.*, **8**, (1976), 11.
39. F. Erni and J. T. Clerc, "Strukturaufklarung organischer Verbindungen durch computerunterstutzen Vergleich spektraler Daten," *Helv. Chim. Acta*, **55**, (1972), 489.
40. D. L. Dalrymple, C. L. Wilkins, G. W. A. Milne, and S. R. Heller, "A Carbon-13 Nuclear Magnetic Resonance Spectral Data Base and Search System," *Org. Magn. Reson.*, **11**, (1978), 535.
41. B. A. Jezl and D. L. Dalrymple, "Computer Program for the Retrieval and Assignment of Chemical Environments and Shifts to Facilitate Interpretation of Carbon-13 Nuclear Magnetic Resonance Spectra," *Anal. Chem.*, **47**, (1975), 203.
42. J. Zupan, M. Penca, M. Razinger, B. Barlic, and D. Hadzi, "KISIK - A Combined Chemical Information System for a Minicomputer," *Anal. Chim. Acta.*, **122**, (1980), 103.
43. V. A. Koptyug, V. S. Bochkarev, B. G. Derendyaev, S. A. Neknoroshev, V. N. Piottukh-Peletskii, M. I. Podgornaya, and G. P. Ul'yanov, "Use of the Computer for the Solution of Structural Problems in Organic Chemistry by the Methods of Molecular Spectroscopy," *Zh. Strukt. Khimii*, **18**, (1977), 440.
44. Y. Katagiri, K. Kanohta, K. Nagasawa, T. Okusa, T. Sakai, O. Tsumura, and Y. Yotsui, "Development of a New File Search System for Nuclear Magnetic Resonance Spectra," *Anal. Chim. Acta*, **133**, (1981), 535.
45. L. A. Gribov and M. E. Elyashberg, "Computer-Aided Identification of Organic Molecules by their Molecular Spectra," *CRC Crit. Rev. Anal. Chem.*, **8**, (1979), 111.
46. H. Abe, T. Yamasaki, I. Fujiwara, and S. Sasaki, "Computer-Aided Structure Elucidation Methods," *Anal. Chim. Acta*, **133**, (1981), 499.
47. J. Lederberg, "Topological Mapping of Organic Molecules," *Proc. Nat. Acad. Sci. (USA)*, **53**, (1965), 134.
48. S. I. Sasaki, H. Abe, T. Ouki, M. Sakamoto, and S. Ochiai, "Automated Structure Elucidation of Several Kinds of Aliphatic and Alicyclic Compounds," *Anal. Chem.*, **40**, (1968), 2220.
49. L. A. Gribov and M. E. Elyashberg, "Symbolic Logic Methods for Spectrochemical Investigations," *J. Molec. Struct.*, **5**, (1970), 179.
50. M. E. Munk, C. S. Sodano, R. L. McLean, and T. H. Hasketh, "Actinbolin. 1. Structure of Actinobolamine," *J. Am. Chem. Soc.*, **89**, (1967), 4158.

51. D. B. Nelson, M. E. Munk, K. B. Gash, and D. L. Herald, "Alanylactonobicyclone. An Application of Computer Techniques to Structure Elucidation," *J. Org. Chem.*, **34**, (1969), 3800.

52. N. A. B. Gray, "Structural Interpretation of Spectra," *Anal. Chem.*, **47**, (1975), 2426.

53. H. R. Henze and C. M. Blair, "The Number of Isomeric Hydrocarbons of the Methane Series," *J. Am. Chem. Soc.*, **53**, (1931), 3077.

54. J. Lederberg, G. L. Sutherland, B. G. Buchanan, E. A. Feigenbaum, A. V. Robertson, A. M. Duffield, and C. Djerassi, "Applications of Artificial Intelligence for Chemical Inference. I. The Number of Possible Organic Acyclic Structures Containing C, H, O and N," *J. Am. Chem. Soc.*, **91**, (1969), 2973.

55. A. M. Duffield, A. V. Robertson, C. Djerassi, B. G. Buchanan, G. L. Sutherland, E. A. Feigenbaum, and J. Lederberg, "Applications of Artificial Intelligence for Chemical Inference. II. Interpretation of the Low Resolution Mass Spectra of Ketones," *J. Am. Chem. Soc.*, **91**, (1969), 2977.

56. A. Buchs, A. B. Delfino, A. M. Duffield, C. Djerassi, B. G. Buchanan, E. A. Feigenbaum, and J. Lederberg, "Applications of Artificial Intelligence for Chemical Inference. VI. Approach to a General Method of Interpreting Low Resolution Mass Spectra with a Computer," *Helv. Chim. Acta*, **53**, (1970), 1394.

57. H. L. Surprenant and C. N. Reilly, "Uniqueness of Carbon-13 Nuclear Magnetic Resonance Spectra of Acyclic Saturated Hydrocarbons," *Anal. Chem.*, **49**, (1977), 1134.

58. A. L. Burlingame, R. V. McPherron, and D. M. Wilson, "Nonheuristic Computer Determination Based Upon Carbon-13 Nuclear Magnetic Resonance Data. Branched Alkanes," *Proc. Natl. Acad. Sci.* USA, **70**, (1973), 3419.

59. J. V. Knop, W. R. Muller, Z. Jericevic, and N. Trinajstic, "Computer Enumeration and Generation of Trees and Rooted Trees," *J. Chem. Inf. Comput. Sci.*, **21**, (1981), 91.

60. R. E. Carhart and C. Djerassi, "Applications of Artificial Intelligence to Chemical Inference. XI. Analysis of Carbon-13 Nuclear Magnetic Resonance Data for Structure Elucidation of Acyclic Amines," *J. Chem. Soc. Perkin Trans. II*, **1973**, 1753.

61. L. M. Masinter, N. S. Sridharan, J. Lederberg, and D. H. Smith, "Applications of Artificial Intelligence for Chemical Inference. XII. Exhaustive Generation of Cyclic and Acyclic Isomers," *J. Am. Chem. Soc.*, **96**, (1974), 7702.

62. R. E. Carhart, D. H. Smith, H. Brown, and N. S. Sridharan, "Applications of Artificial Intelligence for Chemical Inference. XVI. Computes Generation of Vertex Graphs and Ring Systems," *J. Chem. Inf. Comput. Sci.*, **15**, (1975), 124.

63. L. M. Masinter, N. S. Sridharan, R. E. Carhart, and D. H. Smith, "Applications of Artificial Intelligence for Chemical Inference. XIII. Labeling of Objects Having Symmetry," *J. Am. Chem. Soc.*, **96**, (1974), 7714.

64. H. Brown, L. Hjelmeland, and L. M. Masinter, "Constructive Graph Labeling Using Double Cosets," *Discrete Math.*, **7**, (1974), 1.

65. H. Brown and L. M. Masinter, "Algorithm for the Construction of Graphs of Organic Molecules," *Discrete Math.*, **8**, (1974), 227.

66. H. Brown, "Molecular Structure Elucidation," *SIAM J. Appl. Math.*, **32**, (1977), 534.

67. D. H. Smith, "The Scope of Structural Isomerism," *J. Chem. Inf. Comput. Sci.*, **15**, (1975), 203.

68. D. H. Smith, "Applications of Artificial Intelligence for Chemical Inference. Constructive Graph Labelling Applied to Chemical Problems. Chlorinated Hydrocarbons," *Anal. Chem.*, **47**, (1975), 1176.

69. H. Dolhaine, "A Computer Program for the Enumeration of Substitutional Isomers," *Comput. Chem.*, **5**, (1981), 41.

70. R. E. Carhart, D. H. Smith, H. Brown, and C. Djerassi, "Applications of Artificial

Intelligence for Chemical Inference. XVII. An Approach to Computer-Assisted Elucidation of Chemical Structure," *J. Am. Chem. Soc.*, **97**, (1975), 5755.

71. R. E. Carhart, *A Simple Approach to the Computer Generation of Chemical Structures*, Technical Report MIP-R-118, Machine Intelligence Research Unit, University of Edinburgh, 1977.
72. Y. Kudo and S. Sasaki, "Principle for Exhaustive Enumeration of Unique Structures Consistent with Structural Information," *J. Chem. Inf. Comput.Sci.*, **16**, (1976), 43.
73. S. Sasaki, H. Abe, Y. Hirota, Y. Ishida, Y. Kudo, S. Ochiai, K. Saito, and T. Yamasaki, "CHEMICS-F: A Computer Program System for Structure Elucidation of Organic Compounds," *J. Chem. Inf. Comput. Sci.*, **18**, (1978), 211.
74. S. Sasaki, Y. Kudo, S. Ochiai, and H. Abe, "Automated Chemical Structure Analysis of Organic Compounds. An Attempt to Structure Determination by the use of NMR," *Mikrochim. Acta*, **1971**, 726.
75. S. Sasaki, I. Fujiwara, H. Abe, and T. Yamasaki, "A Computer Program System - New CHEMICS - for Structure Elucidation of Organic Compounds by Spectral and Other Structural Information, *Anal. Chim. Acta*, **122**, (1980), 87.
76. T. Oshima, Y. Ishida, K. Saito, and S. Sasaki, "CHEMICS-UBE, a Modified System of CHEMICS," *Anal. Chim. Acta*, **122**, (1980), 95.
77. Y. Kudo and S. Sasaki, "The Connectivity Stack, a New Format for Representation of Organic Structures," *J. Chem. Doc.*, **14**, (1974), 200.
78. H. L. Morgan, "Generation of Unique Machine Descriptions for Chemical Structures," *J. Chem. Doc.*, **5**, (1965), 107.
79. C. A. Shelley, T. R. Hays, M. E. Munk, and R. V. Roman, "An Approach to Automated Partial Structure Expansion," *Anal. Chim. Acta*, **103**, (1978), 121.
80. C. A. Shelley, H. B. Woodruff, C. R. Snelling, and M. E. Munk, "Interactive Structure Elucidation," in *Computer-Assisted Structure Elucidation*, D. H. Smith, ed., American Chemical Society, Washington, DC, 1977, 92, Chapter 7.
81. C. A. Shelley and M. E. Munk, "CASE, Computer Model of the Structure Elucidation Process," *Anal. Chim. Acta.*, **133**, (1981), 507.
82. V. V. Serov, M. E. Elyashberg, and L. A. Gribov, "Mathematical Synthesis and Analysis of Molecular Structures," *J. Molec. Struct.*, **31**, (1976), 381.
83. L. A. Gribov, M. E. Elyashberg, and V. V. Serov, "Computer System for Structure Recognition of Polyatomic Molecules by IR, NMR, UV and MS Methods," *Anal. Chim. Acta*, **95**, (1977), 75.
84. B. Debska, J. Duliban, B. Guzowska-Swider, and Z. Hippe, "Computer-Aided Structural Analysis of Organic Compounds by an Artificial Intelligence System," *Anal. Chim. Acta*, **133**, (1981), 303.
85. C. A. Shelley and M. E. Munk, "Computer Perception of Topological Symmetry," *J. Chem. Inf. Comput. Sci.*, **17**, (1977), 110.
86. C. A. Shelley and M. E. Munk, "An Approach to the Assignment of Canonical Connection Tables and Topological Symmetry Perception," *J. Chem. Inf. Comput. Sci.*, **19**, (1979), 247.
87. C. A. Shelley, M. E. Munk, and R. V. Roman, "A Unique Representation for Molecular Structures," *Anal. Chim. Acta*, **103**, (1978), 245.
88. R. E. Carhart, D. H. Smith, N. A. B. Gray, J. G. Nourse, and C. Djerassi, "GENOA: A Computer Program for Structure Elucidation Utilizing Overlapping and Alternative Substructures," *J. Org. Chem.*, **46**, (1981), 1708.
89. J. G. Nourse, R. E. Carhart, D. H. Smith, and C. Djerassi, "Exhaustive Generation of Stereoisomers for Structure Elucidation," *J. Am. Chem. Soc.*, **101**, (1979), 1216.
90. J. G. Nourse, D. H. Smith, and C. Djerassi, "Computer-Assisted Elucidation of Molecular Structure with Stereochemistry," *J. Am. Chem. Soc.*, **102**, (1980), 6289.

91. J. G. Nourse, "Specification and Enumeration of Conformations of Chemical Structures for Computer-Assisted Structure Elucidation," *J. Chem. Inf. Comput. Sci.*, **21**, (1981), 168.
92. R. E. Carhart, T. H. Varkony, and D. H. Smith, "Computer Assistance for the Structural Chemist," in *Computer-Assisted Structure Elucidation*, D. H. Smith, ed., American Chemical Society, Washington, DC, 1977, 126, Chapter 9.
93. N. A. B. Gray, *STRUCC—Structure Checking Program Manual,* Technical Report, Department of Chemistry, Stanford University, Stanford, CA 94305, 1979.
94. T. H. Varkony, R. E. Carhart, and D. H. Smith, "Computer-Assisted Structure Elucidation. Modeling Chemical Reaction Sequences Used in Molecular Structure Problems," in *Computer-Assisted Organic Synthesis*, W. T. Wipke and J. Howe, eds., American Chemical Society, Washington, DC, 1977, 188, Chapter 9.
95. T. H. Varkony, R. E. Carhart, D. H. Smith, and C. Djerassi, "Computer-Assisted Simulation of Chemical Reaction Sequences. Applications to Problems of Structure Elucidation," *J. Chem. Inf. Comput. Sci.*, **18**, (1978), 168.
96. C. Djerassi, D. H. Smith, and T. H. Varkony, "A Novel Role of Computers in the Natural Products Field," *Naturwissenschaften*, **66**, (1979), 9.
97. L. A. Gribov, M. E. Elyashberg, and M. M. Raikhshtat, "A New Approach to the Determination of Molecular Spatial Structure Based on the Use of Spectra and Computers," *J. Molec. Struct.*, **53**, (1979), 81.
98. J. C. Wenger and D. H. Smith, "Deriving Three-Dimensional Representations of Molecular Structure from Connection Tables Augmented with Configuration Designations Using Distance Geometry," *J. Chem. Inf. Comput. Sci.*, **22**, (1982), 29.
99. L. A. Gribov, M. E. Elyashberg, and V. V. Serov, "On the Solution of One Classical Problem in Vibrational Spectroscopy," *J. Molec. Struct.*, **50**, (1978), 371.
100. L. A. Gribov, V. A. Dementiev, and A. T. Todorovsky, "Calculation of Spectral Absorption Curves for Polyatomic Molecules," *J. Molec. Struct.*, **50**, (1978), 389.
101. C. A. Shelley and M. E. Munk, "Signal Number Prediction in Carbon-13 Nuclear Magnetic Resonance Spectroscopy," *Anal. Chem.*, **50**, (1978), 1522.
102. I. Fujiwara, T. Okuyama, T. Yamasaki, H. Abe, and S. Sasaki, "Computer-Aided Structure Elucidation of Organic Compounds with the CHEMICS System: Removal of Redundant Candidates by 13-C NMR Prediction," *Anal. Chim. Acta*, **133**, (1981), 527.
103. J. T. Clerc and H. Sommerauer, "A Minicomputer Program Based on Additivity Rules for the Estimation of 13-C NMR Chemical Shifts," *Anal. Chim. Acta*, **95**, (1977), 33.
104. N. A. B. Gray, C. W. Crandell, J. G. Nourse, D. H. Smith, M. L. Dageforde, and C. Djerassi, "Computer Assisted Structural Interpretation of Carbon-13 Spectral Data," *J. Org. Chem.*, **46**, (1981), 703.
105. N. A. B. Gray, J. G. Nourse, C. W. Crandell, D. H. Smith, and C. Djerassi, "Stereochemical Substructure Codes for 13C Spectral Analysis," *Org. Magn. Reson.*, **15**, (1981), 375.
106. C. W. Crandell, N. A. B. Gray, and D. H. Smith, "Structure Evaluation Using Predicted C-13 Spectra," *J. Chem. Inf. Comput. Sci.*, **22**, (1982), 48.
107. G. Schroll, A. M. Duffield, C. Djerassi, B. G. Buchanan, G. L. Sutherland, E. A. Feigenbaum, and J. Lederberg, "Applications of Artificial Intelligence for Chemical Inference. III. Aliphatic Ethers Diagnosed by their Low-Resolution Mass Spectra and Nuclear Magnetic Resonance Data," *J. Am. Chem. Soc.*, **91**, (1969), 7440.
108. A. Lavanchy, T. Varkony, D. H. Smith, N. A. B. Gray, W. C. White, R. E. Carhart, B. G. Buchanan, and C. Djerassi, "Rule-Based Mass Spectrum Prediction and Ranking: Applications to Structure Elucidation of Novel Marine Sterols," *Org. Mass Spectrom.*, **15**, (1980), 355.
109. D. H. Smith, B. G. Buchanan, W. C. White, E. A. Feigenbaum, J. Lederberg, and C. Djerassi, "Applications of Artificial Intelligence for Chemical Inference. X. INTSUM - A

Data Interpretation and Summary Program Applied to the Collected Mass Spectra of Estrogenic Steroids," *Tetrahedron*, **29**, (1973), 3117.

110. B. G. Buchanan, D. H. Smith, W. C. White, R. Gritter, E. A. Feigenbaum, J. Lederberg, and C. Djerassi, "Applications of Artificial Intelligence for Chemical Inference. XXII. Automatic Rule Formation in Mass Spectrometry by Means of the Meta-DENDRAL Program," *J. Am. Chem. Soc.*, **98**, (1976), 6168.

111. D. H. Smith and R. E. Carhart, "Structure Elucidation Based on Computer Analysis of High and Low Resolution Mass Spectral Data," in *High Performance Mass Spectrometry: Chemical Applications*, M. L. Gross, ed., American Chemical Society, Washington, DC, 1978, 325, Chapter 18.

112. N. A. B. Gray, R. E. Carhart, A. Lavanchy, D. H. Smith, T. Varkony, B. G. Buchanan, W. C. White, and L. Creary, "Computerized Mass Spectrum Prediction and Ranking," *Anal. Chem.*, **52**, (1980), 1095.

113. K. S. Haraki, R. Venkataraghavan and F. W. McLaffery, "Prediction of Substructures from Unknown Mass Spectra by the Self-Training Interpretive and Retrieval System," *Anal. Chem.*, **53**, (1981), 386.

114. B. R. Kowalski and C. F. Bender, "The K-Nearest Neighbor Rule (Pattern Recognition) Applied to NMR Spectral Interpretation," *Anal. Chem.*, **44**, (1972), 1405.

115. P. C. Jurs, B. R. Kowalski, and T. L. Isenhour, "Computerized Learning Machine Applied to Chemical Problems. Molecular Formula Determination from Low Resolution Mass Spectra," *Anal. Chem.*, **41**, (1969), 21.

116. B. R. Kowalski, P. C. Jurs, T. L. Isenhour, and C. N. Reilley, "Computerized Learning Machines Applied to Chemical Problems. Interpretation of Infrared Spectrometry," *Anal. Chem.*, **41**, (1969), 1945.

117. P. C. Jurs, B. R. Kowalski, T. L. Isenhour, and C. N. Reilley, "An Investigation of Combined Patterns from Diverse Analytical Data Using Computerized Learning Machines," *Anal. Chem.*, **41**, (1969), 1949.

118. P. C. Jurs and T. L. Isenhour, *Chemical Applications of Pattern Recognition*, Wiley, New York, 1975.

119. C. F. Bender and B. R. Kowalski, "Multiclass Linear Classifier for Spectral Interpretation," *Anal. Chem.*, **46**, (1974), 294.

120. S. R. Heller, C. L. Chas, and K. C. Chu, "Interpretation of Mass Spectrometry Data Using Cluster Analysis," *Anal. Chem.*, **46**, (1974), 951.

121. H. B. Woodruff, S. R. Lowry, and T. L. Isenhour, "Bayesian Decision Theory Applied to Multicategory Classification of Binary Infrared Spectra," *Anal. Chem.*, **46**, (1974), 2150.

122. G. S. Zander, A. J. Stuper, and P. C. Jurs, "Nonparametric Feature Selection in Pattern Recognition Applied to Chemical Problems," *Anal. Chem.*, **47**, (1975), 1085.

123. T. J. Stonham, I. Aleksander, M. Camp, W. T. Pike, and M. A. Shaw, "Classification of Mass Spectra Using Adaptive Digital Learning Networks," *Anal. Chem.*, **47**, (1975), 1817.

124. G. L. Ritter, S. R. Lowry, C. L. Wilkins, and T. L. Isenhour, "Simplex Pattern Recognition," *Anal. Chem.*, **47**, (1975), 1951.

125. K. Varmuza, *Pattern Recognition in Chemistry* (Lecture Notes in Chemistry 21), Springer-Verlag, Berlin, 1980.

126. T. M. Mitchell and G. M. Schwenzer, "Applications of Artificial Intelligence for Chemical Inference. XXV. A Computer Program for Automated Empirical 13-C Rule Formation," *Org. Magn. Reson.*, **11**, (1978), 378.

127. B. Pettersson and R. Ryhage, "Mass Spectral Data Processing. Computer Used for Identification of Organic Compounds," *Arkiv fur Kemi*, **26**, (1967), 293.

128. B. Pettersson and R. Ryhage, "Mass Spectral Data Processing. Identification of Aliphatic Hydrocarbons," *Anal. Chem.*, **39**, (1967), 790.

129. L. R. Crawford and J. D. Morrison, "Computer Methods in Analytical Mass Spectrometry. Empirical Identification of Molecular Class," *Anal. Chem.*, **40**, (1968), 1469.

130. L. R. Crawford and J. D. Morrison, "Computer Methods in Analytical Mass Spectrometry. Development of Programs for Analysis of Low Resolution Mass Spectra," *Anal. Chem.*, **43**, (1971), 1790.

131. J. F. O'Brien and J. D. Morrison, "Computing Methods in Mass Spectrometry. Programming for Aliphatic Amines and Alcohols," *Aust. J. Chem.*, **26**, (1973), 785.

132. D. H. Smith, "A Compound Classifier Based on Computer Analysis of Low Resolution Mass Spectral Data," *Anal. Chem.*, **44**, (1972), 536.

133. J. H. van der Maas and T. Visser, "A Systematic Approach to the Structure Elucidation of Carbon-Hydrogen and Hydroxy Compounds by Means of Raman Spectroscopy with Laser Sources," *J. Raman Spectrosc.*, **2**, (1974), 563.

134. H. B. Woodruff and M. E. Munk, "A Computerized Infrared Spectral Interpreter as a Tool in Structure Elucidation of Natural Products," *J. Org. Chem.*, **42**, (1977), 228.

135. N. A. B. Gray and T. O. Gronneberg, "Program for Spectrum Classification and Screening of GC/MS Data on a Laboratory Computer," *Anal. Chem.*, **47**, (1975), 419.

136. H. B. Woodruff and G. M. Smith, "Computer Program for the Analysis of Infrared Spectra," *Anal. Chem.*, **52**, (1980), 2321.

137. H. B. Woodruff and G. M. Smith, "Generating Rules for PAIRS: A Computerized Infrared Spectral Interpreter," *Anal. Chim. Acta*, **133**, (1981), 545.

138. G. M. Schwenzer and T. M. Mitchell, "Computer-Assisted Structure Elucidation Using Automatically Acquired 13-C NMR Rules," in *Computer-Assisted Structure Elucidation*, D. H. Smith, ed., American Chemical Society, Washington, DC, 1977, 58, Chapter 5.

139. J. E. Dubois, M. Carabedian, and B. Ancian, "Elucidation Structurale Automatique par RMN du Carbone 13: Methode DARC-EPIOS. Recherche d'une Relation Discriminante Structurale Deplacement Chimique," *Comptes Rend. Acad. Sci.* (Paris), **290**, (1980), 369.

140. J. E. Dubois, M. Carabedian, and B. Ancian, "Elucidation Structurale Automatique par RMN du Carbone 13: Methode DARC-EPIOS. Description de l'Elucidation Progressive par Intersection Ordonne de Sous-Structures," *Comptes. Rend. Acad. Sci.* (Paris), **290**, (1980), 372.

141. D. H. Smith, N. A. B. Gray, J. G. Nourse, and C. W. Crandell, "The DENDRAL Project: Recent Advances in Computer Assisted Structure Elucidation," *Anal. Chim. Acta*, **133**, (1981), 471.

142. G. Szalontai, Z. Simon, Z. Csapo, M. Farkas, and Gy. Pfeifer, "Use of IR and 13-C NMR Data in the Retrieval of Functional Groups for Computer Aided Structure Determination," *Anal. Chim. Acta*, **133**, (1981), 31.

143. N. A. B. Gray, "Computer Analysis of Carbon-13 NMR Data," *J. Artif. Intel.*, **22**, (1984), 1.

144. M. M. Cone, R. Venkataraghavan, and F. W. McLafferty, "Molecular Structure Comparison Program for the Identification of Maximal Common Substructures," *J. Am. Chem. Soc.*, **99**, (1977), 7668.

145. Committee on Chemical Information, Division of Chemistry and Chemical Technology, National Research Council, *Chemical Structure Information Handling. A Review of the Literature 1962–1968*, National Academy of Sciences, Washington, DC, 1969, (Publication No. 1733).

146. J. E. Ash and E. Hyde, eds., *Chemical Information Systems*, Halsted, New York, 1975

147. CAS ONLINE, Chemical Abstracts Service, Marketing Division - ONF, P. O. Box 3012, Columbus, OH, 43210.

148. Lockheed Information Systems, Organization 5208, Building. 201, 3251 Hanover St., Palo Alto, CA 94304.

149. Systems Development Corporation, 2500 Colorado Ave., Santa Monica, CA 90406.

150. Bibliographic Retrieval Services, Inc., Corporation Park, Building 702, Scotia, NY 12302.

151. QUESTEL, Inc., 1625 Eye St., NW, Suite 818, Washington, DC 20006.

152. Molecular Design, Ltd., 1122 B St., Hayward, CA 94541.

153. E. J. Corey and W. T. Wipke, "Computer-Assisted Design of Complex Organic Syntheses," *Science*, **166**, (1969), 178.

154. E. J. Corey, "General Methods for the Construction of Complex Molecules," *Pure Appl. Chem.*, **14**, (1967), 19.

155. E. J. Corey, "Computer-Assisted Analysis of Complex Synthetic Problems" *Quart. Rev.*, **25**, (1971), 455.

156. A. J. Thakkar, "The Coming of the Computer Age to Organic Chemistry: Recent Approaches to Systematic Synthesis Analysis," *Topics Curr. Chem.*, **39**, (1973), 3.

157. M. Bersohn and A. Esack, "Computers and Organic Synthesis," *Chem. Rev.*, **76**, (1976), 269.

158. W. T. Wipke, G. I. Ouchi, and S. Krishnan, "Simulation and Evaluation of Chemical Synthesis—SECS: an Application of Artificial Intelligence Techniques," *J. Artif. Intel.*, **11**, (1978), 173.

159. H. L. Gelernter, A. F. Sanders, D. L. Larsen, K. K. Agarwal, R. H. Boivie, G. A. Spritzer, and J. E. Searleman, "Empirical Explorations of SYNCHEM," *Science*, **197**, (1977), 1041.

160. J. Haggin, "Computers Shift Chemistry to More Mathematical Basis," *Chem. Eng. News*, **1983** (May) 7.

161. A. K. Long, S. D. Rubenstein, and L. J. Joncas, "A Computer Program for Organic Synthesis," *Chem. Eng. News*, **1983** (May) 22.

162. E. J. Corey, W. T. Wipke, R. D. Cramer, and W. J. Howe, "Techniques for Perception by a Computer of Synthetically Significant Structural Features in Complex Molecules," *J. Am. Chem. Soc.*, **94**, (1972), 431.

163. E. J. Corey, R. D. Cramer, and W. J. Howe, "Computer Assisted Synthetic Analysis for Complex Molecules. Methods and Procedures for Machine Generation of Synthetic Intermediates," *J. Am. Chem. Soc.*, **94**, (1972), 440.

164. J. B. Hendrickson, "A Systematic Characterization of Structures and Reactions for Use in Organic Synthesis," *J. Am. Chem. Soc.*, **93**, (1971), 6847.

165. J. B. Hendrickson, "Systematic Synthesis Design. III. The Scope of the Problem," *J. Am. Chem. Soc.*, **97**, (1975), 5763.

166. J. B. Hendrickson, "A Systematic Organization of Synthesis Reactions," *J. Chem. Inf. Comput. Sci.*, **19**, (1979), 129.

167. J. B. Hendrickson and E. Braun-Keller, "Systematic Synthesis Design 8. Generation of Reaction Sequences," *J. Comput. Chem.*, **1**, (1980), 323.

168. I. Ugi and P. Gillespie, "Representation of Chemical Systems and Interconversion by *be* Matrices," *Agnew. Chem. International Ed.*, **10**, (1971), 914.

169. J. Blair, J. Gasteiger, C. Gillespie, P. D. Gillespie, and I. Ugi, "CICLOPS—A Computer Program for the Design of Synthese on the Basis of a Mathematical Model," in *Computer Representation and Manipulation of Chemical Information*, W. T. Wipke, S. Heller, R. Feldmann, and E. Hyde, eds., Wiley, New York, 1974, 129, Chapter 6.

170. J. Dugundji, P. Gillespie, D. Marquarding, I. Ugi, and F. Ramirez, "Metric Spaces and Graphs Representing the Logical Structure of Chemistry," in *Chemical Applications of Graph Theory*, A. T. Balaban, ed., Academic Press, New York, 1976, 107, Chapter 6.

171. E. J. Corey, H. W. Orf, and D. A. Pensak, "Computer Assisted Synthetic Analysis. The Identification and Protection of Interfering Functionality in Machine Generated Synthetic Sequences," *J. Am. Chem. Soc.*, **98**, (1976), 210.

172. E. J. Corey and W. L. Jorgensen, "Computer Assisted Synthetic Analysis. Generation of

Synthetic Sequences Involving Sequential Functional Group Interchanges," *J. Am. Chem. Soc.*, **98**, (1976), 203.

173. E. J. Corey, W. J. Howe, and D. A. Pensak, "Computer Assisted Synthetic Analysis. Methods for Machine Generation of Synthetic Intermediates Involving Multi-step Look-ahead," *J. Am. Chem. Soc.*, **96**, (1974), 7724.
174. E. J. Corey and A. K. Long, "Computer Assisted Synthetic Analysis. Performance of Long Range Strategies for Stereoselective Olefin Synthesis," *J. Org. Chem.*, **43**, (1978), 2208.
175. E. J. Corey, A. K. Long, J. Mulzer, H. W. Orf, A. P. Johnson, and A. P. W. Hewett, "Computer Assisted Synthetic Analysis. Long Range Search Procedures for Antithetic Simplification of Complex Targets by the Application of Halolactonization Transforms," *J. Chem. Inf. Comput. Sci.*, **20**, (1980), 221.
176. E. J. Corey, A. P. Johnson, and A. K. Long, "Computer Assisted Synthetic Analysis. Techniques for Efficient Long Range Retrosynthetic Searches Applied to the Robinson Annulation Process," *J. Org. Chem.*, **45**, (1980), 2051.
177. H. Gelernter, N. S. Sridharan, A. J. Hart, F. W. Fowler, and H. Shue, "The Discovery of Organic Synthetic Routes by Computer," *Topics Curr. Chem.*, **41**, (1973), 113.
178. M. Bersohn, "Automatic Problem Solving Applied to Synthetic Chemistry," *Bull. Chem. Soc. Jap.*, **45**, 1972, 1897.
179. E. J. Corey, W. T. Wipke, R. D. Cramer, and W. J. Howe, "Computer Assisted Synthetic Analysis. Facile Man-Machine Communication of Chemical Structures by Interactive Computer Graphics," *J. Am. Chem. Soc.*, **94**, (1972), 421.
180. W. T. Wipke, "Computer-Assisted Three Dimensional Synthetic Analysis," in *Computer Representation and Manipulation of Chemical Information*, W. T. Wipke, S. R. Heller, R. J. Feldmann, and E. Hyde, eds., Wiley, New York, 1974, 147, Chapter 7.
181. W. T. Wipke and A. Whetston, *Computer Graphics*, **5**, (1971), 10.
182. J. Gasteiger and C. Jochums, "EROS—A Computer Program for Generating Sequences of Reactions," *Topics Curr. Chem.*, **74**, (1978), 93.
183. *Handbook of Artificial Intelligence*, A. Barr and E. A. Feigenbaum, eds., William Kaufmann, Los Altos, California, 1982.
184. *Building Expert Systems*, F. Hayes-Roth, D. A. Watermann, and D. B. Lenat, eds., Addison-Wesley, Reading, MA, 1983.
185. W. T. Wipke and T. M. Dyott, "Simulation and Evaluation of Chemical Synthesis. Computer Representation and Manipulation of Stereochemistry," *J. Am. Chem. Soc.*, **96**, (1974), 4834.
186. A. Esack and M. Bersohn, "Computer Manipulation of Central Chirality," *J. Chem. Soc. Perkin Trans. I*, **1975**, 1124.
187. H. W. Davis, *Computer Representation of the Stereochemistry of Organic Molecules*, Birkhauser, Basel, 1976.
188. W. T. Wipke and P. Gund, "Simulation and Evaluation of Chemical Synthesis. Congestion: a Conformation Dependent Function of Steric Environment at a Reaction Center," *J. Am. Chem. Soc.*, **98**, (1976), 8107.
189. E. J. Corey and N. F. Feier, "Computer Assisted Synthetic Analysis. A Rapid Computer Method for the Semiquantitative Assignment of Conformation of Six-Membered Rings. 1. Derivation of Preliminary Conformation Description of the Six-Membered Ring," *J. Org. Chem.*, **45**, (1980), 757.
190. E. J. Corey and N. F. Feier, "Computer Assisted Synthetic Analysis. A Rapid Computer Method for the Semi-quantitative Assignment of Conformation of Six-Membered Rings. 2. Assessment of Conformational Energies," *J. Org. Chem.*, **45**, (1980), 757.
191. F. Choplin, C. Laurenco, R. Marc, G. Kaufmann, and W. T. Wipke, "Synthese Assistee par Ordinateur en Chemie des Composes Organophophores," *Nouv. J. Chem.*, **2**, (1978), 659.

192. P. Gund, E. J. J. Grabowski, D. R. Hoff, G. M. Smith, J. D. Andose, J. B. Rhodes, and W. T. Wipke, "Computer Assisted Synthetic Analysis at Merck," *J. Chem. Inf. Comput. Sci.*, **20**, (1980), 88.

193. T. D. Salatin and W. L. Jorgensen, "Computer Assisted Mechanistic Evaluation of Organic Reactions. 1. Overview," *J. Org. Chem.*, **45**, (1980), 2043.

194. B. L. Roos-Kozel and W. L. Jorgensen, "Computer Assisted Mechanistic Evaluation of Organic Reactions. 2. Perception of Rings, Aromaticity and Tautomers," *J. Chem. Inf. Comput. Sci.*, **21**, (1981), 101.

195. R. D. Stolow and L. J. Joncas, "Computer Assisted Teaching of Organic Synthesis," *J. Chem. Ed.*, **57**, (1980), 868.

IV

SPECTRUM MATCHING APPROACHES

IV.A. INTRODUCTION

Most constituents in an extract from an organism will be known compounds. Such constituents may be ubiquitous to all organisms of a given class or may have already been discovered and identified in the course of previous investigations of closely related organisms. If such standard constituents can be reliably and rapidly identified, research efforts can be more usefully and effectively concentrated on novel compounds.

Standard constituents are typically "recognized" by matching their observed spectral properties with data obtained from previous, more definitive studies of reference compounds. Although trivial in principle, such a matching process is complicated by many factors, including:

- The intrinsic variability in the spectral data characterizing a compound.
- The possibility (probability) of contamination of an unknown constituent's spectrum by "signals" from other components.
- The need to minimize computational costs.
- The need to minimize the total quantity of reference data that must be stored.
- The limits of the discriminatory power of any chosen subset of spectral that is used to characterize each reference compound.

- The limits to which any spectrum matching measure can actually capture structurally relevant similarities.
- The (un)reliability of the reference data available.
- The intrinsic limitations to the extent to which a match on any one type of spectral data can be used to reliably identify a compound.

Exact matches of spectral data cannot be required. Instrumental effects can produce significant variations even in spectra recorded under "ideal" conditions. Thus, for example, a reference mass spectrum of a compound recorded on an electric-magnetic sector instrument will show an intensity pattern noticeably different from that of a spectrum recorded on a simple quadrupole instrument (because of the latter's somewhat lower sensitivity at higher masses). Furthermore, even if reference spectra could be obtained under ideal conditions, the data characterizing an unknown are always imperfect. Most spectra of components in a mixture from an organism will be acquired on some instrumentation incorporating both a chromatographic separator and a spectrometer such as a gas chromatographer-mass spectrometer (GC–MS), a liquid chromatographer-mass spectrometer (LC–MS), or a gas chromatography-Fourier transform-infrared spectrometer (GC–FT–IR). Chromatographic resolution is rarely perfect; most spectra will be contaminated by traces of previously eluting components, coeluting minor components, and general column bleed. Further, because of the way in which components elute from a GC column, the concentration of a component in a spectrometer source will typically vary during the course of spectrum acquisition. Generally, such variations in concentration will induce distortions in a recorded spectral pattern.

If spectra had been reproducible, searches for identical matches could have been computationally inexpensive (even if somewhat expensive in terms of file storage). Given reproducible spectral signatures, the "search" method of choice would probably use "hashing" or "scatter storage" techniques.[1] By applying an appropriately chosen mathematical function (a "hashing" function) to the spectral data for an unknown, one could generate a numeric value that could then be used as a pointer to where such data should be stored in some large file. At that point in the file one would either find nothing, indicating that the unknown compund was novel to the reference file, or should find an identical data record and a citation for the appropriate reference compound. (Hashing functions are rarely capable of generating unique file addresses for each data set; instead, a certain proportion of collisions occur where distinct data yield the same hash key. Such collisions only slightly complicate the mechanics of a hash-based search.)

Unfortunately, the irreproducibility of spectral data renders inapplicable the hash-key approach. Because observed and reference spectra cannot be expected to be identical, the designer of a compound recognition system must first determine some criterion of "adequate" spectral similarity and then devise an appropriate strategy for accessing a file of reference spectra to find those references whose spectra might match with that of an unknown. Whereas a hash-key search requires only a few accesses (ideally, just one access) to be made into the reference file for retrieval of relevant data, searches using spectrum similarity matching must process a substantial fraction of—if not the entire set of—reference data records. Since computational costs are proportional to the number of reference data records examined, it is important to minimize both the number of reference records accessed and the amount of processing for each record.

One approach to minimizing overall processing costs is to exploit human intelligence to guide the search for reference compounds. A chemist, with some background knowledge of the spectral technique used and some general information concerning an unknown, can usually select features in an unknown's spectrum that are likely to be highly characteristic of its structure; with experience, it is often possible to make reliable selection even in the presence of noise or contamination of spectral data with signals from other constituents. Although complete spectra cannot be expected to be exactly reproducible, it is reasonable to assume that the most apparent features of a compound's spectrum should be reasonably consistent. The features selected by the chemist for an unknown can be compared with those previously identified as characteristic of each reference compound. Searches of this type generally employ "inverted" files. Rather than search through all the reference records to find those reference compounds characterized by features identical to those of an unknown, one uses features chosen from the spectrum of an unknown as keys for directly indexing into the reference file. The reference data are organized so that, as well as a main file with compound citations and complete spectral records, there are indices listing all compounds characterized by any particular spectral feature. A list of reference compounds, each characterized by the same combination of spectral features as an unknown, can be obtained by intersecting the reference lists associated with each particular feature considered in isolation. The search process itself is typically interactive, requiring detailed human control.

Detailed human analysis and control is impractical in many circumstances. For example, a 1-hr capillary GC-MS run may result in the collection of mass spectra for 100 or more different components; it is simply not feasible for a chemist to attempt to inspect and perform interactive searches for each of these spectra in turn. Most spectral data characterizing unknowns must be processed through some more fully automated procedure. Essentially, the data characterizing each reference structure are either matched with an abbreviated set of features chosen for an unknown, or the characteristic features from reference spectra are sought in the complete spectral record obtained for an unknown. Such searches are inevitably costly; consequently, various devices have been used to attempt to reduce computational overheads. The standard approach has been to attempt to minimize the processing time accorded to each reference compound by recognizing and abandoning obviously inconsistent reference spectra at the earliest possible stage and thus only computing detailed matching coefficients for plausible reference compounds. Alternatively, procedures may be used to try to "cluster" the reference file so that essentially similar reference spectra are grouped into particular subregions of a single file or, possibly, into distinct subfiles. The same "clustering" algorithm applied to the spectral data for an unknown identifies those subfiles that must be searched.

There is a twofold need to minimize the amount of data used to characterize each reference compound: (1) overall processing costs do increase as the volume of data being transferred and manipulated is increased, and (2) there may well be limitations on available storage capacity. Many search systems are intended for laboratory computers or other relatively small computer configurations. On such systems, file storage capacity is typically limited and thus restrictive of the number of reference compounds that may be represented. On larger machines there are fewer intrinsic limitations on file size, but restrictions may still be necessitated by file storage charges.

All types of spectra can be viewed as comprising a set of signals (IR absorption maxima, magnetic resonance signals, m/z values, etc.) characterized by position, intensity, and possibly shape (e.g., peak width in IR, multiplicity in ^{13}C). The total data storage required can be reduced by either selecting only a subset of signals in a spectrum or eliminating detail from the description of each signal; these two types of data reduction are generally used in conjunction. The choices as to which reference spectral features to abstract and as to the details that are recorded for each selected feature determine the form of the spectral matching procedure and, ultimately, the retrieval capability and discriminatory power of the entire search process. Essentially any reasonable spectrum abbreviation process will allow a search system to retrieve reference data for compounds that are present in a small test file. Of greater consequence are the abilities to both discriminate between a number of references with basically similar structures and spectra and retrieve related structures for those compounds not yet included in the search files. These desired abilities are somewhat conflicting; the more effective search systems tend to specialize in either *identification* or *finding analogous structures*.

The performance of a file search system depends directly on the number of reference compounds in the file and on the completeness and accuracy of the data characterizing these compounds. Substantial collections of mass[2,3], IR[4,5] and ^{13}C NMR[6] spectra are all in the public domain; in addition, there exist more limited collections of ^{1}H NMR[7,8] and UV[8] spectra. Many companies maintain proprietary spectral collections; in some cases, such as the BASF ^{13}C spectral collection,[9] these may be larger than the more publicly accessible collections such as the NIH–EPA ^{13}C file.[6] Reviews of the status of various spectral search systems have all appeared recently and include some commentary on the various available collections.[2,4,10] In development of such data bases, conflicts arise between the need to expand the reference files and maintain accuracy in the data. Currently, there are problems with most of the spectral collections. Thus many of the ^{13}C spectra in the NIH–EPA collection are unassigned, with no matchings of resonances to constituent atoms, and still other spectra lack necessary multiplicity data. The largest IR spectral collection is unsatisfactory for it contains only limited spectral data and suffers from inconsistency in spectral encoding[5,4,11]. Recent researches on limited-size IR files, containing much more comprehensive spectral information, suggest that problems in the use of the major IR collections may be related to the inadequacy of the information they contain.[12] The mass spectral collections are possibly the most satisfactory. Yet, even here, there are obviously problems inherent in having a collection of only about 40,000 spectra, for this represents possibly 1% of known compounds. In his recent review,[2] Martinsen provides some rather disappointing statistics with regard to the fraction of biologically and environmentally important compounds included in current mass spectral files.

Along with the problems of limited size and restricted data in the reference files there is a third problem: erroneous data. In a simple retrieval system, references for which the data are erroneous will be effectively inaccessible although, of course, they will occasionally give rise to spurious inappropriate matchings. Erroneous reference data are of greater consequence in file-based systems that attempt to abstract substructural information through an analysis of the forms of those reference structures with spectra matching that of the unknown. For example, there now exist systems for identifying plausible substructures by examining groups of atoms whose assigned resonances have been matched in a ^{13}C file search system.[13,10] Such systems are markedly dependent on the correct assignments of resonances to atoms in the reference structures. There are known problems of misassigned spectra in the major collections of ^{13}C data.[10] One of the more important current research areas is the development of computer methods for helping to maintain the accuracy of spectral data in reference files.[14–16]

When file search methods are used to recognize structures, it is implicitly assumed that a sufficient degree of matching between two spectra constitutes proof of identity. It must be borne in mind that such an assumption is not always valid. Consider, for example, a system processing gas-chromatographic low-resolution mass spectral (GC–LRMS) data of B-ring unsaturated sterols. The principal fragmentations, giving rise to the dominant features in the spectra, will be those of the steroid skeleton itself. Different patterns of substitution on or unsaturation in the steroid nucleus result in clearly distinct fragmentation patterns that can be reliably exploited through even the simplest of file search approaches. In such a system, therefore, the skeletal type of the steroid may be readily recognized. However, only small and rather subtle mass spectral differences result from minor changes in the branching patterns of steroid sidechains. Careful examination might, for example, reveal correlations between side-chain branching patterns and consistent differences in the relative intensities of particular groups of minor ions. Such spectral differences are often too subtle to be captured by normal spectrum abbreviation and matching procedures. Consequently, standard spectral processing methods may incorrectly identify a novel compound, incorporating, say, a standard steroidal nucleus with an unusual side chain, as a standard reference compound with the same nucleus but a more typical side chain. (The chance of misidentifications can usually be greatly reduced by matching on more than one type of spectral data.) As in all other cases, uncritical acceptance of the output of computer file search programs is liable to lead to occasional errors.

IV.B. FILE ORGANIZATIONS AND SEARCH STRATEGIES

IV.B.1. Sequential Files and Conventional Searches

The simplest procedures for spectral based compound identification rely on searching through a sequential file. All data for each reference compound are stored in a single record within this multirecord file. Each such compound record is read from the file in turn and the reference data compared with the spectral data characterizing the unknown. The comparison process provides a basis for computing some numeric measure, or score, expressing the overall similarity or dissimilarity of the unknown and the particular reference spectrum. Details of reference compounds, whose spectra yielded the best matching scores, are kept during the course of the search process. This information can then be used to provide a ranked listing of possible structural identifications for the unknown together with some estimate of plausibility as derived from their scores.

The data on file for each reference compound are typically less than complete spectra. The degree of abbreviation of reference spectra depends on the type of data. For ^{13}C data, the spectra consist of a small number of signals of equal structural significance. The most obvious, although not necessarily the most economic and discriminating representation of a reference ^{13}C spectrum, is as a table of shifts and multiplicities for all resonances. Other ^{13}C spectral data (intensities $J_{C\text{-}H}$, $J_{C\text{-}C}$, T_1 times) would most likely, not be included in a search file because such data are much less commonly available and are of less immediate value for structure elucidation. The data characterizing an IR spectrum could be the absorptions at each wavelength over a chosen spectral range[17,18] (or possibly the equivalent time domain data for an FT–IR spectrum[19]). Such a representation of an IR spectrum incorporates all data—peak shape, peak width, absorption strength, and so on. More typically, the files for IR search systems incorporate much less information. In part this is a consequence of the fact that by the time that computer-based identification methods became topics for active research, there already existed a data base of tens of thousands of IR spectra. The data in this ASTM file is limited to indicating whether a peak maximum occurs within any 0.1-μm interval. The relatively crude digitization and the lack of intensity data have, to some degree, limited the practical utility of this large collection. Nevertheless, it has been the basis for most IR search systems. Although a number of studies have been reported on systems making use of limited intensity data (see the discussion by Rasmussen and Isenhour[18]), most of these studies have been limited to relatively small files.

It is with LRMS that one most clearly confronts the problem of very many spectral signals with widely varying degrees of structural significance and specificity. As with IR, by the time computer methods for handling mass spectra began to be developed, there already existed substantial files of reference spectra; however, these mass spectral files were complete spectral records with all m/z and intensity pairs. Given that searches using complete spectral records would be costly and that different mass spectral peaks conveyed different amounts of structural information, it was inevitable that many attempts would be made to find effective procedures for selecting and then matching on subsets of mass spectral data. A substantial fraction of the papers published on computer applications to chemical structure elucidation have been concerned with choosing particular subsets of mass spectral data and showing the effectiveness of the chosen subset as a method for identifying selected spectra.

The records in the file characterizing each reference compound are necessarily abbreviated spectra containing only those features required by the matching process. There is no intrinsic requirement that the data characterizing unknowns be similarly abbreviated prior to the search. Quite obviously, problems will be encountered in applying any spectral abbreviation procedures to the data obtained for an unknown. Abbreviation procedures typically involve operations such as thresholding against some specified minimum intensity or selecting the most intense signals in the overall spectrum or within specified spectral subregions. These operations are inevitably impacted by signals due not to the unknown, but to some impurity. Matching algorithms exploiting intensities may well be further confounded by varying concentrations of sample during spectrum acquisition. Even given programs that attempt to "clean up" a spectrum,[20,21] the abbreviated spectrum of a compound run as an unknown is likely to deviate from that obtained when the pure compound is analyzed to obtain a reference spectrum. Nevertheless, the basic approach of most search systems is to apply the same abbreviation procedure to the spectrum of the unknown as to the spectra of reference compounds and then match on the abbreviated spectra. The matching coefficient used in such searches is a measure of the correspondence of the two abbreviated spectra. Insofar as the abbreviated spectra capture structurally relevant information, these matching procedures will identify reference compounds with structures similar to that of the unknown.

An alternative approach is to utilize the complete spectrum of the unknown in the matching process. The data characterizing the reference are then regarded as specifications of constraints that must be satisfied by the spectrum of the unknown for that particular reference to be accepted as a potential match. Since the matching procedure is driven entirely by data relevant to the reference compound, it is certainly easier to ignore spurious signals (due to impurities) in the unknown spectrum, and it can be easier to make allowance for other spectral distortions. The approach of seeking characteristic data for the references in the complete spectra of unknowns was exploited in a number of small early mass spectral systems[22-25] and became popularized, under the somewhat quaint name of "reverse search," through the work of Abramson[24] and McLafferty[26].

The difference between conventional "forward" and "reverse" searches is sometimes overemphasized. The search processes are basically identical with records on reference compounds, read in turn from a sequential file, being matched against data characterizing an unknown. In one class of search, the objective is to *reveal structural similarity*. Spectral features considered to be structurally informative are predefined and used as the basis of the abbreviation procedures. Thus, for mass spectra, one might define the structurally significant features as being the number of m/z peaks, the values for the ion series, and the most intense peaks in specified mass ranges. Since there is typically only a limited set of spectral features used to characterize all spectra, the appropriate values can be computed for all these features in all reference spectra and for the unknown. Thus one requires only the abbreviated spectrum of the unknown in the matching process. The scoring of spectral similarity requires some measure of a "distance" between abbreviated spectra.

In the second class of search systems, including all "reverse searches", the objective is to try to *identify the structure*, whose spectrum has been acquired, assuming that it is in fact a standard compound with reference data in the library. Determination of identity is most effective if the matching features can be specific to each reference. Such search systems thus depend on both a procedure for selecting the most reliably unique features in reference spectra, when these are added to the file, and a method for searching for these features in a recorded spectrum. McLafferty's Probability Based Matching (**PBM**) system for mass spectra represents the most highly developed search system of this form.[23,26] In **PBM**, statistics on ion abundancies and the frequencies of occurrence of ions at different m/z values, as drawn from a large data file, are used to determine what data characterize each reference structure entered into the file. These statistics are also employed to give weighting factors used for evaluation of a score expressing how well recorded data express the characteristic intensity pattern of a reference (at the same time, making allowance for the fact that a recorded spectrum may be due to a mixture).

The calculation of any form of overall matching score for pairs of spectra is costly, and it is advantageous if one can avoid computing a precise score for those reference structures that are not reasonable candidate matches. *Prefilters*, which are included in most search systems, are simple tests that eliminate most inappropriate references without the need even to begin to compute a matching score. Typical prefilters used in mass spectral searches include constraints such as requiring that the base peak in the reference spectrum have a specified minimum intensity in the spectrum of the unknown, requiring comparable numbers of peaks in the spectra or requiring similar intensity distributions in ion series spectra (obtained by summing the intensities of all peaks taking their m/z values modulo 14). The "forward" search system of Biemann[27] uses a sequence of such tests to filter out as many as possible reference spectra prior to detailed matching. Prefilters are also important in "reverse" searches, where they usually take somewhat simpler forms such as checking that some key ion has at least a specified minimum intensity in the recorded spectrum. Reported results suggest that an elaborate sequence of filters, such as those in Biemann's mass spectral search program, can eliminate as many as 99% of the reference spectra.[28]

In the Clerc systems [29,30] the prefilter concept is, in effect, elaborated and becomes the basis for the scoring procedure. The Clerc systems involve matching binary attribute vectors computed for both unknown and reference spectra. The scoring procedure assigns differing weights to the four possible cases—both unknown and reference spectra possessing a particular attribute, both lacking that attribute and either case where the attribute is identified for just one of the spectra. The relative discriminating ability of particular attributes depends on the differences in these weights. If the unknown has a particular attribute encoded, the relevant weighting difference is that between the value given if both spectra exhibit that attribute and the value for when the unknown, but not the reference, possesses the relevant attribute. The appropriate weighting differences can be calculated just once from the particular attribute encoding of the unknown spectrum; these differences determine the relative discriminating abilities of the encoded spectral attributes. These relative values can be used to determine an optimal order of comparison steps, an ordering specific to the particular attributes encoded for the unknown. Comparison of unknown and reference attribute vectors is performed in this defined order. The comparison process is abandoned if a sufficient matching score is not realized after a defined number of comparison steps. This ability to abandon profitless comparisons does allow for significant savings in overall computational costs.

Reliable discrimination among a series of similar structures, with similar spectra, can be improved if some "orthogonal" data can be used to supplement the spectral comparisons. A typical example would be the use of chromatographic retention indices to supplement mass spectral data in systems for analyzing GC–MS data.[31,22] A retention index value measured for an unknown can be used to define a range of acceptable retention indices; spectral comparisons can be restricted to those compounds whose retention indices fall within this range. Other orthogonal data constraints include those based on molecular composition or on estimated molecular weights. A typical commercial system, providing searches on files of infrared data, offers a variety of such constraints including restrictions on boiling-point and melting-point ranges, molecular weight ranges, refractive indices, and so on. Tests using orthogonal data are conveniently incorporated into the prefiltering process.

The only problem introduced by prefilter constraints is that if these are too stringent, they may prevent the search system from identifying related reference structures for unknowns not yet represented in the search file. For example,[31] the first step in the biodegradation of some drug might involve oxidation to a more polar metabolite; if GC retention indices were used to *exclude* references, rather than simply as an aid to evaluating a match, the spectrum of a novel metabolite might never be matched against spectral data for the original drug and thus might fail to be identified. Such difficulties can be avoided if the search scheme provides an option whereby the prefilter constraints can be relaxed or completely omitted. Then most processing can utilize the prefilters to reduce computational costs, but any spectra that are not identified during normal processing can be reanalyzed using only the spectrum similarity measure.

Searches based on combined spectral data can also result in improved discrimination. Most existing systems essentially only intersect results from different single spectrum searches and are generally limited by the quality and size of any one of their spectral files. Recently, research has begun into finding more effective mechanisms for exploiting data from combined GC/MS–GC/FT–IR instrumentation.[32]

IV.B.2. "Indexed-Sequential" and Related Files

The major problem with standard sequential files is the cost associated with processing the entire file. "Prefilters" do reduce the cost of processing each record, but it would be preferable if simple prefilter type tests could be used to restrict the parts of the reference file that must be read and processed. One simple approach to restricting that subset of the file to be searched sequentially is to arrange that the file be ordered on some primary key—say, for example, the molecular weight. Together with the main file, a subsidiary index file is maintained. This index defines the set of appropriate records in the main file for either specific values or subranges of values of the primary key. If the records with the main spectral data are of equal length, the data in the index file can simply specify record number; more typically, spectral records will be of variable length and the index will have to define the byte number or block–byte location of the records. The index has to define the first and last record in the main file associated with each index entry, although one of these two limits can often be left implicit.

Computer operating systems commonly provide explicit support for some form of indexed–sequential file. Two or more levels of indices may be provided, with the first-level index file defining a crude partitioning of the primary sort key and containing pointers into additional index files wherein a more precise partitioning of key values is used to gain access to the main file. These operating systems allow the applications programmer to directly create a file, with multiple-level indices, and provide mechanisms for the convenient addition and deletion of new records in such files. On more limited operating systems, such as those on most laboratory mini- and microcomputers, there may be no explicit provision for indexed files. However, so long as the operating system does permit access to an arbitrary point in a file, simple indexed files are readily created. If one foregoes provision for convenient addition and deletion of records, an indexed file can be created simply by sorting the main file on the indexing key and processing the sorted file to develop an index containing record numbers and byte-block numbers for suitable ranges of key value.

The key used to index the file is most often a specific molecular property, such as molecular weight or GC retention index. Then, prior to the spectral matching search it is necessary to measure or estimate the appropriate key value or value range for an unknown. It is sometimes more convenient if the indexing key represents a value computed from the complete spectrum. An example of such a computed value used to index a file is Dromey's "series displacement index" (SDI).[33] This index is related to the ion series filters used in a number of search systems. The SDI abstracts a single number that relates to the relative intensities of the different ion series; values for the SDI range from around 50 to about 700. The file of reference spectra is ordered by SDI values; search of the reference file is restricted to that subset of records with SDI values within 50 units of the value obtained from the spectrum of the unknown. Typically, the use of the SDI index limits the required sequential search to a subset representing only about 22% of the total reference file. Subranges of values for the Series Displacement Index correlate moderately well with particular chemical classes. Thus, for example, alkanes, ethers, and thiophenes typically have SDI values of 150 or less, sugars, esters, and acetals have SDIs of around 200, whereas various aromatic classes occur in the SDI range from 400 to 600. Thus a search restricted to a subrange of SDI values examines predominately spectra of compounds of similar chemical class.

Gronneberg et al.[25] proposed the use of a simple compound classifier as a preprocessing step to file-based spectrum recognition. The approach was investigated in the context of mass spectral identification of important organic geochemicals. Mass spectra were classified by using the ion series method due to Smith[34] in which spectra are classified according to the results of comparisons of their data with prototype ion-series spectra for each reference class. In Gronneberg's the file search system, different subfiles were used to hold the spectra of compounds in the various defined classes. The results of the classification step were used to select the subfile, or subfiles, that had to be searched. More recent proposals for clustered files of mass spectra are due to Domokos et al.[35]

Zupan[36,37] has investigated the use of classification procedures as a possible preprocessing step for, or even as an alternative to, file searches on IR data. Results reported have been concerned mainly with the development of multilevel hierarchical classifiers.

IV.B.3. Inverted Files

The advantage of the various indexed file organizations is that a given (or computed) key can be used to select that subsection of the entire reference file that contains the only reference data that are really relevant. Subsequently, of course, that subset of the file has to be searched in a normal sequential manner. The benefits from limiting the sequential search that can be realized through the use of a single key can be magnified by using two or more keys in conjunction. Elaboration of this concept leads to the ideas of "inverted files" and search procedures devised around such files.

We can choose to index the entries for each reference compound in the main file according to the spectral features used to characterize their spectra. Thus, for example, if ^{13}C shifts were rounded to the nearest 1 ppm, one could have index files identifying all structures with resonances at each shift value from 0 to 240 ppm for example. The entries in each of these indices would comprise lists of record numbers for relevant structures. Structures with specified combinations of features would be obtained by appropriate list intersections. More data can be held in the indices. For ^{13}C it might be appropriate to hold the multiplicity of the signal; an intensity would be appropriate for a mass spectral file. Such additional data permit more precise search requests.

The most widely utilized inverted file systems are those included in the NIH-EPA CIS;[38,39,6] other large inverted files include the BASF ^{13}C data base.[9] The CIS has inverted files for both mass spectral and ^{13}C data. The mass spectral collection can be accessed according to the m/z values and intensities of fragment ions and also by molecular weight. Searches of the ^{13}C collection can be made according to chemical shifts of resonances (required to match within specified tolerances) that can be restricted by multiplicity. (Multiplicity constraints are ignored for those spectra in the collection that lack the relevant data.) Both mass spectral and ^{13}C searches can be further constrained by restrictions on molecular composition. The literature contains many examples illustrating how searches of inverted files can quickly and conveniently lead to the identification of standard compounds whose spectra are included in the files.[6,38]

Such searches are, however, less satisfactory when the unknown is not included in the reference file. For example, the results for searches made using ^{13}C data for the prostaglandin **IV-1** or the kaurenoid **IV-2** give widely differing results depending on the particular resonance shifts initially selected.[10] In both cases shifts for alkene carbons in the test compounds differed by some 3–

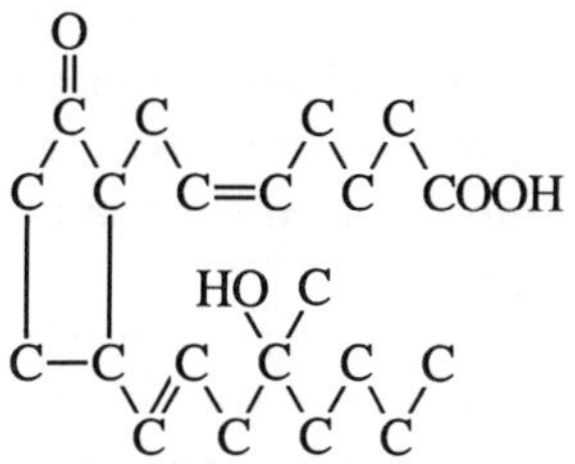

IV-1

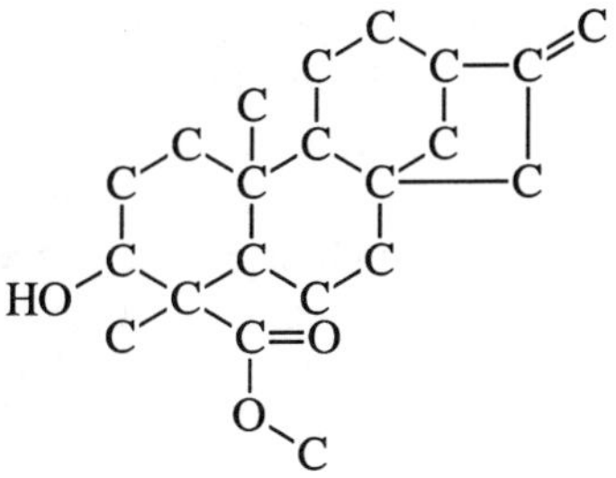

IV-2

5 ppm from the shifts for the corresponding atoms in closely related structures actually present in the reference files. These shift differences are due to changes in either the nature or relative configurations of substituent groups beta or gamma to the alkene carbons. The alkene resonances are among the more distinctive for these compounds and, consequently, should generally be favored as search keys.[9] The use of these alkene resonances inherently excludes relevant structures from the set of references retrieved; the particular reference structures that are retrieved depend on the other resonance shifts used along with those for the alkene carbons. If there is real uncertainty as to the structure of an unknown, the variety of references, as retrieved by slightly differently specified searches, can prove confusing.

IV.C. ABBREVIATED SPECTRA AND MATCHING COEFFICIENTS

As noted earlier, it is mainly in the context of mass spectral analysis that the problem of spectrum abbreviation is of real consequence. A complete mass spectrum may comprise 100-m/z-intensity pairs requiring more than 400 bytes of storage. However, in the spectrum of a sterane, for example, there might be numerous peaks of low or moderate intensity in the mass range 40–100 amu that, resulting from nonspecific multistep fragmentations, convey little or no structural information. The presence of a sterane skeleton might well be recognizable from a few key fragment ions, and a few additional, usually high-mass, ions will distinguish the spectrum of one sterane from that of another. Since the majority of ions in such a spectrum are unnecessary for identification, they can be discarded. Once sets of data have been chosen to characterize each of two spectra, some algorithm must be used to compute the similarity of these data. The problems are, of course, knowing for an arbitrary spectrum which ions to discard and which to select to form the abbreviated spectrum and then, in computing matches, making the most effective use of such data as are used to form the abbreviated spectrum.

The earliest methods used for mass spectral abbreviation relied on the selection of the 8 or 10 most intense peaks to characterize the spectra of both reference compounds and the unknown structure. Inherently, such selection procedures are of only limited efficacy for higher molecular-weight compounds that are often ascribed merely a set of low-mass fragments of little structural significance of specificity. Several methods for selecting ions chosen from throughout the entire spectra region have been investigated in attempts to overcome this problem. The method of Biemann, selecting the two most intense peaks in each 14 amu region of the spectrum,[27] constitutes the prototypical example of such selection procedures; numerous minor variations have been reported and have been reviewed by Chapman.[40]

Intensity data are used in the selection of the ions to characterize a spectrum in all search methods; subsequent use of intensity data varies. Some search systems make no further use of intensities; in such systems, similarity scores simply measure the number of ions common to two abbreviated spectra. More commonly, intensities are used either directly or indirectly. Indirect use of intensity data occurs in systems where relative intensity information is represented through an ordering of the ions used to characterize each spectrum, and this relative ordering is exploited during computation of a matching score. The most elaborate procedures record intensity data in the abbreviated spectra and use these intensities in calculation of a match. Any improved discriminatory power resulting from the use of intensities is obtained at the cost of a significant increase, possibly doubling, of the storage needed for reference data.

For a typical compound in the molecular-weight range 250–300 amu, selection of the two most intense peaks might yield some 20 peaks. Even if only masses are stored, this corresponds to something around 40 bytes, or 300+ bits of storage. The entire spectrum could be recorded in the same number, or maybe fewer bits. If each bit position is taken to correspond to an m/z value, the presence of a peak of greater than a specified intensity can be represented by encoding a "1" bit; the absence of a peak, or the existence of a peak with less than the threshold intensity, is represented by a zero. A single intensity threshold, for encoding peak presence, could be used over the entire spectral range; a more effective approach might determine a specific threshold for each m/z value through an analysis of statistics defining relative frequencies of ions of particular abundancies and m/z values for a wide range of reference compounds. There has been considerable interest in binary-coded mass spectra because of their ability to encode a complete spectrum in comparable or less storage than is required for a conventional abbreviated spectrum. Where the amount of storage used is considered the primary factor, it is possible to further reduce the number of bits needed to represent a spectrum in the files. Statistical analyses can, in effect, identify and discard bits that correspond to m/z values for which ions occur so rarely (or frequently) that their encoding does not convey any structural information.[41,42] A number of slightly different functions exist that measure the distance between two binary vectors.[43-46] Such functions can be used to compute the matching score when comparing the vector encoding the spectrum of an unknown with that for a reference compound.

To achieve the best performance at *retrieving analogous structures*, there is evidence that the features used for matching should not be simple spectral signals but should be computed properties expressing structure-related aspects of the overall spectrum. The use of computed spectral properties is exemplified in the systems for MS, ^{13}C and combined spectral data developed by Clerc[29,30] and in the **STIRS** system developed at Cornell.[47,48] Clerc characterizes both reference and unknown spectra using a binary (boolean) vector whose elements flag the presence or absence of particular spectral attributes. For example, in his system for analyzing ^{13}C data, Clerc has one particular element in the attribute vector representing whether the ratio of signals from hydrogen-bearing carbons to total carbons exceeds 0.2 for all those resonances in the range 124–160 ppm. The truth or falsity of this 1-bit feature conveys considerable chemical structural information concerning the relative proportions of substituted and unsubstituted alkene and aromatic carbons. A set of a hundred or so such feature variables can, if well defined, convey far more structurally relevant information than can the same number of bits used to represent resonances at particular shifts.

Furthermore, the matching process, working through these features, can concentrate on structurally relevant spectral patterns rather than on the correspondence of specific chemical shifts in ^{13}C or m/z intensities in mass spectra. The structural analogs in the reference file may all differ from some unknown structure in the placement of particular substituent groups. Differences in substitution can result in systematic changes in the specific ^{13}C shifts for a substantial number of the constituent carbons and, even if not inducing different fragmentation pathways, can produce systematic changes in the masses of the most intense fragment ions. Such spectral differences will undoubtedly thwart a conventional spectrum matching procedure. However, the spectra of both the unknown and the analogous reference structures may yield very similar attribute vectors. Consequently, a search using the computed attributes can reveal relevant structural analogs even when no individual spectra features are in correspondence. Certainly, in tests on the NIH–EPA's ^{13}C data base, the Clerc search scheme consistently retrieved informative structural matches in cases where the file did not contain a spectrum of the test structure and where searches based on matchings of specific resonance shifts had not yielded useful information.[10]

The **STIRS** system employs data classes that again relate to the underlying spectral processes rather than to specific m/z values. The **STIRS** data classes include ion series, masses of neutral fragments lost from the molecular ion, and "characteristic ions" selected from various defined mass ranges. The various ion series used in **STIRS** include the modulo-14 spectra, amine, and aromatic series; such ion series alone can provide a quite effective indication of general structural type and indeed are used as the basis of certain simple structure classification programs.[34] Many substituent groups give rise to characteristic neutral losses and/or specific low-mass fragment ions. Larger structural components, such as fused carbocyclic ring systems of standard skeleta, result in consistent patterns of fragment ions over a wide mass range. The **STIRS**'s data classes capture much of these characteristic structure-determined features in mass spectra.

In order to develop the "neutral loss" data for an unknown, **STIRS** must be given the molecular weight. In a complete structure elucidation problem one must know the molecular composition, and hence the molecular weight, and consequently there is no difficulty in satisfying this requirement. The **STIRS** system is also applied in more general contexts in order to obtain some suggestion of structural form for compounds whose exact structures are not required and whose compositions and molecular weights are unknown. A number of algorithms for estimating probable molecular weights through analysis of low-resolution mass spectral fragmentation patterns have been proposed.[49,50] The algorithms developed specifically for **STIRS** appear to be fairly effective and can obviate the need for additional molecular-weight determination measurements.

As discussed below, **STIRS** is primarily intended as an aid to spectral interpretation by substructure identification; **STIRS**'s success in such applications depends on its ability to retrieve structurally related compounds for those unknowns not represented by spectra in the data files. Some specific data illustrating the relative performance of a simple matching procedure, exploiting similarities of individual ion-intensity pairs, and one based on elaborate spectral features, are provided by the studies of a k-nearest-neighbor matching scheme and **STIRS**.[51] Both **STIRS** and the k-nearest-neighbor procedure were used to classify test spectra according to whether their search results indicated the presence of specified substructures. The presence of particular substructures was predicted on the basis of their frequency in the most closely matched reference structures. In the k-nearest-neighbor matching schemes, the presence of a substructure would be predicted if it were found in three of the five closest matching spectra. (Other voting schemes were also investigated.) The **STIRS** substructural predictions utilize statistics on the frequency of occurrence of substructures in the **STIRS** reference files; such statistics, defined for each of a repertoire of about 200 standard substructures, determine whether the presence of a substructure in a given proportion of the closest matched references constitutes sufficient evidence to predict its presence in the unknown.

These two methods were tested according to their ability to reliably predict the presence of 20 different substructures in 500 test spectra. The evaluation took into account both the proportion of constituent substructures of test compounds that were successfully detected and the number of false-positive predictions where a substructure was incorrectly proposed for a test compound. The **STIRS** system gave consistently better results than the k-nearest-neighbor scheme. The number of substructures correctly identified—the *recall*—can, of course, be increased by accepting weaker evidence, but the number of false-positive identifications also increases in this case. At any chosen level of recall, **STIRS** gave fewer false positives and thus its predictions were more trustworthy. Its greater success can be ascribed in part to its more sophisticated statistically based prediction criteria. Much of its improved performance is, however, a result of its use in matching of features chosen to reflect the important mass spectral processes. The k-nearest-neighbor matching was based on a simple distance measure using intensity differences between unknown (U) and reference (X) spectra at all m/z values:

$$[\Sigma(U_i - X_i)^2]^{0.5}$$

Such a quadratic metric emphasizes differences in intensity at specific m/z values; as such, it might be a reasonable measure of spectral identity. Any substantial deviations in an intensity pattern are accorded more emphasis than the many small variations that might be expected as a result of differing instrumentation. However, this measure cannot well accommodate differences in, say, the spectra of two homologous compounds in which corresponding key fragment ions may all be shifted by 14 amu. The **STIRS** ion series and neutral loss data classes can readily accommodate such spectral differences and still express the underlying similarity of the spectra.

The various methods for abbreviating mass spectra and computing matches between spectra, as proposed prior to about 1978, have been comprehensively reviewed by Chapman.[40] More recent proposals have been reviewed by Martinsen.[2]

IV.D. EVALUATION OF SEARCH SYSTEMS

McLafferty's group has consistently attempted to provide precise measures of the retrieval performance of their computer-based systems for structure identification[26] and substructure determination through the retrieval of analogous compounds[48]. The measures proposed by the McLafferty group are adaptations, to the chemical context, of the concepts of *recall* and *precision* as devised by Salton for assessing the performance of automated document retrieval systems.[52,53] McLafferty and co-workers have considered a number of overall measures of system performance that combine the separate recall and precision data.[54,52] With the exception the **PBM** and **STIRS** studies, the reporting of performance statistics for compound identification systems has generally been somewhat unsatisfactory. Typically, the only data provided are in the form of a table showing the percentage of searches for which the spectrum of a reference structure was ranked 1st, 2nd,..., *n*th when matching against a test spectrum of the same compound. Differences in reporting style, differences in size of test set, differences in size of reference files all contribute to making very difficult any comparative assessment of the merits of the various schemes for spectrum abbreviation and computation of matching coefficients. Although many search systems have been reported, there have been few detailed studies of comparative performance of different search systems for similar identification tasks.[28,41,44,55,56]

Rasmussen and Isenhour[28] recently presented one of the more comprehensive comparative studies of mass spectral search systems that has actually been performed. In this work, a library of some 17,000 reference spectra and a set of 40 test spectra were used in tests to evaluate search systems based on different spectral abbreviation and matching procedures. Relative success rates for different search systems were expressed graphically, through curves showing the proportion of test spectra for which the appropriate reference was ranked 1st, 2nd,..., *n*th among the matching references. Systems studied include those employing binary-encoded spectra and those using abbreviated spectra with m/z and intensity data. The best results obtained used complete spectra with intensities. Systems using m/z and intensity values for ions selected from throughout the spectrum performed almost as well while requiring significantly less computation and data storage. The exact form of the matching coefficient was found to have only marginal influence on performance. Generally inferior results were obtained from binary-coded spectra. The importance of intensity data for reliable discrimination was further revealed by the greater success rates with the use of binary spectra encoding only the masses of the most intense two peaks in each 14 amu region than with the use of complete spectra encoding all peaks exceeding a standard intensity threshold. (These studies did not extend to the use of distinct intensity thresholds statistically optimized for each m/z value; it is possible that such an encoding might yield a discrimination performance more closely comparable with searches using intensities directly.)

As well as assessing *identification* performance, it is also useful to be able to measure the success of a search system at *retrieving similar structures*. An immediate problem arises as to how, objectively, one might assess structural similarity between a test unknown and the references retrieved through the matching of its spectrum. Objectivity requires some form of algorithmic analysis of computer-compatible structural representations. This important area of research has been somewhat neglected thus far. One of the few studies to have appeared is that of Zupan et al.[57]. Zupan's analysis is based on a comparison of the Wiswesser linear notations (WLNs) for the test search compound with the WLNs of the retrieved reference structures. The comparison of WLNs yields a similarity coefficient that takes into account the number of common symbols and relative displacements of these symbols in the WLN strings.

Zupan et al.[57] have illustrated the use of these WLN-based similarity coefficients in a comparative assessment of file based spectrum matching procedures using IR, mass, and ^{13}C spectra. The search systems evaluated were similar; each involved (1) testing a bit-vector representation of reference spectra against a definition of peak-no-peak regions for the unknown and then, (2) scoring on matches in intensity, shape or multiplicity, and position of the principal reference peaks. The distributions of structural similarity coefficients obtained in these trials showed that searches using ^{13}C and mass spectra were more satisfactory than IR based searches. In the IR searches the retrieved references showed a wide spread of values for the structure similarity coefficient, with some references having matching spectra but only slight structural similarity. Differences in the quality of the spectral files used were identified as major factors in determining the relative performances of the various searches investigated.[57]

Although of value as one of the few efforts to find an appropriate structural similarity measure for assessing file searches, this approach is fundamentally flawed. The priority rules for WLN notation allow similar structures to receive very different encodings. In a counter-example presented by Zupan,[58] one finds that the codes for structures **IV-3** and **IV-4** are U1VR and L66 BVYT&J CU1.

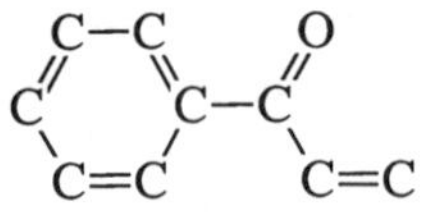

```
                                      C—C     O
    C—C     O                        //  \\   //
   //  \\   //                      C     C—C
  C     C—C                          \   /   \
   \   /   \                          C=C     C=C
    C=C     C=C                          \   /
                                          C—C
```

IV-3 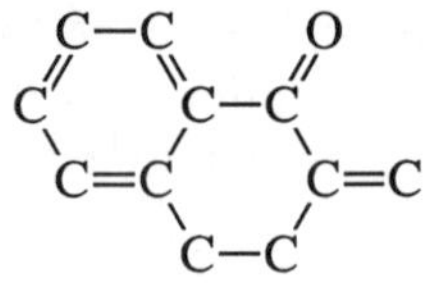 IV-4

Rasmussen and Isenhour[28] have devised an interesting empirical method for assessing the relative effectiveness of different search systems as used to try to retrieve a set of structurally similar compounds. Their method relies on a propagation technique; a test spectrum, selected from the reference file, is used to "seed" an expanding search. First, this spectrum is used to retrieve the n nearest matches in the file; these will consist of the original spectrum and the spectrum of $n - 1$ additional compounds. The spectra of these additional compounds are then all used to probe the reference file using a given search algorithm to produce, for each in turn, its n nearest matches. A list of compounds, found through the successive searches, is built up; the spectra of all new compounds added to the list in one step of this expansion are used as search probes in the next expansion step. If the search method is only marginally effective at selecting related structures, increasingly diverse compounds will be retrieved at each subsequent step and the set of matched compounds will gradually expand to incorporate the entire library. If the search method is able to limit the matches to a homogeneous subset of compounds, the size of the set of matched compounds will grow much more slowly. With an effective search, some of the compounds retrieved in any one search step will already have been identified as potential matches and, consequently, will already be included in the matching set. Rasmussen and Isenhour illustrate how an effective search scheme may in fact produce a closed set to which no further compounds are added by subsequent searches. The relative performance of different search schemes can be assessed by considering the size of the matched set of compounds after a specified number of search steps and the diversity in structure of the compounds thus retrieved.

Rasmussen and Isenhour have presented some illustrative applications of this technique. These involved the evaluation of a number of different mass spectral search schemes. The search schemes considered included the use of complete spectra, modulo-14 spectra, and binary spectra. The results of these studies, with complete spectra resulting in a more homogeneous and limited set of matching references, provide further evidence that intensity data are desirable in an effective mass spectral search system.

IV.E. SPECTRUM INTERPRETATION AND SUBSTRUCTURE IDENTIFICATION

The largest compilations of spectra contain only a few tens of thousands of entries; inevitably, many compounds that will need to be identified will not be represented in the files. And yet, an unknown structure will probably contain a number of relatively large substructural units that are also present in reference compounds whose spectra are included in the file. Matching functions for "forward" searches are designed to retrieve reference structures exhibiting spectral patterns similar to, although not necessarily identical with that of an unknown. Consequently, such search systems should in general be capable of retrieving structures similar to that of the unknown (insofar as similarity of substructural form results in similarity of spectral features).

Few search systems attempt any analysis or interpretation of their results. Compounds with matching spectra are simply listed by name or index number; it is the chemist who has to evaluate the results and thus determine whether a compound has been correctly recognized or, if not, to determine the structural relevance of the various alternative reference structures retrieved as partial matches. There are, of course, many cases where exact structure identification is in fact unnecessary and knowledge of the basic structural type, as inferred from the similar structures retrieved, is sufficient. In still more cases, the chemist may be capable of deriving the correct structure by combining the suggestions present in the retrieved structures with other available information. However, cases remain where exact identification is required and the chemist is unable to guess the structure on the basis of the search results. Hopefully structural information can be abstracted from the search results and exploited through more elaborate structure elucidation procedures in such cases.

There are basically two approaches to obtaining useful substructural constraints through a search system. One may analyze the retrieved reference structures to identify common substructural elements or directly correlate substructures and subspectra in the reference file. A list of plausible substructures, possibly associated with explicit probability ratings, can be derived by either approach. Systems capable of thus abstracting substructural information through file searches have been reported for both mass spectral and ^{13}C NMR data.

McLafferty and co-workers have undertaken the most elaborate current project of this type in their **STIRS** system for the analysis of low-resolution mass spectral (LRMS) data.[47,48,54] The first step of its analysis is a file search based on matching using its detailed spectral features. This search yields a number of sets of reference structures that match best with respect to both an overall matching factor and individual matching factors such as those for neutral losses. The substructure identification procedure then proceeds through an analysis of these best matching reference compounds. The reference file contains data defining the presence of particular substructures in the reference molecules; several hundred different substructures are known to the system.[48,54] If a particular substructure occurs in several of the structures retrieved when matching the spectrum for an unknown, it is reasonable to infer that the unknown also incorporates that substructure. The **STIRS** system employs statistics of the reference file to determine whether the presence of a substructure in a given proportion of retrieved references is likely to be structurally significant.

Obviously, a more or less ubiquitous group, such as methyl, will be found in many retrieved reference structures irrespective of any spectral evidence for the unknown containing a methyl. However, a less common group, such as aryl-F, is much less likely to appear with high incidence among the retrieved references on a purely random basis. Given statistics regarding the frequency of substructures among the reference compounds, it is simple to estimate the frequency of random occurrence of that substructure in any set of, say, 15 reference compounds selected from the file. Then it is possible to estimate the the probability that any specific number of the 15 best matched reference structures should, solely by chance, happen to incorporate a given substructure. These probabilities allow theoretical confidence levels to be defined. Dayringer et al.[54] give an example analysis for phenyl groups. Given that about 28% of the reference structures contain phenyls, it is possible to determine that the chance of finding 10 phenyl containing compounds among the top-ranked 15 is about 1 in 200 on a purely random drawing basis. Consequently, if 10 retrieved structures do indeed contain phenyls, a greater than 99% confidence in that substructure's presence has been attained. A chosen theoretical confidence level determines the number of occurrences of a substructure necessary before its presence is predicted. Thus **STIRS** will suggest methyl presence, at 96% confidence, only when 14 of the top 15 structures contain methyl, whereas only 10 of the top-ranked 15 need contain an aryl-F for this substructure to be proposed. The perfomance of **STIRS** in identifying some 600 substructures has recently been analyzed.[48] The substructures now used in **STIRS** have been selected from an initial set of 7000 possible substructures as being those most reliably identified on the basis of mass spectral correlations.

Working through its list of substructures, **STIRS** can correctly predict about three substructures for each unknown, along with perhaps one incorrect prediction.[48] Naturally, **STIRS** restricts its predictions to positive statements based on evidence indicating the presence of a particular substructure; the competitive nature of mass spectral fragmentation processes makes it inappropriate to infer the absence of a group from the absence fragment ions (because all the ion current may be in ions resulting from fragmentations induced by some other, more strongly directing, group). In principle, substructures identified by **STIRS** could be used in automated structure elucidation systems. However, **STIRS** is primarily a tool for obtaining structural information about an unknown given only LRMS data and an estimated molecular weight. Much more comprehensive data are necessary for complete structure elucidation studies. Certainly, the molecular composition would have to be known, there would be additional spectral data, and, most probably, considerable general structural information inferred from details of the unknown compound's origin and the chemical extraction procedures applied. A fair number of **STIRS** standard substructures can be identified much more readily through complementary spectral data. Obvious examples include the various groups incorporating methyl or phenyl systems in partially specified environments (which could be identified through magnetic resonance spectroscopy) and the predictions for carbonyls and ethers (which could be made more directly from IR absorptions). The presence of other groups (e.g., trimethyl silyl ethers) would usually be known given the history of the chemical extraction and processing of an unknown. Other structural features (e.g., styryl, cinnamoyl, phenylazo, indolyl, steroidal, $C_4H_4O_5$ sugar) might well be known through details of the origin of a compound.

The **STIRS** analysis is not restricted to its list of standard substructures. The **STIRS** system will retrieve the most closely analogous structures that it can find in its files for a given unknown; thus a novel substituted diterpenoid would typically cause the retrieval of related diterpenoid systems. The structures retrieved can be analyzed to identify their common features such as, in this example, any skeletal features common to all the retrieved diterpenoids. Thus evidence can possibly be obtained indicating where a novel structure differs from known compounds in its sites of substitution. This kind of analysis can yield more information than can be derived from an analysis based on standard substructures chosen for showing the best prediction performances in tests on a wide variety of molecules. In the original version of **STIRS**[47] it was the task of the user to examine the WLN notations for retrieved structures and thus identify common features (naturally, this suffers from the same problems as noted earlier with reference to the WLN comparison functions described by Zupan et al.[57,58]). More recently, an automated procedure for structure comparison has been devised.[59] This procedure can identify the largest common substructure of a given pair of structures and can be applied systematically to the retrieved reference structures to determine their common features. This aspect of **STIRS** has received less emphasis than the analyses based on standard substructures, and the effectiveness of such a system has still to be assessed.

There are preliminary data now available for another system relying on a moderately elaborate mass spectral search followed by common substructure abstraction from retrieved references.[60] The mass spectral search system developed at the Novosibirsk Institute utilizes matching on specific mass positions in an abbreviated spectrum and on relative mass values (i.e., an indirect representation of neutral losses). Coded representations of the structures of the highest-ranked matching spectra are analyzed to identify their maximum common substructures. These substructures, together with details of their frequency of occurrence among the top-ranked structures, are presented to the user of the search system. The paper by Lebedev et al.[60] includes a number of specific examples together with percentage success rates achieved over a small number of tests.

Correlations between subspectra and substructures are much more clearly defined for magnetic resonance, and other spectroscopies, than they are in mass spectrometry. In principle, substructure identification systems should be more readily developed for these techniques. From the initial results so far reported, there appear to be some problems.

Shelley and Munk[13] have described a system that typifies the type of analysis that can be undertaken wherever specific spectral signals can be associated with particular atoms (or possibly larger substructures). Their paper describes the use of a spectrum interpretation program in association with a small data base containing ^{13}C NMR spectra of some 900 compounds. Their program is designed for analyzing ^{13}C NMR data but their basic approach should also be applicable to ^{1}H NMR data (provided it is possible to identify the "signal" associated with each proton). The data base is required to contain a representation of the reference molecule (as a form of connection table, see Chapter VII) and details of the assignments of the resonances in the spectrum to the appropriate atoms. Resonances from the observed spectrum of the unknown are matched, within appropriate tolerances on chemical shifts, with resonances in the reference spectrum. The program tags those atoms in its model of the reference structure whose resonances have been matched. If the tagged atoms thus identified form a connected substructure of sufficient size, the program saves a representation of that substructure (containing only tagged carbon atoms and neighboring heteroatoms). The program records each distinct substructure thus obtained together with a record of those reference compound(s) where it was found. At the completion of the search, the chemist may interactively examine the substructures and reference structures.

A problem with ^{13}C NMR data is that fortuitous matchings of resonances can result in identification of misleading substructures. Such misleading results are not particularly overt when only small data bases are employed; however, as the number and diversity of reference compounds are increased, so are the number of erroneous substructural identifications. Frequently, the substructures proposed are almost correct but differ from the correct structural form by implying different substitution patterns on a more or less correctly identified skeleton. Some examples are apparent in the results of Shelley and Munk,[13] as, for instance, in the substructures retrieved for the steroidal benzoate **IV-5**. Along with some totally spurious substructures, a number of steroidal fragments are found for this structure, but all except one of these contain erroneous implications concerning the substitution pattern of the steroidal skeleton. Thus, for example, substructure **IV-6** is suggested on the

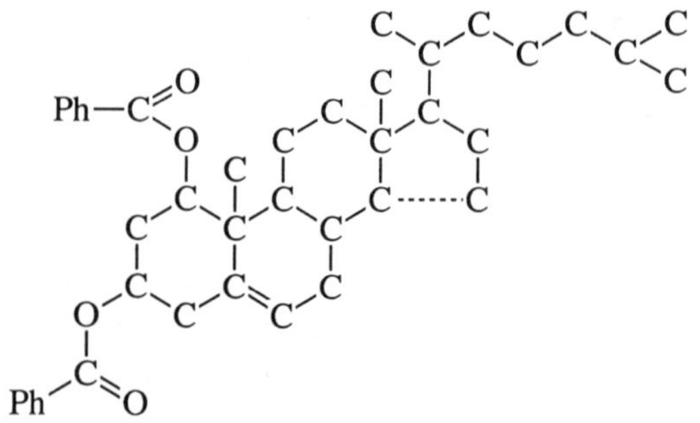

IV-5

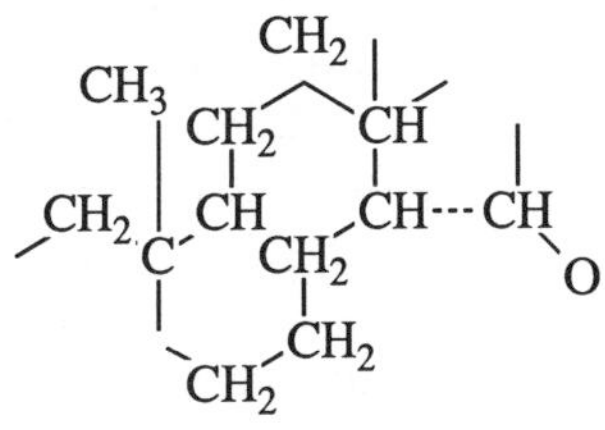

IV-6

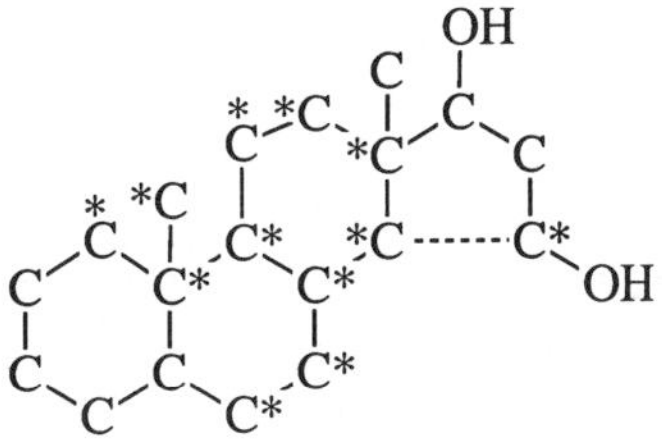

IV-7

basis of matchings with resonances for the tagged atoms in the reference structure **IV-7**; this substructure implies that the "unknown" steroid should be B-ring saturated, D-ring substituted with no substituent group at C(1).

Such problems have been analyzed in some detail elsewhere in relation to results obtained with a rather similar substructure search system working with a somewhat larger and more diverse ^{13}C NMR data base.[10] The results of such studies indicate that it is common for a chance matching of resonances to result in the retrieval of a seemingly plausible but, in fact, highly misleading substructure. Careful analysis by the chemist is required before such substructures can be used with confidence in a structure identification system. Although inspection of retrieved substructures often suffices to eliminate many spurious matches, some errors, such as the proposal of substructure **IV-6** for structure **IV-5**, are not readily perceived as such in processing of true unknowns. In addition, without other constraints, a search against a large and diverse data base generally yields substantial numbers of possible substructures for the chemist to consider.

Carbon-13 NMR data have also been used in systems that employ files of subspectra and substructures.[61,62] Such files are relatively costly to construct, for the chemist must identify "standard" substructural fragments (e.g., a common side chain, substituent group, or part of a ring system) in reference compounds and assemble a file containing these substructures and associated resonances. Ultimately, such file search approaches entail the same problems as do the more flexible systems such as that of Shelley and Munk.[13] Random matchings of resonances will cause the retrieval of incorrect substructures. Substructures and subspectra do, however, have a secondary use in searching of files for matching spectra. One can further constrain a spectrum matching procedure by requiring that all the resonances of defined subspectra be duly matched. Such a constraint can improve the reliability of searches that work through the matching of a specified number of signals. The use of subspectra constraints does involve some presumptions regarding where, on some standard skeleton, differences in substitution are to be permitted while still regarding structures as similar. The subspectrum constraint approach was devised by Bremser et al.[61,62] solely for ^{13}C; probably the method has somewhat wider applicability. Thus, for example, subspectral constraints could be defined for a mass spectral search system. The ion patterns resulting from specific substructural components of a reference molecule could be identified and these subspectra used to additionally characterize that compound. The search system could then be constrained by the requirement that, before a reference structure be counted as relevant, at least one of its defined subspectra be perfectly matched. It is questionable whether any benefits of improved discrimination would outweigh the increased costs of developing the more elaborate files and the biasing introduced by presumptions as to what constituted appropriate substructures and subspectra.

Substructure–subspectra files are essential components in a number of the "knowledge-based" interpretive systems discussed subsequently. These interpretive programs differ from the retrieval systems, discussed in this chapter, in that the file search step (or equivalent procedure) is only the first stage of a much more extensive analysis aimed at identifying consistent interpretations for different spectral features.

IV.F. DATABASE INTEGRITY

It is self-evident that successful identifications are dependent on the use of correct data to characterize the reference compounds in the file. The majority of spectral collections are simply compendia of arbitrary compounds whose spectra have been compiled from various diverse sources. Problems of reliability of data have been reported for most of these collections.[3,4,10] data.

Problems of spectral quality can be tackled in at least two ways. In one approach the reference data collection can be limited to certified spectra, guaranteed to be of the compounds specified, and acquired under carefully controlled conditions. Alternatively, or even better as a complementary approach, computer methods can be used to examine new submitted data for consistency with existing information and explore the self-consistency of data already in the data base.

There are now continuing projects to establish certified collections of quality IR and mass spectral data. Now that most known collections of spectra have already been added to the NIH–EPA–MSDC spectral library, further expansion of the files is being concentrated on acquiring reliable spectra of compounds of particular relevance to the U.S. Environmental Protection Agency (EPA), (which sponsors the CIS).[3,15] The Coblentz Society has established a collection of certified IR data.[63]

Verification of the self-consistency of a spectrum or its consistency with existing information in a data base is basically a matter of "predicting" spectral properties and examining the given data in the light of these predictions. This approach has been most widely applied to mass spectral data. An MS spectral quality index has been developed that may be used to select the best of any duplicate set of spectra of a given compound or identify any highly dubious data. In this system, developed by Speck et al.[14], the prediction process is relatively limited. Given the molecular weight and composition of a compound, it is possible to make crude predictions concerning the total number of ion peaks that might be expected, isotope patterns, and features in the molecular ion region. Observed spectra that show discrepancies from such predictions, such as impurity ions at higher mass than the molecular ion or ions that correspond to "illogical" losses from the molecular ion, are assigned a reduced quality index. The data in the NIH–EPA–MSDC mass spectral collection have been checked with respect to this quality index, and duplicate, lower quality spectra have been discarded.[15]

There exist more elaborate models for mass spectrum prediction (see Chapter XI). At their current state of development, these methods are not adequate for the task of certifying large collections of spectra of arbitrary structures. Some of these methods for spectrum prediction use fragmentation rules that are derived through detailed analyses of specific spectra and structures in some limited chemical class; the derived rules are then applicable only to the analysis of similar structures.[64–66] The only generally applicable mass spectrum prediction method is **DENDRAL**'s half-order theory.[67] Unfortunately, this prediction model cannot provide reliable intensity estimates. Furthermore, it has a tendency to overpredict a spectrum; in its exhaustive search of the possible fragmentations of a given structure, the half-order theory may detect plausible fragmentation mechanisms that could conceivably lead to many ions other than those actually observed in a compound's true spectrum. Any impurity ions that, by chance, occurred at the same mass as one of these extra predictions would prove perfectly acceptable to the half-order model.

Elaborate methods for spectrum prediction and validation have been applied to ^{13}C NMR data. The prediction-validation procedures for ^{13}C NMR data rely on consistency of correlations between an atom's chemical shift and its configurational substructural environment.[10,16,68] In the system developed at Stanford,[16] for example, one component of the ^{13}C data base is a file containing an entry for each carbon atom for which a specific resonance has been assigned in some reference compound. Such entries consist of the shift value and a code defining the constitutional and configurational environment of the resonating atom out to some maximum distance (typically four bonds radius). By encoding the substructural environment of an atom in a given structure and then using the code to access the data base, one can obtain the shift values assigned to atoms in similar substructural environments as found in previously analyzed reference structures. These retrieved shift values establish a range of resonance shifts wherein the signal due to the specified atom should be observed. If, for example, a pair of resonances have been misassigned for some new compound, both misassignments are likely to be detected as lying outside of the predicted ranges for their corresponding atoms. The chance of incorporation of a misassigned spectrum in the data base can be substantially reduced through the application of such checks. It is also possible to run periodic checks, on the entire data base, in which substructures associated with anomalously broad shift distributions can be identified, and thus erroneous assignments may be detected.[16]

IV.G. CONCLUSIONS

Computerized file searches will undoubtedly remain the method of choice for compound identification; the algorithms are simple and can be reasonably fast. In most circumstances, fully automated spectrum matching searches are likely to prove more useful than interactive feature-selection searches. If a compound is not present among the available references, it is possible that an interactive feature-selection search will fail to yield any useful information, whereas a spectrum matching procedure might at least identify basically similar structures. Automated searches are also more appropriate in applications where large numbers of compounds must be identified with the minimum of intervention by the chemist. Simplified spectrum matching procedures can be implemented on the mini- and microprocessor-based systems incorporated into most analytical instruments; in favorable circumstances, the analysis of data acquired for one unknown can proceed simultaneously with the acquisition of data for other samples.

Success at identification, and discrimination among closely related structures, is probably most conveniently achieved by using specific spectral signals, weighted according to their discriminating ability as determined by statistics drawn from the complete reference file. A good current example is provided by **PBM**.[26] Elaboration of indexed file organizations, based on simple spectrum classification techniques,[25,33,35,36] can improve search performance. The classification step can identify that subset of the file that must be searched sequentially, as the small computational cost of the classification step is more than repaid through savings on data transfers and comparisons in the sequential searches. The use of classification functions in a prefilter step may also make it possible to enhance the discriminating power of the searches. The weights used in the matching procedure can be based not on statistics derived for the entire reference file, but on statistics relating solely to the set of compounds within the defined chemical class.

Success at retrieving structural analogs for compounds not yet represented in the files is probably best realized through searches where the matching criteria are based on computed spectral features as in **STIRS** and the systems devised by Clerc.

There are a number of areas where substantial further development of file-based systems is possible. More extensive investigations of retrieval performance are desirable. The "propagating search" scheme of Rasmussen and Isenhour[28] provides a useful empirical probe of performance; however, more direct measures, based on comparison of the correct structure of an "unknown" and the structures of reference compounds retrieved, are necessary. The existing algorithms,[57] utilizing WLN notations for structures, are not satisfactory because the priority rules in WLN (for encoding rings etc.) can result in the representation of similar structures by very different sequences of code symbols. The **STIRS** system suggests two alternative approaches to assessing structural similarity. If the search file contains details of constituent substructures for each reference compound and the various matching procedures being tested all yield substructural suggestions, comparative performance could be assessed on the basis of the reliability of the substructural predictions for a set of test spectra. A possibly more accurate measure of retrieval performance of different searches might be based on predictions of the largest common substructures developed through an analysis of retrieved reference compounds.[59,60,69]

Given a general mechanism for assessing relative retrieval performance, further studies of different procedures for selecting characteristic features of reference spectra would be again justifiable.

Another worthwhile area for research on file-based systems is in development of methods for detecting and eliminating errors in the reference data. Basically, this involves the development of quick procedures for "spectrum prediction" and finding anomalies when comparing predicted properties for a structure with those provided as data. Existing systems, as reported for ^{13}C NMR data[16,68], illustrate the forms of processing appropriate when specific substructure–subspectra correlations exist. It should be possible to adapt these approaches to IR or ^{1}H NMR data. Care has to be exercised with regard to the detail of the substructural descriptions employed. Thus a substructure description lacking stereochemical detail is inappropriate for ^{13}C because it does not allow any expression of those structural differences that induce large "steric" shifts, and, consequently, the same substructure becomes correlated with markedly different spectral signals. For ^{1}H NMR it might well prove necessary to incorporate some form of conformation description in substructures. Whereas subspectrum–substructure compilations make verification of some kinds of data reasonably facile, the corresponding analysis for mass spectra or any other basically global property of a molecule remains a rather more difficult task.

Given the limited size of existing data collections, it is inevitable that many searches must fail to yield identifications simply because relevant reference data are unavailable. Such failures are, of course, the ultimate justification for all the more elaborate methods described subsequently. Although the more sophisticated search schemes can provide substructural suggestions, there are problems in respect to the automated use of these suggestions in subsequent programs for candidate structure generation. A system, like **STIRS**, can suggest a number of substructural constituents for an unknown; plausibility ratings can sometimes be provided for each such suggestion. It doesn't matter that some of these suggested substructures are potentially overlapping, a sophisticated structure generation procedure, such as **GENOA**, can deal with such ambiguities. Nor does it matter that some suggestions will occasionally prove incorrect, all partially assembled structures built on incorrect substructures will eventually be discarded by the structure generation scheme when subsequent conflicting structural constraints are employed. The problem is that each such suggestion defines a completely separate structural problem to be analyzed by the structure generation procedures. Thus the suggestion that a particular unknown probably incorporates, say, a 1,4-disubstituted phenyl and an acetoxy group really results in the definition four distinct structural problems. The unknown may indeed incorporate both groups, or either group alone, or maybe neither of the suggested groups. The plausibility ratings for suggested groups may determine a particular optimum order for investigating each of these structural problems. Ultimately, however, each of the problems will have to be developed in order to guarantee that all possible solutions are indeed considered.

It is possible to proceed with structure generation assigning to different partial structures scores based on the plausibility ratings of incorporated substructures. Partial structures incorporating several less plausible components accumulate poorer overall scores than those based on more plausible components or those incorporating some probable and some less probable components. The overall scores can be used to filter the growing set of partial structures with poorly ranked candidates being discarded.[70] This approach is of only marginal utility even in cases where all the component scores are based on similar spectral analyses, such as all the scores being based on a particular MS interpretation scheme. More generally, there would be problems in determining how commensurate are those substructural scores that are inferred from a MS analysis scheme, a ^{13}C interpreter, and IR spectrum analyzer.

The effective use of substructural constraints inferred through spectral analysis is possible only when either the inferences are certain (the unknown either must or must not contain a suggested component) or when a group of suggested components form an exhaustive set of interpretations for some spectral feature. In such cases, it is possible to use a sophisticated structure generator such as **GENOA** to develop all alternative solutions in parallel.

REFERENCES

1. D. E. Knuth, *The Art of Computer Programming. 3. Sorting and Searching*, Addison-Wesley, Reading, MA, 1975, 506–549.
2. D. P. Martinsen, "Survey of Computer Aided Methods for Mass Spectral Interpretation," *Appl. Spectrosc.*, **35**, (1981), 255.
3. G. W. A. Milne and S. R. Heller, "NIH/EPA Chemical Information System," *J. Chem. Inf. Comput. Sci.*, **20**, (1980), 204.
4. C. L. Fisk and G. W. A. Milne, "The Status of Infrared Data Bases," *J. Chromatogr. Sci.*, **17**, (1979), 441.
5. J. T. Clerc and J. Zupan, "Computer Based Systems for the Retrieval of IR Spectral Data," *Pure Appl. Chem.*, **49**, (1977), 1827.
6. D. L. Dalrymple, C. L. Wilkins, G. W. A. Milne, and S. R. Heller, "A Carbon-13 Nuclear Magnetic Resonance Spectral Data Base and Search System," *Org. Magn. Reson.*, **11**, (1978), 535.
7. Y. Katagiri, K. Kanohta, K. Nagasawa, T. Okusa, T. Sakai, O. Tsumura, and Y. Yotsui, "Development of a New File Search System for Nuclear Magnetic Resonance Spectra," *Anal. Chim. Acta*, **133**, (1981), 535.

8. V. A. Koptyug, V. S. Bochkarev, B. G. Derendyaev, S. A. Nekoroshev, V. N. Piottukh-Peletskii, M. I. Podgornaya, and G. P. Ul'yanov, "Use of a Computer for the Solution of Structural Problems in Organic Chemistry by the Methods of Molecular Spectroscopy," *Zh. Strukt. Khimii*, **18**, (1977), 440.
9. W. Bremser, H. Wagner, and B. Franke, "Fast Searching for Identical C-13 NMR Spectra via Inverted Files," *Org. Magn. Reson.*, **15**, (1981), 178.
10. N. A. B. Gray, "Computer Assisted Analysis of Carbon-13 NMR Spectral Data," *Prog. Nucl. Magn. Res. Spectrosc.*, **15**, 1982, 201.
11. J. T. Clerc, R. Knutti, H. Koenitzer, and J. Zupan, "Conversion of a Conventional IR Library into Computer Readable Form," *Fresenius' Z. Anal. Chem.*, **283**, (1977), 177.
12. S. R. Lowry and D. A. Huppler, "Infrared Spectral Search System for Gas Chromatography Fourier Transform Infrared Spectrometry," *Anal. Chem.*, **53**, (1981), 889.
13. C. A. Shelley and M. E. Munk, "Computer Prediction of Substructures from Carbon-13 Nuclear Magnetic Resonance Spectra," *Anal. Chem.*, **54**, (1982), 516.
14. D. D. Speck, R. Venkataraghavan, and F. W. McLafferty, "A Quality Index for Mass Spectra," *Org. Mass Spectrom*, **13**, (1978), 209.
15. S. R. Heller, G. W. A. Milne, and R. J. Feldmann, "Quality Control of Chemical Data Bases," *J. Chem. Inf. Comput. Sci.*, **16**, (1976), 232.
16. M. R. Lindley, N. A. B. Gray, D. H. Smith, and C. Djerassi, "Computerized Approach to the Verification of Carbon-13 Nuclear Magnetic Resonance Spectral Assignments," *J. Org. Chem.*, **47**, (1982), 1027.
17. L. A. Powell and G. M. Hieftje, "Computer Identification of Infrared Spectra by Correlation Based File Searching," *Anal. Chim. Acta*, **100**, (1978), 313.
18. G. T. Rasmussen and T. L. Isenhour, "Library Retrieval of Infrared Spectra Based on Detailed Intensity Information," *Appl. Spectrosc.*, **33**, (1979), 371.
19. L. V. Azarraga, R. R. Williams, and J. A. de Haseth, "Fourier Encoded Data Searching of Infra Red Spectra," *Appl. Spectrosc.*, **35**, (1981), 466.
20. J. E. Biller and K. Biemann, "Reconstructed Mass Spectra, a Novel Approach for the Utilization of Gas-Chromatography-Mass Spectrometry Data," *Anal. Lett.*, **7**, (1974), 515.
21. R. G. Dromey, M. J. Stefik, T. C. Rindfleisch, and A. M. Duffield, "Extraction of Mass Spectra Free of Background and Neighboring Component Contributions from Gas Chromatography-Mass Spectrometry Data," *Anal. Chem.*, **48**, (1976), 1368.
22. C. C. Sweeley, N. D. Young, J. F. Holland, and S. C. Gates, "Rapid Computerized Identification of Compounds in Complex Biological Mixtures by Gas Chromatography-Mass Spectrometry," *J. Chromatogr.*, **99**, (1974), 507.
23. F. W. McLafferty, R. H. Hertel, and R. D. Villwock, "Probability Based Matching of Mass Spectra," *Org. Mass Spectrom.*, **9**, (1974), 690.
24. F. P. Abramson, "Automated Identification of Mass Spectra by the Reverse Search," *Anal. Chem.*, **47**, (1975), 45.
25. T. O. Gronneberg, N. A. B. Gray, and G. Eglinton, "Computer Based Search and Retrieval System for Rapid Mass Spectral Screening of Samples," *Anal. Chem.*, **47**, (1975), 415.
26. G. M. Pesyna, R. Venkataraghavan, H. E. Dayringer, and F. W. McLafferty, "Probability Based Matching Using a Large Collection of Mass Spectra," *Anal. Chem.*, **48**, (1976), 1362.
27. H. S. Hertz, R. A. Hites, and K. Biemann, "Identification of Mass Spectra by Computer-Searching a File of Known Spectra," *Anal. Chem.*, **43**, (1971), 681.
28. G. T. Rasmussen and T. L. Isenhour, "The Evaluation of Mass Spectral Search Algorithms," *J. Chem. Inf. Comput. Sci.*, **19**, (1979), 179.
29. F. Erni and J. T. Clerc, "Strukturaufklarung organischer Verbindungen durch computerunterstutzen Vergleich spektraler Daten," *Helv. Chim. Acta*, **55**, (1972), 489.

30. R. Schwarzenbach, J. Meile, H. Konitzer, and J. T. Clerc, "A Computer System for Structural Identification of Organic Compounds from C-13 NMR Data," *Org. Magn. Reson.*, **8**, (1976), 11.
31. H. Nau and K. Biemann, "Utilization of Automatically Assigned Retention Indices for Computer Identification of Mass Spectra," *Anal. Lett.*, **6**, (1973), 1071.
32. S. S. Williams, R. B. Lam, D. T. Sparks, T. L. Isenhour, and J. R. Haas, "Data Evaluation in Combined Gas Chromatography-Infrared Spectroscopy-Mass Spectrometry," *Anal. Chim. Acta*, **138**, (1982), 1.
33. R. G. Dromey, "Simple Index for Classifying Mass Spectra with Applications to Fast Library Searching," *Anal. Chem.*, **48**, (1976), 1464.
34. D. H. Smith, "A Compound Classifier Based on Computer Analysis of Low Resolution Mass Spectral Data," *Anal. Chem.*, **44**, (1972), 536.
35. J. Domokos, E. Pretsch, H. Mandli, H. Koenitzer, and J. T. Clerc, "Cluster-analyse Massenspektrometricher Daten," *Fresenius' Z. Anal. Chem.*, **304**, (1981), 241.
36. M. Penca, J. Zupan, and D. Hadzi, "Hierarchical Preprocessing of Infrared Data Files," *Anal. Chim. Acta*, **95**, (1977), 3.
37. J. Zupan, "Hierarchical Clustering of Infrared Spectra," *Anal. Chim. Acta*, **139**, (1982), 143.
38. S. R. Heller, "Computer Techniques for Interpreting Mass Spectrometry Data," in *Computer Representation and Manipulation of Chemical Information*, W. T. Wipke, S. Heller, R. Feldmann, and E. Hyde, eds., Wiley, New York, 1974, 175, Chapter 8.
39. S. R. Heller, H. M. Fales, and G. W. A. Milne, "A Conversational Mass Spectral Search and Retrieval System - II. Combined Search Options," *Org. Mass Spectrom.*, **7**, (1973), 107.
40. J. R. Chapman, *Computerized Mass Spectrometry*, Academic Press, London, 1978.
41. S. L. Grotch, "Computer Identification of Mass Spectra Using Highly Compressed Spectral Codes," *Anal. Chem.*, **45**, (1973), 2.
42. G. van Marlen and A. Dijkstra, "Information Theory Applied to the Selection of Peaks for Retrieval of Mass Spectra," *Anal. Chem.*, **48**, (1976), 595.
43. S. L. Grotch, "Matching of Mass Spectra When Peak Height is Encoded to One Bit," *Anal. Chem.*, **42**, (1970), 1214.
44. S. L. Grotch, "Computer Techniques for Identifying Low Resolution Mass Spectra," *Anal. Chem.*, **43**, (1971), 1362.
45. S. L. Grotch, "Automatic Identification of Chemical Spectra. A Goodness of Fit Measure Derived from Hypothesis Testing," *Anal. Chem.*, **47**, (1975), 1285.
46. H. B. Woodruff, S. R. Lowry, G. L. Ritter, and T. L. Isenhour, "Similarity Measures for the Classification of Infrared Data," *Anal. Chem.*, **47**, (1975), 2027.
47. K. S. Kwok, R. Venkataraghavan, and F. W. McLafferty, "Computer-Aided Interpretation of Mass Spectra. III. A Self-Training Interpretive and Retrieval System," *J. Am. Chem. Soc.*, **95**, (1973), 4185.
48. K. S. Haraki, R. Venkataraghavan, and F. W. McLaffery, "Prediction of Substructures from Unknown Mass Spectra by the Self-Training Interpretive and Retrieval System," *Anal. Chem.*, **53**, (1981), 386.
49. R. G. Dromey, B. G. Buchanan, D. H. Smith, J. Lederberg, and C. Djerassi, "Applications of Artificial Intelligence for Chemical Inference. XIV. A General Method for Prediction of Molecular Ions in Mass Spectra," *J. Org. Chem.*, **40**, (1975), 770.
50. I. K. Mun, R. Venkataraghavan, and F. W. McLafferty, "Computer Prediction of Molecular Weights from Mass Spectra," *Anal. Chem.*, **53**, (1981), 179.
51. S. R. Lowry, T. L. Isenhour, J. B. Justice, F. W. McLafferty, H. E. Dayringer, and R. Venkataraghavan, "Comparison of Various K-Nearest-Neighbor Voting Schemes with the Self-Training Interpretive and Retrieval System for Identifying Molecular Substructures from Mass Spectra," *Anal. Chem.*, **49**, (1977), 1720.

52. F. W. McLafferty, "Performance Prediction and Evaluation of Systems for Computer Identification of Spectra," *Anal. Chem.*, **49**, (1977), 1441.
53. G. Salton, *Automatic Information Organization and Retrieval*, McGraw-Hill, New York, 1968.
54. H. E. Dayringer, G. M. Pesyna, R. Venkataraghavan, and F. W. McLafferty, "Computer-Aided Interpretation of Mass Spectra. Information on Substructural Probabilities from STIRS," *Org. Mass Spectrom.*, **11**, (1976), 529.
55. R. J. Mathews and J. D. Morrison, "Comparative Study of Methods of Computer-Matching Mass Spectra," *Austral. J. Chem.*, **27**, (1974), 947.
56. R. J. Mathews and J. D. Morrison, "Comparative Study of Methods of Computer-Matching Mass Spectra. II.," *Austral. J. Chem.*, **29**, (1976), 689.
57. J. Zupan, M. Penca, D. Hadzi, and J. Marsel, "Combined Retrieval System for Infrared, Mass, and Carbon-13 Nuclear Magnetic Resonance Spectra," *Anal. Chem.*, **49**, (1977), 2141.
58. J. Zupan, "Problems in Data Retrieval Systems for Analytical Spectroscopy," *Anal. Chim. Acta*, **103**, (1978), 273.
59. M. M. Cone, R. Venkataraghavan, and F. W. McLafferty, "Molecular Structure Comparison Program for the Identification of Maximal Common Substructures," *J. Am. Chem. Soc.*, **99**, (1977), 7668.
60. K. S. Lebedev, V. M. Tormyshev, B. G. Derendyaev, and V. A. Koptyug, "A Computer Search System for Chemical Structure Elucidation Based on Low-Resolution Mass Spectra," *Anal. Chim. Acta*, **133**, (1981), 517.
61. W. Bremser, M. Klier, and E. Meyer, "Mutual Assignment of Subspectra and Substructures—A Way to Structure Elucidation by C-13 NMR Spectroscopy," *Org. Magn. Reson.*, **7**, (1975), 97.
62. W. Bremser, "The Importance of Multiplicities and Substructures for the Evaluation of Relevant Spectral Similarities for Computer Aided Interpretation of C-13 NMR Spectra," *Z. Anal. Chem.*, **286**, (1977), 1.
63. Coblentz Society, P. O. Box 9952, Kirkwood, Missouri 63122.
64. D. H. Smith, B. G. Buchanan, W. C. White, E. A. Feigenbaum, J. Lederberg, and C. Djerassi, "Applications of Artificial Intelligence for Chemical Inference. X. INTSUM—A Data Interpretation and Summary Program Applied to the Collected Mass Spectra of Estrogenic Steroids," *Tetrahedron*, **29**, (1973), 3117.
65. B. G. Buchanan, D. H. Smith, W. C. White, R. Gritter, E. A. Feigenbaum, J. Lederberg, and C. Djerassi, "Applications of Artificial Intelligence for Chemical Inference. XXII. Automatic Rule Formation in Mass Spectrometry by Means of the Meta-DENDRAL Program," *J. Am. Chem. Soc.*, **98**, (1976), 6168.
66. A. Lavanchy, T. Varkony, D. H. Smith, N. A. B. Gray, W. C. White, R. E. Carhart, B. G. Buchanan, and C. Djerassi, "Rule-Based Mass Spectrum Prediction and Ranking: Applications to Structure Elucidation of Novel Marine Sterols," *Org. Mass Spectrom.*, **15**, (1980), 355.
67. N. A. B. Gray, R. E. Carhart, A. Lavanchy, D. H. Smith, T. Varkony, B. G. Buchanan, W. C. White, and L. Creary, "Computerized Mass Spectrum Prediction and Ranking," *Anal. Chem.*, **52**, (1980), 1095.
68. J. E. Dubois and J. C. Bonnet, "The DARC Pluridata System: The ^{13}C NMR Data Bank," *Anal. Chim. Acta*, **112**, (1979), 245.
69. T. H. Varkony, Y. Shiloach, and D. H. Smith, "Computer-Assisted Examination of Chemical Compounds for Structural Similarities," *J. Chem. Inf. Comput. Sci.*, **19**, (1979), 104.
70. N. A. B. Gray, A. Buchs, D. H. Smith, and C. Djerassi, "Computer-Assisted Structural Interpretation of Mass Spectral Data," *Helv. Chim. Acta*, **64**, (1981), 458.

V

PATTERN RECOGNITION APPROACHES

V.A. INTRODUCTION

When searches of files of reference spectral data fail to identify a compound, elucidation of its structure will depend on procedures for generating and evaluating all possible structural forms. Such generation procedures are combinatorial and, unless a problem is well constrained, can produce innumerable candidates. The constraints, defining constituent subassemblies of atoms and restrictions on the bonding among these subassemblies, must be derived primarily by the chemist applying rules for spectral interpretation. As discussed in the Chapter VI, conventional spectral interpretation rules can, to a degree, be incorporated into spectral analysis programs. However, there are limitations to such "knowledge-based" interpretive schemes. Usually, such schemes are computationally costly. Different types of interpretive program have to be developed for analyzing each different type of spectral data; it may, indeed, be necessary to develop substantially different procedures for analyzing the same kind of spectral data as obtained with different classes of compound. Furthermore, such automation is possible only in cases where well-defined sets of spectral interpretation rules are already extant. Yet, there remain many spectral-structural correlations for which there are no strong models and no well-defined interpretive rules; given empirical methods for exploiting such correlations it might be possible to derive valuable additional structural constraints.

For example, conventional methods for mass spectral interpretation do not permit one to establish whether a compound incorporates a heteroatom in a ring system, nor would one normally be able to infer the existence of a steroidal nucleus from a relatively cursory examination of the pattern of resonance shifts in a compounds ^{13}C spectrum. Yet, both such correlations have been identified and exploited through empirical classification procedures working on (limited) sets of structures and spectra.[1,2] The task of a combinatorial structure generation procedure would obviously be greatly reduced if it could indeed include 19 carbon atoms, and at least four degrees of unsaturation, into a standard skeleton and could then further restrict the placing of heteroatoms so that all of these were incorporated into acyclic appendages on this skeleton.

Direct correlations between spectral data and structural features obviate the need for any intermediate interpretation steps and any intermediate concepts. If substructures can be inferred directly from mass spectral data, there is no need to invoke rules describing the fragmentation mechanisms appropriate to a particular class of compounds. Analyses based on finding consistent interpretation of individual resonances can be bypassed when reliable correlations exist between substructures and patterns of ^{13}C resonances.

Direct correlations can be expressed in the form of classification functions that range over the set of possible classes and whose domains are the spectral measurements made on the compounds. Thus we can conceive of a "heteroatom-in-ring" function, $F_{\text{ring-het}}$, with the range "compound with a ring-heteroatom, compound without a ring-heteroatom" and a domain as the measured intensities of ions over a range of some 150 amu.

Such functions attribute no physical significance to individual spectral measurements nor to any relationships between individual measurements. The values found for different spectral properties simply define the values of the elements of a vector. A mass spectrum may be represented by a vector of 150 elements with each element having the intensity recorded at a particular m/z; a ^{13}C spectrum may be represented by a 200-element vector whose elements are encoded as 1s and 0s according to whether resonances were observed in particular 1-ppm shift ranges. The classification functions involve just simple arithmetic manipulations of the values of these vector elements. These vectors are regarded as defining the position, in some hyperdimensional pattern space, of a point representing a compound. Such classification functions typically involve computation of some form of generalized distance between one point and another reference point; an alternative, although essentially equivalent form of classification, function tests a weighted sum of its argument values against a specified threshold. Different types of spectral data may all be interpreted in a uniform manner, and it may even be practical for a single interpretive function to combine multisource spectral data.

In simple cases, where compounds are to be classified on the basis of two or three measurements of spectral properties, it is practical to plot the data points and, by inspection, determine whether a particular structural attribute appears to be correlated with the measured property values. Of course, the physical basis for any such simple correlation is likely to be known, and, if the relationship is significant and reliable, interpretation rules for analyzing the data will probably exist. With multidimensional data, using measurements on 100 or more spectral channels (m/z values, resonance shifts, IR absorption ranges, etc.), direct visualization of correlations is impractical. Alternative approaches must be devised either to uncover spectral-property–structural-attribute relationships present in the data or determine whether the data are consistent with some postulated relationship.

In one approach (usually called *display* methods), the points representing compounds are projected from the multidimensional space onto a visualizable two- or three-dimensional model. Clusterings of points, representing molecules showing similarities in their spectra, may then be apparent. The clusterings thus revealed may suggest some novel relationship or confirm some postulated spectral-structural correlation.[3,4] Data characterizing an unknown compound can similarly be projected onto the two- or three-dimensional model; the relationship between the projected image of the unknown with the clusters of points representing known compounds can be used by the chemist as a basis for classification. As an alternative, standard mathematical procedures can be used to examine the available reference data and reveal any clusterings.[5] Essentially, these algorithms identify those subgroups–clusters of reference compounds for which, according to some appropriate distance measure, the intragroup distances are significantly smaller than intergroup differences. Although possibly of value as aids to discovering new spectral–structural relationships, neither the *display* nor the *clustering* approaches have proved of much immediate value in structure elucidation.

In contrast to the predominantly *unsupervised* automatic clustering and display approaches, most applications of pattern recognition methods to structure elucidation are *supervised*. In supervised methods, the desired classes are predefined. Each reference spectral record can be tagged with its appropriate class. These classified reference data can be used directly when analyzing unknowns. "Distances" between the point representing an unknown and the points representing known, tagged reference compounds can be estimated. If it is assumed that the distance metric being used really captures some structurally significant aspect of spectral similarity, an unknown should be structurally similar to, and can thus be assigned to the same class as its nearest reference point. Although this *nearest-neighbor* criterion does provide the basis for quite effective classification performance, most pattern recognition studies are concerned with alternative approaches.

An unknown can be classified by relating its position to the positions, not of individual reference points, but instead, of the clusters representing each class. The data used to characterize the position and form of such clusters can vary widely in degree of detail. The most exacting of pattern recognition procedures require details of the probability densities—the distributions of values found for each spectral property in all the classes. These distributions are expressed as parameterized functions, typically gaussian distributions. Clusters, representing each user-defined class, are identified in terms of their mean and a covariance matrix. Once the clusters have been determined from the available reference data, the probability density for each class at a point in the pattern space representing an unknown can be computed and used as the basis for classification. The major problem with such *parametric* methods is the amount of reference data needed to derive statistically significant probability distributions for the classes. Only a few attempts have been made to use parametric methods for identifying structural attributes; most of these studies have been restricted to the special case of binary data.[5]

Most pattern recognition methods for structural interpretation of spectra rely on less complete, *nonparametric* descriptions of the various classes. Usually, the classification problem is simplified to the case of two-class binary classification. Multiclass categorizers can be created through appropriate combinations of binary classifiers. The typical binary classifier employs a *decision plane* in the hyperdimensional pattern space to separate the region containing points in class α and from the region with points from class β. An unknown is classified according to the side of this plane on which it lies; this calculation involves simply taking a weighted sum (*val*) of its coordinate values (spectral measurements, x_i):

$$val = \Sigma w_i x_i$$

The value of this sum determines the class appropriate for the point defined by the coordinate vector $(x_1, x_2, \ldots, x_d)$ (where d represents the dimensionality of the feature space, i.e. the number of features recorded in the spectrum). The coefficients w_i in this summation are derived from the position of the decision plane. This position, in hyperdimensional pattern space, is determined by using the available reference data; these data are said to constitute a *training set* for the procedure that finds the decision plane.

Often an attempt is made to develop a function that can correctly classify the spectra of *all* compounds in the training set (i.e., a 100% *recognition rate*). Several different algorithms can be used to determine the position of an optimal decision plane for a given training set of classified spectra. These algorithms work by adjusting the position of an initial, possibly arbitrarily placed, decision plane. In some algorithms the spectral data for each known reference structure need be analyzed only once. In other procedures, adjustment of the form of the classification function necessary to accommodate one misclassified member of the training set may result in previously correctly identified references being misclassified by the new function. In such cases the available reference data must be repeatedly reanalyzed until an optimal classification performance is attained. The procedures for developing classification functions can be said to "learn" to recognize compounds of known classes. The performance of such a classifier can then be assessed according to its success rate at classifying additional spectra of known test compounds (*prediction rate*).

It is the various nonparametric, supervised pattern recognition approaches that have been most commonly proposed for application in structure elucidation. Research efforts have concentrated on developments of different procedures for deriving the decision planes, on evaluations of pre-processing techniques that may usefully be applied to spectral data prior to classification and on investigations of feature selection mechanisms. Preprocessing includes normalization, scaling, and,—sometimes—transformation of the original spectral data. The performance of classifiers is usually enhanced by appropriate spectral normalization; scaling is also necessary, particularly in situations where different types of spectral data are to be combined in a single classifier and need to be made commensurate. More elaborate preprocessing steps have also been considered, including taking the Fourier, Haddamard, or Walsh transforms of the original spectra. These transforms create new features that are linear combinations of the original spectral measurements. Functions using transformed data have sometimes been reported as possessing better classification performance. Feature selection is used, in main, either to reduce the number of raw data to more manageable proportions thus making classification more economical, or to investigate which original spectral features are the most important to the classification task.[6] Feature selection can be combined with the data transformation process through mechanisms whereby the feature with the greatest discriminatory power is selected, correlations between this and other features then measured and used to generate new, transformed features orthogonal to those selected. The entire feature-selection-orthogonality transform being reapplied to realize a set of independent features with maximum discriminatory power.

The treatment of pattern recognition methods given in Section V.B is relatively cursory. There are two justifications for this. First, these methods have been subject to several recent reviews. Kryger[7] has published a brief general overview of the methods, and a much more comprehensive treatment is available in Varmuza's text.[5] Chapman's book[8] contains a detailed study of applications of pattern recognition specifically for mass spectral analysis. Much of the earlier work in this field is collected and reviewed in the book by Jurs and Isenhour.[6] Second, as noted both by Kryger[7] and Varmuza[5], the majority of studies of chemical applications of pattern recognition techniques have been concerned with the introduction and demonstration of new pattern recognition procedures and investigations of the (usually marginal) effect of different preprocessing techniques for combining the data. Despite the relatively large effort expended over almost 15 years, there are still relatively few examples of practical, routine uses of pattern recognition techniques in structure elucidation.[9]

V.B. PATTERN RECOGNITION METHODS FOR STRUCTURE ELUCIDATION

In many ways, the k-nearest neighbor (k-n-n) algorithm represents the simplest classification procedure; this algorithm is, of course, just a rather conservative form of file search. A normal sequential search procedure is used to match the spectral data for an unknown with reference data and thus to determine its "nearest neighbors" based on some simple distance metric measured over the entire spectra. Rather than asserting that the unknown *is* the reference compound with the most closely matching spectrum, as would be done in a file search, this classification procedure simply assigns the unknown to the same chemical class as its closest match (1-n-n). The decision regarding class membership can be based on data for more than one neighbor, and various voting schemes, possibly incorporating some form of inverse distance weighting, can be devised under which the unknown is assigned to that class including the majority of its closest three, five, and subsequent neighbors.

The k-n-n algorithm establishes a kind of norm against which other pattern classifiers can be assessed. Because no elaborate training procedures are necessary, any newly acquired reference data can be simply added to the reference file and then exploited immediately in order to improve subsequent classification performance. The k-n-n approach is, inherently, a multiclass categorizer (whereas many other simple classification functions in their basic forms allow distinction of only two classes). No special assumptions or requirements need to be made about how the points corresponding to different classes need to be distributed. The k-n-n algorithm will work as effectively in identification of a member of a class that is best conceived of as comprising an island of points in a sea of nonclass members, as when discriminating between two well-separated, well-defined clusters. The algorithms for k-n-n are simple and their interpretation intuitively obvious. The performance of a nearest-neighbor classifier can be theoretically related to the expected "Bayes risk";[5] the maximum risk of the 1-n-n method is twice the Bayes risk, which represents the minimum risk for misclassification of any method either parametric or nonparametric.

The disadvantage of k-n-n is its computational cost. Although we are assuming that similar structures will give rise to similar spectra that, as measured by an appropriate metric, are clustered closely together in some multi-dimensional space, we are making no use of any such clustering effect. All data for known references are used for classification of any unknown. Most of the comparisons between the unknown and reference data from well-defined classes are redundant, as can be illustrated through consideration of the simplified two-dimensional example shown in Figure V-1. Rather than measure the distance of point x, representing an unknown, from the points representing each of the known reference compounds (the as and bs), we could rely on a single distance measurement from the mean–center-of-gravity A of the α class. (The true mean vector for the class is, of course, unknown; an estimated mean, based on available reference data, has to be used. Provided a reasonable-sized, representative set of compounds is used to estimate the mean, the difference between true and estimated value should not be significant). If x is within a specified maximum distance of A then it can be assigned to the α class.

More typically, the data are envisaged as comprising two or more well-

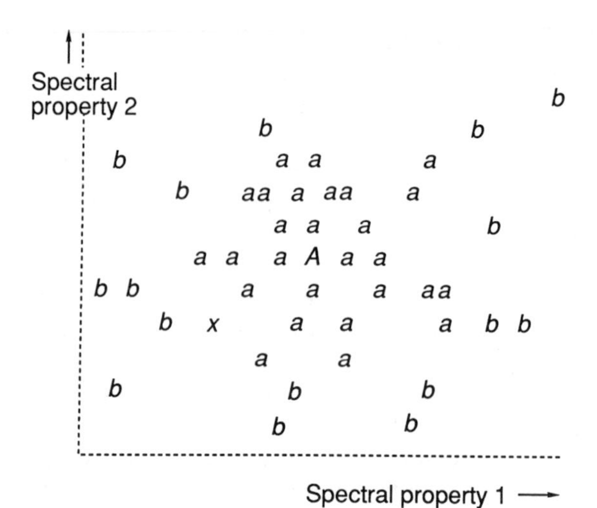

FIGURE V-1. A simple two-dimensional classification problem. The points a and b represent the positions of known reference compounds from the two classes α and β. A marks the mean of the α class. The point x represents the unknown that is to be classified.

defined classes as illustrated in Figure V-2. Here, an effective classification criterion could be be based on distance from the class means or centers-of-gravity. The unknown is assigned to the class with the nearest center of gravity. Various distance measurements are possible; the euclidean distance is satisfactory for most data. The points representing the unknown and the reference compounds are all defined in terms of d dimensional vectors. Distances between points are given by summations of absolute differences, or squares of differences, in each coordinate dimension. (Obviously, the scheme extends easily to multiple categories.)

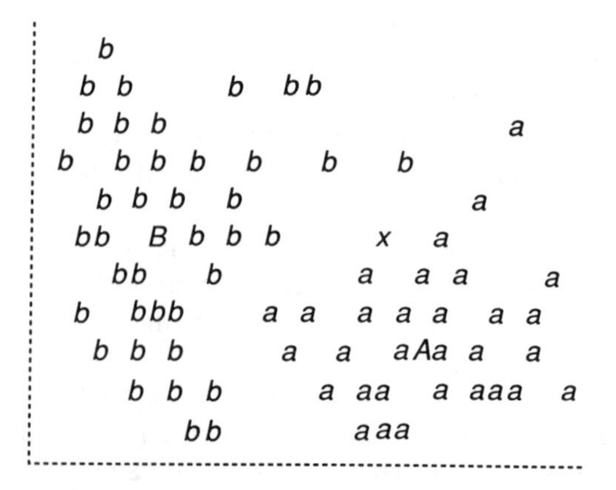

FIGURE V-2. A simple two-dimensional classification problem where both classes are well-defined with class members distributed around some class mean.

The square of the euclidean distance D_1 from unknown X to mean A of α class is expressed as:

$$D_1^2 = \Sigma(X_i - A_i)^2$$

Similarly, the square of the distance D_2 from unknown X to mean B of β class is:

$$D_2^2 = \Sigma(X_i - B_i)^2$$

The difference in the (squares of) distances is:

$$Y = D_1^2 - D_2^2$$

$$Y = |A|^2 - |B|^2 - 2\Sigma A_i X_i + 2\Sigma B_i X_i$$

$$Y = |A|^2 - |B|^2 - 2\Sigma(A_i - B_i)X_i$$

$$Y = |A|^2 - |B|^2 - 2\Sigma W_i X_i$$

If

$Y < 0$ then X in α class

$Y > 0$ then X in β class

Another expression of the same basic classification procedure is to use the scalar products of the vector which defines the position of the unknown point x and the vectors defining the positions of the class means. The unknown is assigned to that class with which it has the largest scalar product.

The scalar product, S_1, between the unknown $\boldsymbol{X}$ vector and the mean $\boldsymbol{A}$ of α class

$$S_1 = \Sigma A_i X_i$$

The scalar product, S_2, between the unknown $\boldsymbol{X}$ vector and the mean $\boldsymbol{B}$ of β class

$$S_2 = \Sigma B_i X_i$$

The difference in scalar products

$$S_1 - S_2 = \Sigma(A_i - B_i)X_i$$

If the class mean vectors are all normalized to a constant length, $|A| = |B|$, these two approaches give identical classification results.

Whether formulated in terms of distances or (normalized) mean vectors, the use of this classification criterion in a two-class problem is just equivalent to determining where the unknown point lies with respect to the symmetry plane of the class centers. How well such a symmetry plane classifier works obviously depends on the cohesiveness of the clusters and their degree of overlap. Whereas reasonable performance might be attained with data such as those illustrated in Figure V-3, less satisfactory results are likely if both classes are diffusely distributed or if one class is relatively diffuse and the other cohesive.

Intuitively, one expects that better classification performance could be realized if the inclination of the decision plane and its point of intersection with the line of centers could both be optimally adjusted. In the example illustrated in Figure V-3, moving the dividing plane slightly toward the β-cluster will yield a higher overall classification success rate. It may, as in the case of the data shown in Figure V-2, be impossible to achieve a 100% classification success rate because the clusters overlap to a small degree. However, the position of the dividing plane can be adjusted so that as many as possible points, representing molecules from the training set, are correctly classified. The process for positioning the decision plane can be biased to reflect any differences in the relative importance of detecting either of the two classes. If, for example, it were more important to detect α-class molecules than β-class ones, it would be appropriate to weight the points so that, when assessing the performance of a particular decision plane, one misclassified α compound would be as detrimental as, say, 10 misclassified β compounds.

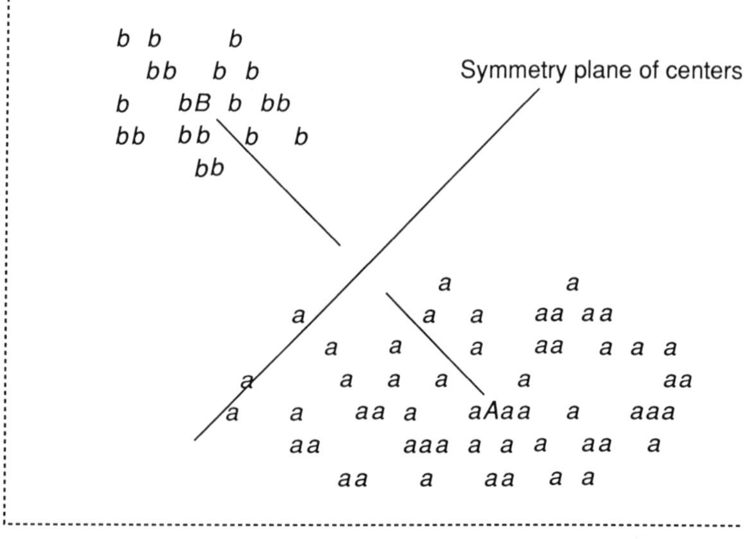

FIGURE V-3. Most simple "distance"-measuring classification functions are equivalent to tests determining compound class according to where a point lies relative to the symmetry-plane of the class centers.

In a simple case, such as this two-dimensional example, a reasonable performance classifier can obviously be found by starting with the symmetry plane separating the cluster centers and moving its point of intersection with the line of centers or its inclination until at least a local minimum is found for the number of misclassified training compounds. It is convenient to arrange for such a dividing plane to pass through the center of the coordinate system, for this allows simplification of some of the calculations necessary to determine on which side of the dividing plane a particular point lies. It is always possible to arrange that the plane pass through the coordinate origin simply by augmenting the coordinate system by one additional dimension; Varmuza's text includes informative illustrations showing how this may be done.[5]

In a trivial two-dimensional case, an optimal division plane for a set of training structures can be identified by visual inspection. In higher dimensions, simple procedures can find a good division plane by small incremental shifts in the inclination of an initial symmetry plane. Less *ad hoc* procedures for finding the best linear classifier are, however, desirable. As is noted later, there are in fact statistical procedures for determining the "best" position for a discriminant plane subject to certain assumptions about the distributions of measurements in each dimension. Although such statistical methods are well-defined and not overly complex, a greater emphasis has been accorded to *empirical* procedures for finding discriminant functions for chemical data.

There exists an iterative algorithm that is guaranteed to find *a* dividing plane separating two nonoverlapping clusters of points. This *linear learning machine* (LLM) algorithm involves a simple iterative loop.[6] In each cycle of this loop, reference spectra are classified according to their positions relative to the current candidate decision plane. The program proceeds analyzing each successive reference spectrum in turn as long as the derived classifications are consistent with those originally assigned by the investigator. If a spectrum is misclassified, the position of the decision plane is adjusted so that a correct result is obtained for the data that caused the error. The program then restarts its analysis, reevaluating all reference data with respect to the new decision plane. This learning process terminates when either all data are correctly classified or a limit on the number of iterations is exceeded. (A version of the algorithm, in FORTRAN-IV, is included in the book by Jurs and Isenhour.[6]) Lack of convergence of the iterative cycle may indicate that the classes are, in fact, overlapping.

There are several prescriptions defining how the position of the decision plane should be defined and modified during the training process.[5] The standard convention is to arrange that the decision plane pass through the origin of the coordinate space (by augmenting the property vectors with an extra dimension); this decision plane is defined by its normal vector $\boldsymbol{W}$. This normal vector is the weight vector used for classification of a pattern vector $\boldsymbol{X}$:

$$s = \Sigma w_i x_i$$

If

$$s < 0 \quad \text{then } X \text{ is assigned to } \alpha \text{ class}$$

$$s > 0 \quad \text{then } X \text{ is assigned to } \beta \text{ class}$$

If a particular vector $\boldsymbol{X}$ is misclassified, a new weight vector is derived according to the formula

$$W' = W + cX$$

$$c = -\frac{2s}{X \cdot X}$$

This correction factor arranges that the originally misclassified point now lies within its correct region. In fact, it lies just as far from the new decision plane as it had previously lain from the old plane, but now it lies on the correct side. The vector used to start the iterative cycle is usually arbitrarily initialized with all elements equal to +1 or -1.

As noted, the learning procedure terminates as soon as all patterns in the training set are correctly classified. Although having a 100% recognition rate, the resulting decision surface may not yield the optimal prediction performance. Problems with outliers, compounds whose spectral properties are not typical of their class, are inherent in the LLM approach.

The example data shown in Figure V-4 illustrate how outliers can adversely effect performance. The indicated clusters represent the regions that should contain 90% of examples of each class, and the plotted points designate instances from a limited training set. An outlier, such as point $b1$, can result in the selection of a decision plane that will yield a prediction performance inferior to that of a simple classifier based on the symmetry plane of the cluster centers. Such problems are unavoidable if the classifier is to be based on a simple linear decision plane and trained to a 100% recognition rate. Such outliers are likely when the classes are not really perfectly separable but, instead, overlap to at least a small degree. Any arbitrarily selected subset of reference compounds may then prove linearly separable, but only at the cost of forcing the decision plane to a nonoptimal position.

The LLM algorithm can yield poor decision planes even in those cases where the classes are nonoverlapping and can indeed be linearly separated. Thus, as illustrated by the data in Figure V-5, the LLM algorithm could terminate on finding the decision plane indicated. Although positioned so as to yield 100% recognition on the limited training set data, this decision plane will exhibit a relatively poor prediction success rate.

A number of empirical approaches have been devised that reduce the adverse affects of LLM's convergence to a poor decision plane. One such mechanism involves training with a *dead-zone*.[6] With dead-zone training, it is

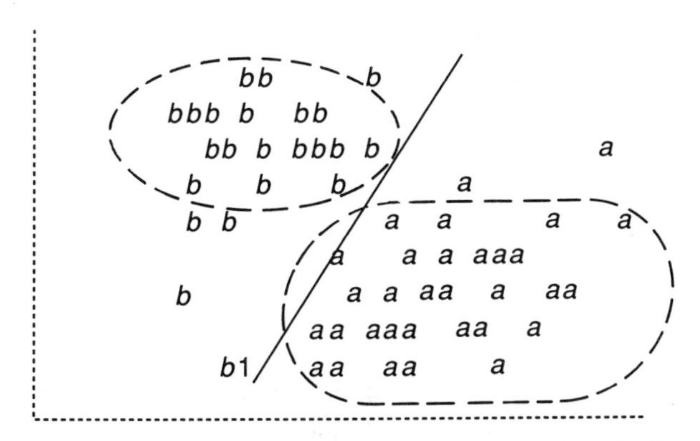

FIGURE V-4. "Outlier" points among the training set data can result in a classifier that, if trained to 100% recognition, has a nonoptimal prediction success rate.

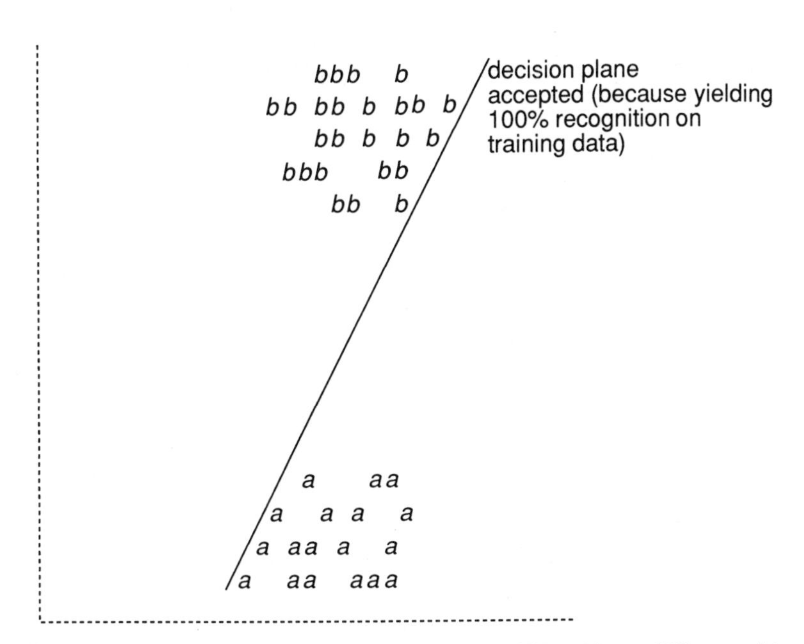

FIGURE V-5. Termination of training as soon as the classifier achieves 100% recognition can result in a nonoptimal classification plane.

not sufficient for a point to lie on the correct side of the decision plane. Classifications are determined by

$$s = \Sigma w_i x_i$$

If

$$s < Z \quad \text{then } X \text{ is assigned to } \alpha \text{ class}$$

$$s > Z \quad \text{then } X \text{ is assigned to } \beta \text{ class}$$

In effect, the decision plane is broadened to a band of width 2Z; to be classified, a point must lie within either the α or β region and at least a distance Z from the decision plane. Training is continued until the decision plane has been shifted so that this condition is satisfied for all the members of the training set. As illustrated in Figure V-6, such dead-zone training will improve the position of the decision plane, although still not necessarily yielding an optimal plane.

Another proposed approach uses a set of separately trained classification functions. Different classification functions can be developed from a single training set of data through the use of different initializations (e.g., one weight vector might be trained from an initial vector all of whose elements equal +1, and a second vector could be trained from an initial vector with elements equal to -1). Alternatively, different subsets of the available training data may be used

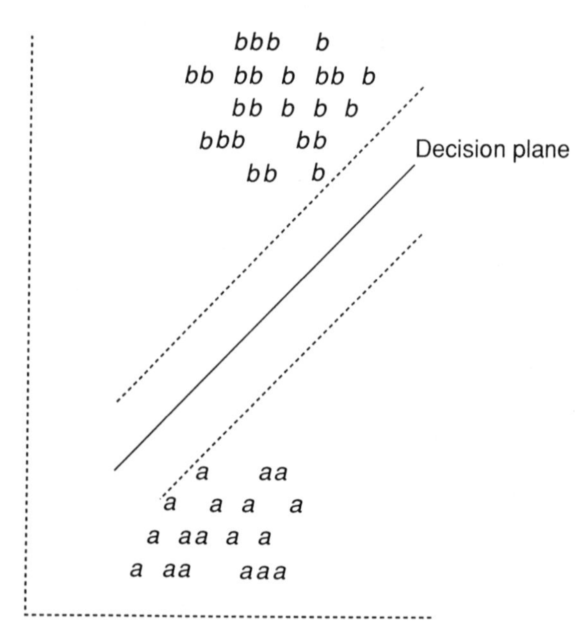

FIGURE V-6. Training with a "*dead-zone*" reduces the chance of accepting an unsatisfactory decision plane.

to develop weight vectors. Predictions are then based on the combined use of these separate classification functions. The class of an unknown is predicted by each function, and its final class designation is decided on the basis of the majority vote of this *committee machine*.[6] The prediction improvements thus attained are not dramatic. Furthermore, the advantages of computational simplicity and speed claimed for pattern recognition schemes are significantly lessened by the necessity of first having to train multiple classifiers and then use several such functions when processing an unknown.

Often, poor choice for the final position for the dividing plane is a consequence of starting with an arbitrary initial vector. The LLM algorithm does not require arbitrary initialization. The use of the symmetry plane of the two class centers as the initial plane will typically both greatly increase the rate of convergence of the iterative algorithm and reduce the chance of terminating with a badly positioned final separating plane.

The fact that the data, as obtained for a set of training spectra, are linearly separable is, of itself, of no great significance (save that it permits convergence of the LLM algorithm for developing classification functions). Linear separability does not in itself establish the existence of some as yet uninterpreted spectral–structural relationship. As discussed later, perfect linear separability may be achievable with a small training set of data in a high-dimensional system for a totally arbitrary classification scheme lacking any physical reality. But when classification is attempted by use of only a limited number of spectral features, it is frequent for the classes to overlap to some degree. It is then impossible to achieve a 100% recognition rate, and the LLM algorithm will not terminate. Although overlapping of the data clusters for the different classes makes perfect recognition and prediction impossible, there may still be a significant degree of spectral-structural correlation. Whereas nearest-neighbor methods are likely to yield the best results for such data, linear discriminant functions can perform reasonably provided the degree of overlap is not too great. Alternative algorithms for deriving linear classification functions are necessary to handle these really more practical cases of imperfectly separable data.

Varmuza has discussed the use of *linear regression* and *simplex* algorithms for deriving weight vectors.[5] A number of applications of the linear regression algorithm were reported when pattern recognition approaches were first introduced for spectral analysis.[10,11] In some cases the method was used to derive multiclass categorizers.[6] More recently, this method has been applied with a fair degree of success to the analysis of mass spectral data for steroids.[12] The simplex algorithms have been more consistently investigated; most of the work on the simplex algorithm and its variants has been done by Wilkins and co-workers.[13–15] The simplex algorithm treats the problem of selecting a weight vector as an optimization task in which the coefficients of the weight vector are adjusted to yield the maximum value for a chosen response function. This response function is basically a measure of the classifier's recognition success rate; the response function incorporates additional performance measures because the success rate is discrete, and the optimization procedure requires a continuous function. Although sometimes rather costly to develop, these simplex-derived classifiers generally appear to achieve higher prediction success rates than does the simple symmetry plane or an LLM classification function.

The "linear regression algorithms" are closely related to those standard statistical procedures known as *discriminant analysis*. Detailed presentations of discriminant analysis are available in many comprehensive statistics texts (such as that of Mardia et al.[16]). Simpler treatments are also available; one example is the text by Davis[17] (which contains illustrative applications on geochemical data). The objective in discriminant analysis is exactly that needed in our spectral classification task; an individual is to be assigned to one of a predefined set of classes on the basis of a set of measurements characterizing that individual.

In discriminant analysis, the classification procedure again involves simply testing a linear weighted sum of feature measurements found for an unknown against a threshold value, the result of the test determining the appropriate class attribution. The weights used in this function are determined, in part, from the difference of the class means as in the simplest linear classifier considered. In the discriminant analysis method, however, the effects of the variance in each feature dimension and the covariance between different pairs of features are taken into account.

The influence of differences in variance in different dimensions can possibly be visualized from the example in Figure V-7. For the data shown there, the dispersion in the horizontally plotted dimension x_1 exceeds that in the vertical dimension x_2; the variances for the α and β classes are about the same in each individual dimension. Intuitively, one recognizes that a feature for which the measurements are relatively broadly dispersed, with a large standard deviation, is likely to be less effective when one discriminates between two classes. The weighting accorded to such a feature should be reduced from that suggested by any difference in the class means. The position of the decision plane is, in effect, adjusted so that it lies slightly more closely parallel to the axis of the more disperse feature. Thus one expects that allowance for the different degrees of variance for the data of Figure V-7 should result in a less steep decision plane intersecting with the x_1 axis at a small, or –ve, x_1 coordinate value. Such intuitive expectations are borne out by the mathematical analysis.

The mathematical development of a discriminant function can presume that the groups are samples from a multinormal distribution with different means and the same covariance matrix (both of which must be estimated from available sample data); the discriminant function thus derived gives the *maximum likelihood classification* based on the presumed distributions. Alternatively, as with the *Fisher discriminant*, a more empirical analysis can be used to choose a form for a classifier without any particular presumptions concerning the distributions. It happens that, for the two-class problem of particular interest, both maximum likelihood and Fisher's discriminant derivations yield the same decision rule.

Mathematically, the optimal discriminant function is defined as

$$\text{If} \quad a'[x - 1/2\,(A + B)] > 0 \quad \text{then } x \text{ in } \alpha \text{ class}$$

i.e., $\Sigma w_i x_i >$ constant => x in α class)

where x = vector of unknown being classified
A = mean vector for class α
B = mean vector for class β
$a' = [S]^{-1}$ (A – B)
S = pooled estimate for variance–covariance matrix.

The variable S is given by

$$[S] = \frac{[S_A] + [S_B]}{n_A + n_B - 2}$$

where n_A: the number of class α members in the training set
n_B: the number of class β members in the training set

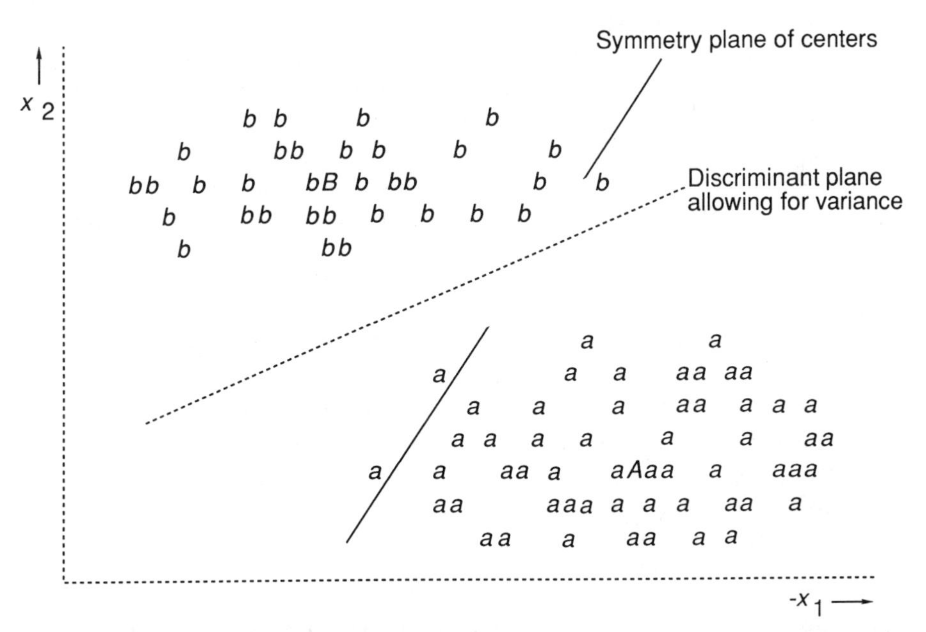

FIGURE V-7. If the class variances differ among the measures of spectral attributes, the symmetry plane of centers no longer constitutes the best discriminant plane. The best discriminant plane will lie more parallel to the axis with greater variance.

Elements from the one-class estimates for the variance-covariance matrices $[S_A]$ and $[S_B]$ are calculated according to the formula

$$S_{A_{jk}} = \Sigma x_{ij} x_{jk} - \frac{\Sigma x_{ij} \Sigma x_{jk}}{n_A}$$

The calculations necessary to derive the discriminant function are not complex. The data of the training set must be read and sums, sums of squares, and sums of pairwise products of feature measurements accumulated. These sums of products determine the elements of the (The *linear regression analysis* described by Varmuza[5] differs from this discriminant analysis procedure mainly with respect to details of the calculation of the terms corresponding to elements of the pooled variance-covariance matrix.) The variance–covariance matrix must be inverted. The weighting coefficients for the discriminant function are determined by the product of the inverted matrix and the vector of differences in class means. The threshold value used in the discriminant test is determined from the product of the derived weight vector and the vector representing the average of the two-class means.

The discriminant function thus developed provides the optimal decision plane given the estimates of class means and variances derived from the available data. This decision plane is not constrained to separate the classes. If there are outliers, it is quite possible for a linear function with 100% recognition to be developed by LLM while the "optimal" decision plane based on the statistical analysis misclassifies some of the training data. Although not achieving perfect recognition on such data, the statistical linear discriminant function should possess a higher prediction success rate.

These simple linear discriminant functions can be further elaborated. Through more sophisticated analyses, it is possible to make allowance for the fact that the different classes may have different intrinsic probabilities of occurrence. It is also possible to bias the procedure to reflect differences in the relative importance of misclassifying instances of the majority and minority classes.

Davis's text includes example FORTRAN-IV programs for simple discriminant analysis;[17] although the mathematical formalism may be more complex than that for LLM, the actual programs are not substantially more elaborate. Sophisticated discriminant analysis programs are available in most standard statistical packages. A few applications of discriminant analysis methods for the classification of spectral data have been reported; one such example is in the work of Rasmussen et al.[18]

V.C. PROBLEMS OF SIMPLE PATTERN CLASSIFICATION FUNCTIONS

The essence of many of the studies on supervised pattern recognition methods for spectral analysis is first to assume that a spectral–structural relationship

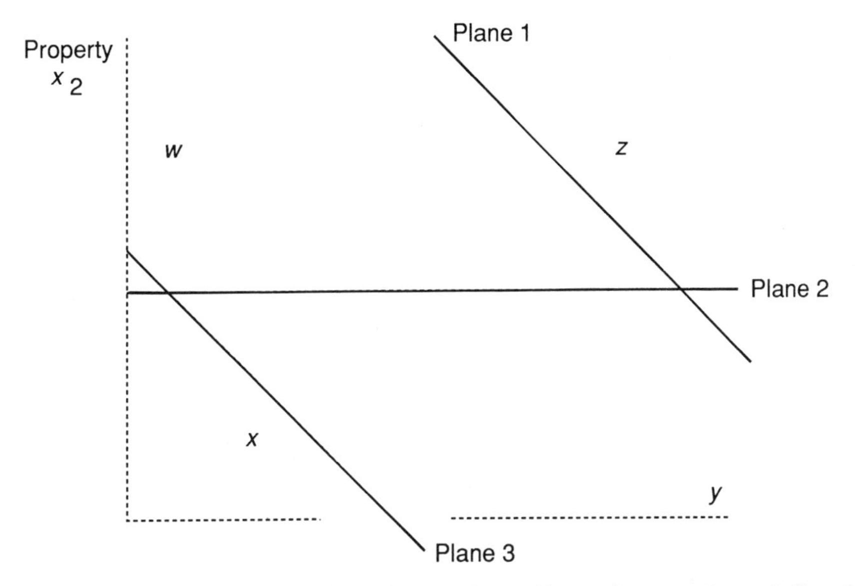

FIGURE V-8. Almost any grouping of compounds into arbitrary classes can be made linearly separable in a space of sufficiently high dimensionality.

exists, then to impose this classification on a set of training data, and finally to seek out a linear decision plane, in the multidimensional space, which will separate points representing compounds in the different classes. If such a plane exists, and thus a classifier with a 100% recognition rate can be derived, it is concluded that the presumed spectral-structural relationship has indeed been established. Unfortunately, this is not necessarily true.

Consider the trivial case of four compounds '*w*, *x*, *y* and *z*' which are to be classified into two classes α and β on the basis of two physical measurements. There are several ways of drawing planes (i.e., lines in this simplified case) separating different groupings of these four compounds, three such planes are indicated in Figure V-8. Obviously, some classification schemes for the data points, totally lacking in any physical basis, may be found to yield linearly separable groupings in such a small set of data. Thus it may well appear that some obscure property of the compounds is related to measurable properties through some as yet unknown relationship.

With chemical data, there are typically 100 or more measurements defining the positions, in multidimensional space, of the points representing each compound. There is a significant chance that in such high-dimensional spaces any labeling of points with either one of two class designations will yield a linearly separable set. For n points, there are 2^n possible labelings. Each such labeling represents a different *dichotomy* or assignment of points to classes. If the points are distributed over a d dimensional space, a mathematically defined fraction $f(n,d)$ of the possible labelings will be linearly separable. This fraction, as a function of n and d, is given by

$$f(n,d) = 1 \qquad\qquad n \leqslant d + 1$$

$$f(n,d) = \frac{2}{2^n} \sum_{i=0}^{d} \binom{n-1}{i} \qquad n > d + 1$$

When $n = 2(d+1)$, about half of the dichotomies are linearly separable, that is, for any arbitrarily chosen labeling of points with classes, there is a 50:50 chance that linear separability can be achieved. For high-dimensional systems $d > 100$, the curves representing the fraction $f(n,d)$ are steep, almost step function in form. If the size of the training set employed is less than about twice the dimensionality, then $f(n,d)$ is almost 1. Therefore, irrespective of any real spectral-structural relationship it is almost certain that linear separability should be achievable with the training set data. As the ratio $n/(d+1)$ is increased beyond 2, the fraction of possible dichotomies that are linearly separable drops markedly, and with $n>3d$ there is only a very small chance of linear separability. With training sets of sufficient size, the discovery of a linear separating plane would constitute strong empirical evidence for a spectral-structural relationship. (More usually, as the size of the training set increases relative to the dimensionality of the problem, it is found to be increasingly more difficult to find a separating plane and achieve convergence in the LLM procedure.[5]) Numerous reported studies have had $n<2d$; for example, many of the mass spectral classifiers have been developed for training sets of 200 spectra characterized by 155 m/z-intensity measurements. With such data, successful convergence of the LLM algorithm and discovery of a separating plane does not prove the existence of any spectral-structural relation.

The real import of these results concerning the number of linear dichotomies is that they show how the use of large numbers of features to characterize each compound increases the risk that a totally spurious hypothesized spectral-structural correlation *may appear consistent* with a limited training set of data.[19,20] It may be quite possible to find a linear discriminant function with a 100% recognition rate. However, this linear separability is not itself sufficient to establish real class differences and hence the postulated spectral–structural relationship. A relationship may exist,[21] probably on the basis of differences between the classes in just a few key features. However, a more precise statistical analysis would be required before such a relationship could be accepted.

There are statistical measures for estimating the validity of a discriminant function. Taking a slightly simplified view, these statistical measures can be considered as estimates of the chance that any apparent differences between two defined classes are not in fact intrinsic but are merely artifacts arising from random sampling of examples from a single distribution.[16,17] If these statistical measures show that there is only a very low probability of the observed class differences arising by chance, the hypothesized spectral-structural relationship can be accepted. Few, if any, of the reports on spectrum classification have contained any discussion of the statistical validity of the functions developed.

Inspection of Figure V-8 will show that the only desired classification groupings for which no linear classifier can be found are those wherein x and z comprise one class, say, the α class, with w and y in the other. This particular grouping of points into classes would correspond to a situation in which the two properties measured were closely related. The α class would include all compounds for which the property values were either both large or both small; the β class would include compounds characterized by one large and one small property. The mean values for the properties considered individually are approximately the same for both classes. Simple classification rules, such as one testing the absolute value of the difference in the two measures, can be defined for such classes. Other distributions for points can be envisaged wherein, in two dimensions, the α class points are all within a limited radius of the class mean whereas the β class points are also distributed around the same mean but at greater radii; again, simple discriminant tests can be defined, for this case, in terms of radial distances. However, classification criteria using absolute values or radial distances cannot be expressed in the form of a single test against a limit value for the sum of individual weighted measurements for the different properties. A simple linear classifier can only exploit features where the means for the two classes differ.

The fact that the only effective discriminant features are those for which the class means differ can be used from the outset to help determine which features should be included in a classification function. An estimate of the value of any feature can be estimated in terms of a ratio between the difference in class means and the breadth of class distributions, expressed as standard deviations, in that feature dimension. Standard statistical methods for feature selection exploit such measures. It is not necessary to resort to more empirical approaches such as comparing weights in discriminant vectors derived by using different subsets of the reference spectra.[6]

The essential requirement for a feature to serve as a useful element in a discriminant function is a difference in its mean value as measured for the two separate classes. Unfortunately, we don't know the true class means and must rely on estimates based on the sample, training set data. A mean based on a very small sample can be substantially in error. For example, if only five spectra are available to characterize a compound class, there exists a small chance (about 2.5%) that the estimated mean will be in error for any individual feature by a full standard deviation. If there are more than 32 features used to characterize each compound, there will be at least a 50% chance that the estimated mean, for a class with so few examples, will be substantially incorrect in one or other feature dimension.

Some reported applications of pattern recognition techniques have used training sets consisting of 200 or fewer spectra of which less than 10 might correspond to minority class examples. In such systems, it is common for 100 or more different spectral measurements to be used to characterize each compound. If a proposed classification scheme has no physical basis, then in each spectral dimension the two "classes" as defined by the training set data, will represent simply different random samples from a single distribution. However, there is a significant chance that in a number of dimensions the mean estimated for the supposed minority class will differ substantially from the single true mean. There will appear to be class differences. Such differences will be picked up by whatever procedure is used to develop a discriminant function and will allow for the derivation of a classification function with a high, if not perfect, recognition capability. The resulting classification function will be accepted on the basis of its successful recognition abilities, and its predictive performance will be assessed. Typically, the frequencies of occurrences of members of the different classes will be similar in the training and prediction sets. Under such circumstances, the classification function may apparently achieve a high overall prediction success rate.

The apparent paradox of high predictive ability for a function based on a supposed but nonexistent spectral–structural relationship is resolved by recognizing that the overall success rate is not an appropriate measure of classifier performance. Success rates estimated for the individual classes would show marked differences, with the classification function having little or no ability to detect members of the supposed minority class among the prediction set.[19]

Assessment and comparison of prediction performances constitute a problem with much published work on pattern recognition procedures. Commonly, the only performance statistic given is an overall averaged prediction success rate derived through tests on a small set of test data. These overall prediction success rates are then used to assess the relative merits of different classifiers. The small size of the test sets used to determine these prediction rates is in itself such as to make comparisons of relative performance of dubious statistical significance.[19] More importantly, the overall prediction rate provides a very poor measure of the quality of a classification function. There have been a number of attempts at defining a single overall measure of classifier performance. A good measure would be one that manages to indicate the degree to which application of the classifier really increases the structural information about a compound while at the same time is independent of the composition of the prediction test set used. Varmuza has reviewed the problems in formulating such a useful measure of classifier performance.[5]

An alternative, more physically based description may help to illustrate the problems that can arise when overly small training sets are used to develop a classification function. Consider, for example, the development of a linear classifier that is to detect phenyl compounds from binary ^{13}C spectra (encoding peak–no-peak data at each of 200 1-ppm shift ranges). Phenyls will be evidenced by shifts in the characteristic Csp^2 region particularly, in the case of simple phenyl compounds, by shifts in the range 110–150 ppm. Any procedure used to develop a classifier will quickly accord positive weights to such spectral features. However, shifts in this region will also be found in the spectra of any alkenes, dienes, and other similar compounds that may be present among the training set data. Such structures will be misclassified as phenyls unless other spectral features are accorded appropriate negative weights.

These negatively weighted features must correspond to resonance shifts that *among the spectra of the training set* happen to characterize one or more alkenes or dienes and yet are not found in the phenyl compounds. The spectral features satisfying these requirements may not bear any relationship to the structural aspects that are intended to be classified. Thus it might happen that all the alkenes of the training set incorporate a couple of alkyl methylenes with resonances at low fields, with shifts in, say, the region 45–55 ppm, whereas none of the training phenyl compounds contain such resonance signals. The low-field methylene resonances will be selected as the feature by which alkenes can be successfully distinguished from phenyls and a linear classifier, with a 100% recognition rate, might readily be developed.

Inevitably, the prediction performance of such a classifier will be disappointing. Alkenes that lack low-field methylenes will be accepted as phenyl compounds; phenyl containing compounds that, fortuitously, do incorporate methylenes with low-field resonances will pass undetected. In part, such difficulties reflect a poor choice for the classes; a linear classifier could be developed that would reliably discriminate "alkene or phenyl" compounds from "not (alkene or phenyl)" compounds, but such a classifier would have less analytic purpose. More generally, such problems are due to the highly restricted form of these "CNPE" (Complete Neglect of Previous Experience) classification functions. Through the application of knowledge of chemistry, a trivial phenyl-nonphenyl classifier, with a performance comparable to an elaborately developed linear function, can be based on a single test of the form: "Are there resonances for six or more carbons in the range 110–150?" (a test that will miss few phenyls and reject all simple mono- and diolefins).

V.D. APPLICATIONS TO STRUCTURE ELUCIDATION

Varmuza's text contains an exhaustive compilation of the various reported chemical applications of pattern recognition techniques.[5] This compilation includes work on structure-activity relationships, environmental chemistry, applications in electrochemistry and chromatography, and the structural interpretation of the spectra of organic compounds.

The majority of papers reporting applications in spectral interpretation have described the use of linear classifiers for two-class problems. The simple symmetry plane of class centers (possible expressed as distances from class means, or in terms of scalar products with mean vectors) has been employed occasionally. More frequently, the decision plane has been derived through LLM, simplex, or other algorithms. The k-n-n approach has also been fairly widely employed. In general, the k-n-n approach has proved to constitute the basis for a satisfactory, if expensive, classification scheme. (With k-n-n, care has to be exercised in the choice of an appropriate distance metric particularly when unnormalized, e.g., binary, spectral data are used.)

The overall prediction performances for linear classification functions reported in the literature are typically at, or above, 90%. Although usually rather different rates are obtained for minority and majority classes, it is convenient for simplicity to assume 90% success rates for both classes. These prediction success rates measure the probability that the classifier will produce the correct response for a given spectrum. This is not the probability of greatest interest to the analyst. What the analyst needs to know is the probability of a compound really being in class α given that a particular classifier identifies α as its appropriate class.

Such *a posteriori* probabilities are not normally included in the statistics given to characterize classification functions. The reason is that these probabilities depend on the *a priori* probabilities of occurrence of members of the two classes. If the training set of data is sufficiently large and diverse—say, a file of 25,000 mass spectra, then it *may* be reasonable to assume that the frequencies of occurrence of compounds from various classes as observed in this training set are indeed representative. This is basically what McLafferty and co-workers do in estimating *recall–precision* performances of file search systems for predicting constituent substructures.

Some examples will illustrate how the *a priori* probabilities determine the *a posteriori* probability of a class. Consider first a system where compounds from the two classes are equally probable (with the *a priori* probabilities of both α and β compounds being 0.5). Then in a thousand tests there should be 500 α compounds of which 90%, (i.e., 450), should be correctly identified as α. About 10% (i.e. 50) of the β class will also, but erroneously, be identified as α. The probability of a compound identified as an α actually belonging to the α class is then 0.9. More typically, one is interested in detecting a relatively rarely occurring structural feature—say, one present in only 10% of all compounds. The *a priori* probabilities of the classes are now very disparate; α class has a probability of 0.1, and β of 0.9. Working again with 1000 test examples, one might detect 90 of the 100 αs and also get 90 βs falsely identified as αs. Here, there is only a 50% chance of the suggested α classification being correct. If the distributions of frequencies of occurrence of members from each class are still more disparate (e.g., 5:95), the greater proportion of minority class identifications will in fact be erroneous.

Many of the reported linear classifiers have been developed to detect structural features that had *a priori* probabilities of 0.1 or less, even among the test compounds of the prediction set. With 90% success rates, there is only at best a 50:50 chance of a compound classified as being in the minority class actually possessing the relevant structural feature. In such circumstances, it is obviously unwise to use such results to constrain a structure generator.

Furthermore, a substantial number of the classification functions used to illustrate pattern recognition systems are essentially irrelevant to the real problems involved in structure determination. It may, for example, be true that simplex-derived linear classifiers can achieve 80% or higher overall success rates in detection of methyls, phenyls, and carbonyls from ^{13}C shift patterns. Quite apart from the fact that more conventional ^{13}C spectrum interpretation schemes should realize at least equal performances on such problems, real structure elucidation studies exploit multiple spectral data. A success rate of approaching 100% at discrimination of carbonyl from noncarbonyl compounds should be achievable through a simple inspection of their IR spectra. Methyls should be clearly evidenced in the ^{1}H NMR spectra. Usually, in real structural problems more complete ^{13}C data, with resonance multiplicities, would be available, in which case the number of methyls would be explicit. It is possible that no single spectral feature would be adequate for unambiguous establishment of the presence of a phenyl moiety; however, very strong evidence can be adduced through almost any combination of UV, IR, ^{1}H NMR, ^{13}C NMR or MS data.

Pattern recognition studies of mass spectra in particular have suggested many more novel spectral–structural correlations. Thus the presence of ring heteroatoms, the size of the largest ring, the size of the largest "clump" (largest number aggregate of carbon atoms bonded together), and the number of "clumps" have all been determined from MS data through the use of linear learning machine classifiers. If any such novel correlations could indeed be clearly established, they might provide valuable additional constraints for a structure generator. However, these classifiers are in practice of no value. Such classifiers are typically developed using small training sets of 200 spectra. The dimensionality of the data used to define each training compound is often sufficiently high that almost any arbitrary dichotomy of the training set would yield a system for which a 100% recognition rate could be achieved by a linear classifier. Even where there are spectral properties positively correlated with the desired structural attribute, it is very probable that any negatively correlated features will be purely an artifact of the limited training data and thus inevitably will confound prediction. The reported prediction rates for such classifiers are often in the range 85–95% (or higher); however, since only the overall success rates are given, these data are insufficient for assesment of the utility of the classifiers. In most cases such functions will yield a higher prediction success rate for the majority (less interesting?) class than for the minority class.[22] Finally, while mass spectral-substructural correlations may be useful for establishing the presence of a substructure, it is doubtful that, given the competitive nature of fragmentation processes, lack of required mass spectral patterns constitutes evidence for the absence of a specific substructure.

There have been only a few attempts to use the results from pattern classifiers to direct a structure generation procedure. In one such study, Abe and Jurs[23] demonstrated a system for structure elucidation of small hydrocarbon molecules, with 10 or fewer carbon atoms, in which Sasaki's structure generation procedure[24] was combined with a set of linear pattern classification functions. The spectral data used logarithmically scaled mass spectral intensity measurements, mostly representing intensities of individual ions, but with a few measurements representing intensities of specific neutral losses. A total of 32 linear classifiers were developed to recognize the presence of some 18 structural attributes including methyls, ethyls, isopropyls, acetylene groups, and benzene rings. Multiple classifiers were developed to determine the numbers of particular groups; thus, for example, five different methyl classifiers were used to determine whether a compound contained specific numbers of methyl groups. In a significant number of cases, training to 100% recognition could not be achieved as a result of either slow convergence of the iterative feedback algorithm or inseparability of the classes. The derived classification functions were tested and found to average a prediction success rate of about 90%. Predictions from these classifiers were used to provide the structure generator with minimum and maximum constraints on the numbers of particular groups. The generation procedure first determined sets of substructures consistent with these constraints and the molecular composition and then proceeded to derive all possible unique molecular structures.

A reanalysis of the training set data resulted in 142 compounds being either uniquely identified or at least included in a small set of generated structures consistent with the inferred mass spectral constraints. Because many of the classifiers could not be trained to 100% recognition, the remaining 58 training spectra were misclassified by one or more functions, and the correct structure could not be generated for these. Only brief details were given of results obtained on nontraining set spectra. In at least 32 of the 177 cases tested the correct structure was included among a set of five or fewer candidates. The problem of achieving correct results from all the linear classifiers was noted. Considered individually, prediction success rates of 90–95% may seem high. However, if results from large numbers of classifiers have to be combined then the chance of achieving complete success does fall. With 32 functions, as needed even in this restricted case of hydrocarbons with 10 or fewer carbon atoms, the chance of all being correct is $(p)^{32}$, p representing their average success rate. Even with p as high as 0.95 (95% success rate), this value is only 0.19 (19% chance of all results correct).

V.E. CONCLUSIONS

The results from pattern recognition methods are as difficult, if not more difficult, to use in practical structure generation algorithms than substructures inferred through **STIRS** and other interpretive file search procedures. As shown by the results presented by Abe and Jurs, it is unrealistic to assume that the predictions will all be correct and, therefore, appropriate for constraining a structure generator. The statistical nature of the predictions has to be recognized. With **STIRS** and similar programs, the substructure predictions are accorded various confidence estimates. These, as noted earlier, could be used either to determine an optimal order for exploring possible solutions or for assigning scores to partial structures incorporating differing combinations of the predicted substructures. However, few pattern recognition classification procedures give any measure of confidence in their predictions; without such confidence estimates, even such limited use of the predictions becomes less practical.

The results from a conventional pattern classification procedure will generally be inferior to those obtained from an interpretive file search system. File-based substructure recognition procedures exploit computed properties that relate to the physical processes giving rise to the spectral signals. Thus a count of the number of resonances in some spectral subrange, or a ratio of the counts of singlet and doublet resonances in a spectral region, are both examples of the type of computed feature, based on chemical knowledge, that would typically be exploited in an interpretive file-based system. Pure pattern recognition methods attempt instead to delineate relationships between the spectral patterns and the "obscure property" (chemical classification) without the use of chemical knowledge or prejudices. It is difficult to capture many known spectral–structural relationships through functions restricted to being a weighted sum of individual spectral measurements.

The k-n-n, and the file-based methods, utilize all available data characterizing known reference compounds. Pattern classifiers, based on decision planes, attempt to compress the information present in these data into a single weight vector. Although this allows for much more rapid and economical classifications, it obviously will increase the rate of misclassification.

Yet, the linear discriminant functions, preferably those derived by statistically valid procedures, still have potential. They are computationally simple and rapid. One possible application of binary classifiers would be in screening spectral data from large numbers of multicomponent samples to detect a small minority of compounds that require more detailed analysis; many routine biochemical analyses do involve just such screening processes. Such an application requires that a classifier detect all instances of compounds from the minority class of interest. In such circumstances it is appropriate to bias the training procedures so that the detection rate for the minority class is enhanced even at the cost of more false-positive identifications and a lower overall success rate.

Multicategory classifiers are of potential value as preprocessors that select more detailed, class-specific spectrum matching or spectrum interpretation procedures. One such application, noted in Chapter IV, would have a simple classifier select subsets of a file that must be searched in a file-based compound identification system. Another, hypothetical, example would involve a simple multicategory pattern classifier as a selector that identifies an appropriate set of compound class-specific mass spectrum interpretation rules.

REFERENCES

1. P. C. Jurs, B. R. Kowalski, T. L. Isenhour, and C. N. Reilley, "Computerized Learning Machines Applied to Chemical Problems. Molecular Structure Parameters from Low Resolution Mass Spectra," *Anal. Chem.*, **42**, (1970), 1387.
2. H. B. Woodruff, C. R. Snelling, C. A. Shelley, and M. E. Munk, "Computer Assisted Interpretation of Carbon-13 NMR Spectra Applied to the Structure Elucidation of Natural Products," *Anal. Chem.*, **49**, (1977), 2075.

3. B. R. Kowalski and C. F. Bender, "Pattern Recognition. A Powerful Approach to Interpreting Chemical Data," *J. Am. Chem. Soc.*, **94**, (1972), 5632.
4. B. R. Kowalski and C. F. Bender, "Pattern Recognition. II. Linear and Nonlinear Method of Displaying Chemical Data," *J. Am. Chem. Soc.*, **95**, (1973), 686.
5. K. Varmuza, *Pattern Recognition in Chemistry* (Lecture Notes in Chemistry 21), Springer-Verlag, Berlin, 1980.
6. P. C. Jurs and T. L. Isenhour, *Chemical Applications of Pattern Recognition*, Wiley, New York, 1975.
7. L. Kryger, "Interpretation of Analytical Chemical Information by Pattern Recognition Methods," *Talanta*, **28**, (1981), 871.
8. J. R. Chapman, *Computerized Mass Spectrometry*, Academic Press, London, 1978.
9. C. L. Wilkins, "Interactive Pattern Recognition in the Chemical Analysis Laboratory," *J. Chem. Inf. Comput. Sci.*, **17**, (1977), 242.
10. P. C. Jurs, B. R. Kowalski, T. L. Isenhour, and C. N. Reilley, "Computerized Learning Machine Applied to Chemical Problems. Multicategory Pattern Classification by Least Squares," *Anal. Chem.*, **41**, (1969), 695.
11. C. F. Bender and B. R. Kowalski, "Multiclass Linear Classifier for Spectral Interpretation," *Anal. Chem.*, **46**, (1974), 294.
12. H. Rotter and K. Varmuza, "Computer Aided Interpretation of Steroid Mass Spectra by Pattern Recognition Methods. Part III. Computation of Binary Classifiers by Linear Regression," *Anal. Chim. Acta*, **103**, (1978), 61.
13. G. L. Ritter, S. R. Lowry, C. L. Wilkins, and T. L. Isenhour, "Simplex Pattern Recognition," *Anal. Chem.*, **47**, (1975), 1951.
14. T. R. Brunner, C. L. Wilkins, T. F. Lam, L. J. Soltzberg, and S. N. Kaberline, "Simplex Pattern Recognition Applied to Carbon-13 NMR Spectroscopy," *Anal. Chem.*, **48**, (1976), 1146.
15. S. L. Kaberline and C. L. Wilkins, "Evaluation of the Super Modified Simplex for Use in Chemical Pattern Recognition," *Anal. Chim. Acta*, **103**, (1978), 417.
16. K. V. Mardia, J. T. Kent, and J. M. Bibby, *Multivariate Analysis* (Probability and Mathematical Statistics, A Series of Monographs and Textbooks), Academic Press, London, 1979.
17. J. C. Davis, *Statistics and Data Analysis in Geology*, Wiley, New York, 1973.
18. G. R. Rasmussen, G. L. Ritter, S. R. Lowry, and T. L. Isenhour, "Fisher Discriminant Functions for a Multilevel Mass Spectral Filter Network," *J. Chem. Inf. Comput. Sci.*, **19**, (1979), 255.
19. N. A. B. Gray, "Constraints on Learning Machine Classification Methods" *Anal. Chem.*, **48**, (1976), 2265.
20. A. J. Stuper and P. C. Jurs, "Reliability of Non-parametric Linear Classifiers," *J. Chem. Inf. Comput. Sci.*, **16**, (1976), 238.
21. G. L. Ritter and H. B. Woodruff, "Dimensionality and the Number of Features in Learning Machine Classification Methods," *Anal. Chem.*, **49**, (1977), 2116.
22. C. P. Weisel and J. L. Fasching, "Deceptive Correct Separation by the Linear Learning Machine," *Anal. Chem.*, **49**, (1977), 2114.
23. H. Abe and P. C. Jurs, "Automated Chemical Structure Analysis of Organic Molecules with a Molecular Structure Generator and Pattern Recognition Techniques," *Anal. Chem.*, **47**, (1975), 1829.
24. S. Sasaki, Y. Kudo, S. Ochiai, and H. Abe, "Automated Chemical Structure Analysis of Organic Compounds. An Attempt to Structure Determination by the use of NMR," *Mikrochim. Acta*, **1971**, 726.

VI

"KNOWLEDGE-BASED" SPECTRUM ANALYSIS

VI.A. INTRODUCTION

Chemists' conventional approaches to spectral interpretation are based on models, possibly simplified models, of the physical processes underlying spectral resonances and absorptions. These physical models permit specific spectral signals to be related to particular molecular components. Usually, there will be several factors that together determine the detailed characteristics of a spectral signal, with generally some form of hierarchical relationship defining their relative importance. An initial analysis of a spectral signal may identify the presence of a specific type of atom or bond in the molecule; more detailed analysis of the form of the same spectral signal may determine aspects of the larger environment of that atom or bond.

The simplest expression of the physical rules, relating spectral and structural features, takes the form of a correlation chart detailing *spectral-feature–substructure* relationships. One approach to spectral interpretation involves hypothesizing of a particular substructural component, using the correlation chart to determine required spectral features and confirming, or refuting, the presence of the hypothesized substructure according to whether the required features are instantiated in the spectrum. An alternative approach is more data directed; the observed spectral features are used to index into the correlation chart and thus obtain various alternative interpretations.

For example, ^{13}C NMR spectra are typically interpreted by considering the multiplicities and chemical shifts of the various resonances. These interpretations are expressed in the form of substructures defining the chemical environment of the resonating atoms. A chemist seeking to determine whether a compound contains a >CH–O– group would search for a doublet resonance in a limited region of the ^{13}C spectrum. In a data-directed approach, the observation of a doublet resonance with a shift of, say, 75 ppm would yield a number of alternative structural interpretations, most incorporating some form of >CH–O– group.

Such approaches to spectral interpretation contrast markedly with typical pattern recognition methods. In a pattern recognition procedure, the determination of the presence of >CH–O– would not focus on its specific signals. Instead, the complete spectral data would be subject to some mathematical transform, a resulting number would be compared with a cutoff value determined from previous reference data, and the result of this comparison would be taken to establish the presence, or absence, of the >CH–O– substructure.

VI.A.1. Interpretation Rules

Chemists' conventional approaches, exploiting chemical and spectral knowledge as encoded in the form of either simple *spectral-feature–substructure* rules or slightly more general procedural forms can be used as the basis for computer-aided interpretation procedures. A computer program for spectral interpretation must possess, at the very least, a set of *spectral-feature–substructure* rules that can be applied to specific data. These rules are employed in either some data-directed or hypothesis-directed manner so as to identify possible constituent substructures for some unknown molecule. Generally, more elaborate capabilities are required in an *effective* spectrum interpretation program.

When attempting structural elucidation based on spectral data, a chemist typically uses much general chemical knowledge in addition to the spectral data immediately at hand. Knowledge of the source of a natural product and the extraction procedures used provides valuable constraints that aid in spectral interpretation. Data that would otherwise be ambiguous or uninterpretable can frequently be resolved given partial information either concerning the likely skeletal structure of a compound or specifying known component fragments. Chemists utilize different spectral data in combination, with structural features suggested by one spectrum being confirmed, or refuted, through the examination of other data. Furthermore, chemists' spectral analysis procedures are intimately coupled with the conscious (or subconscious ?) mechanisms for building up hypothesized structures.

Similar capabilities would be required for programs to achieve a performance at the structural interpretation of spectra comparable to that of skilled chemists. In these respects, most current spectral interpretation programs are restricted. Usually, a particular program will process only one specific form of spectral data; if several types of spectra are processed, the analysis performed is akin to intersecting results such as could be obtained by considering each spectrum in isolation. Few current programs exploit known structural constraints to simplify their spectral analysis. A number of programs do entail some form of structural analysis in order to find self-consistent sets of substructural interpretations. Although possibly involving quite elaborate computations, such intercomparisons of substructures are less flexible, and less sophisticated, than those that can be achieved by a skilled chemist.

The majority of current programs use some hypothesis-driven approach, checking spectral data for evidence for the presence of some limited set of standard substructures. The difference between hypothesis- and data-driven strategies is not of major consequence at this stage in the development of computerized spectral interpretation systems. Primarily, this is an efficiency consideration and, as such, can await demonstration of the practical feasibility and utility of any automated method of spectral interpretation. The current hypothesis-driven systems tend to employ more limited sets of smaller substructures than do current data-driven systems. Consequently, these hypothesis-driven systems have less need for detailed procedures for analyzing possible substructural interpretations and are more likely to be limited in the extent of their analyses.

The hypothesis-driven programs work in terms of a predefined set of 200 or so standard substructures. These substructures are small; most often they will comprise a central atom and its immediate bonded neighbors or may define a bond with qualifying data identifying conjugation or polar substituents. Each such substructure is correlated with some defined spectral pattern such as an IR absorption or a chemical shift. These spectral features characterizing substructures are normally rather generalized descriptions of broad spectral ranges wherein required resonances or absorptions should occur. Such a set of initial substructures is, in effect, screened against the recorded spectral data. The program discards those substructures whose required spectral patterns are not instantiated. The result of the analysis is a subset of the standard substructures that, considered individually, are consistent with the data characterizing the unknown.

Sometimes the suggestions produced by the systems are not definitive; instead, each suggested substructure is accorded some plausibility rating. Thus a strong IR absorption near 1750 cm^{-1} can be taken to constitute definitive evidence for some carbonyl-based substructure, but several alternative elaborations are possible. With the use of other IR data it may be possible in a particular case to conclude that a γ-lactone seems a more plausible interpretation of an absorption than does a vinyl-ester group. However, such detailed structural inferences must be qualified with some form of plausibility rating.

Many spectral interpretation systems can proceed no further than such a first-order analysis based on subspectral–substructural correlations. A system relying solely on IR data is obviously limited to inferences concerning bonds that will be active in the IR spectrum; since not all bonds in a molecule will show characteristic and unique absorptions, no complete description of a molecule can be generated.

However, a more complete analysis may be possible if the set of initial substructures can be considered as *spanning the range of all possibilities* for a molecule. The substructures remaining after the subspectral–substructural correlation analysis then constitute the permitted components for an unknown. The presumption that the structure must be assembled from some selection of these remaining substructures makes further analysis appropriate.

Although individually consistent with the spectral data, all the remaining substructures will not necessarily be incorporated into the final molecular structure. Frequently, several of these substructures constitute alternative interpretations of some specific spectral feature; the actual structure may include only one of these alternatives. Other constituent substructures may have been correctly identified, but a simple correlation analysis may not have been sufficient to determine the number of instances of occurrence of these substructures within the unknown molecule.

VI.A.2. Derivation of Mutually Consistent Interpretations

The identification of mutually consistent sets of permitted components can be achieved through a procedure for finding those combinations of permitted substructures that are compatible with overall composition constraints. Each such combination of substructures then defines a distinct problem to be passed to a subsequent structure generation procedure.

More elaborate processing may sometimes be possible. As in the example of Szalontai et al.,[1] a particular ^{13}C resonance might be found to admit interpretation as a carbon in a vinyl ether (i.e., $=C^{*}H{-}O{-}$); such an interpretation could be valid only if some other resonance could be interpreted as due to a carbon either in the environment $>C^{*}{=}CH{-}O{-}$, $-C^{*}H{=}CH{-}O{-}$, or $H_2C^{*}{=}CH{-}O{-}$. If a particular set of substructures, consistent with the molecular composition, were created that had a $=CH{-}O{-}$ substructure but none of the matching interpretations, a structure generator might explore many partially assembled structures but would inevitably fail to create any complete candidates.

Consistency checks on the interpretations for different spectral features can eliminate combinations of substructures that, although consistent both with individually considered spectral features and overall composition constraints, would not yield any connected molecules if passed to a structure generation procedure. In simple examples, such as that of Szalontai, the checks involve simply the comparison of each individual substructure's requirements, for neighbors of particular atom type, hydrogen-substitution and hybridization, with the bonding offered by other substructures.

Of course, with some spectral data it is possible to relate a precisely defined spectral pattern to an elaborately specified substructure. If substructural descriptions are really limited to the one-bond environment of a central carbon atom, the interpretation of ^{13}C spectra is necessarily limited to a consideration of at most a few dozen distinct substructures and their expected resonances (of defined multiplicity and broad range of possible shift values). Interpretations using substructures representing a two-bond atom environment can distinguish a larger number of cases, each related to a more narrowly prescribed range of shift values.

The number of substructures that must be considered rises exponentially with the increasing detail of their definitions. Systems attempting to exploit precise spectral-structural relationships must deal, therefore, with large numbers of possibilities. Rather than 200 or so simple substructures, the spectral interpretation system may define tens of thousands. Such systems are, consequently, usually data-directed. The initial search for possible substructural interpretations for the data is driven by a process that uses observed spectral features to index into a *spectral-feature–substructure* correlation table and thus to retrieve possible interpretations.

Of course, without additional constraints, the only effect of expanding the set of initial substructures employed by the system is to increase the inherent ambiguity of any spectral feature. Something of the extent of ambiguity is illustrated by the data in Figure VI-1. These data show some of the various substructures characterized by a quartet resonance at 20.75 ± 0.25 ppm. An interpretation system for ^{13}C data that relied on one-bond substructural models would derive only one interpretation for this spectral feature (in this case, a methyl bonded to another carbon atom CH_3–C). If the two-bond environment of the resonating methyl is considered, many substructural interpretations can be derived. Attempts at interpretation in terms of still more detailed substructures, describing three-bond and four-bond environments, must contend with hundreds of distinct possible substructural interpretations for each different resonance in a spectrum.

To be of practical value, such spectrum interpretation systems must possess effective mechanisms for eliminating those substructures that cannot be combined into complete candidate structures. Simple substructure intercomparison processes, such as those used by Szalontai et al.[1], must be elaborated. Substructure elimination procedures will necessarily involve detailed searches for mutually consistent interpretations of different spectral features. These procedures should also be capable of exploiting any known structural constraints.

Later, we consider problems inherent in a scheme for validating interpretations by substructure intercomparisons. Ignoring any problems for now, one can see that many possible substructural interpretations of a particular resonance could be easily excluded on the basis of requirements for mutual consistency among the interpretations selected for the different resonances in a complete spectrum. For example, none of the interpretations found for the 20.75-ppm quartet involved either an ethyl group (CH_3–CH_2–) or a vinyl methyl group of the form CH_3–CH=. So, unless some other methyl resonance

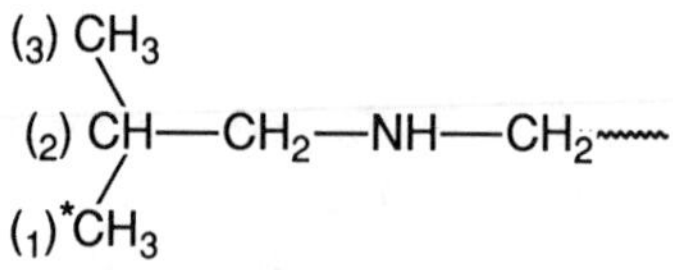

VI-1

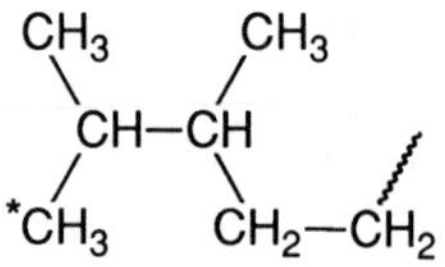

VI-2

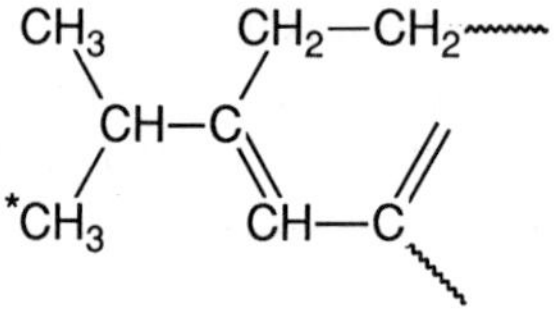

VI-3

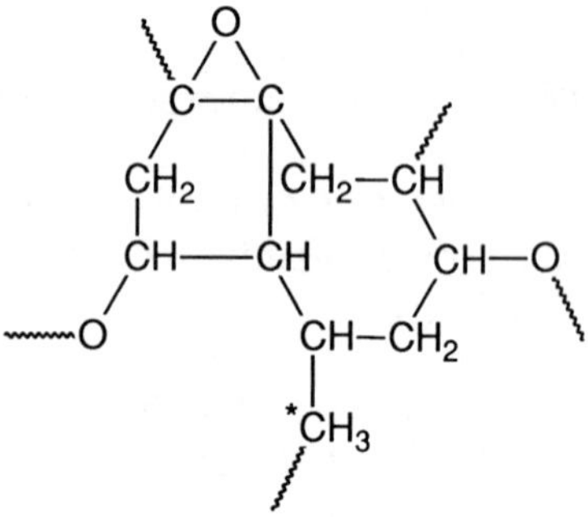

VI-4

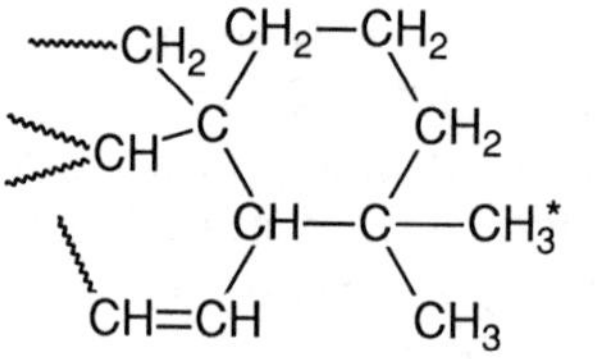

VI-5

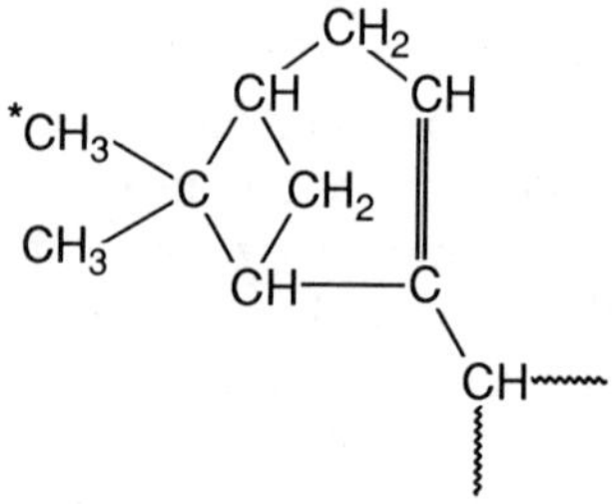

VI-6

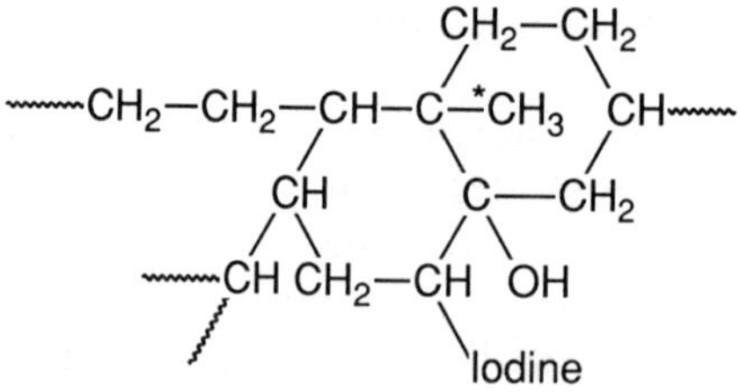

VI-7

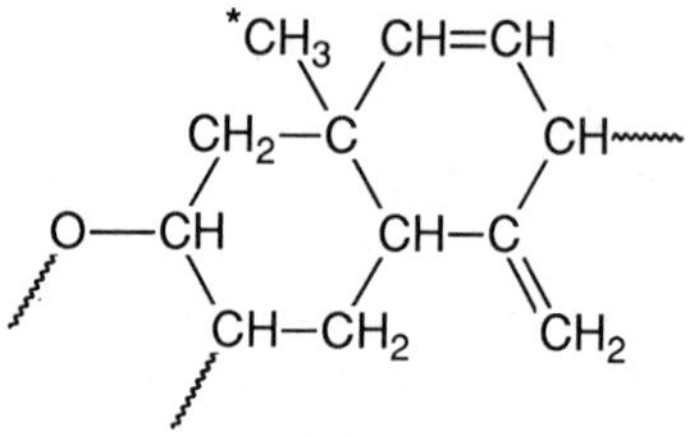

VI-8

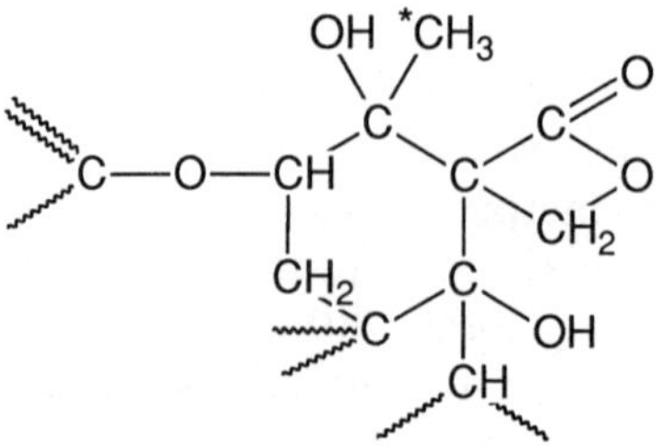

VI-9

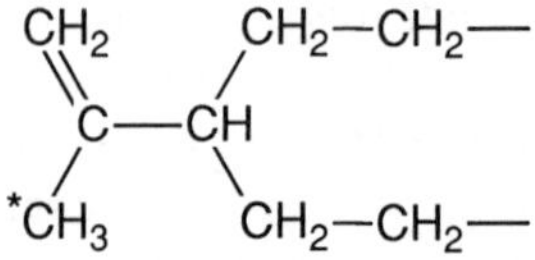

VI-10

FIGURE VI.1.

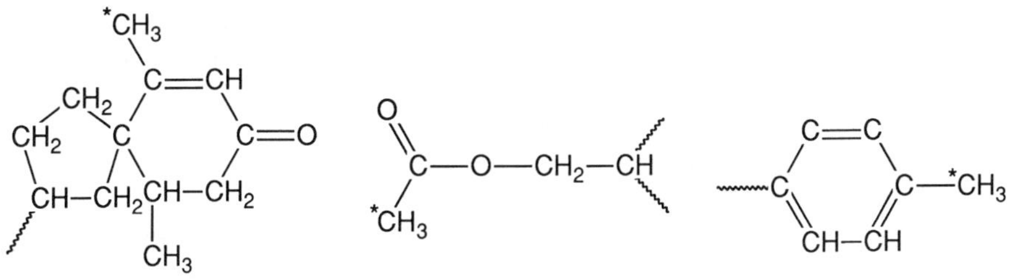

VI-11 VI-12 VI-13

FIGURE VI-1. Some possible interpretations for a 20.75-ppm quartet resonance.

in the spectrum did admit such interpretations, it would be possible to eliminate any substructures initially correlated with $-CH_2-$ or $-CH=$ resonances that had these atoms bonded to a $-CH_3$.

Further analysis of, for example, substructure **VI-1** reveals that this can be a valid interpretation of a quartet at 20.75 ppm in the spectrum of one compound [i.e., C(1) in **VI-1**] only if there is another methyl resonance yielding an identical topological interpretation [i.e., C(3) in **VI-1**] and a doublet resonance that can be interpreted in terms of a $>CH-$ bonded to two methyls and a $-CH_2-$ [i.e., C(2) in **VI-1**]. Analysis at a three-bond radius would establish the requirement for the –NH– group (because no other $(CH_3)_2>CH-CH_2-$ system results in a methyl resonance near 20.75 ppm). If this perceived requirement for a $>CH-CH_2-NH-$ system proves incompatible with constraints derived for other atoms, it would be possible to eliminate substructures of the type $(CH_3)_2>CH-CH_2-$ not only for the 20.75-ppm quartet, but possibly also for other methyl and methine resonances.

Constraints from other chemical and spectral data may also help to eliminate inappropriate substructural interpretations of individual spectral features. For example, if aromatics could be excluded, substructure **VI-13** could be eliminated as a possible interpretation for a 20.75-ppm methyl resonance. More elaborate analyses are possible. For instance, if a compound's molecular skeleton is known or is limited to one of a few possibilities, alternative substructural interpretations for particular spectral features can be checked through a procedure that attempts to map them onto the known skeleton. A substructure may be eliminated if it can neither be mapped onto the skeleton nor assembled from residual atoms of the molecular composition.

VI.A.3. The Interpreter Program

Automated spectral interpretation programs may be employed as stand-alone systems or integrated into complete structure elucidation systems. The results from the interpreter can consist simply of a list of substructure identifiers printed for the benefit of the chemist. These same identification data may also be expressed in terms of a file of substructure definitions in a form appropriate for use in a structure generator program.

In most automated spectral interpretation systems there is no need to manipulate representations of substructures or structures; substructure identifications, for use by candidate structure generators, can simply take the form of references to entries in a library file containing appropriate computer-compatible representations (see Chapter VII).

Checks for mutually compatible substructural interpretations of different spectral features do require some form of substructure representation. In most cases the requirements are limited and easily satisfied by some tabular summary of the constituent atom types, rings, and other features of the substructure. The most elaborate analyses do require tests to determine whether a substructure can be fitted to some general molecular skeleton; these tests involve graph matching procedures similar to those detailed in Chapter VIII.

Once having decided on the particular type(s) of data to be analyzed, the implementor of an automated spectral interpretation scheme confronts two main problems. First, a method must be devised for identifying the possible substructural interpretations of given spectral (and/or chemical) data. This entails the development of systems allowing for a chemist to define and encode the relevant spectral-chemical knowledge, and for a computer to utilize these encoded data. Second, a method for searching for mutually consistent sets of alternative interpretations, as found for a compound's various spectral features, may have to be formulated. Examples of current attempts to solve these two problems are presented in the following sections.

VI.B. IDENTIFICATION OF CONSTITUENT SUBSTRUCTURES

The first, or only, component of a spectrum interpretation program will be a function that identifies plausible substructural interpretations for the available spectral data. In the simplest cases, the various possible substructures may be enumerated and their spectral characteristics defined. This approach is obviously convenient for data such as ^{13}C spectra where a single spectral feature (a signal of defined shift and multiplicity) can be directly correlated with a specific substructure. Possible constituent substructures may then be found by a simple program that searches through the set of alternative interpretations identifying those consistent with observed spectral data.

Figure VI-2 illustrates typical "rules" for simple spectral interpretation. Given such *spectral feature–substructure* rules, spectra can be analyzed through the use of a simple *rule interpreter* program. The spectral data characterizing a compound are held in memory, the rule interpreter reads the various rules, attempts to match their *"premise parts"* (i.e., descriptions of required spectral patterns) to the available data and, if successful in matching, executes the rules *"action parts"* by asserting the (possible) presence of the corresponding substructure.

With other data, such as mass and IR spectra, it may be appropriate to define more elaborate classification schemes. Typically, such schemes define

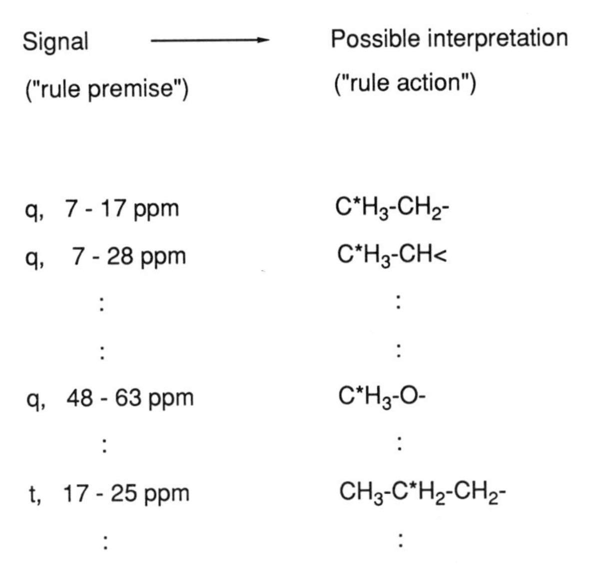

FIGURE VI-2. Simple spectrum-substructure rules for interpreting ^{13}C magnetic resonance spectra. If a recorded spectrum exhibits a resonance of appropriate multiplicity in any of the defined spectral regions, the corresponding substructure may be taken as constituting a possible interpretation.

treelike multilevel classifications. Thus, as shown schematically in Figure VI-3, one might first seek to establish the existence of an aldehyde group from its carbonyl and C-H band absorption bands in the IR spectrum. Subsequently, if some form of aldehyde is evidenced, one might wish to distinguish, on the basis of the exact position of the carbonyl band, between saturated and unsaturated aldehydes.

Of course, such a scheme can be formulated for a rule interpreter. Thus one can conceive of rules of the form

If: (IR bands in 2750-2660) and
(IR bands in 1760-1660)
then: infer aldehyde with confidence *X*
:
:
:
If: (confidence in aldehyde > 0) and
(IR bands in 1740-1715)
then: infer saturated aldehyde with confidence *Y*

The various classes in the hierarchy can be referenced in the premise parts of subsequently applied inference rules. Although this is possible, it does require either a more sophisticated rule interpreter or some subsidiary system for

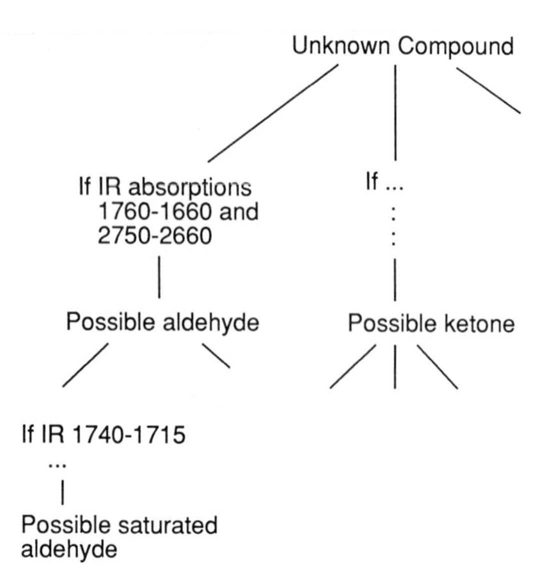

FIGURE VI-3. Schematic representation of a simplified hierarchical scheme for identifying carbonyl compounds by their IR absorptions.

organizing the rules (so that the plausibility of "aldehyde" has been assessed prior to the application of the rule concerning "saturated aldehyde"). It is often more convenient to specify explicitly the structure of a hierarchical classification scheme rather than leave it implicit in relationships between discrete rules. Consequently, such classification schemes often take the form of "programs" in which the flow of control through various sequences of spectral tests is explicitly defined.

The various possible methods for representing the spectral interpretation rules range from "hard coding" as FORTRAN subroutines through to essentially table-driven rule interpreters.

VI.B.1. Hard Coding of Interpretive Rules

In a number of the early studies, the rules for spectral interpretation were indeed expressed through FORTRAN subroutines or equivalent program constructs.[2–10] A good example of this approach is provided in the work on the interpretation of raman spectra by van der Maas and Visser.[8] As illustrated in this work, detailed "chemists' flowcharts" can be defined that prescribe a systematic search of the raman spectra for the absorption bands indicative of particular groups. A fragment of one such flowchart, adapted from an example by van der Maas and Visser, is shown in Figure VI-4.

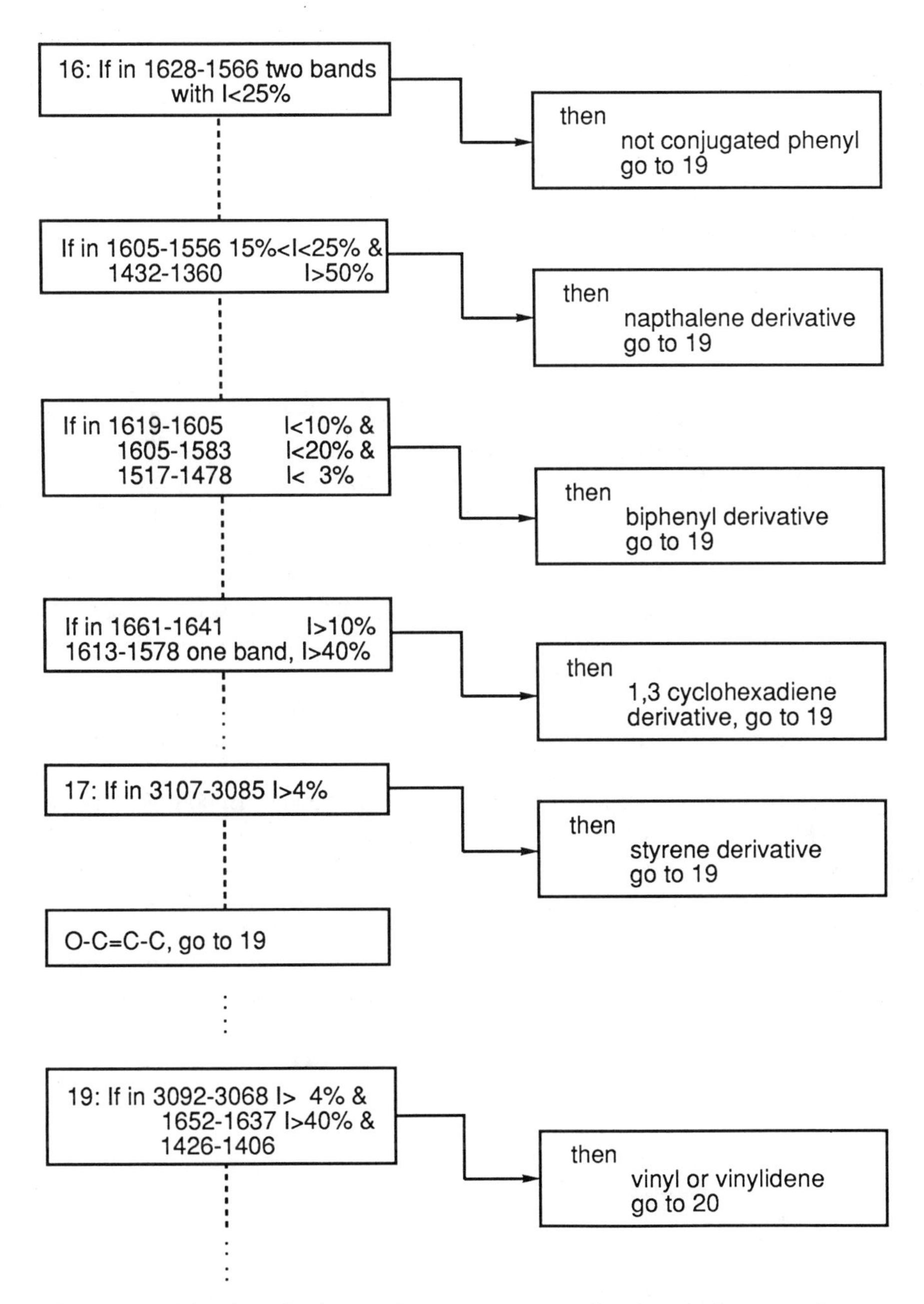

FIGURE VI-4. Flowchart for interpreting raman spectra of carbon, hydrogen, and oxygen compounds. (Adapted from J. H. van der Maas and T. Visser, *J. Raman Spectrosc.*, **2** (1974), 563; John Wiley & Sons.)

Once constructed, these flowcharts can readily be converted into simple FORTRAN code:

```
   :
16 IF(ABSORB(1628,1566,2,0,25)) GOTO 19
   IF(ABSORB(1605,1556,1,15,25).AND.ABSORB
(1436,1360,1,50,100)) .

   :
   :
```

In their work van der Maas and Visser did develop FORTRAN programs implementing, initially, this type of spectrum classification scheme. In subsequent studies, the basic approach was considerably refined and extended to include IR as well as raman spectral data.[11] Other analyses of IR data[10,12] and mass spectral data (of specific compound classes[2–7]) have been presented.

Similar approaches to the interpretation of multisource spectral data have also been attempted.[9] In one typical program, the code of the interpretation component of the program was built around tests for the different spectral features characteristic of each of a limited set of functional groups. An example of such a test is shown in Figure **VI-5**.

By the application of the encoded tests to recorded data, the constituent groups of simple molecules could be derived and then used in a simplified structure generation procedure.

There are, however, disadvantages to these approaches. Encoding interpretation rules as FORTRAN–ALGOL subprograms eliminates the distinction between *rule* and *rule interpreter* and this has at least two adverse consequences: (1) it becomes difficult to amend or extend rules; and (2) the rules are no longer in a form amenable to simple analysis by other programs, and, consequently, it may well prove difficult to generate explanations for decisions based on the use of particular combinations of rules, or to examine sets of rules to detect inconsistencies.

```
If   IR (ragged, medium, 3200, 2500) and
     IR (broad, strong, 1740, 1670) and
     proton (13,9)
then: begin
     mark—spectral—features—accountable
     identify—constituent (''-CO2H'')
     end
```

FIGURE VI-5. Simple programmed code for identifying the presence of a $-CO_2H$ group from its characteristic spectral features.

VI.B.2. Spectrum Interpretation "Languages" and Systems

Examination of a typical spectrum analysis routine as written in FORTRAN, or in some ALGOL dialect, will reveal that it consists primarily of elaborate conditional tests for various spectral properties. These conditional tests evaluate evidence for particular groups and, simultaneously, define the flow of control of the program. Through this control over the flow of the program, various (usually hierarchical) classification networks are implicitly defined. Given the disadvantages of hard-coded spectral analysis schemes and the highly stylized and limited structure of such code, it is obvious that an alternative formulation is both desirable and possible.

Typically, the spectral analysis consists of searching in various restricted spectral regions for particular patterns of signals. In a "hard-coded" interpreter of multisource spectral data, one will find routines for searching for an absorption, of given relative intensity and shape, within a specified region of the IR spectrum, and possibly some other routine for comparing the total intensities for two series of ions in a compound's mass spectrum. Such routines define the basic operations comprising some spectrum interpretation "language."

In the simplest classification schemes,the class of a compound is determined by the result of a single test checking for one, or more, defined spectral patterns. A spectrum interpretation language–system must at least provide a method for creating a composite test for class membership out of a set of calls to several individual functions that each check for specific spectral features.

More elaborate compound classification schemes involve multiple levels of classification. A compound may, for example, first be identified as containing an aldehyde group, and then further tests may be applied to the spectral data to differentiate between saturated and unsaturated aldehydes. A spectrum interpretation system must, therefore, allow the chemist to define the relationships between classes. After the successful testing for the spectral features of some general class, the interpretation process must be able to select and apply the spectral tests characterizing each of the more precisely defined subordinates to that class.

VI.B.2.a. Subsystems for Defining and Applying Spectrum Classification Schemes

These spectral interpretation systems normally comprise two components: (1) there must be a subsystem through which a chemist can specify a desired classification scheme; and (2) there must be a subsystem for applying an encoded classification scheme to subsequently acquired data.

The first subsystem must provide a means for identifying the spectral features characteristic of a particular compound class. A "menu" of spectral tests can be offered; the chemist need only select the appropriate test, (e.g., *search for IR band*) and provide appropriate parameters as required by the system (e.g., *SHAPE = ragged, STRENGTH = medium, START OF SPECTRAL WINDOW (in cm*^{1}*) = 3200, END WINDOW = 2600*).

As new classes are introduced by the chemist, the definition subsystem can "request" details as to how these are related to existing, previously defined classes. Procedures for displaying the structure and content of existing classification schemes should be included, and there must be provision for subsequent amendment and combination of schemes.

As well as enabling the chemist to specify classification schemes, this first component of the overall system "translates" or "compiles" the specified schemes. This translation converts from a representation convenient to the chemist into a representation more convenient for a computer.

The second component of the overall system performs the actual analysis of acquired spectral data. This second component is usually a "table-driven" program. The "tables" contain the computer-compatible representation of the classification scheme as defined by the chemist. A contrived example, based on the earlier example of Figure VI-5, is illustrated in Figure VI-6.

The code testing for the spectral attributes of some particular compound class would be represented as a set of entries in the table. Calls to functions testing for particular spectral features, such as IR bands or ^{1}H NMR resonances, would be represented by the identifier numbers assigned to the appropriate functions. Encoded arguments, normally all represented as integers, would follow each function identifier. Each such spectral test function would have some fixed number of arguments; successive tests could be placed in immediately following locations within the table. The end of the sequence of tests would be marked by some special function identifier.

The table would also contain information encoding the structure of the classification scheme. This information would take the form of pointers relating the various sections of coded tests in the table. At the head of the tests defining any one class there would be a pointer to the tests describing the first of any alternative classes. For instance, at the start of the $-CO_2H$ test sequence there could be a pointer linking through to a test for features characteristic of ketones, from where there might be a further link to aldehydes. At the end of the test sequence of each class, there would be a pointer linking to the first of any sequence of alternative more detailed subclass descriptions. These various pointers could be represented by the index numbers of the table locations at which the relevant class definitions were stored.

A particular spectral analysis system may employ a single elaborate table. For example, a system could use a table that contained definitions of the IR absorption patterns characteristic of all common functional groups. Other spectral analysis systems have been designed to process more restricted data. Such systems use separate tabular classification schemes devised for specific classes of molecules.

The run time interpretation program would typically be a simple function for the exhaustive exploration of tree structures such as those implicitly

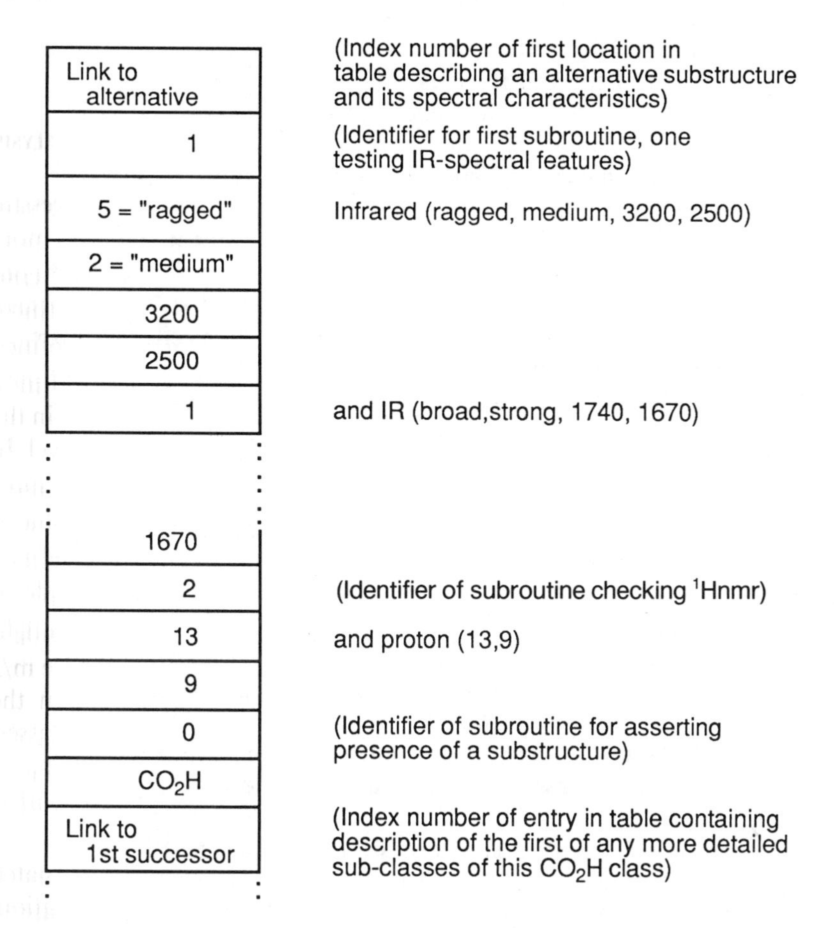

FIGURE VI-6. Representation of a spectrum classification scheme in a tabular form as might be used by an interpretation program.

defined through the links in such a table. (Such tree-searching functions are most readily defined recursively; iterative variants are possible.) An initial pointer would identify the spectral tests characterizing the first defined class. The program would save the pointer to any alternatives for subsequent use and proceed to apply the spectral tests defined. If the observed data did not satisfy the spectral tests, processing of the current class could be abandoned; search of an entire branch of the tree structure could thus be curtailed. If the data did satisfy the requirements for some class, that classification could be proposed (possibly with some plausibility estimate as computed by the spectrum testing functions).

If some subclasses are defined, the pointer to the first of these could be followed. The entire analysis sequence would then be repeated, with the spectral tests being applied to determine whether the observed data were consistent with those of the more precisely defined subclass. Again, successful matching of required spectral features would result in a further, now more precise classification. The search deeper into the tree structure would continue, in a recursive manner, until no more detailed subclasses were defined.

Usually, the spectral characteristics of a particular subclass can be defined independently of the features used to characterize its parent class. Sometimes, it is advantageous if the results of matching spectral features at one stage in the classification scheme can be subject to further analysis at some deeper level. In a mass spectrum classifier, therefore, it might be necessary to identify the most intense fragment ion in some ion series; subsequently, the specific m/z value of this ion may be relevant for discrimination among more precisely defined subclasses. Such requirements are fairly readily satisfied through simple, if rather *ad hoc*, extensions to the simplest tree-searching procedures. One might deal with the mass spectral case through the concept of a "key" ion whose m/z value, when defined at one level of the hierarchy, could be used in the subsequent tests for the characteristic spectral features of subordinate classes. More exacting approaches are also possible. Techniques, for matching of strings and parsing of languages, known as "augmented transition networks" are relevant to such problems.[13,14]

Search along one branch of the tree terminates either with a failure to match required features in observed data or with a successful proposed classification. On termination of the search along one branch, the spectrum classification program can "back up" to the last stage at which a pointer to alternatives was saved. These alternative classifications can also be explored by the same overall procedure.

The run time spectrum classification program will thus find all possible interpretations of some given spectral data. If the spectral tests involve computation of some measure of plausibility, the resulting plausibilities can be used to rank order the possible interpretations of the data.

VI.B.2.b. Example Systems

Such table-driven spectrum analysis systems differ mainly in their first component, that is, the subsystem that they provide to interface to chemists. Interactive menu-based programs provide one form of interface. One such system was devised to classify GC–MS data of organic chemicals extracted from marine sediments.[15] In that system, the chemist used an interactive program to define the characteristic mass spectral features of the various compound classes of interest that would be expected to be found in samples from a known source. The interactive program encoded the tests for spectral features in a tabular form and saved the complete encoded classification scheme on a file. Subsequently, the file with the encoded classification scheme could be loaded by the run-time analysis routines and used to classify mass spectral data as they were acquired during the course of a GC–MS run.

When using the interactive program to define a classification scheme, the chemist would first identify a main compound class such as *sterane* or *triterpane*. The interactive program would then display a menu of possible spectral tests such as (1) searches for a possible molecular ion, (2) comparisons of intensities of groups of ions at various specified mass ranges, and (3) comparisons of intensities of ions at m/z values defined by specified neutral losses from a molecular ion. The chemist could select the appropriate test and specify the necessary parameters. After indicating that all the spectral characteristics of a given class had been specified, the chemist would be asked to list any subclasses that were to be defined. The interactive program would then obtain details of the spectral characteristics of each of these subclasses in turn, and link the records on subclasses back to the main class to which they were subordinate. The order in which subclasses would be characterized by the chemist was determined by a simple recursive and backtracking scheme, similar to that used by the run-time analysis program.

Such an approach is adequate for strictly hierarchical, tree-structured classification schemes. Some applications may require more flexible control structures. For example, it might be convenient for the same sequence of spectral tests to be invoked at different points in some classification procedure (thus implicitly involving some form of "subroutine" construct). Other tests might be only conditionally apposite as, for instance, where a solvent might obscure particular absorptions. In such cases it may be convenient to provide some form of conditional programming construct (*if*...*then*...*else*...) in the "chemical–spectral interpretation language."

Chemical programming languages are applied primarily in systems for planning chemical syntheses. In the synthesis programs, such as **LHASA** and **SECS**, knowledge of chemical transformations is encoded in special languages such as ALCHEM.[16] Elaborate conditional constructs, and possibly also subroutines, are necessary to express modifications to a chemical reaction such as may be induced by neighboring functionality, ring strain, or other forms of steric congestion. There have also been some applications of such languages to spectral interpretation. The most elaborate such application is in the ***Program for the Analysis of IR Spectra*** (PAIRS) system.[17]

The **PAIRS** interpreter for analyzing spectra is a fairly conventional table-driven system such as outlined above. It contains routines for testing for spectral absorptions in specified ranges and updating measures of evidence ("probabilities") for particular functional groups. The tables used to drive the interpreter contain numeric keys that identify each routine and its required arguments. Additionally, the tables contain the information defining the control structure of the "program" being executed. A program for **PAIRS** is basically an arbitrarily deeply nested sequence of *if*...*then*...*else*... constructs. The conditional clauses can involve tests for spectral features in data recorded for an unknown, or for composition constraints, or for experimental details (e.g., solvent used for sample). The various *then* and *else* clauses can involve further conditionals or can cause adjustment of functional group probabilities.

The **PAIRS** tables are created by a "compiler" from original source rules defined, by the chemist, in the ***C****omputer* ***O****riented* ***N****otation* ***C****oncerning* ***I****nfrared* ***S****pectral* ***E****valuation* (CONCISE) language. An example of one of **PAIRS**'s interpretation subprograms, that for determining the presence of various aldehyde groups, is shown in Figure VI-7. This example is taken from the set of spectral interpretation rules that are provided with the **PAIRS** programs as distributed by the Quantum Chemistry Program Exchange, Bloomington, Indiana 47405 (Catalog No. QCPE 426). Woodruff and Smith[18] have provided a detailed discussion of how such rules may be created, expressed in the CONCISE language, tested, and refined.

The example in Figure VI-7 illustrates some of the ways in which the **PAIRS** system goes beyond simple hierarchical classification. This particular piece of CONCISE code evaluates evidence for three different groups; these are the main ***ALDEHYDE*** class and its two subclasses ***ALDEHYDE-UNSATURATED*** and ***ALDEHYDE-SATURATED***. In this example, most of the coded tests for spectral patterns are in common and the "probabilities" for the two aldehyde subclasses are initially set from the probability associated with the main ***ALDEHYDE*** class. The last two tests in the code modify these subclass probabilities in accord with particular composition and spectral data. Many more elaborate adjustments of relative ratings of different subclasses are allowed in the CONCISE language. Evidence found in support of one subclass can be used to adjust the probability either of some other subclass or of the main class to which they are subordinate.

In addition to allowing for such comparisons of relative probabilities of different subclasses of the same main structural type, **PAIRS** also allows for interdependencies between the probabilities for classes defined in different code sections. For instance, the rating accorded to an ***ALDEHYDE*** depends in part on the system's assessed probability for ***ACID*** groups (as expressed in tests around line 17). Acids and aldehydes both do, of course, exhibit absorptions in the "carbonyl" regions around 1800–1650 cm^{-1}; in addition, the hydrogen bond absorptions of an acid are likely to obscure H–CO absorptions of the aldehyde. Effective use of the **PAIRS** system requires explicit consideration of such potential interferences. Of course, the total set of rules given to the interpreter must be organized so that the evaluation of evidence for an ***ACID*** group will have been completed prior to the tests for ***ALDEHYDE***.

VI.B.2.c. Problems of Obscured or Missing Data

An important element in the design of such interpretive procedures is the provision of a simple, consistent method of dealing with missing data. The problems of missing data are readily apparent with IR spectra. Depending on the medium used when acquiring IR data, different spectral regions may be obscured. In the **PAIRS** system such solvent effects must be expressed in the rules for assessing evidence for a particular class. Thus in the ***ALDEHYDE*** code the tests expressed in lines 56–68 require that a check be made for an absorption in the 2800–2850 cm^{-1} region, provided the spectrum was **not** run in oil (which would obscure this region).

```
1    $
2    $==============================================
3    $
4    $
5     ALDEHYDE C 1 O 1 H 1 UNSAT 1 2000
6    $
7    $
8     IF ANY INTENSITY 7 TO 10 SHARP TO BROAD PEAKS ARE IN
9     RANGE 1765 TO 1660
10   $
11    THEN BEGIN
12    IF ANY INTENSITY 1 TO 10 SHARP TO BROAD PEAKS ARE IN
13    RANGE 2750 TO 2680
14   $
15    THEN BEGIN
16    SET ALDEHYDE TO 0.25
17    IF ACID GREATER THAN 0.90
18   $
19    THEN BEGIN
20    SUBT 0.15 FROM ALDEHYDE
21    DONE
22   $
23    ELSE BEGIN
24    IF ANY N EXIST
25   $
26    THEN BEGIN
27    IF ANY INTENSITY 4 TO 10 BROAD PEAKS ARE IN RANGE 2750 TO 2300
28   $
29    THEN BEGIN
30    SUBT 0.15 FROM ALDEHYDE
31    DONE
32    DONE
33    DONE
34   $
35    ELSE BEGIN
36    IF ANY INTENSITY 1 TO 10 SHARP TO BROAD PEAKS ARE IN
37    RANGE 2679 TO 2660
38   $
39    THEN BEGIN
40    SET ALDEHYDE TO 0.05
41    DONE
42    DONE
43    IF ANY INTENSITY 1 TO 10 SHARP PEAKS ARE IN RANGE 2750 TO 2660
44   $
45    THEN BEGIN
46    ADD 0.25 TO ALDEHYDE
47    DONE
48   $
49    ELSE BEGIN
```

FIGURE VI-7. Example of the rules for identifying the presence of an aldehyde, from IR absorptions, as expressed in CONCISE—the interpretive language of the **PAIRS** system.

```
50      IF ANY INTENSITY 1 TO 5 AVERAGE PEAKS ARE IN RANGE 2750 TO 2660
51     $
52      THEN BEGIN
53      ADD 0.25 TO ALDEHYDE
54      DONE
55      DONE
56      IF SPECTRUM RUN IN OIL
57     $
58      THEN BEGIN
59      CONTINUE
60      DONE
61     $
62      ELSE BEGIN
63      IF ANY INTENSITY 1 TO 10 SHARP TO BROAD PEAKS ARE IN
64     RANGE 2850 TO 2800
65     $
66      THEN BEGIN
67      ADD 0.25 TO ALDEHYDE
68      DONE
69      DONE
70      IF ANY INTENSITY 6 TO 10 SHARP TO BROAD PEAKS ARE IN
71      RANGE 1739 TO 1681
72     $
73      THEN BEGIN
74      ADD 0.25 TO ALDEHYDE
75      DONE
76      IF ALDEHYDE GREATER THAN 0.00
77     $
78      THEN BEGIN
79      IF ANY INTENSITY 7 TO 10 SHARP TO BROAD PEAKS ARE IN
80      RANGE 1739 TO 1715
81     $
82      THEN BEGIN
83      SET ALDEHYDE-SATURATED TO ALDEHYDE
84      MULT ALDEHYDE-SATURATED BY 0.90
85      DONE
86     $
87      ELSE BEGIN
88      SET ALDEHYDE-SATURATED TO ALDEHYDE
89      MULT ALDEHYDE-SATURATED BY 0.10
90      DONE
91      IF AT LEAST 2 UNSAT EXIST
92     $
93      THEN BEGIN
94      IF ANY INTENSITY 7 TO 10 SHARP TO BROAD PEAKS ARE IN
95      RANGE 1714 TO 1660
96     $
97      THEN BEGIN
98      SET ALDEHYDE-UNSATURATED TO ALDEHYDE
```

FIGURE VI-7. (*Continued*).

```
 99      MULT ALDEHYDE-UNSATURATED BY 0.90
100      DONE
101     $
102      ELSE BEGIN
103      SET ALDEHYDE-UNSATURATED TO ALDEHYDE
104      MULT ALDEHYDE-UNSATURATED BY 0.10
105      DONE
106      DONE
107      DONE
108      DONE
109     $
110      FINISH
111     $
```

FIGURE VI-7. (*Continued*).

Although it is always possible to require explicit consideration of solvent effects (and any other factors that might distort or obscure expected spectral patterns), a more general solution is preferable. A spectrum interpretation program, such as the **PAIRS** interpreter, really combines some assessed spectral evidence with a strength of inference scoring factor (probability), given in a rule for identifying a substructure. If the spectral evidence is uncertain, the strength of inference of the rule should be appropriately reduced.

The rule interpreter should assess the strength of evidence for a particular required spectral pattern being instantiated in the recorded data and use this assessment to qualify the strength of inference. The final confidence rating for a substructure could be taken as the product of the strength of inference given in a rule and the assessed confidence for correct identification of the spectral pattern defined in that rule's premise. If a clearly defined absorption at around 2700 cm^{-1} contributes 0.2 to the belief rating of an aldehyde, therefore an uncertain absorption—possibly due to a contaminant or maybe an overtone band, might be accorded a confidence of 0.5 and thus would contribute only 0.1 to the final belief rating for an aldehyde group.

If the spectrum were acquired in a solvent known to absorb strongly in a particular spectral region, absorptions due to the sample being analyzed could not be detected with confidence. Spectral interpretation rules, whose premise parts depend on the detection of spectral patterns in obscured regions, consequently contribute nothing to the final belief ratings of substructures. A zero confidence rating in the validity of the premise part results in a zero overall strength of inference. If the rule interpreter assesses the strength of evidence for spectral features, one can simply reflect modifying factors, such as obscuring solvent bands, into the original spectral data. It is then no longer necessary to make explicit allowance for modifying solvent influences in the rules used to identify each individual substructure.

Some other common problems of spectral interpretation schemes are alleviated through this approach of assessing the strength of evidence for a required spectral feature and accordingly modifying the strength of inference of a rule. Most spectral interpretation rules specify a spectral "window" where absorptions or resonances must occur. Compounds incorporating the relevant substructure will exhibit spectral bands whose exact position within the spectral region will depend on additional structural influences of secondary importance. It is obviously desirable to define these spectral windows fairly narrowly so that subtle spectral differences, due to detailed structural influences, can be captured by the interpretation scheme. However, the spectral windows should not be defined so narrowly that they do not encompass the signals observed from example compounds that contain the relevant substructure within a somewhat unusual larger environment.

In most rule schemes problems are encountered with "end effects" on the windows. A saturated aldehyde that fortuitously gives a peak absorption at 1740 cm^{-1} would not be identified as such by the **PAIRS** rules; however, any further extension of the allowed region for aldehyde carbonyl absorptions would result in the inappropriate identification of lactones and other groups as possible saturated aldehydes. Any such identifications would be accorded a confidence equal to that of a compound whose carbonyl absorption was at the very center of the region characteristic of saturated aldehydes.

The actual absorptions, or resonance shifts, associated with a substructure can be presumed, in the lack of more detailed data, to be normally distributed about some mean value. This distribution of peak positions can be defined in terms of mean and estimated standard deviation. The distribution of peak positions as found in reference compounds incorporating a substructure may be presumed to be universally valid. The probability of a particular observed signal belonging to this distribution can then be assessed (from the presumed normal distribution curve). This probability can then be used as the strength of evidence for the validity of the rule premise. This approach avoids any abrupt end effects. An absorption at 1740 or even 1755 cm^{-1} can still be interpreted as satisfying a requirement for an absorption centered around 1725 cm^{-1}; however, the 1740 cm^{-1} absorption is accorded a lower confidence rating than another centered at 1720 cm^{-1}. As the confidence rating in a rule premise is reduced, so is the final confidence rating for the corresponding substructure. Absorptions at 1720 cm^{-1} can be taken as strong evidence for aldehydes; those at 1740 cm^{-1} still admit such interpretation, but with less confidence.

VI.B.2.d. Assessment of Substructural Plausibilities

An ubiquitous problem of heuristic inference systems is illustrated by the fractional "probabilities" used in the **PAIRS** IR analysis program. These are not mathematical probabilities; they are not combined according to the laws of probability and it is quite possible for a probability exceeding 1 to be accorded to a specific substructure.

If sufficient statistics were available, a genuine probability-based scheme for assessing evidence could be constructed. It would be necessary to know the conditional probabilities of cooccurrence of different spectral features. For example, a probability-based system would need to have a measure of the probability of observing an absorption at around 2700 cm^{-1}, given that a carbonyl absorption at around 1720 cm^{-1} has been observed. Two such conditional probabilities would have to be known: one for aldehydic and one for nonaldehydic compounds. Given sufficient data defining conditional probabilities, Bayesian inference schemes could be used to determine the probability of a particular substructure given certain observed data.

It is rare to have such detailed statistics. Typically, more empirical inference schemes are employed. For the most part, these schemes neglect the various conditional probabilities for the cooccurrence of various spectral features. Each datum is considered as providing some independent evidence for or against a particular hypothesis. There have been a number of studies of the problems of how a measure of strength of evidence may be associated with some datum, and of how different evidence may be combined. These problems have been addressed in the context of various "Expert Systems" dealing with medical and geologic data. Shortliffe's model of confirmation,[19] based on "Confidence Factors," could serve as an initial model for developing a scheme for combining plausibility estimates for substructures suggested by uncertain spectral data. Although simple to implement, the Shortliffe confidence factor model has a number of limitations. Various alternative models of plausible inference have been proposed; Quinlan has recently reviewed several possible models for plausible inference.[20]

VI.B.3. Rule-Based Approaches

The simplest, but in many ways the most versatile interpretation programs exploit what are in essence simply correlation charts, that is, sets of rules relating substructures and spectral features. The spectral features employed are usually simple. An example might be a spectral feature comprising a single ^{13}C resonance of defined shift and multiplicity; observed data on an unknown would have to exhibit an appropriate resonance, of closely similar shift, for the corresponding substructure to be accepted. Another similar scheme, appropriate for more general substructures, would have each substructure characterized by a set of regions in IR, ^{1}H NMR and ^{13}C NMR spectra. With this scheme, the interpretation program would suggest those substructures each of whose spectral regions encompassed some signal observed in the data characterizing an unknown.

Such spectrum-substructure rules represent reference data that can be manipulated by the interpretation program. Other programs may also manipulate the rules and can, therefore, ensure that rules for related classes are consistent.[21] One of the major advantages of the rule-based systems is this ability to examine and test the encoded spectral knowledge.

VI.B.3.a. Identification of Small Standard Substructures

The interpretation procedures of Sasaki's **CHEMICS** system[22-25] and Gribov's **STREC**[26] system both use correlation charts. Many of the more recently developed programs employ similar spectrum–substructure rule sets.[1,27,28]

Although restricted to molecules containing only carbon, hydrogen and oxygen, the **CHEMICS** scheme is typical. The **CHEMICS** scheme uses tables that define those features in the IR, ^{1}H NMR and ^{13}C NMR characteristic of each of a set of 189 standard substructures known to the system. These substructures are mostly small, typically containing a single carbon or hetero atom of defined hybridization and with a specified number of substituent hydrogens. Some substructures are slightly more complex; examples are those substructures comprising, in effect, the two methyls of a *gem*-dimethyl group. By successive comparison of the unknown's recorded spectra with the requirements as specified in these tables, increasing numbers of these standard components may be eliminated from further consideration.

Details of **CHEMICS** substructure–spectral rules have been published.[22,29] A part of the rule set relating to ^{13}C NMR data is summarized in Figure VI-8. The **CHEMICS** interpreter identifies both a substructure and some constraints on its larger environment. Thus, given appropriate resonances in the ^{1}H NMR and ^{13}C NMR spectra, the **CHEMICS** interpreter can infer that the unknown must incorporate a methoxy group, with the added restriction that it form a part of an ester, rather than ether, structure. (Limitations of **CHEMICS**'s structure generator, specifically its inability to handle overlapping substructures like CH_3–O–CO– together with –O–CO–C(sp^2) or –O–CO–C(sp^3), necessitate this form of qualified substructure identification).

Appropriate definitions of the spectral patterns characteristic of each group are obtained from spectral data of known reference compounds. Details of the construction of the **CHEMICS**'s tables have been published.[29] If resonance shifts or absorption bands can be assigned to specific atoms or bonds of standard structures represented in computer-compatible form, the process of building up rule sets can be largely automated.[21,30]

VI.B.3.b. Inferring Larger Substructures

The same general approaches may be adopted in systems attempting to identify larger substructural components. The spectral attributes of the substructures are tabulated and organized in some data base. An interpretive program matches these reference data to the spectra recorded for the unknown and, through this matching, identifies possible substructural components of the unknown . (If the set of substructures and subspectral rules is too large to fit into computer memory, this matching process may use some simple file search mechanism.)

The attempted identification of large constituent substructures engenders new problems for the overall interpretive system. Commonly, the substructures thus identified will not be discrete but will overlap one another to some

Substructure	(Larger environment)	Signal required in range:
⋮	⋮	⋮
CH_3-O-	$CH_3-O-C(=O)-$	50.34–52.53
⋮	⋮	⋮
CH_3-O-	$CH_3-O-C(sp^2)$	56.68–61.51
⋮	⋮	⋮
CH_3-O-	$CH_3-O-C(sp^3)$	49.95–60.60
⋮	⋮	⋮
CH_3-CO-	$CH_3-CO-O-$	19.81–23.39
⋮	⋮	⋮
CH_3-CO-	CH_3-CO-C (aromatic)	22.95–31.79
⋮	⋮	⋮

FIGURE VI-8. Carbon-13 interpretation "rules" as used in ***CHEMICS*** (adapted from Ref. 29). If the spectral pattern correlated with a substructure is not observed, that substructure may be eliminated from further consideration.

arbitrary degree. Furthermore, several alternative large substructures may be correlated with similar spectral patterns.

In some interpretive systems, the resolution of ambiguities and overlaps is left entirely to the user. The output from such systems comprises just a small set of names of the seemingly plausible substructures. More complete analyses are necessary whenever there are likely to be very large numbers of alternative interpretations or when the results of the interpretive procedures are to be employed directly in some automatic structure generation process. In these situations the identification of plausible substructures represents only a first, relatively minor, step in the overall interpretive process. The major part of the interpretation program will be concerned with combining the substructures inferred from different spectral features.

Often, an unknown may be presumed to incorporate a standard skeleton with substituents (or multiple bonds) whose nature and position must be determined. In such a case, knowledge of the form of the skeleton can be helpful in resolving among various alternative interpretations for some spectral features. The interpretive system can eliminate those substructures that cannot fit anywhere onto the molecular skeleton. Restrictions on the placement of substituent groups can be derived by exploring all possible successful matches, onto the skeleton, of the remaining alternative substructures. In other cases, such with as ^{13}C NMR data, additional constraints may be derived through an intercomparison of the substructures correlated with individual spectral features. These additional constraints eliminate many of those substructures initially selected by the spectral matching process.

Substructural Interpretation of Mass Spectra. Some attempts have been made to interpret mass spectra. Much of the original work on Heuristic **DENDRAL** concerned mass spectral interpretation. Heuristic **DENDRAL** used a rule interpreter, in LISP, working with rules represented as LISP data structures. These spectrum–substructure rules implicitly defined various hierarchical classification schemes. The substructures, were used directly in a subsequent structure generation procedure. The analyses attempted by Heuristic **DENDRAL** were limited to simple monofunctional compounds.[31,32] The rules could employ unambiguous spectral patterns. The substructures comprised the functional group together with its immediate bonded neighbors; such substructures constituted substantial parts of the simple molecules studied. The approach of Heuristic **DENDRAL** has recently been extended; details of the interpretation rules, as used in this work, are limited.[33]

The **STIRS**[34] program was discussed in Chapter IV. One development of **STIRS**[35,36] can identify the presence of any of 200 standard substructures. Some of these standard substructures are substantial moieties such as steroid fragments and fused aromatic ring systems. The **STIRS** spectral matching procedure works on arbitrary compounds and can, in principle, suggest the presence of substructures other than those in its standard set of 200.[37]

The **STIRS** program is atypical in that the relevant substructure-spectral relationships are not predefined but are, in a sense, rediscovered through each application of **STIRS**'s matching procedure to the mass spectrum of an unknown. The **STIRS** program suggests the presence of a substructure in an unknown if it occurs, at significantly greater than chance frequency, among those reference compounds matched to the unknown's spectrum. However, **STIRS** is limited to suggesting possible constituent substructures of an unknown. The program does not attempt to identify or resolve alternative, ambiguous interpretations. The **STIRS** user must determine which of the suggested interpretations are mutually compatible and must identify, and resolve, possible overlaps among the various proposed substructures. Data derived through **STIRS** are not immediately usable in constraining automatic structure generators.

Preliminary studies of more direct substructure-mass-spectral pattern relationships have also been reported.[38] These developments focused on the task of detailed interpretation of mass spectra of compounds of a known class. For such problems, conventional interpretation schemes, as employed by human spectroscopists, typically infer the presence of particular skeletal fragments and substituent groups from observed patterns of ions in the spectrum of a compound.

Ion patterns, as used to characterize specific skeletal fragments or other standard substructures, normally comprise a principal fragment resulting from some initial molecular cleavage and some secondary ions resulting from further characteristic fragmentations of, and neutral losses from this principal fragment. Usually, one can describe such a pattern in terms of the relative intensities of ions at specified m/z values; occasionally, the pattern may be more conveniently defined in terms of neutral losses from the molecular ion. An automated mass spectral interpretation scheme can readily exploit such simple substructural–spectral-pattern relationships.

An automated interpretation system requires a detailed data base of substructures and their characteristic ion patterns. Construction of such a data base is relatively costly, necessitating the detailed analysis of the spectra of many known compounds of similar structure, in order to identify workable subspectral-substructural relationships. Once constructed, such a data base can be used to assist in the analysis of a subsequently isolated compound incorporating a similar skeleton.

The interpretation process starts by identifying those reference substructures whose ion patterns are at least partially extant in the unknown's spectrum. Then, the quality of match must be assessed and, finally, possible substructures proposed. These substructures must then be combined and used to build up candidate molecular structures for the unknown.

In the reported study,[38] the chemist retained much of the initiative and responsibility for interpretation. The chemist had to select ions, from the unknown's spectrum, and give these to the program to start each interpretive step. In response, the program presented those standard substructures, characterized by that ion, together with details of their complete spectral patterns. The chemist was further responsible for assessing the match between the observed spectrum and these specified patterns. Plausibility scores of the substructures were derived from this assessment.

The published examples concern a system that uses a data base describing the characteristic fragmentations of different sterol nuclei and side chains.[38] The applications were to the identification of new marine sterols. In this context, it was reasonable to presume that all unknowns would incorporate one of a set of standard sterol skeletons. Substructures, derived through the spectral analysis, could be matched onto these prototype skeletons; results of this matching procedure yielded suggestions as to where multiple bonds and substituents were placed. Thus evidence could be accumulated showing that some specific unknown incorporated a Δ-5 or Δ-7 sterol skeleton if several of its observed ion patterns could be rationalized in terms of fragments involving unsaturated B rings.

More completely automated systems for substructural interpretation of mass spectra can be conceived. An exhaustive search through the substructure/ion-pattern data base could be used to identify possible substructures. If it were really necessary to index into the data base, some selection algorithm could be used to select suitable indexing ions from the spectrum of the unknown. (Appropriate algorithms could be adapted from those used to select the ions for characterizing reference compounds for a compound identification file search system.) Scores accorded to substructures could be determined from some measure of consistency of observed and specified ion–intensity patterns. Although some improvements could be obtained, there are limits as to what can be achieved through a system relying on low-resolution mass spectral data.

Any data base of substructures and ion patterns will be restricted. Only a few substructures will be correlated with any given ion or pattern of ions. If an ion pattern is matched to the unknown's spectrum, corresponding reference substructures are retrieved. These substructures will be proposed by the interpreter system as a set of alternatives to be used to constrain a subsequent generation procedure. Of course, it is possible that, in the spectrum of some particular unknown, a matched ion pattern results from some completely different fragmentation of some non-standard skeleton. Then, none of the reference substructures retrieved would be correct. If the correct substructure is not included among those used to constrain the generation procedure, the correct structure cannot be derived. Consequently, in such interpretive schemes it is necessary to allow for the possibility that none of the standard alternatives is actually correct. This can be achieved by allowing for an "or none of these" alternative in any set of substructures used as constraints.

This requirement for an "or none of these" alternative greatly reduces the general efficacy of such interpretive systems. In cases where the interpretation is really a matter of identifying the correct skeletal structure from among a limited set of choices,[38] the "or none of these" constraint is merely an irritant. However, if it is not possible to limit the skeletons, a conclusion such as "(1) Δ-5 sterol is likely, (2) Δ-7 sterol is possible, (3) Δ-5–Δ-7 sterol is improbable, (4) it could be something completely different" really means that **no** constraints have been derived. At best, such interpretation results suggest an order in which to explore different possible structural types. The unconstrained problem, resulting from the "or none of these" alternative, in fact subsumes all the others and will eventually yield the same structural candidates as the more constrained problems (along with a great many other structural forms).

Essentially, ion patterns in a low-resolution mass spectrum are of too limited specificity. The relative intensities, as expected for the ions comprising some pattern, may be distorted by isobaric ions due to other fragmentations, and consequently the pattern may not be recognized. It is also quite possible for a series of unrelated fragmentation processes to result in ions that, fortuitously, match a pattern specification. Additional data, such as metastables, are necessary to establish that the observed ions are related by the postulated fragmentation processes. When a chemist interprets mass spectra, such limitations of ion patterns are borne in mind and the results are always appropriately qualified. An automated system will all too readily be treated as a "black box" and its suggested structural interpretations accepted uncritically. Reliable interpretation, as essential in any automated system, requires more exacting data. Later, we consider the possibility of using MS–MS data as the basis for a reliable, automated interpretation scheme.

Substructural Interpretation of ^{13}C NMR Spectra. Several schemes have been proposed for the analysis of ^{13}C data. All involve large sets of rules relating chemical shifts to substructures defining the local chemical–stereochemical environment of the resonating carbon atom. The first step in these interpretation programs is some form of search for reference substructures whose spectral features match the observed data to within some specified tolerance. The earliest systems, by Bremser[30,39] and Jezl and Dalrymple[40], were only file search systems. These search systems could identify substructures that could account for any particular resonance selected from the spectrum of an unknown. These substructure identifications were intended to be helpful suggestions for a chemist analyzing an already partially interpreted spectrum. Jezl and Dalrymple outlined a general approach to spectral interpretation using substructural information as derived from different resonances.[40]

Subsequent developments have seen both the introduction of stereochemistry[41] and the development of programs for combining the substructural data derived from all the resonances in the spectrum (thus realizing the suggestions of Jezl and Dalrymple).[42,43] Details of known skeletons, or of postulated sets of alternative molecular skeletons, can be used to help in the interpretation of the ^{13}C data;[44] more general analyses involve searching for consistent interpretations of the various resonances.[42,43] Other developments have been based on multiresonance-substructure relationships.[45–47] Again a major part of these systems is an elaborate search procedure for finding consistent interpretations for different groups of ^{13}C resonance signals. It is for these ^{13}C interpretation programs that elaborate methods for *searching for consistency* have been developed. These consistency checks are reviewed in Section VI.C.

VI.C. THE SEARCH FOR CONSISTENCY

A requirement for some procedure to find mutually consistent sets of interpretations for various spectral data arose first in the systems of Sasaki et al.[22] and Gribov et al.[48,49] In these systems, fully automated spectrum interpretation procedures have to identify the substructural building blocks to be used in a subsequent structure generation program.

In the first phase of these systems, some subset of the defined standard substructures consistent with the data are identified. However, there are seldom constraints on either the minimum or maximum number of occurrences of any one of these selected substructures. A second phase of analysis entails selection of all combinations of these structures that make up the molecular formula and that can account for all observed absorptions and resonances. Of course, such selection procedures must also verify requirements such as single carbon resonances not being assigned to more than one atom in the set of substructures selected. Programs for solving these selection problems can be based on methods for solving sets of logical equations.[48–51] The approach used in **CHEMICS** has been illustrated with some detailed examples.[29]

More elaborate methods for finding consistent substructural interpretations for spectral data have been developed specifically for ^{13}C data. The first stage of any ^{13}C interpretation programs results in large numbers of possible substructures. In each, a central carbon is correlated with a resonance that has a shift and multiplicity similar to that of some observed resonance in the spectrum of the unknown. As well as specifying the environment of the central carbon, these substructures specify, at least in part, the environments of its neighbors. There are, therefore, redundant data that can be used as the basis for elaborate consistency checks. A particular substructure constitutes a valid interpretation of one resonance only if the environments that it defines for neighboring atoms are consistent with the interpretations found for other resonances.

Consistent interpretations for the various resonances must be identified by what is basically a search process. Two different search algorithms have been used in experimental programs. One search algorithm involves a depth-first search that identifies those sets of substructural interpretations that can yield complete structures. The other search method attempts to establish constraints on the possible substructural interpretations of each resonance and, consequently, restrictions on the bonding of the corresponding carbons.

VI.C.1. Depth-First Search for Consistent Substructural Interpretations

The depth-first search is the more obvious method. In this method, search starts with one particular substructural interpretation being chosen for one of the resonances in the spectrum. This substructure would define at least the one-bond, α neighbors of the resonating atom (in a practical system, the defined structural environment would normally extend through at least the atom's β neighbors). For instance, one might select **VI-14** as the initial choice of substructural interpretation for a quartet resonance at 22.8 ppm:

$$C^{*}H_3\text{–}CH< \;\leftarrow\; (\text{q, 22.8 ppm})$$

VI-14

The search procedure must then find doublet resonances that could correspond to the methine bonded to the methyl group in **VI-14**. The spectrum might contain a number of doublets with various shifts, such as 43.9, 55.8, and 80.6 ppm. Each of these might admit interpretations involving a methine bonded to a methyl group (e.g., **VI-15–VI-18**)

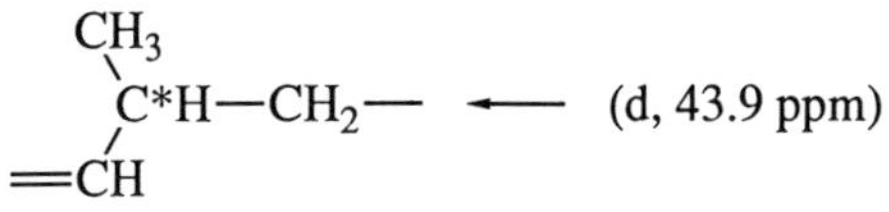

```
    CH3
      \
      C*H—CH2—   ←—   (d, 43.9 ppm)
     /
=CH
```

VI-15

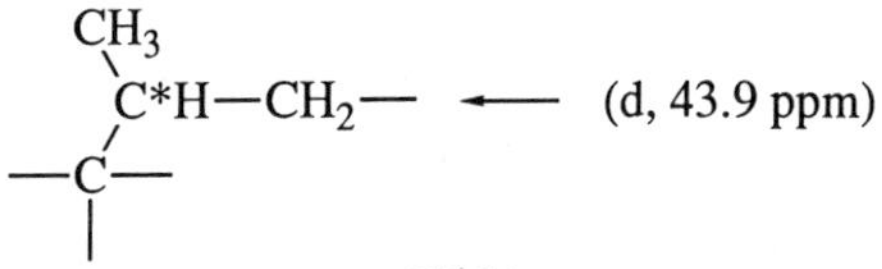

```
    CH3
      \
      C*H—CH2—   ←—   (d, 43.9 ppm)
     /
 —C—
   |
```

VI-16

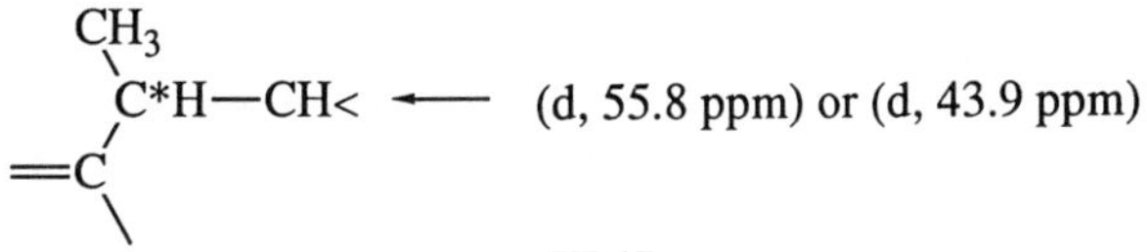

```
  CH3
    \
    C*H—CH<   ←—   (d, 55.8 ppm) or (d, 43.9 ppm)
   /
=C
  \
```

VI-17

```
    CH3
      \
      C*H—CH<   ←—   (d, 80.6 ppm)
     /
 —O
```

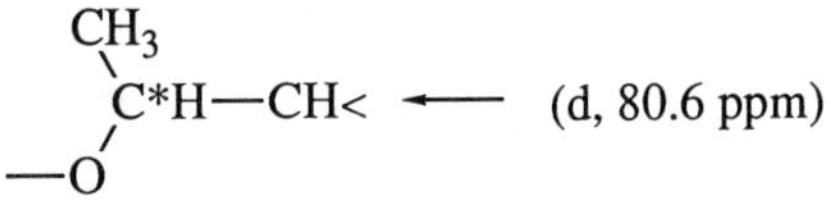

VI-18

The methine carbon atom in **VI-14** might correspond to any of the three observed doublet resonances. Thus, starting from the presumed substructural interpretation of the (q, 22.8 ppm) signal as being due to **VI-14**, there would be three main paths for development. Each path would involve selection of a different one of the observed doublet resonances as corresponding to the methyl's neighboring methine. In general, for each choice of methine resonance, there would be a number of possible substructural interpretations. Thus a branching tree of possible consistent substructural interpretations would be developed as illustrated in Figure VI-9.

Each of these paths would have to be explored in turn. Search might proceed first along the path where the (d, 43.9 ppm) resonance was attributed to the neighboring methine of substructure **VI-14**. Each possible alternative interpretation of this resonance (**VI-15**, **VI-16**, etc) would need to be considered. Some of these choices might prove incompatible with the observed spectral data. Exploration along the path through substructure **VI-15** could be curtailed if, for example, the ^{13}C data did not exhibit any doublet resonance that could be interpreted as being due to a olefinic –CH= carbon bonded to a methine. In

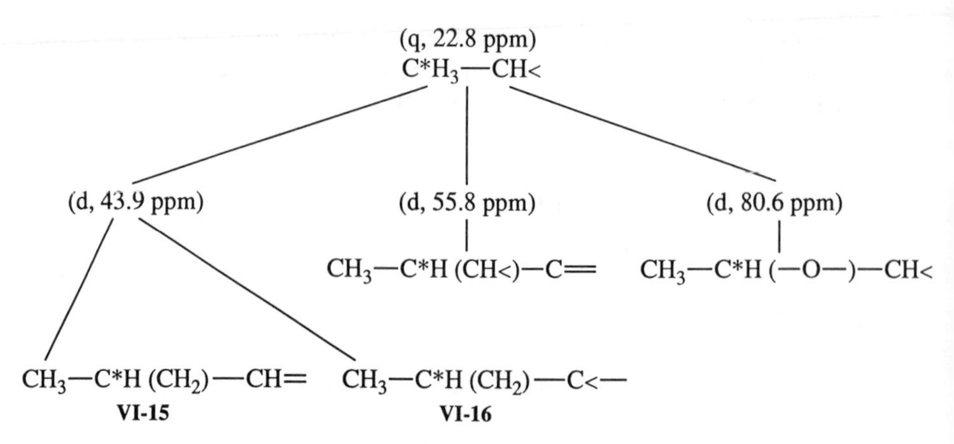

FIGURE VI-9. Developing a depth-first search for consistent substructural interpretations.

the event of such a failure, the search procedure could backtrack to the last point at which there existed unexplored alternative pathways. In the example of Figure VI-9, search would resume with **VI-16** being considered.

Usually, such a tree structure will show a high degree of branching at each level. Thus the next step along the path through substructure **VI-16** would entail the identification of possible resonances that could correspond to the neighboring $-CH_2-$ methylene. The observed spectrum might exhibit several different triplet resonances in, for example, a range from 18 to 64 ppm. Each of these observed triplet resonances would have to be considered as possibly corresponding to the $-CH_2-$ of substructure **VI-16**. Each triplet resonance might be associated with several alternative substructures incorporating $-CH_2-CH<$ bonds. The various possible substructures for each triplet resonance would have to be analyzed in turn. Thus, emanating from the point where substructures **VI-14** and **VI-16** had been combined, there would be many alternative branches to new partially assembled structures at the next level of the search tree. The search procedure would have to explore these alternatives in some systematic manner, perhaps starting with the first substructure identified as representing a possible interpretation for the triplet resonance of smallest shift.

At the next level of search it would be appropriate for the search procedure to attempt to identify singlet resonances that could be due to the quaternary carbon neighboring the methine in **VI-16**. Again, there would be in general be several possible singlet resonances in the spectrum, each having many alternative substructural interpretations. Each choice of resonance, and each choice of substructure for that resonance, would define a new branch of the tree that would have to be explored further. Many branches would lead to dead ends. These dead ends would arise at points where incompatibilities were detected between the bonding requirements of the various atoms as combined into some partially assembled candidate structure.

Such search procedures are costly. Many unsuccessful probes proceed deep down particular branches of the search tree before being abandoned and causing the search to be backed up to a point where unexplored alternatives remain. Exploration of the various branches in such a tree structure can ultimately lead to final candidate structures. However, such searches do not constitute a very effective approach to structure generation.

Although the depth-first search approach has some intrinsic problems, it has been adopted in two experimental systems for the analysis of ^{13}C data.[45–47] Both of these experimental systems involve elaborate substructures and detailed tests restricting the combinations of substructures that are considered. These elaborations help to restrict the number of paths that must be searched.

The system due to Schwenzer and Mitchell[45] has been described in the greater detail. The actual Schwenzer-Mitchell study was limited to analysis of ^{13}C spectra of acyclic alkanes and amines; only resonance shifts, and not multiplicities, were used. However, their basic approach can be generalized.

Schwenzer and Mitchell[45] employed ^{13}C interpretation rules of a more elaborate form than the simple (*shift* + *multiplicity*)–*substructure* relationships considered thus far. Their substructures extended beyond the immediate α environment of the central carbon atom (not necessarily extending to the same bond radius along all bonds from the central atom). The shift range for the central atom was supplemented by additional data, "support predictions" characterizing the shifts of certain of the neighboring carbons. Example interpretation rules, similar in concept to those used in the analysis of ^{13}C spectra of alkanes and amines, are shown in Figure VI-10.

Substructure **VI-19**, as shown in Figure **VI-10**, would be saved as a possible interpretation of a (q, 22.8 ppm) resonance only if there were also a doublet resonance in the region 30–46 ppm. In a search building on **VI-19** as an

```
          CH2-
         /
C*H3—CH                 <=>    (q, 10-23 ppm) & support
        |                       (d, 30-46 ppm)
      —C—
        |

    VI-19

          CH2—CH2
         /
CH3—C*H                 <=>    (d, 42-45 ppm) & support
       |                        (q, 10-23 ppm) & (t, 14-45 ppm)
CH3—C—C=                        & (s, 34-55 ppm)
       |   \
       CH2
      /

    VI-20
```

FIGURE VI-10. Example multi-resonance rules for the structural interpretation of ^{13}C spectra.

interpretation for a 22.8-ppm quartet, only doublet resonances in this range need be considered. Thus only the substructural interpretations of the doublet resonance at 43.9 ppm would be employed; one example substructural interpretation is shown as substructure **VI-20**. The support predictions associated with **VI-20** would limit the choice of candidate triplet and singlet resonances that might correspond to the methine's $-CH_2-$ and >C< neighbors. The high degree of branching, characteristic of the original simple search tree, can be reduced through this use of additional data characterizing substructures.

In the Schwenzer-Mitchell study, additional heuristics were employed in an attempt to further improve the efficacy of the search procedure. The substructures, identified as possible interpretations for each resonance in the spectrum, were in effect ranked. Substructures inferred from rules with narrow shift ranges were preferred over those associated with broad shift ranges. A resonance with only a few possible interpretations was considered to represent a better starting point for subsequent searches than another resonance with many alternative interpretations. Substructures with narrow ranges for support predictions were preferred over those wherein the support predictions were poorly defined and thus would not effectively constrain the next building step.

The objective of using these heuristics was to constrain and guide the search process. A constrained search should lead, more rapidly, to a candidate structure. A constrained search should also detect, sooner, any inconsistencies due to the selection of an inappropriate substructure as an interpretation of any one of the resonances. In the Schwenzer–Mitchell system, the interpretation process was terminated when some predefined number of possible structures had been generated.

The Schwenzer–Mitchell pilot study was limited to the spectra of small acyclic alkanes and amines and has not been developed further. Those investigators' heuristic approach, of selecting and combining the most plausible of substructures until some predefined number of structures had been generated, is reasonable in a pilot project but inappropriate in practical structure elucidation. All possible candidate structures for an unknown are required; consequently, exhaustive search and candidate generation must be performed.

Another depth-first search approach to the interpretation of ^{13}C spectra has been briefly outlined by Dubois et al.[46,47] Again, the procedure is based on selection of one alternative interpretation for some spectral feature and using this as the starting point for a recursive search. At each successive step in the search process, an existing partially assembled structure is extended. The structural environment of one of the peripheral atoms, of the current partially assembled structure, is elaborated using information on possible substructural environments as inferred from the ^{13}C data.

The method of Dubois et al., like that of Mitchell and Schwenzer, uses data from more than one resonance shift when attempting to identify possible substructures. Each of these substructures defines a spherical environment of one atom. Substructures are characterized by the difference between two

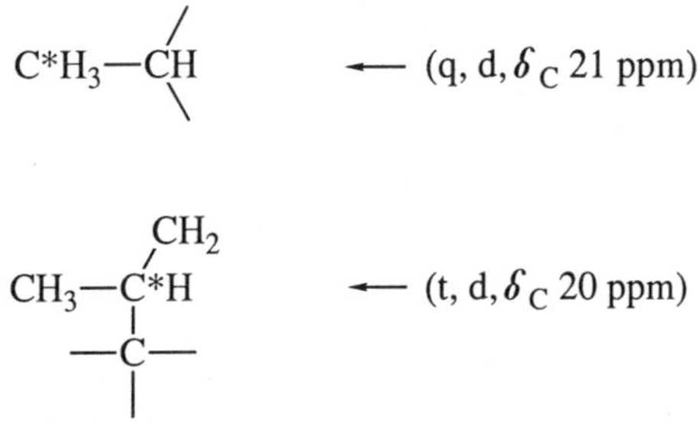

$$C^*H_3\text{—}CH\!<\quad \leftarrow (\text{q, d}, \delta_C\ 21\ \text{ppm})$$

$$CH_3\text{—}C^*H(CH_2)\text{—}C\!<\quad \leftarrow (\text{t, d}, \delta_C\ 20\ \text{ppm})$$

FIGURE VI-11. Rules for ^{13}C interpretation based on differences in chemical shift values.

chemical shifts; these are the shift of the central atom and that of one of its α neighbors. Multiplicities are not required but, obviously, could be employed to improve selectivity in interpretation. Examples of possible rules relating substructures and shifts (and multiplicities) are given in Figure VI-11.

Possible component substructures are identified by use of differences in shifts between pairs of observed resonances. Then a depth-first search scheme is employed to explore out from some chosen initial substructure. Only a brief review of Dubois's system has been published with no examples of complete spectral interpretation. The published description of the system includes a diagram summarizing statistics that illustrate how the use of a pair of resonances, as opposed to a single resonance, can greatly reduce the number of substructures that need to be analyzed.

Even with elaborate substructures characterized by several shifts, the depth-first search is costly. In this search method, it is difficult to take advantage of substructural constraints inferred from other data or derived through some more global analysis of the ^{13}C data themselves. In effect, these systems use an inefficient structure generation method to determine whether a particular set of substructural interpretations can possibly yield a connected structure.

VI.C.2. Iterative Constraint Refinement Search for Consistent Substructural Interpretations

The second method for searching for consistent interpretations is based on the concept of *constraint refinement*. Initially, the only constraints on the interpretations found for any given spectral feature would be those derivable from molecular composition. Analyses of the interpretations derived in the initial step will, in general, yield restrictions on the bonding of the constituent atoms in a molecule. These restrictions typically eliminate some of those possible substructural interpretations that were obtained initially. A reanalysis of the remaining substructural interpretations can reveal new, more precise bonding constraints. These refined constraints on bonding serve to eliminate yet more of the possible substructures. The entire procedure may be repeated until no further structural information can be derived.

The results of such an analysis will rarely define a unique structure. Rather, the results take the form of restrictions on the bonding of some, or all, of the molecule's constituent atoms. These restrictions may take the form of requirements for bonds, or prohibitions on the existence of bonds between specific atoms; sometimes, extensive multiatom substructures, or sets of alternative substructures, can be identified.

This approach has been explored in an experimental program for the partial interpretation of ^{13}C resonance spectra. This experimental program was devised to yield substructural constraints that could subsequently be used in the Stanford **GENOA** structure generation system. The first step in the program's interpretive procedure consists of reading a data base of four-bond substructure-shift combinations and abstracting those that have chemical shifts matching, within the specified tolerances, any of the observed resonances of the same multiplicity. The retrieved substructures are analyzed to determine the atom types of the α (one-bond) and β (two-bond) neighbors of the resonating atom. The counts of neighbors of different atom types are checked against the limits derived from the composition, and inappropriate substructures are eliminated. Acceptable substructure–shift combinations are then written to a new file. In addition, the analysis of the substructures retrieved for the atom associated with some particular observed resonance can give more precise limits on the numbers of its α and β neighbors of various atom types.

A ^{13}C spectrum typical of those that the interpretive program is intended to analyze is shown in Table VI-1. These spectral data, together with the molecular composition $C_{20}H_{32}O_2$, provide a few gross constraints on the type of substructure that might be correlated with any one of the resonance lines. Thus it is immediately evident that the molecule consists of two methyls (CH_3–s), nine alkyl methylenes (–CH_2–s), four methines (>CH–s), three alkyl quaternary carbons (>C<s), a vinylic methylene (CH_2=), a vinylic quaternary carbon (>C=) and two hydroxyl groups (–OHs). Substructures proposed as interpretations of resonance lines must be restricted to those comprised solely of these subunits. Other constraints on bonding derive directly from considerations of valence and atom type; thus, there certainly cannot be bonds between methyl and hydroxy groups, and the two vinylic carbons must comprise a >C=CH_2 system. However, even with these composition constraints there are numerous possible substructures for each resonance. The numbers of substructures, distinct at the two-bond environment and matching to within 1 ppm (0.75 ppm for methyls), are given in Table VI-1 for each resonance.

Analysis of the substructures matching the observed resonances, to within required tolerances, provides numerous constraints on the bonding of their associated atoms. Thus, for example, it is found, on examination of all their possible interpretations that both methyls must be α to a methine >CH–, or to a quaternary alkyl carbon >C<, or a a vinylic carbon >C=. The only atoms that allow interpretations involving α-hydroxy groups are the –CH_2–, associated with the triplet at 64.3 ppm, and the >CH– with the doublet at 80.6 ppm. The two hydroxy groups can, unequivocally, be bonded to these atoms. Only the methine at 43.9 ppm yields interpretations of the form CH_3–CH<. The vinylic carbon at 155.4 ppm is identified as being in one of the two substructural environments represented by substructures **VI-21** and **VI-22.**

```
      |                      \ /
     —C—                      CH
     /                       /
CH2=C                   CH2=C
     \                       \
      CH2—                    CH2—

    VI-21                   VI-22
```

TABLE VI-1. Spectral Data for a Compound Extracted from the Medicinal Herb *Stachys lanata*[52] [a]

Resonance	Type	Shift	Multiplicity	Number of Two Bond Environments	
				Initial	Final
1	>C=	155.4	s	4	1
2	>C<	43.9	s	38	2
3	>C<	42.7	s	44	3
4	>C<	39.6	s	48	2
5	>CH–	80.6	d	15	3
6	>CH–	55.8	d	32	3
7	>CH–	55.8	d	32	3
8	>CH–	43.9	d	41	1
9	$CH_2=$	103.1	t	5	3
10	$-CH_2-$	64.3	t	7	3
11	$-CH_2-$	48.8	t	11	2
12	$-CH_2-$	41.3	t	48	6
13	$-CH_2-$	38.7	t	70	8
14	$-CH_2-$	38.5	t	76	8
15	$-CH_2-$	33.0	t	95	11
16	$-CH_2-$	27.6	t	80	10
17	$-CH_2-$	20.1	t	31	8
18	$-CH_2-$	18.3	t	28	6
19	$-CH_3$	22.8	q	25	2
20	$-CH_3$	18.3	q	23	2
21	–OH				
22	–OH				

[a] The column labeled "Number of Two-Bond Environments" gives the number of different possible substructures that, to with 1.0 ppm (0.75 ppm for $-CH_3$s), match with the observed resonance and are consistent with the constraints of the molecular composition and available atom types.

TABLE VI-2. Constraints Derived for Atoms Corresponding to Each Resonance[a]

Resonance	Type	Shift	Atoms Permitted in α Environment	Allowed Neighbors
1	>C=	155.4	=CH_2, –CH_2–, >CH–, or >C<	9, 10–18, 5–8 2–4
2	>C<	43.9	–CH_3, –CH_2–, >CH–, or >C<	19, 20, 10–18, 5–8, 3, 4
3	>C<	42.7	–CH_3, –CH_2–, >CH–, >C<, or >C=	19, 20, 10–18, 5–8, 2, 4, 1
4	>C<	39.6	–CH_3, –CH_2–, >CH–, >C<, or >C=	19, 20, 10–18, 5–8, 2, 3, 1
5	>CH–	80.6	–OH, –CH_2–, >CH–, or >C<	21, 10–18, 6–8, 2–4
6	>CH–	55.8	–CH_2–, >CH–, >C<, or >C=	10–18, 5, 7, 8, 2–4, 1
7	>CH–	55.8	–CH_2–, >CH–, >C<, or >C=	10–18, 5, 6, 8, 2–4, 1
8	>CH–	43.9	–CH_3, –CH_2–, >CH–, >C<, or >C=	19, 20, 10–18, 5–7, 2–4, 1
9	CH_2=	103.1	>C=	1
10	–CH_2–	64.3	–OH, –CH_2–, >CH–, >C<, or >C=	22, 11–18, 5–8, 2–4, 1
11	–CH_2–	48.8	>CH–, >C<, or >C=	5–8, 2–4, 1
12	–CH_2–	41.3	–CH_2–, >CH–, >C<, or >C=	10, 11, 13–18, 5–8, 2–4, 1
13	–CH_2–	38.7	–CH_2–, >CH–, >C<, or >C=	10–12, 14–18, 5–8, 2–4, 1
14	–CH_2–	38.5	–CH_3, –CH_2–, >CH–, >C<, or >C=	19, 20, 10–13, 15–18, 5–8, 2–4, 1
15	–CH_2–	33.0	–CH_3, –CH_2–, >CH–, >C<, or >C=	19, 20, 10–14, 16–18, 5–8, 2–4, 1
16	–CH_2–	27.6	–CH_3, –CH_2–, >CH–, >C<, or >C=	19, 20, 10–15, 17, 18, 5–8, 2–4, 1
17	–CH_2–	20.1	–CH_3, –CH_2–, >CH–, >C<, or >C=	19, 20, 10–16, 18, 5–8, 2–4, 1
18	–CH_2–	18.3	–CH_3, –CH_2–, >CH–, >C<, or >C=	19, 20, 10–17, 5–8, 2–4, 1
19	–CH_3	22.8	>CH–, >C<, or >C=	5–8, 2–4, 1
20	–CH_3	18.3	>CH–, >C<, or >C=	5–8, 2–4, 1
21	–OH		>CH–	5
22	–OH		–CH_2–	10

[a]These constraints are based solely on features found in the substructures matching the resonance lines to within the required tolerances. Only the atom types allowed as α (one-bond) neighbors are noted. The analysis also identifies limits on the number of atoms of each type that are permitted. Thus, resonance 12, triplet at 41.3, could correspond to a –>C–C*H_2–C<–; but, resonance 13, triplet at 38.7, does not allow two >C<s adjacent to the –CH_2–. In addition to the α restrictions, some restrictions on the atoms' β (two-bond) environments are also derived by the program.

For other resonances, such as the triplet at 20.1 ppm, the constraints implied by the retrieved substructures are weaker; for this particular–CH_2–, the minimum-maximum limits neither require nor prohibit neighbors of any particular atom type. Yet, even here, some restrictions are obtained through reductions of the maximum limits allowed for the numbers of alpha and beta neighbors of a given type. The constraints defined by the types of substructure matching each resonance are summarized in Table VI-2, along with their implications regarding bonding.

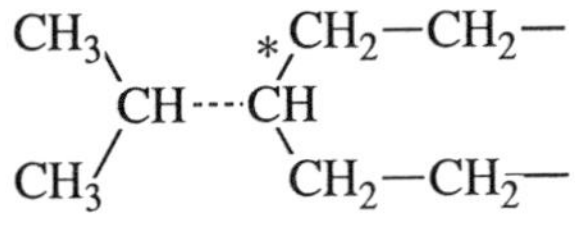

```
CH3\        *CH2—CH2—
    CH····CH
CH3/         \CH2—CH2—
```

VI-23

The constraints thus abstracted from the substructures associated with individual resonances have implications that must be propagated through the entire set of atoms. Many of the additional implied constraints derive directly from a simple one-bond analysis. Thus the absence of ethyl (CH_3–CH_2–) substructures amongst the interpretations found for the quartet resonances allows the α–CH_3 maximum limit to be reduced to zero for the –CH_2–s associated with the triplets at 18.3, 20.1 27.6, 33, and 38.5 ppm (despite the fact that all these resonance can be interpreted as due to –CH_2–s in ethyl groups). Similarly, the types of substructure found for the >C= resonance at 155.4 ppm imply that interpretations of methyls in vinylic environments are invalid. Less direct implications include the fact that neither methyl can be β to an hydroxy group, and that substructure **VI–23** can now be established as a being inadmissible for the 43.9 ppm doublet resonance. (All these inferences can be derived from the data given earlier.) The results of such an analysis are summarized in Table VI–3 in terms of the new α environments and allowed neighbors.

These derived constraints eliminate some of substructures previously considered for the various resonances. For the data given in Table VI-1, about 10% of the substructures initially retrieved can be eliminated just on the basis of alpha and beta limits, leaving over 2000 possible four-bond substructures incorporating about 400 distinct two-bond environments. More detailed analysis, involving the assembly of candidate substructures and testing against more specific bonding constraints, permits further reductions in the number of possibilities. At this stage, for example, the interpretations for the singlet at 43.9 ppm include, amongst others, the one-bond environments **VI-24** and **VI-25**. Consequently, the derived limits on –CH_3s and –CH_2–s are 0–2 and 0–3. Such limits imply, incorrectly, that substructure **VI-26** is also a valid interpretation for this resonance in this molecule. As a consequence of this overgeneralization for substructures correlated with the 43.9 ppm resonance, substructure **VI-27** is accepted as a valid interpretation for the triplet resonance at 41.3 ppm.

TABLE VI-3. Constraints Derived for the Various Atoms by Propagation of Implications of Initial Constraints Shown in Table VI-2

Resonance	Type	Shift	Atoms Permitted in α Environment	Allowed Neighbors
1	>C=	155.4	$=CH_2$, $-CH_2-$, >CH–, or >C<	9, 3, 4, 6–8, 10–18
2	>C<	43.9	$-CH_3$, $-CH_2-$, >CH–, or >C<	19, 20, 10–18, 5–8, 3, 4
3	>C<	42.7	$-CH_3$, $-CH_2-$, >CH–, >C<, or >C=	19, 20, 10–18, 5–8, 2, 4, 1
4	>C<	39.6	$-CH_3$, $-CH_2-$, >CH–, >C<, or >C=	19, 20, 10–18, 5–8, 2, 3, 1
5	>CH–	80.6	–OH, $-CH_2-$, >CH–, or >C<	21, 10–18, 6–8, 2–4
6	>CH–	55.8	$-CH_2-$, >CH–, >C<, or >C=	10–18, 5, 7, 8, 2–4, 1
7	>CH–	55.8	$-CH_2-$, >CH–, >C<, or >C=	10–18, 5, 6, 8, 2–4, 1
8	>CH–	43.9	$-CH_3$, $-CH_2-$, >CH–, >C<, or >C=	19, 20, 10–18, 5–7, 2–4, 1
9	CH_2-	103.1	>C=	1
10	$-CH_2-$	64.3	–OH, $-CH_2-$, >CH–, >C<, or >C=	22, 12–18, 5–8, 2–4, 1
11	$-CH_2-$	48.8	>CH–, >C<, or >C=	5–8, 2–4, 1
12	$-CH_2-$	41.3	$-CH_2-$, >CH–, >C<, or >C=	10, 13–18, 5–8, 2–4, 1
13	$-CH_2-$	38.7	$-CH_2-$, >CH–, >C<, or >C=	10, 12, 14–18, 5–8, 2–4, 1
14	$-CH_2-$	38.5	$-CH_2-$, >CH–, >C<, or >C=	10, 12, 13, 15–18, 5–8, 2–4, 1
15	$-CH_2-$	33.0	$-CH_2-$, >CH–, >C<, or >C=	10, 11–14, 16–18, 5–8, 2–4, 1
16	$-CH_2-$	27.6	$-CH_2-$, >CH–, >C<, or >C=	10, 11–15, 17, 18, 5–8, 2–4, 1
17	$-CH_2-$	20.1	$-CH_2-$, >CH–, >C<, or >C=	10, 11–16, 18, 5–8, 2–4, 1
18	$-CH_2-$	18.3	$-CH_2-$, >CH–, >C<, or >C=	10, 11–17, 5–8, 2–4, 1
19	$-CH_3$	22.8	>CH–, >C<	8, 2–4
20	$-CH_3$	18.3	>CH–, >C<	8, 2–4
21	–OH		>CH–	5
22	–OH		$-CH_2-$	10

The implication in substructure **VI-27** that there be a >C< with two methyls and two methylenes can then be identified as incompatible with the allowed two-bond environments of all the three different >C<s; consequently, this interpretation of the 41.3 ppm triplet can be eliminated.

```
 \    CH3                      CH2—
  CH—C—CH3               \     |
 /    |                   CH—C—CH2—
      CH                 /     |
     / \                       CH2—

    VI-24                    VI-25

       CH3                              CH3
       |                                |
 —CH2—C—CH3       —CH2—CH2—CH2—C—CH2—
       |                                |
       CH2—                             CH3

    VI-26                    VI-27
```

For this example, the analysis based on considering only the two-bond environments of the atoms eventually converged successfully. The final constraints identified for the alpha neighbors and bonding are summarized in Table VI-4.

The spectral data in Table VI-1 pertain to a compound isolated from the medicinal herb *Stachys lanata*.[52] The data in Table VI-1 include the final number of two-bond substructural environments derived for the various resonances at the point where a two-bond analysis converged. The two-bond environments for two of the atoms are determined exactly, the >C= associated with the 155.4-ppm singlet and the >CH– with the 43.9-ppm doublet. Several other atoms are limited to only two or three possible two-bond interpretations. Several examples of inferred alternative substructural interpretations are shown as **VI-28–VI-36**. Significant ambiguity remains only for some of the $-CH_2-$s where there are 10 or more alternative substructures. Constraints on the one-bond environment of atoms are obtained for all atoms save the $-CH_2-$ associated with the triplet at 27.6 ppm for which the limits on alpha neighbors remain as 0–2 for each of $-CH_2-$, >CH– and >C<.

The final unique two-bond environment for line 1 in Table VI-4 (155.4 ppm, s) is

```
          \
           CH—CH2
          /     |
  CH2=C*        |
          \     |
           CH2—C—
                |
```

VI-28

TABLE VI-4. Constraints Derived for the Various Atoms at Completion of Iterative Analysis

Resonance	Type	Shift	Atoms Permitted in α Environment	Allowed Neighbors
1	>C=	155.4	=CH$_2$,-CH$_2$-, >CH	9, 11, 8
2	>C<	43.9	-CH$_3$,-CH$_2$-, >CH-	19, 20, 10-16, 5-7
3	>C<	42.7	-CH$_3$,-CH$_2$-, >CH-	19, 20, 10-16, 5-7
4	>C<	39.6	-CH$_3$,-CH$_2$-, >CH-	19, 20, 10, 12-16, 5-7
5	>CH-	80.6	-OH,-CH$_2$-, >C<	21, 15-16, 2-4
6	>CH-	55.8	-CH$_2$-, >C<	16-18, 2-4
7	>CH-	55.8	-CH$_2$-, >C<	16-18, 2-4
8	>CH-	43.9	-CH$_2$-, >C=	13-16, 1
9	CH$_2$=	103.1	>C=	1
10	-CH$_2$-	64.3	-OH,-CH$_2$-, >C<	22, 15, 16, 2-4
11	-CH$_2$-	48.8	>C<, >C=	2, 3, 1
12	-CH$_2$-	41.3	-CH$_2$-, >C<	13-18, 2-4
13	-CH$_2$-	38.7	-CH$_2$-, >CH-, >C<	12, 14-18, 8, 2-4
14	-CH$_2$-	38.5	-CH$_2$-, >CH-, >C<	10, 12, 13, 15-18, 8, 2-4
15	-CH$_2$-	33.0	-CH$_2$-, >CH-, >C<	10, 11-14, 16-18, 5, 8, 2-4
16	-CH$_2$-	27.6	-CH$_2$-, >CH-, >C<	10, 11-15, 17, 18, 5-8, 2-4
17	-CH$_2$-	20.1	-CH$_2$-, >CH-	12-16, 18, 6, 7
18	-CH$_2$-	18.3	-CH$_2$-, >CH-	12-17, 6, 7
19	-CH$_3$	22.8	>C<	2-4
20	-CH$_3$	18.3	>C<	2-4
21	-OH		>CH-	5
22	-OH		-CH$_2$-	10

Alternative two-bond environments (**VI-29** and **VI-30**) for line 2 in Table VI-4 (43.9 ppm, s) are

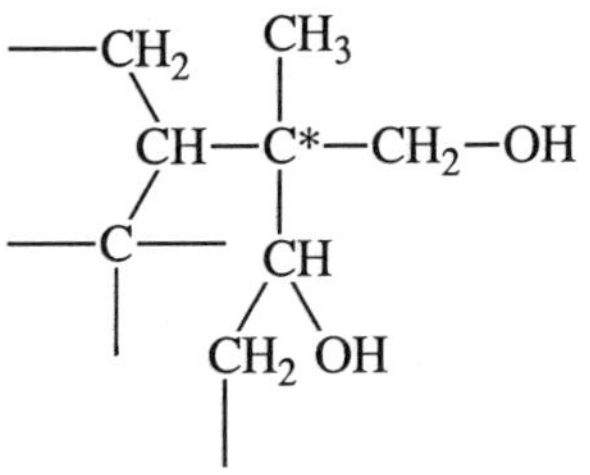

```
                                        \
                                       —C    CH2
                                       / \  /
—CH2    CH3                               CH
    \    |                                 |
    CH—C*—CH2—OH                      CH2—C*—CH2—CH2—
    /    |                             |   |
—C—      CH                            CH  CH2
  |     /  \                            \  /
       CH2  OH                           C
        |                                ||
```

VI-29 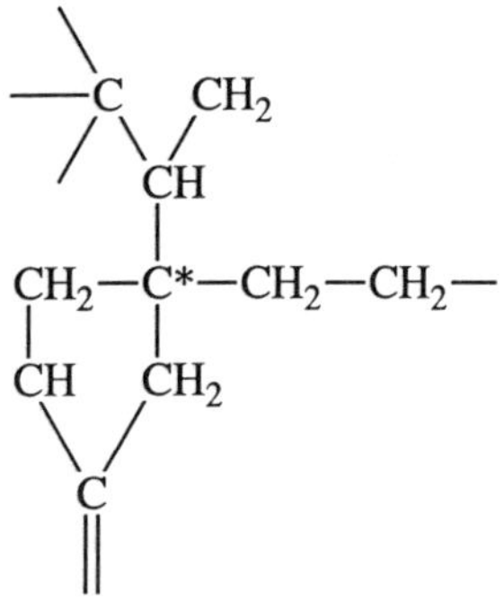 **VI-30**

Alternative two-bond environments (**VI-31**-**VI-36**) for Line 12 in Table VI-4 (41.3 ppm, t) are

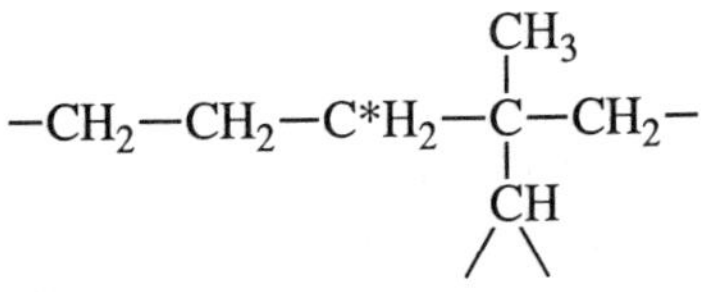

```
                        CH3                                   CH3 /
                         |                                     | /
—CH2—CH2—C*H2—C—CH2—                  —CH2—CH2—C*H2—C—CH
                         |                                     |  \
                        CH                                    CH
                       /  \                                  /  \
```

VI-31 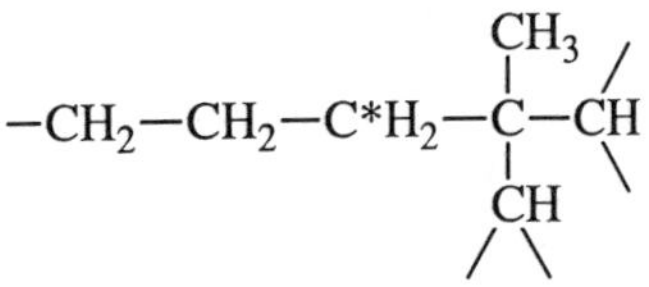 **VI-32**

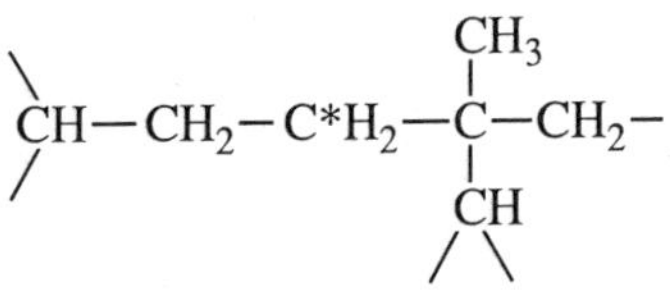

```
\                 CH3                  \                  CH3 /
 CH—CH2—C*H2—C—CH2—                 CH—CH2—C*H2—C—CH
/                  |                   /                  |   \
                  CH                                     CH
                 /  \                                   /  \
```

VI-33 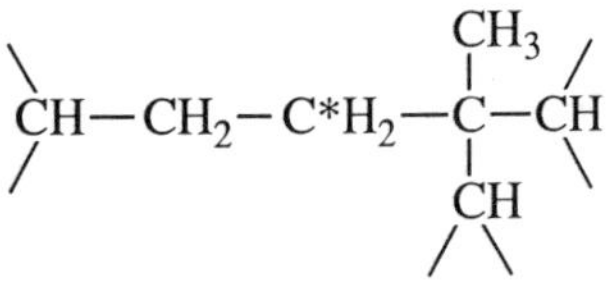 **VI-34**

```
                                                      |
                                               CH2—C—
                                              /      |
                                          *CH2       |
\                 CH2—                        \      |
 CH—CH2—C*H2—C—CH2—                        C—CH
/                  |                          /  \
                  CH                       CH3    CH—
                 /  \                             |
```

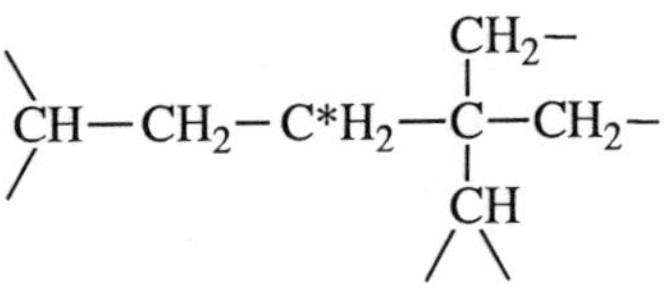

VI-35 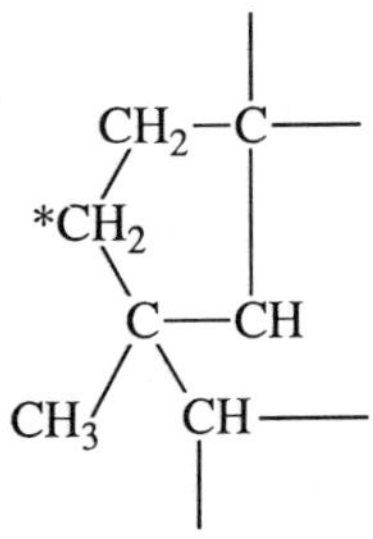 **VI-36**

The specific two-bond substructures, the sets of alternative two-bond substructures for resonances with six or fewer alternatives, and the one-bond environments found for the resonances were passed to the **GENOA** program and used to constrain the structure generation process. Seven constitutional isomers resulted. Only one of these isomers incorporated a standard diterpene skeleton; this candidate structure (**VI-37**) does represent the correct constitution of the unknown diol that had been established to be *ent*-3β,19-dihydroxy-kaur-16-ene.[52]

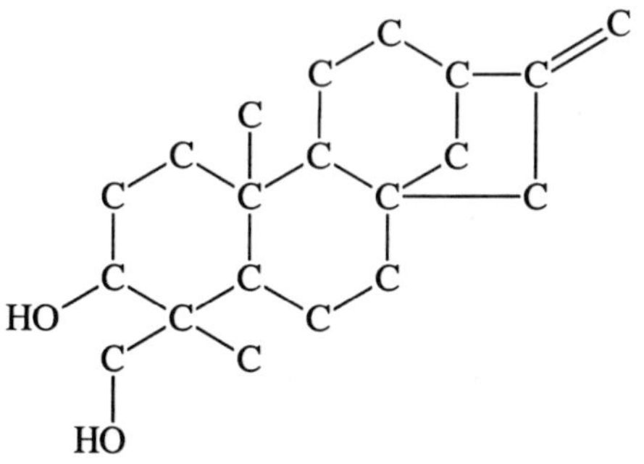

VI-37

The success of the interpretation for this structure was due in part to the fact that every one of its atom environments was correctly represented, at least out to a two-bond radius, in the substructural templates of one or more rules in the data base. The rules in the data base relevant to atoms in the B, C, and D rings were derived mainly from previously characterized kaurenes; the rules used when analyzing the A-ring atoms were derived mostly from labdane and pimarane diterpenes.

In a few favorable cases, like the kaurenoid diol (**VI-37**) it is possible to determine structure from composition and ^{13}C data alone. However, this result is exceptional. More usually, the composition and basic spectral constraints would permit many more interpretations for each resonance line. Several different combinations of interpretations for different features might typically be found; thus the inferences that could be drawn from the ^{13}C would prove far less effective at constraining the structure generator.

It would certainly be atypical to possess only ^{13}C data. The ^{13}C experiment is time-consuming and costly. Infrared and ^{1}H NMR spectra are almost invariably acquired and analyzed first. Results from other spectral studies and assumptions concerning the biological origin of a compound, can provide valuable constraints to aid in interpreting the ^{13}C spectrum. For example, the two methyl groups of the kaurenoid-diol appear in the ^{1}H NMR spectrum as 3H singlets at 0.99 and 1.22 ppm; these observations imply that both of these methyl groups must be bonded to quaternary alkyl carbons. This information would have eliminated from the outset all substructures of the form CH_3-CH_2-, $CH_3-CH<$, or $CH_3{>}C{=}$, thus leading to a more rapid convergence of the iterative procedures.

The interpretation program is designed to accept additional constraints that can be inferred from other sources. Constraints can be expressed either in terms of substructures defined through a standard structure editor, or more simply as restrictions on the neighbors and next-nearest neighbors of atoms associated with particular ^{13}C resonances. Elaborate substructural constraints are particularly useful in problems that involve determination of the position of substituents on some standard skeleton. For example, one could analyze the data for the kaurenoid assuming, on the basis of evidence from cooccurring compounds, that it incorporated a kaurenoid skeleton.

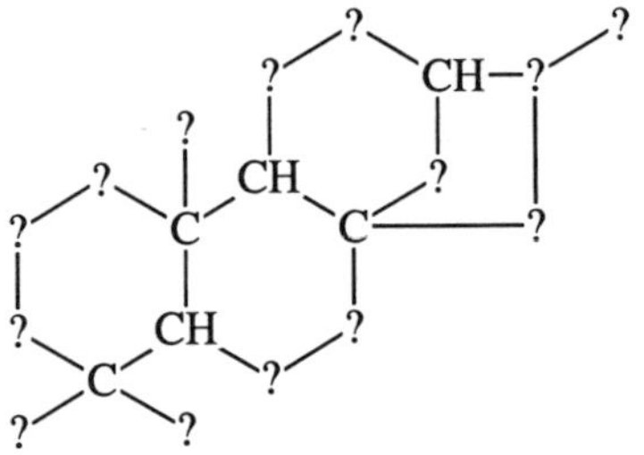

VI-38

A basic C_{20} kaurene skeleton could be taken from a library of standard skeletons and given to the ^{13}C interpretation program prior to the initial search of the substructure–shift data base. The interpretation program analyzes the skeleton and compares its requirements for particular substituted carbons with the multiplicities of observed signals. For these data, it is possible to immediately establish some bonding constraints. Since only three quatemary alkyl carbon resonances are observed, these must match onto C(4), C(8), and C(10) of the skeleton. There must be >CH– methines at C(5), C(9), and C(13). These initial constraints on the kaurene are shown in structure **VI-38**. [Bond orders are not defined in the standard skeletons taken from libraries; unless explicitly contraindicated, any bond can have a bond order of greater than 1. The current interpretation program has some limitations and, for instance, cannot establish that the vinylic resonances at 155.4 ppm and 103.1 ppm must correspond to atoms C(16) and C(17) of the skeleton].

Several general bonding constraints on the various resonances can be derived. No quaternary alkyl >C< can have another alkyl >C< as an alpha neighbor; each must have at least one >C< as a beta neighbor. No methine can have have three bonded >C< neighbors, and no methylene can have two >C<s. (It is evident that two >CH–s must each have two alpha >C<s; however, this constraint cannot be used at this stage because there are no known correspondences of resonances and skeletal atoms.) These constraints can be employed in the initial selection of possible substructural interpretations matching the observed shifts. The selection is more effective, particularly for the >C< carbons for which the average number of different possible initial two-bond

TABLE VI-5. The Presumption of a Known Skeleton Allows the Interpretation Program to Constrain the Selection of Possible Substructures that Match with Observed Resonances[a]

Resonance	Type	Shift	Multiplicity	Number of Two-Bond Environments Initial
2	>C<	43.9	s	22
3	>C<	42.7	s	17
4	>C<	39.6	s	20
	⋮			
9	CH_2=	103.1	t	5
10	$-CH_2-$	64.3	t	7
	⋮			
13	$-CH_2-$	38.7	t	67
14	$-CH_2-$	38.5	t	72
	⋮			

[a]The column labeled "Number of Two-bond Environments" gives the number of different possible substructures that, to within 1.0 ppm (0.75 ppm for–CH_3s), match with the observed resonance and which are consistent with the constraints of the molecular composition, available atom types and the requirement for a kaurenoid skeleton. These values should be compared with those given in Table VI-1.

environments is halved (to about 20 possible choices for each resonance). Details of the number of substructures retrieved are given in Table VI-5.

Analysis of the substructures retrieved for each resonance line provides additional bonding constraints on the associated atoms. These new constraints can be used in a further process of matching atoms onto the given kaurenoid skeleton. The more restricted matching process limits the possible distribution of substituents on the skeleton; the results of the matching step are shown as structure **VI-39**. The matching step also yields further restrictions on allowed

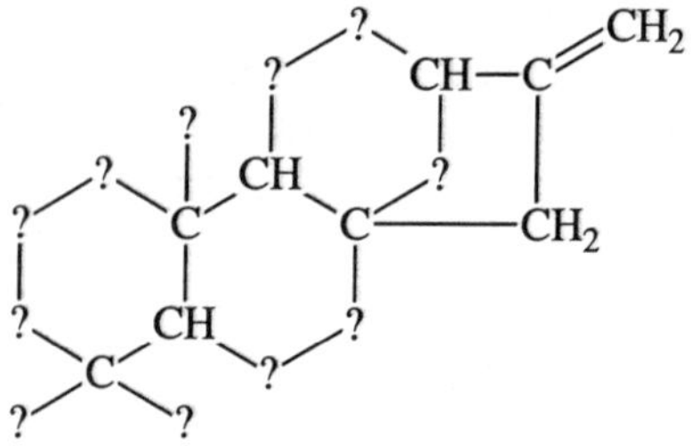

VI-39

TABLE VI-6. Constraints Derived for Atoms Corresponding to Each Resonances[a]

Resonance	Type	Shift	Atoms Permitted in α Environment	Allowed Neighbors
1	$>C=$	155.4	$=CH_2, -CH_2-, >CH-$	11–15, 8, 9
2	$>C<$	43.9	$-CH_3, -CH_2-, >CH-$	19, 20, 10–18, 5–7
3	$>C<$	42.7	$-CH_3, -CH_2-, >CH-$	19, 20, 10–18, 5–7
4	$>C<$	39.6	$-CH_3, -CH_2-, >CH-$	19, 20, 10–18, 5–7
5	$>CH-$	80.6	$-OH, -CH_2-, >CH-$, or $>C<$	21, 11–18, 6–8, 2–4
6	$>CH-$	55.8	$-CH_2-, >CH-, >C<$	12–18, 5, 2–4
7	$>CH-$	55.8	$-CH_2-, >CH-, >C<$	12–18, 5, 2–4
8	$>CH-$	43.9	$-CH_2-, >CH-, >C=$	11–18, 5, 1
9	$CH_2=$	103.1	$>C=$	1
10	$-CH_2-$	64.3	$-OH, >C<$	22, 2–4
11	$-CH_2-$	48.8	$>CH-, >C<$ or $>C=$	8, 2–4, 1
12	$-CH_2-$	41.3	$-CH_2-, >CH-, >C<$, or $>C=$	13–18, 5–8, 2–4, 1
13	$-CH_2-$	38.7	$-CH_2-, >CH-, >C<$, or $>C=$	12, 14–18, 5–8, 2–4, 1
14	$-CH_2-$	38.5	$-CH_2-, >CH-, >C<$, or $>C=$	12–13, 15–18, 5–8, 2–4, 1
15	$-CH_2-$	33.0	$-CH_2-, >CH-, >C<$, or $>C=$	12–14, 16–18, 5–8, 2–4, 1
16	$-CH_2-$	27.6	$-CH_2-, >CH-, >C<$	12–15, 17, 18, 5–8, 2–4
17	$-CH_2-$	20.1	$-CH_2-, >CH-, >C<$	12–16, 18, 5–8, 2–4
18	$-CH_2-$	18.3	$-CH_2-, >CH-, >C<$	12–17, 5–8, 2–4
19	$-CH_3$	22.8	$>C<$	2–4
20	$-CH_3$	18.3	$>C<$	2–4
21	$-OH$		$>CH-$	5
22	$-OH$		$-CH_2-$	10

[a]These constraints are based on features found in the substructures matching the resonance lines to within the required tolerances and derived by attempting to match substructures onto the presumed kaurenoid skeleton.

bonding of the atoms associated with each resonance; the restrictions are summarized in Table VI-6. These restrictions can then be used in the next iterative step of checking for matching substructures in the reference file. The use of data defining a known skeleton, or simply limited data defining some structural fragments and functional groups, can help the ^{13}C analysis to converge more quickly.

VI.D. COMMENTARY

VI.D.1. Problems

The primary objective of computer-aided structural analysis is the *guaranteed generation of all possible structural candidates for some unknown*. Therefore, strictly, the only structural constraints that should be applied during the generation procedure are those established, beyond reasonable doubt, from available spectral and chemical data. Acceptable constraints may define specific constituent substructures or may take the form of a set of alternative interpretations *provided the correct substructural interpretation is included among the members of this set of alternatives*. Unfortunately, there are limits to the extent to which inferences can be made with certainty by any current general purpose spectrum interpretation procedure.

VI.D.1.a. The Use of Substructural Plausibility Scores

There are several examples of interpretation programs that can identify sets of alternative substructures each associated with some plausibility estimate. Thus **PAIRS** can suggest substructures and "probabilities" derived from IR data, simple mass spectral interpretation schemes can relate substructures and scores to specified ion patterns, and **STIRS** can identify substructures through its elaborate "match-factor" analysis of spectra. However, the use of such (substructure$_1$, score$_1$) · · · (substructure$_n$, score$_n$) information in constraining a structure generator remains problematical.

An unknown structure may incorporate some relatively unusual substructural feature that gives rise to an observed spectral pattern normally associated with one of some small set of more common substructures. The less common interpretation of the spectral data may not be included among the interpretations available to an automated spectral analysis system. The constraints, as proposed by such a system, would then limit the generator to creating structures incorporating one or other of the more common substructures.

Of course, the correct structure of an unknown cannot be derived if a structure generator is incorrectly constrained. Usually, as with the experimental mass spectrum system discussed above, one must allow for the case "or none these" when enumerating the possible substructural interpretations of some data. Unfortunately, this largely defeats the objective of the constraint determining process. The generator will inevitably create candidate structures from the least constrained alternative that duplicate those incorporating precisely defined, elaborate substructures; these duplicates must then be identified and discarded. The work performed by a candidate structure generator is not in any way reduced by "constraints" that must include an "or none these" alternative.

Even where a set of alternative substructural interpretations may be assumed complete, their plausibility estimates may have little practical utility. The most obvious application of the plausibilities might be as a filter in structure generation. A partially assembled structure can accrue an overall plausibility based on some combination of the plausibilities accorded to its constituent substructures. These overall plausibilities can be used to rank the partially assembled candidates. Candidates with the poorest overall plausibilities, presumably incorporating several disfavored substructures, can be discarded.

Combination of the plausibilities of constituent substructures to yield overall plausibility scores may be appropriate when these various substructure plausibilities can be presumed to be commensurate (having all been based on some similar analysis). With most multisource spectral interpretation systems, the substructures and their plausibilities are likely to be derived through very different procedures. A particular partial structure could, for example, incorporate one substructure accorded a high probability by **PAIRS** with some other substructure assigned a poor plausibility score by some mass spectrum interpreter. These conflicting plausibility data would have to be evaluated and some final overall score derived before an appropriate ranking of the candidate could be determined. It is difficult to conceive of a general system for resolving conflicting structural inferences as evidenced by such differing spectrum interpretation procedures.

Even in cases where partial structures can be ranked according to some plausibility measure, it is usually more appropriate to generate *all* possible structures. Then there is no chance of the correct solution being eliminated by some arbitrary cutoff value applied to crudely estimated plausibilities. The overall plausibilities of complete candidates can, of course, be used to help guide further analysis (see Chapter XI).

VI.D.1.b. Problems with Incomplete Sets of Alternative Interpretations

Missing alternative interpretations can, as noted above, cause problems even in schemes working with sets of alternative substructures and associated plausibility scores. Problems resulting from incomplete sets of substructures are of greater consequence in systems that entail elaborate searches for consistency.

The difficulties that can then result can be illustrated through the analysis of another kaurenoid compound, **VI-40**, also derived from *Stachys lanata*.[52] Compound **VI-40** occurs in the plant as a hydroxy acid and is derivatized to a methyl ester during the extraction process. It happens that the environments of

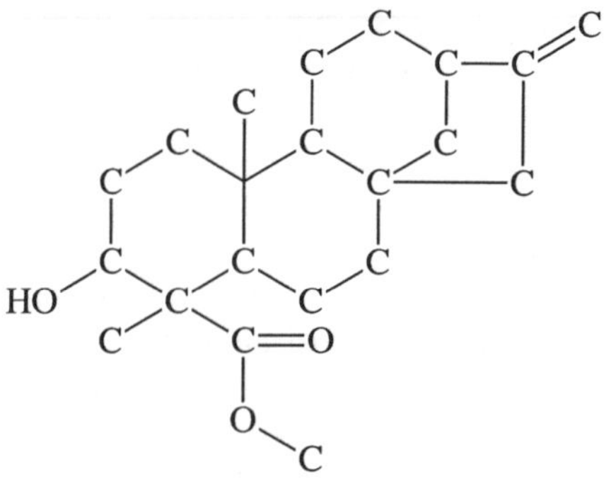

VI-40

atoms C(4) and C(5), represented as substructures **VI-41** and **VI-42**, respectively, are not incorporated in any of the subspectrum–substructure rules available to the ^{13}C NMR interpretation program.

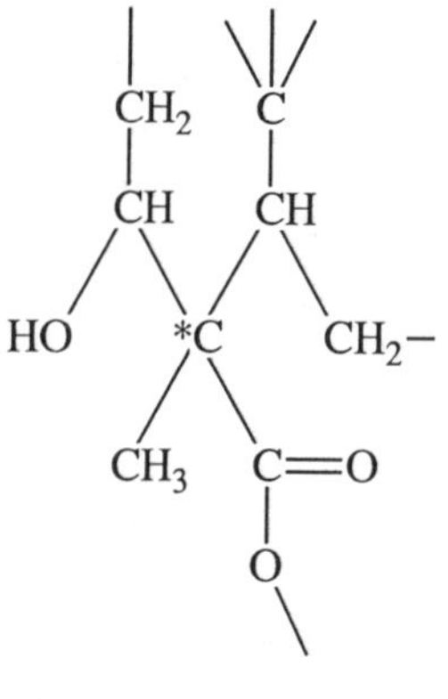

VI-41 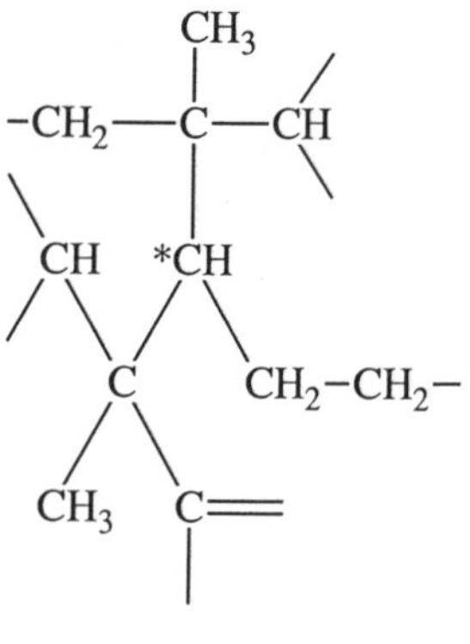 **VI-42**

Since the correct substructural interpretations of some resonances are not available to the program, attempts to interpret the spectrum in a manner similar to that used for the kaurenoid diol must inevitably fail. Since a mutually consistent set of interpretations for the resonance lines is not available to the program, the iterative cross-checking procedures will eventually eliminate all possible interpretations for one or more resonance lines.

In a real analytic application, one would not, of course, know *a priori* which atoms of the unknown structure were in novel substructural environments. The resonance line at which the program first detects an inconsistency among the interpretations is not necessarily that for which the possible alternative interpretations are incomplete. For example, in the case of **VI-40** the inadequacy of the rule set is first apparent in the analysis of the resonance line due to atom C(8). The substructure (**VI-43**) identified as a possible interpretation of this resonance is in fact correct; however, incorrect substructures inferred for other resonances have implied constraints on atoms that prevent **VI-43** from being constructed.

C(8) correct

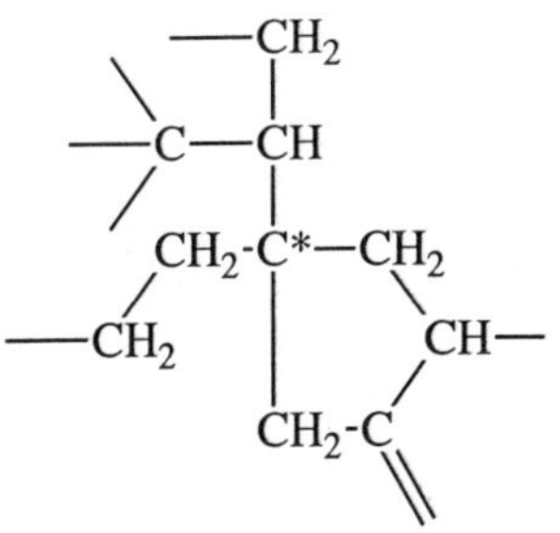

VI-43

The substructural environment derived for the resonance due to C(4) (**VI-44**) is substantially correct, but it prohibits a bond between the hydroxymethine group, C(3), and the quaternary C(4) carbon. Since the substructural interpretation **VI-45** derived for the doublet due to C(3) implies a bond to a quaternary carbon atom, this bond is forced to C(10). Consequently, the interpretations for the C(10) resonance are restricted to those incorporating the $\rangle$C–CH(OH)– substructure and eventually yield substructure **VI-46**; this results in the propagation of further, erroneous restrictions on bonding of the atoms associated with different resonances. Eventually, restrictions derived for $-CH_2-$s make construction of **VI-43** impossible.

C(4) (wrong)

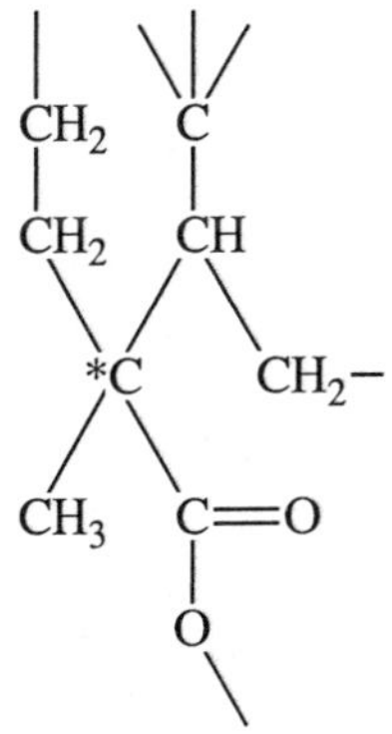

VI-44

C(3) (wrong)

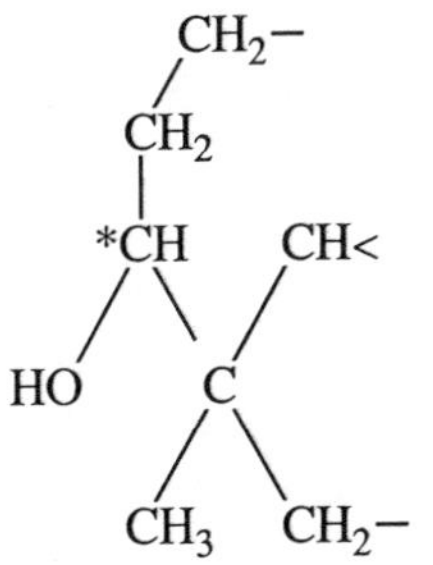

VI-45

C(10) (wrong)

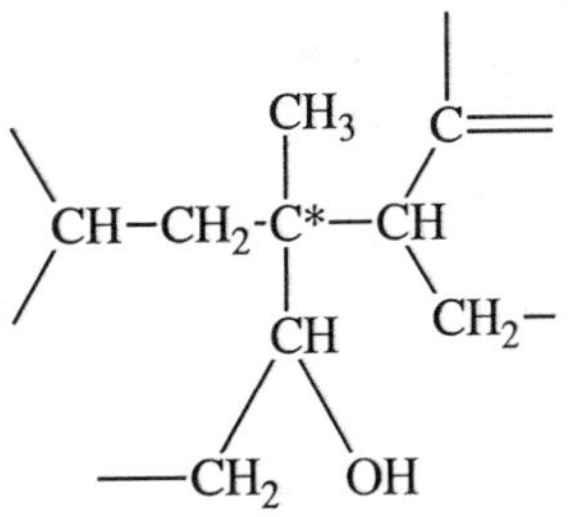

VI-46

If the consistency search does ultimately fail, at least some of the bonding constraints identified for various atoms will be wrong. The program has no means for identifying which derived constraints are correct and which are erroneous. Consequently, it is inappropriate to use any of the inferred bonding constraints in the structure generation process.

VI.D.2. Possible Lines for Development and Improvement

A number of possible developments could lead to improved performance of automated interpretation systems. New spectroscopic techniques offer the prospect of unambiguous data. Such developments in experimental techniques are the most beneficial, but other improvements are also possible. Better use of the interplay of information derived from different spectral (and chemical) evidence could enhance the performance of systems for analyzing data. Alternative approaches to the interpretation task can reduce, although not entirely eliminate, problems engendered by incomplete sets of interpretation rules.

VI.D.2.a. New Spectroscopies

Two new spectral techniques appear to merit particular reference. One of these allows specific bonds to be mapped out from $^{13}C-^{13}C$ couplings; the other allows identification of the structure of individual fragment ions.

Tracing the Carbon Skeleton of an Organic Molecule. The analysis of $^{13}C-^{13}C$ couplings provides a direct method of mapping out the complete structure of an unknown.[53] The experiment involves the use of an appropriate pulse sequence to create double quantum coherence in molecules with two (natural-abundance) ^{13}C nuclei joined by a chemical bond; signals from molecules with only one ^{13}C nucleus are suppressed.[54]

There are a number of variations on the basic experiment; the most useful for direct structure elucidation is a two-dimensional NMR experiment. In this two-dimensional experiment, one dimension spans the normal chemical shift spectrum, and the other measures the double quantum coherence frequency. The results from the experiment can be plotted as a *carbon-carbon connectivity plot*. Signals from each carbon appear as (sets of) doublets, centered about the appropriate chemical shift value; the doublet signals of pairs of bonded carbons are at the same double quantum coherence frequency. From the carbon-carbon connectivity plot, bonded carbon atoms can be unequivocally recognized.

Applications of the methods have been recently reported with both demonstrations on known compounds[55] and actual structure elucidations of unknowns.[56–58] The experiment does require relatively large samples, of several hundred milligrams, and long acquisition times. Some signals, particularly those from carbons with long T_1 times, may still remain obscured by noise. Consequently, it is likely to be rare, although by no means impossible, for the complete structure of an unknown to be determined solely from the one NMR experiment.

Often, however, only limited data may be available defining the connectivities of some subset of the carbon atoms of an organic molecule. However, the definitiveness of these data is such as to provide very powerful constraints for an automated combined spectral-interpretation–structure-elucidation system. The results presented by Lindley et al. provide illustration of this effect.[55] In this study, substructures inferred from the ^{13}C–^{13}C couplings were combined with those derived through conventional analyses of chemical and spectral data. The combined sets of substructures were used as inputs to the **GENOA** program.

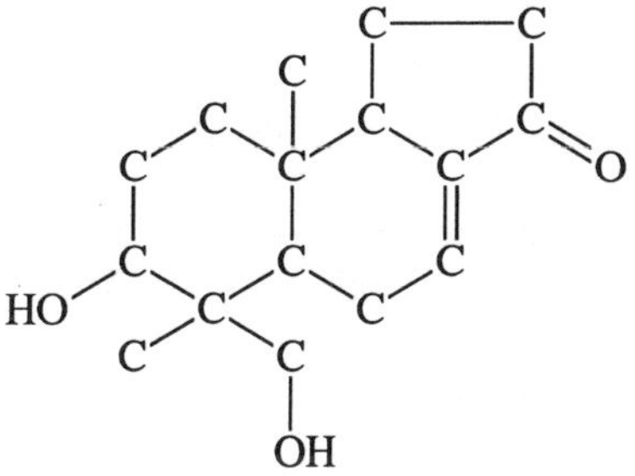

VI-47

In the example given of *iresin* (**VI-47**),[55] the structural inferences derived solely from examination of the ^{13}C and ^{1}H NMR shifts, multiplicities, and interpretable ^{1}H NMR couplings were insufficient to constrain the structural problem. More than 8000 isomers were apparently compatible with the given constraints. A single additional substructural constraint derived from two ^{13}C–^{13}C couplings, specifying the presence of the fragment $-CH_2-CH_2-CH(OH)-$, limited the problem to less than 2000 candidate structures. By analyzing less than 10 ^{13}C–^{13}C couplings, and deriving appropriate constraints for **GENOA**, it was possible to uniquely identify the structure.[55]

In Lindley's second example, the structure of *genipin*, the analysis of the complete results of a two-dimensional NMR experiment established the presence of a substituted cyclopentene system. The combination of this substructure with a methyl ester group, identified from other data, essentially defined the structure of genipin, and only three candidates were produced by **GENOA**. It is worthwhile to reexamine this example, making more limited use of ^{13}C–^{13}C coupling data, and using a slightly different approach to the assembly of identified constituent fragments.

For the purposes of example, the "unknown" compound genipin has the composition $C_{11}H_{14}O_5$. IR, ^{1}H NMR, and ordinary ^{13}C shift and multiplicity data establish the presence of an α,β-unsaturated methyl ester group with a single vinylic proton beta to the carbonyl, shown as substructure **VI-48**. The shifts of the vinylic carbons indicate a bond from the $-CH=$ to an oxygen.

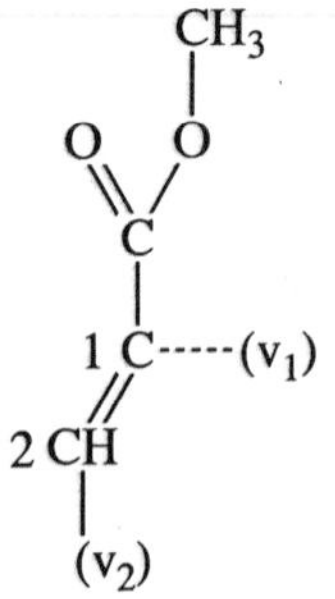

```
          CH3          (δ(C) = 51.1 (q) )
          |
      O   O
       \\ /
        C              (δ(C) = 168.2 (s) )
        |
      1 C-----(v1)     (δ(C) = 110.6 (s) )
       //
    2 CH               (δ(C) = 152.7 (d) )
      |
     (v2)
```

VI-48

There are three methine resonances. One, at δ (C) 96.3, indicates the presence of a methine carbon bonded to two oxygens. There are also two methylene resonances; again, for one, the shift δ(C) 61.0 ppm indicates a bond to an oxygen. There is also a second >C=CH– group. Composition constraints and limits on the number of Csp^2 carbons, establish that there are two –OH groups and an ether oxygen –O–. The structure elucidation problem is thus reduced to establishing connections between these fragments.

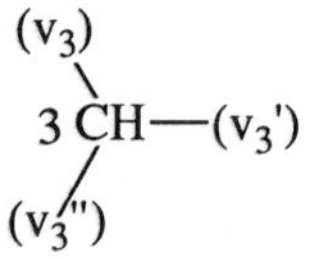

```
   (v3)
     \
    3 CH—(v3')        (δ(C) = 96.3 (d))
     /
  (v3'')
```

VI-49

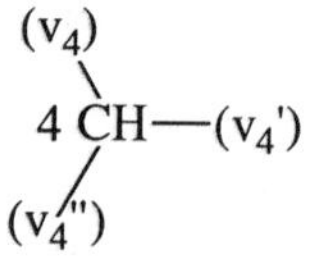

```
   (v4)
     \
    4 CH—(v4')        (δ(C) = 47.8 (d))
     /
  (v4'')
```

VI-50

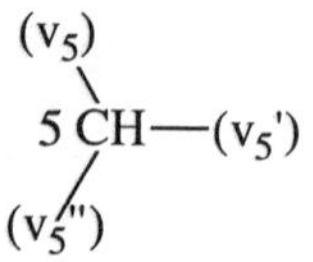

```
   (v5)
     \
    5 CH—(v5')        (δ(C) = 36.4 (d))
     /
  (v5'')
```

VI-51

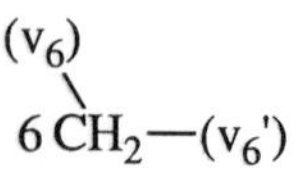

```
   (v6)
     \
   6 CH2—(v6')        (δ(C) =38.9 (t))
```

VI-52

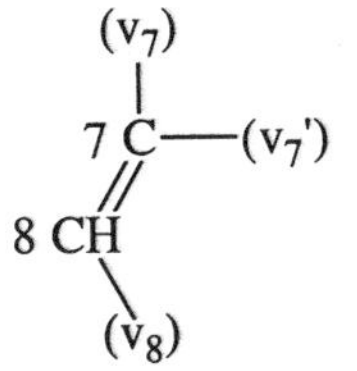

$$\begin{array}{l} \quad (v_7) \\ \quad\ | \\ 7\,C\!-\!\!-(v_7') \\ \ \ // \\ 8\,CH \\ \quad \backslash \\ \quad (v_8) \end{array}$$

$(\delta(C) = 142.6\ (s))$

$(\delta(C) = 129.6\ (d))$

VI-53

$$(v_9)\!-\!CH_2\!-\!(v_9) \quad (\delta(C) = 61.0\ (t))$$

9

VI-54

A practical representation of a structural problem is as a connectivity matrix of known components. It is convenient to simplify this representation by excluding atoms whose bonding is completely specified, such as the atoms comprising the methyl ester moiety in genipin. A connection matrix of atoms, rather than valences, is sufficient if stereochemistry (of the double bonds and other stereocenters) can be neglected at this stage of the analysis. In this example, therefore, a 12 × 12 array will suffice for the connection matrix of the nine incompletely specified carbons and three remaining oxygens. (The array is necessarily symmetrical; only the lower half need be represented in a computer program.) The entries in this array designate whether bonds are known (i.e., "*"—bond order is not indicated in these examples), whether bonds are simply possible, that is neither definitely established nor eliminated ("?"), or where bonds are known to be impossible (".").

The first connectivity matrix (Figure VI-12) incorporates the known structural information that the C(2), the –CH= carbon at δ = 152.7 ppm, must bond to an oxygen. All possible bonds from this atom to other carbons, apart from C(1), have been excluded. Bonds between the remaining –O– atom and –OH groups have been excluded on the premise that perols were implausible. Shift information has been used to eliminate the possibility of some bonds such as those between the oxygroups and the methines at around 30–45 ppm.

Not all structural information is represented in such a connection matrix format. Thus the known requirement that the C(3) methine, δ(C) 96.3, be bonded to two oxygens must be represented in some separate auxiliary data table. However, the connection matrix provides a convenient basic representation for mapping the bonding constraints and exploring their ramifications.

Additional structural constraints can eliminate possible bonds and, in consequence, define other possible bonds as being present. It might, for example, be appropriate to prohibit acetals (hydrated aldehydes) on the grounds of chemical instability and to exclude enols because the ^{1}H NMR spectrum shows no enolic resonance signal. Such prohibitions eliminate the possibility of a bond between C(2) and –OH, thus forcing a bond from C(2) to the remaining –O– oxygen. C(3), the >CH– at δ(C) 96.3 ppm, must also be

	C(1)	C(2)	C(3)	C(4)	C(5)	C(6)	C(7)	C(8)	C(9)	-OH	-OH	-O-
C(1)	.	*	?	?	?	?	?	?	?	.	.	.
C(2)	*	.	.	.	.	.	.	.	.	?	?	?
C(3)	?	.	.	?	?	?	?	?	?	?	?	?
C(4)	?	.	?	.	?	?	?	?	?	.	.	.
C(5)	?	.	?	?	.	?	?	?	?	.	.	.
C(6)	?	.	?	?	?	.	?	?	?	.	.	.
C(7)	?	.	?	?	?	?	.	*	?	.	.	.
C(8)	?	.	?	?	?	?	*	.	?	.	.	.
C(9)	?	.	?	?	?	?	?	?	.	?	?	?
-OH	.	?	?	.	.	.	.	.	?	.	.	.
-OH	.	?	?	.	.	.	.	.	?	.	.	.
-O-	.	?	?	.	.	.	.	.	?	.	.	.

FIGURE VI-12. Initial structural information for *Genipin* represented in terms of a connectivity matrix of atomic fragments. Asterisks indicate definite bonds, question marks designate possible bonds whose presence or absence must be determined to resolve the structural problem, and periods indicate where bonds are impossible.

	C(1)	C(2)	C(3)	C(4)	C(5)	C(6)	C(7)	C(8)	C(9)	-OH	-OH	-O-
C(1)	.	*	?	?	?	?	?	?	?	.	.	.
C(2)	8	.	.	.	.	.	..		.	.	.	*
C(3)	?	.	.	?	?	?	?	?	?	.	*	*
C(4)	?	.	?	.	?	?	?	?	?	.	.	.
C(5)	?	.		?	?	.	?	?	?	.	.	.
C(6)	?	.	?	?	?	.	?	?	?	.	.	.
C(7)	?	.	/	?	?	?	.	*	?	.	.	.
C(8)	?	.	?	?	?	?	*	.	?	.	.	.
C(9)	?	.	?	?	?	?	?	?	.	*	.	.
-OH	.	.	.	.	.	.	.	.	*	.	.	.
-OH	.	.	*	.	.	.	.	.	.	.	.	.
-O-	.	*	*	.	.	.	.	.	.	.	.	.

FIGURE VI-13. Connectivity matrix of atomic fragments following exclusion of acetals and enols.

bonded to this same oxygen and to one of the –OH groups, if its requirements for two oxysubstituents are to be satisfied without engendering any acetal structures.

The –CH_2–, at δ(C) 61.0 ppm, must thus be bonded to the other –OH group (see Figure VI-13). The possible bond between C(1) and C(3) could probably be eliminated at this point (unless one really desired to allow for the possibility of cyclic enolic ethers with a ring size of 4). This C(1)–C(3) bond can, in fact, be definitely excluded, on the basis of ^{1}H NMR data, as is discussed below (see Figure VI-14).

In the case of genipin, it is fairly easy to establish a correspondence between signals in the ^{1}H NMR spectrum and the signals in the ^{13}C spectrum. The proton of the –CH(OH)–O– methine resonates in the ^{1}H NMR spectrum as a doublet, $J = 8$ Hz, at $\delta_H = 4.8$ ppm. The coupling to another proton shown by this signal can be used to exclude the possibility of a bond from the methine carbon to either of the quaternary vinylic carbons. Similarly, the –CH_2–OH protons can be identified with the resonance at $\delta_H = 4.29$ (s, br). The ^{1}H NMR shift and lack of further coupling, other than the slight broadening due to coupling with an –OH proton, both point to a bond from the –CH_2–OH to a quaternary vinylic carbon (See Figure VI-14).

	C(1)	C(2)	C(3)	C(4)	C(5)	C(6)	C(7)	C(8)	C(9)	-OH	-OH	-O-
C(1)	.	*	.	?	?	?	?	?	?	.	.	.
C(2)	*	.	.	.	.	.	.	.	.	.	.	*
C(3)	.	.	.	?	?	?	.	?	.	.	*	*
C(4)	?	.	?	.	?	?	?	?	.	.	.	.
C(5)	?	.	?	?	.	?	?	?	.	.	.	.
C(6)	?	.	?	?	?	.	?	?	.	.	.	.
C(7)	?	.	.	?	?	?	.	*	?	.	.	.
C(8)	?	.	?	?	?	?	*	.	.	.	.	.
C(9)	?	.	.	.	.	.	?	.	.	*	.	.
-OH	.	.	.	.	.	.	.	.	*	.	.	.
-OH	.	.	*	.	.	.	.	.	.	.	.	.
-O-	.	*	*	.	.	.	.	.	.	.	.	.

FIGURE VI-14. Connectivity matrix of atomic fragments following use of ^{1}HNMR coupling data.

	C(1)	C(2)	C(3)	C(4)	C(5)	C(6)	C(7)	C(8)	C(9)	-OH	-OH	-O-
C(1)	.	*	.	?	?	.	?	.	?	.	.	.
C(2)	*	.	.	.	.	.	.	.	.	.	.	*
C(3)	.	.	.	?	?	.	.	.	.	.	*	*
C(4)	?	.	?	.	?	.	?	.	.	.	.	.
C(5)	?	.	?	?	.	*	?	.	.	.	.	.
C(6)	.	.	.	.	*	.	.	*	.	.	.	.
C(7)	?	.	.	?	?	.	.	*	?	.	.	.
C(8)	.	.	.	.	.	*	*	.	.	.	.	.
C(9)	?	.	.	.	.	.	?	.	.	*	.	.
-OH	.	.	.	.	.	.	.	.	*	.	.	.
-OH	.	.	*	.	.	.	.	.	.	.	.	.
-O-	.	*	*	.	.	.	.	.	.	.	.	.

FIGURE VI-15. Connectivity matrix of atomic fragments following mapping in of connections identified from ^{13}C–^{13}C coupling data for C(6).

The structure can be completely resolved by the identification of as few as three C–C bonds from observed ^{13}C–^{13}C couplings. Couplings of the $-CH_2-$ methylene at δ(C) = 38.9, [C(6) (Figure VI–15)] should be detected fairly readily. These couplings in fact show bonds from this methylene to the C(5) methine at δ(C) 36.4 and the C(8) vinylic –CH= at δ(C) 129.6 ppm.

The structure could then be determined from the further bonding of the C(5) methine. If the coupling between this atom and C(1) could be detected, the establishment of a C(1)–C(5) bond forces a C(9)–C(7) bond. The bonding of C(1) to C(5) satisfies the valence requirements of C(1) and thus eliminates the possibility of a C(1)–C(4) bond. Atom C(4) is then left with only three possible

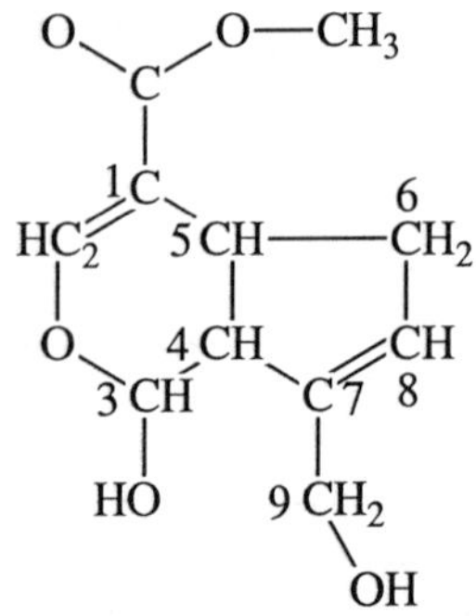

VI-55

bonds; its valence of 3 can be satisfied only if each of these is indeed present. Thus, the structure of genipin (**VI-55**) could be directly derived by using only limited $^{13}C-^{13}C$ coupling data.

Structural Information from Ion Fragmentations. Although a very powerful structure elucidation technique, the carbon–carbon connectivity plot two-dimensional NMR experiment does have the practical disadvantage of requiring hundreds of milligrams of sample and long acquisition times. In contrast, mass spectra can be rapidly acquired from much smaller samples. The problem with mass spectra is, of course, the difficulties associated with their interpretation.

Generally, one attempts to treat mass spectra as a molecular fingerprints and "interprets" spectra by matching them in a file. If the compound was analyzed previously, and its spectrum is in the file, identification is unambiguous. It would be desirable to obtain some structural information even when the compound being analyzed is not present in the file. The unknown is likely to share substructural features with many of the compounds actually present. Using methods similar to those outlined earlier or through McLafferty's **STIRS** system, therefore, one may attempt to identify constituent substructures of an unknown by some analysis of its mass spectrum.

The problem with these methods is that they are not very reliable. There are, however, data suggesting that it may be possible to unambiguously identify substructures in a molecule by some more detailed analysis of mass spectra. A high-resolution mass spectrum of an unknown might exhibit an ion at, say, C_9H_{16}. Although this information has some structural implications, these are not such as to be easily exploited in any structure generator. If this fragment ion from the unknown spectrum could be shown to be identical to the fragment **VI-56** (as previously identified in the mass spectra of some reference compound), however, the structural information is much more explicit and easily exploited.

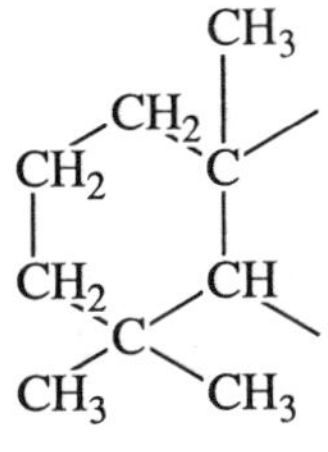

VI-56

This approach to structural analysis requires that one be able to select ions from the spectrum of an unknown and determine their structures. Such an analysis yields a number of (overlapping) substructures taken from the unknown molecule. If the unknown molecule (or its fragment ions as used in the analysis), had undergone rearrangement, however, it would be possible for some of the substructures so identified to be misleading. This approach to structure elucidation depends on molecules fragmenting into distinctive ions without any intervening rearrangement processes.

The structures of the unknown's fragment ions are determined by mass spectral techniques. There are many variations on the basic experiment. All entail an initial fragmentation of the unknown molecule; the fragment ions produced in this step are subject to further study. Specific ions are selected by m/z value by the use of the focusing controls of the first stage mass spectrometer. Typically, the selected ions pass into a reaction chamber where further fragmentation is induced by collision with some inert gas. The processes inducing further fragmentation are relatively low energy and do not cause rearrangement of the original fragment ion.

The fragments resulting from this collision-induced process are analyzed in a further conventional mass spectral scanning step. Just as an ordinary mass spectrum may serve as a fingerprint for a molecule, so may these collision-induced fragmentation spectra serve as fingerprints for individual ions. A library of a few thousand mass spectra of fragment ions could provide much more useful structural information than could a collection of tens of thousands of standard mass spectra of known molecules.

This approach to structure elucidation has been proposed several times.[59,60] There are a few example applications where, for example, the equivalence of MIKES spectra has been used as evidence for a particular skeletal structure for an unknown. The method has not yet received any widespread application. Possibly, this lack of application has been due to the limited availability of appropriate instrumentation and the greater importance accorded to using available instruments for studies of ion–molecule reactions in the gas phase. Suitable triple–quadrupole mass spectrometers and other alternative instruments are now more readily available commercially. Possibly, this will allow for more systematic studies of both the advantages and the limitations of this approach to structure elucidation. Fourier transform mass spectrometers may also be capable of providing equivalent structural information.[61,62]

VI.D.2.b. Improvements to Interpretation Programs

Some improvements to current interpretation programs might be realized through adaptations and extensions of the approaches used in the **DENDRAL** ^{13}C interpreter and underlying the simple example analysis of ^{13}C–^{13}C couplings. These approaches view spectral interpretation as the process of developing restrictions on the bonding of a defined set of fragments.

These fragments are typically atoms augmented by details of hybridization and number of substituent hydrogens. Initially, subject to any obvious valence constraints, bonds can exist between any pair of fragments. Each possible bond can be considered in turn, and any spectral or chemical implications that it may entail can be evaluated against the experimental data.

The establishment or elimination of a few bonds would provide constraints for more elaborate search procedures. All remaining possible two-bond, then three-bond substructural environments could be generated for each constituent carbon atom and used in spectrum prediction and verification procedures. If the "rule base" lacked appropriate substructural models for prediction (as was the case with substructures **VI-44** and **VI-45** considered earlier), verification procedures could simply indicate that the proposed bonding pattern was possible. When detailed predictions could be made, mismatches of predicted and observed spectral patterns would identify unacceptable bonding arrangements.

Similar analyses can apply to many types of spectral data. A detailed data base of bond environments–IR absorptions could be accessed in a manner similar to a data base of atom environments–^{13}C NMR shifts. Predictions of ^{1}H NMR shifts and couplings could also be made and checked. Data on large fragments, such as might be inferred from MS–MS analysis, could be handled through a generalization of the methods used in the **DENDRAL** ^{13}C interpreter to match substructures associated with resonances onto predefined skeletal fragments. The overall spectral analysis cycle would be iterative. Each cycle would commence with generation of remaining possible atom and bond environments. These would then be used in the spectral prediction–evaluation process where unsatisfactory bonding combinations would be saved. These would then be analyzed to determine whether any possible bonds could be definitely eliminated or any specific bonds could be formed. The analysis process would terminate if no such bonding changes could be effected.

In addition to spectral data, it should be possible to exploit various chemical constraints in such an interpretation system. Chemical data are relatively little used in existing automated structure elucidation systems. Munk's **CASE** system does permit certain structural constraints to be expressed in terms of results of standard analytic tests.[63] For instance, in **CASE** one may prescribe a constraint in terms of the number of moles of periodate consumed in a reaction, rather than specify the constraint directly in terms of bonds between carbons with polar substituents [e.g., C(OH)–C(OH), C(OH)–C(NH), and C(OH)–C=O]. This approach may be convenient for chemists, in that the expression of structural constraints in terms of experimental results may be more natural than an equivalent specification in terms of the number of occurrences of some defined substructure. However, such provision for synonymous constraint declarations adds nothing to the overall effectiveness of the interpretive system.

One way in which chemical data could be exploited would be to determine substructures contraindicated by a compound's origin or chemical history. Often, structural constraints that could have been inferred from a structure's chemical history are left unspecified. Only when the chemist sees partial assembled candidates incorporating, say, $-CO_2H$ groups is it remembered that these can be excluded on the grounds that the extraction procedures entailed the derivitization of acids to yield methyl esters.

Procedures checking for consistency with chemical constraints would require data from the chemist concerning the origin of a compound and the extraction and derivitization procedures employed. Tables similar to those used by Corey et al.,[64] could define the sensitivity of various functional groups to the conditions of extraction and workup. Then, whenever bonds were hypothesized by other spectral interpretation procedures, the resulting partially assembled structures could be checked for any resulting violations of chemical constraints.

Sometimes, as in the case of genipin, the application of such spectral and chemical interpretation procedures may yield a unique final structure. More typically, one would expect to identify several overlapping structural fragments. Then candidate structure generation must be completed through the use of one of the more sophisticated structure generators discussed in subsequent chapters.

VI.E. CONCLUSIONS

Existing knowledge-based systems for spectral interpretation are not satisfactory:

- Although there exist a few fully automated systems that attempt to derive all the structural constraints that are to be used in a subsequent candidate structure generation process, it is yet to be demonstrated whether any of these systems is viable in practical application over a wide range of real structural problems.
- It appears inappropriate to employ any form of plausibility ratings to constrain candidate generators. A subsequent evaluation step can better employ those data yielding interpretations defined only in terms of relative plausibilities for various structural features.
- There are systems that can achieve quite exacting and detailed interpretations of specific types of spectral data. However, existing systems are completely dependent on the comprehensiveness of the spectral feature–substructure rule sets that they use. Through the expense of considerable effort at expanding the rule set to include relevant reference data, it is possible to tailor such systems to specific applications.[64] Routine application to arbitrary structural problems will remain impractical until the rule sets are greatly extended.
- Existing systems are limited in their abilities to combine inferences obtained from different data or to exploit known structural constraints to simplify the interpretation task.

Most aspects of knowledge-based approaches, therefore, require further research and development before substantial practical application can be expected. New spectral analysis techniques, such as the analysis of MS–MS data or the exploitation of coupling patterns, may approve amenable to these approaches. The interaction of different spectral analysis procedures should allow for more detailed and exacting interpretations. Ultimately, some of these developments may not prove viable, but it is assumed that much more is achievable than is currently obtained through existing programs.

REFERENCES

1. G. Szalontai, Z. Simon, Z. Csapo, M. Farkas, and Gy. Pfeifer, "Use of IR and 13-C NMR Data in the Retrieval of Functional Groups for Computer Aided Structure Determination," *Anal. Chim. Acta*, **133** (1981), 31.
2. B. Pettersson and R. Ryhage, "Mass Spectral Data Processing. Computer Used for Identification of Organic Compounds," *Arkiv fur Kemi*, **26** (1967), 293.
3. B. Pettersson and R. Ryhage, "Mass Spectral Data Processing. Identification of Aliphatic Hydrocarbons," *Anal. Chem.*, **39** (1967), 790.
4. L. R. Crawford and J. D. Morrison, "Computer Methods in Analytical Mass Spectrometry. Empirical Identification of Molecular Class," *Anal. Chem.*, **40** (1968), 1469.
5. L. R. Crawford and J. D. Morrison, "Computer Methods in Analytical Mass Spectrometry. Development of Programs for Analysis of Low Resolution Mass Spectra," *Anal. Chem.*, **43** (1971), 1790.
6. J. F. O'Brien and J. D. Morrison, "Computing Methods in Mass Spectrometry. Programming for Aliphatic Amines and Alcohols," *Aust. J. Chem.*, **26** (1973), 785.
7. D. H. Smith, "A Compound Classifier Based on Computer Analysis of Low Resolution Mass Spectral Data," *Anal. Chem.*, **44** (1972), 536.
8. J. H. van der Maas and T. Visser, "A Systematic Approach to the Structure Elucidation of Carbon-Hydrogen and Hydroxy Compounds by means of Raman Spectroscopy with Laser Sources," *J. Raman Spectrosc.*, **2** (1974), 563.
9. N. A. B. Gray, "Structural Interpretation of Spectra," *Anal. Chem.*, **47** (1975), 2426.
10. H. B. Woodruff and M. E. Munk, "A Computerized Infrared Spectral Interpreter as a Tool in Structure Elucidation of Natural Products," *J. Org. Chem.*, **42** (1977), 228.
11. T. Visser and J. H. Van der Maas, "Systematic Computer Aided Interpretation of Vibrational Spectra," *Anal. Chim. Acta*, **133** (1981), 451.
12. H. B. Woodruff and M. E. Munk, "Computer Assisted Interpretation of Infrared Spectra," *Anal. Chim. Acta*, **95** (1977), 13.
13. *The Handbook of Artificial Intelligence*, Vol. 1, A. Barr and E. A. Feigenbaum, eds., W. Kaufmann, Los Altos, CA, 1981.
14. P. H. Winston and B. K. P. Horn, *LISP*, Addison-Wesley, Reading, MA, 1981.
15. N. A. B. Gray and T. O. Gronneberg, "Program for Spectrum Classification and Screening of GC/MS Data on a Laboratory Computer," *Anal. Chem.*, **47** (1975), 419.
16. W. T. Wipke, G. I. Ouchi, and S. Krishnan, "Simulation and Evaluation of Chemical Synthesis—SECS: an Application of Artificial Intelligence Techniques," *J. Artif. Intell.*, **11** (1978), 173.
17. H. B. Woodruff and G. M. Smith, "Computer Program for the Analysis of Infrared Spectra," *Anal. Chem.*, **52** (1980), 2321.
18. H. B. Woodruff and G. M. Smith, "Generating Rules for PAIRS—a Computerized Infrared Spectral Interpreter," *Anal. Chim. Acta*, **133** (1981), 545.
19. E. H. Shortliffe, *Computer-based Medical Consultations: MYCIN*, American Elsevier, 1976.

20. J. R. Quinlan, *Consistency and Plausible Inference*, Technical report P-6831, Rand Corporation, Santa Monica, CA 90406, 1982.
21. M. R. Lindley, N. A. B. Gray, D. H. Smith, and C. Djerassi, "A Computerized Approach to the Verification of C-13 NMR Spectral Assignments," *J. Org. Chem.*, **47** (1982), 1027.
22. S. Sasaki, Y. Kudo, S. Ochiai, and H. Abe, "Automated Chemical Structure Analysis of Organic Compounds. An Attempt to Structure Determination by the use of NMR," *Mikrochim. Acta*, **1971**, 726.
23. S. Sasaki, H. Abe, Y. Hirota, Y. Ishida, Y. Kudo, S. Ochiai, K. Saito, and T. Yamasaki, "CHEMICS-F: A Computer Program System for Structure Elucidation of Organic Compounds," *J. Chem. Inf. Comput. Sci.*, **18** (1978), 211.
24. S. Sasaki, I. Fujiwara, H. Abe, and T. Yamasaki, "A Computer Program System - New CHEMICS - for Structure Elucidation of Organic Compounds by Spectral and Other Structural Information," *Anal. Chim. Acta*, **122** (1980), 87.
25. T. Oshima, Y. Ishida, K. Saito, and S. Sasaki, "CHEMICS-UBE, a Modified System of CHEMICS," *Anal. Chim. Acta*, **122** (1980), 95.
26. L. A. Gribov, M. E. Elyashberg, and V. V. Serov, "Computer System for Structure Recognition of Polyatomic Molecules by IR, NMR, UV and MS Methods," *Anal. Chim. Acta*, **95** (1977), 75.
27. M. Farkas, J. Markos, P. Szepesvary, I. Bartha, G. Szalontai, and Z. Simon, "A Computer-Aided System for Organic Functional Group Determinations," *Anal. Chim. Acta*, **133** (1981), 19.
28. B. Debska, J. Duliban, B. Guzowska-Swider, and Z. Hippe, "Computer-Aided Structural Analysis of Organic Compounds by an Artificial Intelligence System," *Anal. Chim. Acta*, **133** (1981), 303.
29. T. Yamasaki, H. Abe, Y. Kudo, and S. Sasaki, "CHEMICS: A Computer Program System for Structure Elucidation of Organic Compounds," in *Computer-Assisted Structure Elucidation*, D. H. Smith, ed., American Chemical Society, Washington, DC, 1977, 108, Chapter. 8.
30. W. Bremser, M. Klier, and E. Meyer, "Mutual Assignment of Subspectra and Substructures - A Way to Structure Elucidation by C-13 NMR Spectroscopy," *Org. Magn. Reson.*, **7** (1975), 97.
31. A. M. Duffield, A. V. Robertson, C. Djerassi, B. G. Buchanan, G. L. Sutherland E. A. Feigenbaum, and J. Lederberg, "Applications of Artificial Intelligence for Chemical Inference. II. Interpretation of the Low Resolution Mass Spectra of Ketones," *J. Am. Chem. Soc.*, **91** (1969), 2977.
32. A. Buchs, A. B. Delfino, A. M. Duffield, C. Djerassi, B. G. Buchanan, E. A. Feigenbaum, and J. Lederberg, "Applications of Artificial Intelligence for Chemical Inference. VI. Approach to a General Method of Interpreting Low Resolution Mass Spectra with a Computer," *Helv. Chim. Acta*, **53** (1970), 1394.
33. C. Damo, C. Dachun, K. Teshu, and C. Shaoyu, "An Artificial Intelligence System for Computer Aided Mass Spectral Interpretation of Saturated Monohydric Alcohols", *Anal. Chim. Acta*, **133** (1981), 575.
34. K. S. Kwok, R. Venkataraghavan, and F. W. McLafferty, "Computer-Aided Interpretation of Mass Spectra. III. A Self-Training Interpretive and Retrieval System," *J. Am. Chem. Soc.*, **95** (1973), 4185.
35. H. E. Dayringer, G. M. Pesyna, R. Venkataraghavan, and F. W. McLafferty, "Computer-Aided Interpretation of Mass Spectra. Information on Substructural Probabilities from STIRS," *Org. Mass Spectrom.*, **11** (1976), 529.
36. K. S. Haraki, R. Venkataraghavan, and F. W. McLaffery, "Prediction of Substructures from Unknown Mass Spectra by the Self-Training Interpretive and Retrieval System," *Anal. Chem.*, **53** (1981), 386.

37. M. M. Cone, R. Venkataraghavan, and F. W. McLafferty, "Molecular Structure Comparison Program for the Identification of Maximal Common Substructures," *J. Am. Chem. Soc.*, **99** (1977), 7668.
38. N. A. B. Gray, A. Buchs, D. H. Smith, and C. Djerassi, "Computer-Assisted Structural Interpretation of Mass Spectral Data," *Helv. Chim. Acta*, **64** (1981), 458.
39. W. Bremser, "The Importance of Multiplicities and Substructures for the Evaluation of Relevant Spectral Similarities for Computer Aided Interpretation of C-13 NMR Spectra," *Z. Anal. Chem.*, **286** (1977), 1.
40. B. A. Jezl and D. L. Dalrymple, "Computer Program for the Retrieval and Assignment of Chemical Environments and Shifts to Facilitate Interpretation of Carbon-13 Nuclear Magnetic Resonance Spectra," *Anal. Chem.*, **47** (1975), 203.
41. N. A. B. Gray, J. G. Nourse, C. W. Crandell, D. H. Smith, and C. Djerassi, "Stereochemical Substructure Codes for ^{13}C Spectral Analysis," *Org. Magn. Reson.*, **15** (1981), 375.
42. N. A. B. Gray, C. W. Crandell, J. G. Nourse, D. H. Smith, M. L. Dageforde, and C. Djerassi, "Computer Assisted Structural Interpretation of Carbon-13 Spectral Data," *J. Org. Chem.*, **46** (1981), 703.
43. N. A. B. Gray, "Computer Analysis of Carbon-13 NMR Data," *J. Artif. Intell.*, **22** (1984), 1.
44. D. H. Smith, N. A. B. Gray, J. G. Nourse, and C. W. Crandell, "The DENDRAL Project: Recent Advances in Computer Assisted Structure Elucidation," *Anal. Chim. Acta*, **133** (1981), 471.
45. G. M. Schwenzer and T. M. Mitchell, "Computer-Assisted Structure Elucidation Using Automatically Acquired 13-C NMR Rules," in *Computer-Assisted Structure Elucidation*, D. H. Smith, ed., American Chemical Society, Washington, DC, 1977, 58, Chapter 5.
46. J. E. Dubois, M. Carabedian, and B. Ancian, "Elucidation Structurale Automatique par RMN du Carbone 13: Methode DARC-EPIOS. Recherche d'une Relation Discriminante Structure Deplacement Chimique," *Comptes Rend. Acad. Sci.* (Paris), **290** (1980), 369.
47. J. E. Dubois, M. Carabedian, and B. Ancian, "Elucidation Structurale Automatique par RMN du Carbone 13: Methode DARC-EPIOS. Description de l'Elucidation Progressive par Intersection Ordonne de Sous-Structures," *Comptes Rend. Acad. Sc.* (Paris), **290** (1980), 372.
48. L. A. Gribov and M. E. Elyashberg, "Symbolic Logic Methods for Spectrochemical Investigations," *J. Molec. Struct.*, **5** (1970), 179.
49. L. A. Gribov, M. E. Elyashberg, and L. A. Moscovkina, "Solution of Spectral Problems by Methods of Symbolic Logic," *J. Molec. Struct.*, **9** (1971), 357.
50. L. A. Gribov and M. E. Elyashberg, "Computer-Aided Identification of Organic Molecules by their Molecular Spectra," *CRC Crit. Rev. Anal. Chem.*, **8** (1979), 111.
51. L. A. Gribov, M. E. Elyashberg, V. N. Koldashov, and I. V. Pletnjov, "A Dialogue Computer Program System for Structure Recognition of Complex Molecules by Spectroscopic Methods," *Anal. Chim. Acta*, **148** (1983), 159.
52. F. Piozzi, G. Savona, and J. R. Hanson, "Kaurenoid diterpenes from *Stachys Lanata*," *Phytochemistry*, **19** (1980), 1237.
53. A. Bax, R. Freeman, and T. A. Frenkiel, "An NMR Technique for Tracing Out the Skeleton of an Organic Molecule," *J. Am. Chem. Soc.*, **103** (1981), 2102.
54. A. Bax, R. Freeman, and S. P. Kempsell, "Natural Abundance ^{13}C-^{13}C Coupling Observed via Double Quantum Coherence," *J. Am. Chem. Soc.*, **102** (1980), 4581.
55. M. R. Lindley, J. N. Shoolery, D. H. Smith, and C. Djerassi, "The Application of the Computer Program GENOA and Two-Dimensional NMR Spectroscopy to Structure Elucidation," *Org. Magn. Reson.*, **21** (1983), 405.
56. R. Freeman, T. Frenkiel, and M. B. Rubin, "Structure of a Photodimer Determined by Natural Abundance ^{13}C-^{13}C Coupling," *J. Am. Chem. Soc.*, **104** (1982), 5545.

57. A. C. Pinto, S. K. do Prado, R. B. Filho, W. E. Hull, A. Neszmelyi, and G. Lukacs, "Natural Abundance ^{13}C-^{13}C Coupling Constants Observed via Double Quantum Coherence. Structure Elucidation by the One and Two Dimensional Experiments of *Velloziolone*—a New Seco Diterpene," *Tetrahedron Lett.*, **23** (1982), 5267.

58. R. Jacquesy, C. Nabone, W. E. Hull, A. Neszmelyi, and G. Lukacs, "Reduction Isomerization in Superacidic Media. Carbon Connectivity Determination Based on Natural Abundance ^{13}C-^{13}C Couplings and a Two Dimensional NMR Experiment," *J. Chem. Soc., Chem. Commun.*, **1982**, 409.

59. M. H. Bozorgzadeh, R. P. Morgan, and J. H. Benyon, "Application of Mass-Analyzed Ion Kinetic Energy Spectrometry (MIKES) to the Determination of the Structures of Unknown Compounds," *Analyst*, **103** (1978), 613.

60. R. A. Yost and C. G. Enke, "Triple Quadrupole Mass Spectrometry for Direct Mixture Analysis and Structure Elucidation," *Anal. Chem.*, **51** (1979), 1251A.

61. C. L. Wilkins and M. L. Gross, "Fourier Transform Mass Spectrometry for Analysis," *Anal. Chem.*, **53** (1981), 1661A.

62. R. B. Cody, R. C. Burnier, and B. S. Freiser, "Collision-Induced Dissociation with Fourier Transform Mass Spectrometry," *Anal. Chem.*, **54** (1982), 96.

63. C. A. Shelley, H. B. Woodruff, C. R. Snelling, and M. E. Munk, "Interactive Structure Elucidation," in *Computer-Assisted Structure Elucidation*, D. H. Smith, ed., American Chemical Society, Washington, DC, 1977, 92, Chapter 7.

64. E. J. Corey, H. W. Orf, and D. A. Pensak, "Computer Assisted Synthetic Analysis. The Identification and Protection of Interfering Functionality in Machine Generated Synthetic Sequences," *J. Am. Chem. Soc.*, **98** (1976), 210.

65. J. Finer-Moore, N. V. Mody, S. W. Pelletier, N. A. B. Gray, C. W. Crandell, and D. H. Smith, "Computer-Assisted Carbon-13 Nuclear Magnetic Resonance Spectrum Analysis and Structure Prediction for the C-19-Diterpenoid Alkaloids," *J. Org. Chem.*, **46** (1981), 3399.

VII

REPRESENTATION OF STRUCTURES AND SUBSTRUCTURES

VII.A. INTRODUCTION

The previous chapters have centered on methods for obtaining structural information about an unknown compound, either in the form of identifications of complete, (hopefully) related, structures or in the form of substructural fragments of the unknown structure. The next steps involve computer aids for analyzing this structural information to determine possible structures for the unknown. At this point it is appropriate to consider how structural information is stored, retrieved, and subsequently used. This chapter introduces several methods for representing a chemical structure in a computer; these methods are fundamental to problems of storage, retrieval, and subsequent structural analysis.

These discussions of representations of structures are restricted to covalently bonded organic compounds. Almost every representational system can accommodate other types of structure, and several do so. However, these extensions are beyond the scope of this presentation. Specific categories of structures excluded from our treatment and references where additional information on such representations can be found include: (1) incompletely specified structures, Markush structures,† and polymers;[1-3] (2) organometallic

†The representations of substructures discussed in this chapter include incompletely specified substructures and Markush forms for generic classes of substructures.

compounds;[2–5] (3) inorganic compounds;[2–5] and (4) catenanes, rotaxanes, and knots.[6]

To be useful for structural analysis, the structure representation used internally by the computer programs must be chosen so that the overall system will satisfy at least the following requirements:

1. It must be simple to input a standard structural diagram, as would be used by a chemist, and convert this into a suitable unambiguous form that can be stored in a computer.
2. It must be possible to manipulate the computer's internal structure representation so as to model chemical transformation, fragmentation processes, and other functions. These manipulations should not entail excessively expensive interconversions to some alternative intermediate representation prior to each modification step.
3. The results of computer manipulations must be displayed back to the chemist, preferably as structural diagrams. Consequently, there must be an unambiguous method of converting the internal computer representations back into interpretable structural diagrams.
4. The representational system must be capable of storage, manipulation, retrieval, and display of all relevant structural descriptions included in the original specification of the structure to the computer.

There are several reviews of methods of computer-compatible structure representations.[1,2,7,8] The National Academy of Sciences report *Chemical Structure Information Handling*[2] groups systems into three categories: (1) fragment codes; (2) linear notations, and (3) tabular and graphic representations.

Fragment codes capture predefined groups of atoms and bonds and simply summarize the substructural content of a structure within the chosen definition. As such, the codes do not allow direct reconstruction of a structure, nor do they allow a general definition of what constitutes a substructure. These limitations preclude their use in most aspects of a structure elucidation system; however, such codes do have important applications in searching large data bases (see Chapter VIII).

The remaining two categories correspond to three systems described by Lynch et al.[8]: (1) linear notations, (2) connection tables, and (3) coordinate representations. The categorization of Lynch and co-workers is more suitable as a subdivision of various systems because it differentiates between the fundamentally different tabular and graphical representations (that were grouped together in the National Academy report[2]).

"Graphical" or *"coordinate"* representations are of value in systems primarily for exploring the shapes of molecules and studying structure–activity relationships that depend, at least in part, on molecular shape. A purely graphical representation would be ill-suited to studies in structure elucidation or synthesis planning. Such studies require that it be simple to test for

particular patterns of bonds between atoms of defined type; it would be impractically laborious to have to reidentify the bonded atoms from the given coordinate data at every structure manipulation step. There are practical uses of coordinate data as a supplement to some other structure representation. Given a coordinate representation of a molecule, it thus might be easier to assess which of two alternative reaction sites in a structure is the more stericaly hindered and, consequently, to predict the outcome of a proposed reaction. However, the most common use of supplementary coordinate data in structure manipulation programs is simply as an aid to creating a satisfactory structural output display.

Linear notations and connection table representations both have wider application. Linear notations are particularly suitable in information retrieval systems involving large data bases of chemical structures. Connection tables are the most suitable structure representation for structure manipulations. The utility of both linear and connection table methods for representing structural connectivity, configuration, and conformation are considered in the following sections.

Before proceeding further with discussion of structure representations, however, it is useful to define several terms. These terms are intuitively understood by chemists, but it is important to define them explicitly for the discussions in subsequent chapters of computer manipulation of the representations. Chemists have detailed mental models of chemical structure, models that allow them to embellish simple representations with additional structural detail that is not formally part of the representation. The definitions of terms used in reference to structure representation must be more precise for use in a computer.

VII.A.1. Structure

Thus far, the word "structure" has been used rather loosely. Chemists may mean many different things when they refer to the "structure" of a molecule; the exact meaning depends on the context of the discussion. In the strictest sense, the word "structure" implies a detailed, geometric representation of a molecule. It implies, at least, knowledge of the coordinates of atoms comprising the structure in three-dimensional space, that is at the level of representation of Dreiding models. Added to these coordinate-based descriptions may be some representation of the Van der Waals radii of each atom, that is, at the level of representation of space-filling models. These models include configurational stereochemistry and allow one to examine conformational stereochemical alternatives assuming that the model is not already locked into a rigid conformation.

Unfortunately, there is seldom sufficient information about a structure to specify such detailed, three-dimensional representations. However, many less detailed specifications of molecular structure are important to computer manipulations, several of which are defined in subsequent sections.

VII.A.2. Substructure

The term "substructure" is used to refer to *any* part of a structure, from a single atom up to, and including, the structure itself. Although it may seem inappropriate to view a structure as a substructure of itself, this definition is relevant to discussions of substructure search and graph matching in Chapter VIII.

The concept of substructure is one that is intuitive to chemists, who generally think in substructural terms when they refer to functional groups, ring systems, and substituents. Generally, substructures are represented by simple combinations of atoms and the bonds that interconnect the atoms. For example, the carbonyl group (**VII-1**) and a benzene ring (**VII-2**) are two simple, common substructures.

```
                        \   /
                         C=C
  \                     /   \
   C=O               —C       C—
  /                    \\   //
                         C—C
                        /   \
```

VII-1 **VII-2**

Such substructures are generally represented on paper with additional features that indicate how the substructure may be bonded within a larger structure. Both **VII-1** and **VII-2** are presented with unused valences filled with bonds with unspecified termini. These bonds are termed *free valences* because they are free to form bonds within a larger structure. Furthermore, it is normally assumed that these free valences represent bonding sites to other, *nonhydrogen* atoms.[9] If, for example, the carbonyl group were known to be part of an aldehyde functionality, the correct representation would be **VII-3** or **VII-4**, depending on whether one chose to represent explicitly the hydrogen atom. In many representational schemes hydrogens are assumed implicitly, as in **VII-3** and **VII-5**.

—C=O —CH=O

VII-3 **VII-4**

Similarly, if the benzene ring were known to be monosubstituted and connected to a larger structure through an ether linkage, the appropriate representation would be **VII-5**.

```
   C=C
  /   \
 C     C—O—
  \\  //
   C—C
```

VII-5

There are several other features of substructures that must be represented to complete our definition of what constitutes a substructure. These features are irrelevant to structures in our definition. There must be formalisms for: (1) hydrogen substitution and bonding sites, as introduced above; (2) variable atom type; and (3) variable bond type.

Furthermore, substructures can, in principle, possess all the stereochemical properties of complete structures. This may not seem obvious, since the excision of a substructure from a structure removes it from the context within which some stereochemical features make sense. However, one must keep in mind the very general definition of a substructure. If, for example, one is referring to a ring system, it is still meaningful to represent in the substructure the configurational stereochemistry at the ring junctions. In fact, for certain applications discussed in Chapter XI, it is *essential* to represent the stereochemistry of substructures where a sufficiently large substructure exists to make such representations meaningful.

VII.A.3. Connectivity of Structures and Substructures

The simplest representational scheme for structures and substructures captures only their *connectivity*, that is the details of the interconnections of the atoms. *This representation possesses no two- or three-dimensional information*. Atom properties such as names, hybridization (and similar), and bond multiplicities can be, and generally are, all represented. Structural connectivity is also referred to as structural *constitution* or structural *topology*; (subsequently, these terms are used interchangeably).

In structure elucidation, establishment of structural connectivity is usually the first goal, with stereochemical detail added as additional information is obtained on a structure. The normal form for presentation of connectivity to chemists is in the form of structural diagrams. Such diagrams (e.g., **VII-5**) often are drawn conveying some two-dimensional information. Ring systems and substituents are drawn with bond angles that approximate their standard values, so that six-membered rings, for example, resemble hexagons. It is important to remember, however, that information on bond angles is *not* present in the actual representation of connectivity; it is added because structural diagrams are easier to interpret when displayed this way. This introduces several problems in converting from an internal representation of connectivity in a computer to a structural diagram interpretable by a chemist. An internal representation that specifies only connectivity will have to be augmented with other data, such as presumed default values for arrangements of rings or at least bond lengths and angles, before a satisfactory structural diagram can be created and presented to the chemist.

Topological representations of structure have proven extremely useful for computational manipulation. There is a direct relationship between this structural representation and undirected graphs in the mathematical discipline of graph theory. The *atoms and bonds* of structural topology correspond to *nodes and edges* of graphs. Thus several theorems of graph theory, including those related to connectivity, symmetry, subgraphs, and isomorphism, are directly applicable to topological structures. Many applications of graph theory to chemical problems have been reviewed recently by Balaban.[10]

Many of the more important computer programs for structure manipulation are rooted in graph theoretic methods for manipulation of structures. As these programs are discussed in subsequent chapters, the intimate relationship between chemical problems and graph theory will be obvious. *The success of such programs is a reflection of the relative ease of performing computations on topological representations with the use of mathematically proven algorithms.* However, one must always keep in mind the inherent limitations of this representation in that the absence of geometric information precludes relating structures to properties that depend on geometry.

VII.A.4. Stereochemistry of Structures and Substructures

All discussions of stereochemistry refer to some implicit or explicit representation of the relative positions of atoms in a structure in three-dimensional space. Stereochemical representations can be regarded as augmentations to structural connectivities to yield a representation that specifies atom connectivity and geometric information. Two types of stereochemistry are usefully distinguished here: *structural configuration* and *structural conformation*.

Configurational stereochemistry includes the configurations of chiral centers and *cis–trans* isomerism about double bonds. This grouping reflects the fact that both require physical interchange (bond breaking) of substituents on chiral centers or double-bonds to convert one enantiomer to its mirror image, or to interchange *cis* and *trans* substitution.

The definition of configuration is intimately involved with atom connectivity; it makes no sense to discuss configuration without defining the constitution of a substructure or structure. However, given the constitution, several configurational stereoisomers may be possible. A representation of configuration is necessary for differentiation among the possibilities.

As in the case of structural connectivity, there are powerful methods of graph theory that allow rigorous specification and enumeration of stereoisomers within the representation of connectivity augmented by stereochemical designations. These methods are discussed in Chapter X.

Again, the inherent limitations of this representation must be acknowledged. No geometric information is represented explicitly; additional chemical information and computer programs are necessary for translation of the designations into geometry.

Conformational stereochemistry implies the specification of details of the geometry of a structure. It can be regarded as the next level of detail in a hierarchy of representation

$$\text{Connectivity} \longrightarrow \text{configuration} \longrightarrow \text{conformation}$$

The *conformation of a molecule* refers to the relative locations of atoms in three-dimensional space. A representation of conformation in a computer will include structural connectivity and configuration, explicitly or implicitly, together with a representation of the relative positions of atoms. The positional information may be explicit, in the form or actual X, Y, Z coordinates of all atoms, or implicit, in the form of torsional angles for every rotatable bond.

VII.A.5. Isomerism of Structures and Substructures

Structural isomers are topologically distinct ways of connecting together atoms to yield structures possessing the same molecular formula. *Configurational stereoisomers* are structures that possess the same topology but differ in the configurations at chiral centers. In the computer programs to be discussed, configurations at double-bonds are included, although strictly speaking these are examples of geometric isomerism. *Conformational stereoisomers* are structures that possess the same topology and configurations but differ in the arrangement of atoms in three-dimensional space.

It is meaningless to discuss the structural isomers of a substructure, because there is no application for such a notion. However, it is meaningful to characterize the stereochemistry of substructures.

VII.B. REPRESENTATION OF CONNECTIVITY

VII.B.1. General Approaches

The main methods for representing structural connectivity in computers are *linear notations* and *connection tables*. Linear notations represent chemical structures with strings of alphanumeric characters (possibly supplemented by other printable characters). Considerable effort has been devoted to developing linear notational systems because of their inherent suitability for storage and retrieval of structural information in computers.† These notations have the advantage of being quite compact, resulting in storage of tens of thousands of structures in a relatively small amount of computer memory or magnetic disk space. However, linear notations have a number of inherent limitations that restrict their use in structure elucidation–manipulation systems.

The primary disadvantage of linear notations is that they impose a formidable barrier between the chemist and the desired structural of linear notations. Chemical nomenclature can be regarded as one type of linear notation and one with which most chemists are familiar. But the rules for nomenclature are highly complex and *ad hoc*. Few chemists have the expertise to translate between structural diagrams and chemical nomenclature without recourse to a

† For the purposes of this discussion, conventional chemical nomenclature is considered as a type of linear notation, although exception has been taken to this view.[11]

compendium of rules. Computer programs have been developed that are capable of translation of structures to names[12] and names to structures,[13–15], but these very large and complicated programs will be successful in this translation in considerably less than 100% of cases.[12,15]

Wiswesser linear notation (WLN)[16] is the most popular of the linear notational systems that represent structures with character strings; the Hayward[17,18] and IUPAC[19] systems represent other examples of such systems. Comparisons of structural representations obtained in these systems have been presented.[2,8] The systems use sets of complex *ad hoc* rules to derive names for structures. Most chemists are incapable of deriving structural diagrams from such names, and vice versa. Computer programs for translation of structural diagrams to names and names to diagrams have been only partially successful, primarily because of the complexity of the rules.[2,8]

A second limitation is that it is very difficult to manipulate linear notations in the computer to derive substructures or search for structural similarities. For example, nomenclature-based substructure search systems have been described,[20] but can retrieve only substructural information that is explicit in the name. Wiswesser line notation has been used as the basis for substructure search systems,[21] but WLN strings often must be converted to a tabular representation for successful search. Problems in such conversions of representations were mentioned above. The **SYNCHEM** synthesis planning programs due to Gelernter et al.[22,23] provide an actual case study of these problems. WLN is used in some parts of the **SYNCHEM** system (e.g., WLN codes are used for the structures defined in the library of available compounds that is checked as **SYNCHEM** evaluates hypothesized precursor structures). Reports on **SYNCHEM** have noted the problems that can arise in interconversion between the WLN representations and various connection tables used elsewhere in the program.[22,23] Most of the applications discussed in subsequent chapters *require* the capability for detailed manipulation of structural representations. Linear notations were developed for information retrieval, not computer-based structural analysis.

Connection matrices, tables and *atom-connectivity* lists provide the most valuable representation for the vast majority of applications in chemical structure elucidation and manipulation. All these methods capture topology in tabular forms that directly represent structural connectivity.

The earliest suggestion of connection tables as a method for representing structural topology is apparently due to Mooers.[24] Ray and Kirsch[25] were the first to adapt Mooers's suggestions for actual representation in a computer. The example of representation of the structure of chloral (**VII-6**) cited in their paper[25] is a good introduction to the content of connection tables. Structure **VII-6** is presented in two forms, one with atom names and the other with atom numbers. (Note that the double-bond is explicitly represented as a numbered "atom," number 2.)

The connection table derived for **VII-6** by Ray and Kirsch is given in Table VII-1.

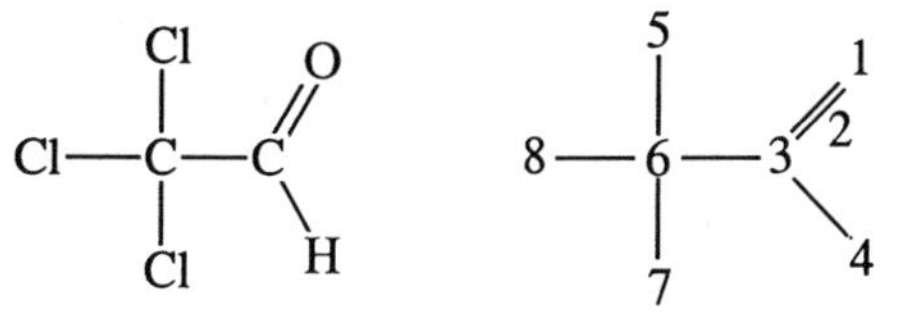

```
      Cl   O              5    1
      |   //              |   //2
Cl—C—C          8—6—3
      |    \              |    \
      Cl   H              7     4
```

VII-6

In this connection table, the hydrogen atom H(4) is represented explicitly, as is the carbonyl double bond. This connection table can be read directly in structural terms. For example, referring to Table VII-1, atom 6, component 6, is connected to atoms 3, 7, 8, and 5 (implicitly by single bonds) and is a carbon atom. Connections tables are usually redundant in that bonds are represented twice. Thus C(6) appears in the "connections" list for C(3) in **VII-6**.

The information presented in Table VII-1 was stored in the computer in eight words. Each word corresponded to one row of the table and contained eight fields, one for the component number, four for the (up to) four connections, and one for the element symbol.[25]

Connection tables can assume many other forms. The particular form chosen depends on factors such as:

1. The amount of structural information that is represented explicitly. Information represented implicitly must be derivable from the representation by any program that utilizes the connection table.
2. The uses to which the connection tables are put. Based on the application, one representation may be more appropriate than another in terms of storage requirements and the amount of decoding of the internal computer representation that must be done.
3. The data structures available in (or imposed by) a specific programming language in which programs are written to manipulate the connection tables.

TABLE VII-1. Connection Table for Chloral (VII-6) as Given by Ray and Kirsch[25]

Component	Connections	Element Symbol
1	2	O
2	1, 3	=
3	2, 4, 6	C
4	3	H
5	6	Cl
6	3, 7, 8, 5	C
7	6	Cl
8	6	Cl

Other forms of connection tables can serve to illustrate these points. These examples use the structure of hydroquinone (**VII-7**), whose numbering is also given to aid in interpretation of the subsequent connection tables. This numbering corresponds to the Chemical Abstracts Service (CAS) Registry III[3] connection table for **VII-7**, given in Table VII-2.[11].

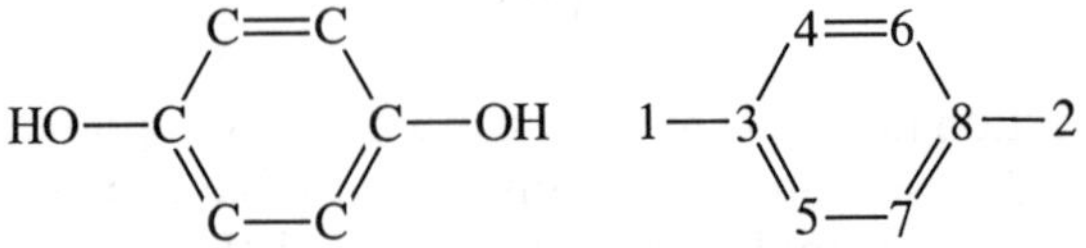

```
      C=C                    4=6
     /   \                  /   \
HO—C       C—OH       1—3       8—2
    \\   //                \\   //
      C—C                    5—7
```

VII-7

The representation for connection tables given in Table VII-2 is based on a particular view of molecular structure by CAS. Structures are broken down into their constituent ring systems and acyclic chains. The first segment of the connection table (Table VII-2) summarizes the topology of structure **VII-7**, consisting of the acyclic portion, two hydroxyl groups whose oxygen atoms are numbered 1 and 2, the ring system identifier and numbering with respect to

TABLE VII-2. CAS Registry III Connection Table for VII-7[11]

```
TOPOLOGY

          ATOM NUMBER   1   2
          CONN
          ELEMENT       O   O
          BOND

          RING CORRESPONDENCE LINK

          RING ID               46T.150A.182
          RING ATOM NOS              1  2  3  4  5  6
          SUBSTANCE ATOM NOS        03 04 05 06 07 08

          LINK GROUP          01 -1 03, 02 -1 08

RING
          RING IDENTIFIER NO = 46T.150A.182

          ATOM NO    1   2   3   4   5   6
          CONN          01  01  02  03  04
          ELEMENT    C   C   C   C   C   C

          BOND          *5  *5  *5  *5  *5

          RING CLOSURE PAIRS  05 *5 06
```

TABLE VII-3. Connection Table Representation of VII-7 in DENDRAL

ATOM#	TYPE	NEIGHBORS	ARTYPE	HYBRID
1	O	3		SP3
2	O	8		SP3
3	C	5 1 4	AROM	SP2
4	C	6 6 3	AROM	SP2
5	C	7 3 3	AROM	SP2
6	C	8 4 4	AROM	SP2
7	C	8 8 5	AROM	SP2
8	C	2 6 7 7	AROM	SP2

VII-7, and the points of attachment of the acyclic portions to the ring ("LINK GROUP," Table VII-2). The attachment list specifies the numbers of the atoms and the bond type forming the connection (e.g., atom 1 connected to atom 3 with a single bond). The second segment specifies the connection internal to the ring itself, the bond type (type "*5" represents an aromatic bond), and the ring closure bonds. This representation is a logical outgrowth of CAS interests in maintaining ring indices for chemical structures. Ring systems are explicitly defined and named in their connection tables, permitting easy access to this information for certain file search techniques.

A rather different connection table for structure **VII-7** is that used in the **DENDRAL** programs[9] (shown as Table VII-3).

The connection table shown in Table VII-3 is closely related to the connection table of Ray and Kirsch in that it merely lists the atoms by name, number and connections. Here, however, multiple bonds are represented by multiple entries in the neighbors list. In addition, a number of other atoms properties are included,[9] in this case the aromatic nature and hybridization of the atoms. Like the CAS connection tables, hydrogen atoms are not represented explicitly. There is no ring system information explicit in the **DENDRAL** connection tables. Rings are defined implicitly by the connectivity; if specific data on rings are required, they must be rediscovered through an analysis of the connectivity performed in the applications programs.

Internally to the computer, the **DENDRAL** programs maintain connection tables as a set of *vectors*. Atom names, properties, and connection lists are stored separately, consuming one computer word for each separate entry. Although inefficient with regard to use of storage in main memory, this representation has the advantage of having all relevant information about the structure readily available for manipulation.†

It is possible to represent connection tables in a variety of matrix forms. An early description of matrix representations was presented by Spialter[26] and

†**DENDRAL** connection tables are compressed into abbreviated character strings for efficient temporary and long-term storage on a magnetic disk. This reduces the storage requirements but means that the strings must be decoded back to connection tables when read back into computer memory.

TABLE VII-4. Atom Connectivity Matrix Representation of VII-7

	1	2	3	4	5	6	7	8
1	O	0	1	0	0	0	0	0
2	0	O	0	0	0	0	0	1
3	1	0	C	1	2	0	0	0
4	0	0	1	C	0	2	0	0
5	0	0	2	0	C	0	1	0
6	0	0	0	2	0	C	0	1
7	0	0	0	0	1	0	C	2
8	0	1	0	0	0	1	2	C

expanded on in a subsequent publication.[5] Spialter defined the *atom connectivity matrix* (ACM) to be a matrix of order n, where n is the number of atoms in a structure. The diagonal elements can be used for atom properties (name, valence, charge). The off-diagonal elements $a_{i,j,i \neq j}$ can be used for the connectivity, generally defined to include bond order, force constants, and other variables. The ACMs can represent hydrogen atoms explicitly or implicitly as may be required by the application.

Table VII-4 illustrates the hydrogen-suppressed ACM for hydroquinone (**VII-7**), where the diagonal elements contain the atom name and the off-diagonal elements the bond order, with 0 representing no connection. Here the atom numbers are included for clarity; they are not formally part of the ACM. (In this example the aromatic system is defined in terms of alternating double and single bonds, although another designation could be chosen to represent the aromatic nature of these bonds.)

Spialter also presented a set of advantages and disadvantages of the method of representation shown in Table VII-4.[5] Many of the disadvantages he cites have been solved since his paper appeared. His suggested solution to some of the disadvantages, the ACM characteristic polynomial, is, in part, wrong. These problems are related to deriving a canonical form for matrices, details of which are presented in Chapter IX, and in manipulating the representation in various ways, details of which are presented in Chapter VIII. There is nothing wrong with the representation per se, other than it is inefficient in terms of computer storage because most off-diagonal elements in matrices representing organic compounds will be zero. (Because the matrix is symmetrical, only the elements in the part of the matrix above the diagonal need to be stored to represent the structure fully and so some savings in storage are possible.) The representation has some computational utility using programs written in languages such as FORTRAN that use arrays as fundamental data structures.

Several other matrix-based systems for representing structural topology have been presented. The *bond-electron* (BE) matrix proposed by Ugi and coworkers[27,28] was developed to characterize systems of molecules undergoing transformation. For this reason, valence electrons are explicitly represented, as are hydrogen atoms. The actual form of a BE matrix closely resembles the ACM (Table VII-4), with the diagonal elements used to store the number of valence electrons rather than atom names. Detailed discussions of BE matrices including presentation of their computational utility have been published.[28,29] Rouvray[30] has presented a thorough discussion of matrix representations of structures, covering a number of additional matrix types and their chemical meaning and utility, particularly with regard to quantum chemical applications.

VII.B.2. Special Requirements of Substructure Representations

Substructures can be represented using methods similar to those just outlined. The representation methods used for substructures do, however, have to be somewhat more flexible. As noted above, substructure definitions must allow for variable bond types, variable atom types, and free valences. Some of these complexities are discussed in the following paragraphs.

Several of the examples used to illustrate these discussion use rather generalized concepts of "possible bonding sites" and so on and really go beyond the context of merely asking questions about whether a substructure is "contained within" a larger structure. These examples are meant to introduce the concept of a substructure as a building block in a structure elucidation system as well as a passive entity used to search against files of structures. The substructural representations of the **DENDRAL** programs are used for illustrative purposes. Problems of substructure representation are, or course, common to all structure manipulation systems and have been solved in a number of essentially similar ways.

In the following examples, many substructures are presented with atom numbers assigned by the computer program used to define the substructures. The actual numbers are irrelevant to the discussions; the same substructure could have been defined using any other numbering of the atoms. Computer systems designed to manipulate substructures use the numberings as internal references to the atoms involved. These numberings bear no relationship to formal chemical numberings; the results obtained by any program using the substructures are invariant to the numberings.

VII.B.2.a. Bonding of Substructures to Other Atoms

The first important consideration in representing substructures in a computer is how to specify conventions that will aid subsequent programs in determining how the substructures might be found, or connected within larger structures. This specification includes not only connections to external atoms—that is, those not included within the substructure, but also connections among atoms in a given substructure. This generality is important to discussion in the subsequent chapters mentioned above. For the examples presented below, the most general interpretation are assumed; bonds may be to any atom including those defined as part of the substructure, assuming that such bonding is allowed by the actual definition, (e.g., when more than one bonding site is available).

Consider as the first example substructure **VII-8**, which is closely related to structure **VII-7**, but with bonding sites, or free valences on C(4) and C(6).

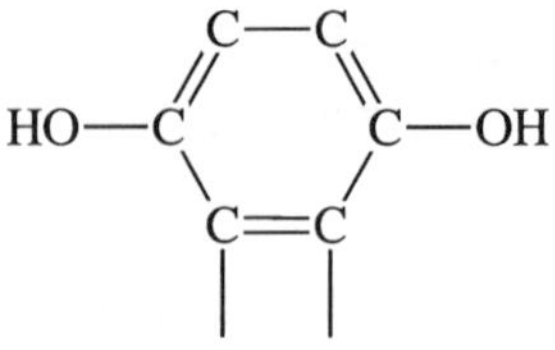

C—C

HO—C C—OH

C=C

VII-8

The connection table in the **DENDRAL** programs for this substructure (Table VII-5) closely resembles the connection table for **VII-7** (Table VII-3), with important differences that help to characterize **VII-8** as a substructure.

First, there are bonding sites explicitly defined at C(4) and C(6) (**VII-8**). These free valences are shown in Table VII-5 as "FVs" in the lists of neighbors of C(4) and C(6), with one FV for each free valence. The "HRANGE" column in the table specifies the minimum to maximum number of hydrogens on designated atoms. The oxygen atoms O(1) and O(2) bear exactly one hydrogen, specifying them as hydroxyl groups. The remaining hydrogen ranges specify that C(5) and C(7) are unsubstituted, whereas the hydrogen range of 0–0 on C(4) and C(6) force them to be bonded to nonhydrogen atoms.

Hydrogen ranges are irrelevant for representation of structures (e.g., Table VII-3) because one can assume for structures that remaining valences, unfilled by entries in the connection table, are filled with hydrogen atoms. Similarly, structures are assumed to be complete, so that free valences have no meaning in structures.

Although bonding information of substructures could be specified in many different ways, Table VII-5 illustrates some important considerations. There

TABLE VII-5. Connection Table in the DENDRAL system for Substructure VII-8

ATOM#	TYPE	NEIGHBORS	HRANGE	ARTYPE	HYBRID
1	O	3	1-1		SP3
2	O	8	1-1		SP3
3	C	5 5 1 4		AROM	SP2
4	C	6 6 3 FV	0-0	AROM	SP2
5	C	7 3 3	1-1	AROM	SP2
6	C	8 4 4 FV	0-0	AROM	SP2
7	C	8 8 5	1-1	AROM	SP2
8	C	2 6 7 7		AROM	SP2

must be a method for designating atoms that participate in bonding to a larger structure. This condition is satisfied by the free valences of **VII-8**. Equally important is some convention for specifying which atoms *might* participate in bonding, because often one is not completely certain. Therefore, substructural connection tables must be able to represent three possibilities for bonding: (1) atoms that *cannot* participate in forming new bonds; (2) atoms that *might* participate in forming new bonds, and a convention for how many (i.e., a range); and (3) atoms that *must* participate in forming new bonds. In **DENDRAL** these three possibilities are treated by assumptions of default values for hydrogen ranges, designated hydrogen ranges on some or all atoms, or a hybrid representation using hydrogen ranges and free valences. These possibilities are best illustrated through some examples. Some understanding of **DENDRAL**'s internal representation of hydrogen ranges is necessary for understanding these examples. In the computer, *all specifications of hydrogen atoms and bonding sites are controlled by hydrogen ranges*. However, the specification of "free valences" on an atom often serves as a more concise way of defining possibilities of bonding (and has the added advantage of corresponding to pictorial descriptions of substructures used routinely by chemists). *All free valence designations are translated in the computer into hydrogen ranges*.

As a first example, consider substructure **VII-9**, whose connection table is given in Table VII-6.

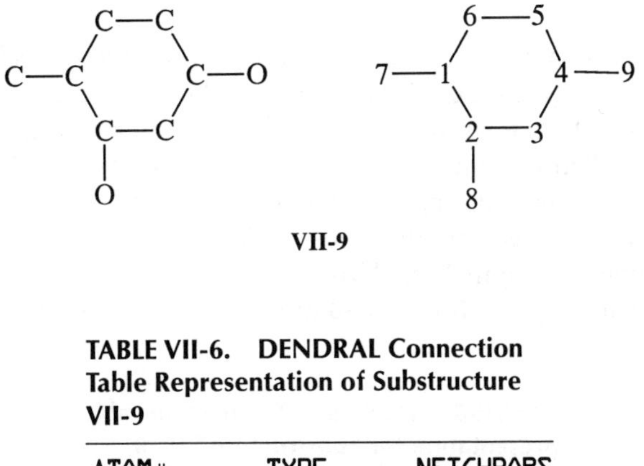

```
     C—C                    6—5
    /   \                  /   \
C—C     C—O           7—1     4—9
    \   /                  \   /
     C—C                    2—3
     |                      |
     O                      8
```

VII-9

TABLE VII-6. DENDRAL Connection Table Representation of Substructure VII-9

ATOM#	TYPE	NEIGHBORS
1	C	7 6 2
2	C	8 3 1
3	C	4 2
4	C	9 5 3
5	C	6 4
6	C	1 5
7	C	1
8	O	2
9	O	4

In Table VII-6 there is no mention of hydrogen ranges, free valences, or any other special atom or bond properties. In fact, **VII-9** resembles a structure. It can be used as a substructure, however, merely by telling **DENDRAL** that it is to be treated as one. This is the most general form of a substructure in **DENDRAL** and it is interpreted according to the default condition that every hydrogen range is from zero to "any," where "any" is limited only by the valence of a particular atom. Thus every atom in **VII-9** is a potential bonding site for connection to any other atom, including hydrogen.

Statements about such a substructure can be made more specific by detailing additional atom properties. Consider as an example that in **VII-9**, C(7) must be a methyl group and O(8) must be an hydroxyl group. The hydrogen range on those two atoms can then be fixed, resulting in substructure **VII-10**, whose connection table is given in Table VII-7.

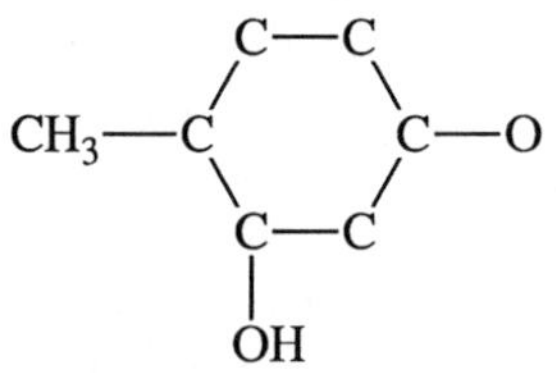

```
           C—C
          /   \
CH3—C          C—O
          \   /
           C—C
           |
           OH
```

VII-10

The connection table presented in Table VII-7, fixes specific hydrogen counts for atoms C(7) and O(8). Other atoms, however, are unrestricted and may still participate in forming interconnections with any other atoms.

A more restricted version of **VII-9** might be one where C(2,3,5,6) and O(9) all possess exactly one bonding site. These restrictions could be expressed by defining hydrogen ranges on all atoms, yielding substructure **VII-11**, whose connection table is given in Table VII-8.

The hydrogen ranges in Table VII-8 force C(2,3,5,6) and O(9) to be bonded

TABLE VII-7. DENDRAL Connection Table Representation for Substructure VII-10

ATOM#	TYPE	NEIGHBORS	HRANGE
1	C	7 6 2	
2	C	8 3 1	
3	C	4 2	
4	C	9 5 3	
5	C	6 4	
6	C	1 5	
7	C	1	3-3
8	O	2	1-1
9	O	4	

TABLE VII-8. DENDRAL Connection Table Representation for Substructure VII-11

ATOM#	TYPE	NEIGHBORS	HRANGE
1	C	7 6 2	1-1
2	C	8 3 1	0-0
3	C	4 2	1-1
4	C	9 5 3	1-1
5	C	6 4	1-1
6	C	1 5	1-1
7	C	1	3-3
8	O	2	1-1
9	O	4	0-0

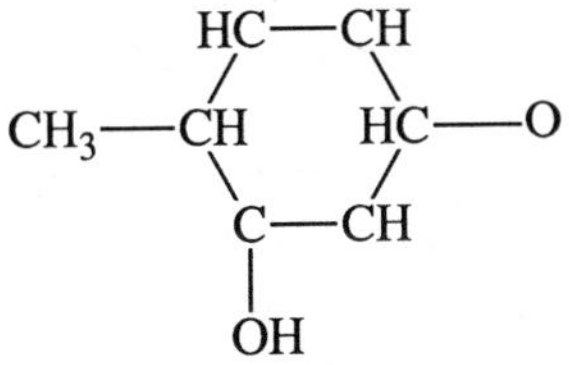

```
          HC—CH
         /     \
$CH_3$—CH       HC—O
         \     /
          C—CH
          |
          OH
```

VII-11

to nonhydrogen atoms. The **DENDRAL** programs allow the use of free valences to designate bonding sites as an alternative to the specification of hydrogen ranges on all atoms. In large substructures with only a few bonding sites, a specification in terms of free valences is usually more concise. If free valences are used, then by default atoms not bearing free valences have any unspecified connections made to hydrogen atoms. Substructure **VII-12** and its corresponding connection table (Table VII-9) result from specification of one free valence on each of C(2,3,5,6) and O(9).

Substructure **VII-12** is equivalent to **VII-11**; in the computer, the connection table for **VII-12** is translated to a representation that specifies hydrogen

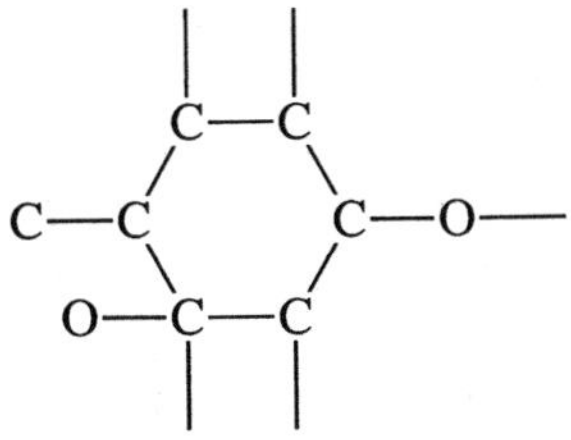

```
         |  |
         C—C
        /    \
     C—C      C—O—
        \    /
       O—C—C
         |  |
```

VII-12

TABLE VII-9. DENDRAL Connection Table Representation of Substructure VII-12

ATOM#	TYPE	NEIGHBORS
1	C	7 6 2
2	C	8 3 1 FV
3	C	4 2 FV
4	C	9 5 3
5	C	6 4 FV
6	C	1 5 FV
7	C	1
8	O	2
9	O	4 FV

ranges that is the same as that obtained from **VII-11**. It is much simpler, however, to define this substructure with free valences as **VII-12**.

Examples **VII-11** and **VII-12** demonstrate how one can define a substructure to possess specific bonding sites to nonhydrogen atoms. Examples **VII-9** and **VII-10** demonstrate how one can define many atoms to be potential bonding sites. The scheme for substructure definitions used in **DENDRAL** can also handle more complex situations such as those where there are a small number of potential bonding sites, some of which may or may not be bonded to hydrogen atoms. Such substructures can be defined either by specifying required hydrogen ranges for every atom or, more conveniently, by appropriate combined use of hydrogen range restrictions on some atoms and free-valence allocations to other atoms. Such advanced use of the **DENDRAL** substructure definition scheme are detailed elsewhere.[31]

VII.B.2.b. Representations for Variable Atom and Bond Type

There are many uses of substructures where it is important to be able to refer to variable atom or bond type. For example, in structure elucidation the available data can sometimes yield information on the topology of a substructure without always precisely specifying the type of every atom and bond. (For instance, a particular carbon in the substructure might be required to bond to a heteroatom where there are possibilities of bonds to either nitrogen or oxygen.) As another example, there are often applications in substructure search systems where one may want to characterize and search for occurrences of a entire class of similar substructures.

Most computer-based systems applied to these problems have conventions for treating variable atom and bond type. Generally, the following types of variability must be captured by a useful substructural representation:

1. *Atom Properties*. Atoms of a given type must be allowed to possess variable properties. An example of variable hydrogen ranges was presented in the previous section. Other properties that may possess

ranges or sets of values include aromatic type, hybridization, isotope, electronegativity, partial charge, and so on.

2. *Atom Type*. A given atom must be allowed to be of several different (specified) types (e.g., halogen) or be "any" type of atom. The latter is essential in representing general topologies (e.g., a six-membered rings) without defining the atoms constituting the ring.
3. *Bond Order*. It is often important to characterize a bond as being of "any" bond order rather than single, double, and so on. This convention simplifies specification of general topologies in the same way as does the "any" atom type, in item 2 above.
4. *Bond Type*. For some applications, particularly substructure search, it is extremely useful to characterize a bond by its order *and* topology (e.g., "ring-double," "chain-triple").

The extent of the requirements for flexibility in representation depend on the intended applications. Substructure search systems require the greatest flexibility and usually include mechanisms for treating all of these four variabilities. For example, any system based on the CAS connection tables[3] has a structural representation that describes all bond and atom types, facilitating development of substructure search systems that access any necessary part of this information using substructure representations that allow variability.[32,33] Several other systems have been presented that possess similar features in specifying substructures.[8,34]

Programs for structure elucidation, such as the **DENDRAL** programs, have somewhat less need for variability in substructure definitions; substructures inferred from spectroscopic or chemical data can generally be specified in some detail. However, these programs still possess a rich "language" for substructural description. This language is exemplified by the definition of substructure **VII-13** whose connection table is presented in Table VII-10.

```
SUBSTRUCTURE TRY:(ARTYPES, HRANGES AND ANYBONDS NOT INDICATED)
POLYATOMS: P2→(N O), P1→(F CL BR I)
BONDS 4-3, 5-4, 6-5, 6-1 AND 2-1 ARE OF TYPE 'ANY'
NON-C ATOMS: 1→(N O), 7→(F CL BR I)
LNODES ARE INDICATED BY AN ATTACHED @
```

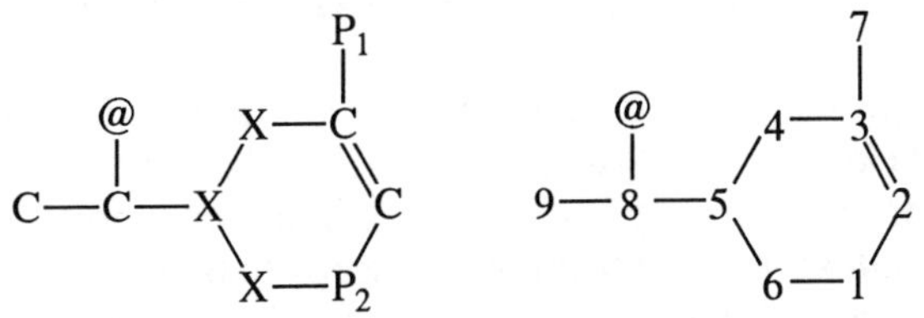

VII-13

TABLE VII-10. DENDRAL Connection Table Representation for Substructure VII-13

ATOM#	TYPE	NEIGHBORS	HRANGE	LNODE	ARTYPE	POLYNAME
1	POLY	6 2			NON-AR	(N O)
2	C	1 3 3	1-1		NON-AR	
3	C	4 7 2 2			NON-AR	
4	X	4 6 8			NON-AR	
5	X	4 6 8			NON-AR	
6	X	5 1			NON-AR	
7	POLY	3				(F CL BR I)
8	C	5 9		1-3		
9	C	8	3-3			

BONDS 4-3, 5-4, 6-5, 6-1 AND 2-1 ARE OF TYPE 'ANY'

Substructure **VII-13** (connection table shown in Table VII-10) specifies a six-membered ring possessing a carbon-carbon double bond [at C(2)-C(3)]. Atom 1, however, is specified as a *polyatom*, that is as an atom potentially with several names, in this case nitrogen or oxygen. Atoms 4, 5, and 6 are all specified as being of type *X*—this usage following from the **DENDRAL** convention for any atom name. In addition, atoms 1–6 are all designated as nonaromatic, C(2) is specified to bear a single hydrogen atom, and the bond order of the indicated bonds (Table VII-10) can be any multiplicity. Thus the ring system of **VII-13** corresponds to a wide variety of ring systems possessing six atoms.

Substructure **VII-13** is rendered more specific by additional atoms that limit the types of substituents on the six-membered ring. Atom 7, connected to C(3) of the double bond, is another polyatom, this time any halogen atom (Table VII-10). The atom 5 substituent corresponds to an alkyl chain terminated with a methyl group [the hydrogen range on C(9) is exactly 3]. This substituent introduces the concept of a variable length chain of atoms, defined by a *link node*, whose minimum to maximum lengths are given as one to three atoms (Table VII-10). Thus this substituent corresponds to substitution at atom 5 with a C_2–C_4 chain. This chain may be arbitrarily branched because nothing was said about the degree of substitution of the atoms comprising the link node, C(8). If the hydrogen range on C(8) were given as exactly 2, the substituent would correspond to an ethyl, propyl, or *n*-butyl group. Link nodes can also be used in ring systems to specify a range of ring sizes.

VII.C. REPRESENTATION OF CONFIGURATIONAL STEREOCHEMISTRY

Systems for representing and manipulating configurational stereochemistry are far less developed than those for topological data. This is true despite the fact that stereochemistry plays a crucial role in understanding the behavior of molecules in contexts ranging from spectroscopy to biological activity. Some of the reasons why configurational stereochemistry is often handled inadequately include:

1. Initial choice of inadequate or incomplete representational systems that cannot be modified to include stereochemistry.
2. The inertia of having large, computer-based systems designed to access large data bases of structural information, all of which operate on topological representations of structures and substructures.
3. The paucity of proven algorithms and programs for analyzing, manipulating, and displaying stereochemistry.

No generally useful method for representing configurational stereochemistry in one of the linear notational systems has ever been developed. These systems were designed to capture structural topology, and it is unlikely that they can be successfully modified to capture stereochemistry. The primary reason for this is that fragments of structures are represented by the notations, whereas configuration is an atom, or atom-centered property. Unless all atoms are described explicitly by a notation, it is extremely difficult to represent configurations.

Conventional chemical nomenclature can, of course, be used to represent stereochemistry using conventional textual descriptions of configurations. Such systems are exemplified by the CAS registry system,[3] which describes stereochemistry with text strings appended to the name and connection table of a structure.[35] The CAS system is useful for structural storage and retrieval using this method for describing stereochemistry. However, a simple example suffices to demonstrate that such systems as currently implemented are useless for any subsequent manipulation of stereochemistry. Suppose that one wanted to query such structural representations for those that possessed a given substructure in a configuration designated as *R* using the Cahn-Ingold-Prelog rules for *R,S* nomenclature of chiral centers.[36] Unfortunately, the designations *R* or *S* depend on the atomic weights of attached substituents. Thus, although the desired substructure can easily be located in structures on the basis of topology, the configuration designations for stereocenters in the sought-for configuration will be arbitrarily *R* or *S* depending on the structural context in which they are found.

Therefore, current systems of linear notation and conventional nomenclature are ill suited for representation of stereochemical information in forms that can be manipulated by computer programs.

In contrast, because configurational stereochemistry can be regarded as an atom property (or a two-atom property in the case of double bonds), connection table representations of structure are ideally suited for capturing configurations. The first general approach to this problem was presented by Wipke[37] and Wipke and Dyott.[38] Their paper presents a detailed discussion of the problems and their solutions, of computer representation and manipulation of chiral center and double-bond stereochemistry. Some details are also given of how configurations can be defined for atoms and double bonds using the graphical structure input subsystems of **SECS**.

Their method relies on defining configuration by ordering the connection table entries (i.e., the *neighbors* list) for the chiral center or the double bond whose stereochemistry is defined. Recognition and manipulation of configuration is handled by algorithms that take advantage of the predefined scheme for ordering. For example, in the computer, inversion of configuration is merely a matter of reordering the list of neighbors by interchanging two of the entries.[37,38]

The crucial advantage of this approach is that it is a completely general and unambiguous method for representing stereochemistry. There is no introduction of *ad hoc* schemes[36] for naming the chirality of a stereocenter and no reference to an arbitrary standard such as tartaric acid. Configuration is defined solely by the list of atom numbers (or bond numbers, depending on the details of the representation) connected to the stereocenter.

It is, of course, true that the atoms of a structure can be numbered arbitrarily. The sets of numbers accorded to the neighbors of a particular stereocenter in two different representations of the same structure could, consequently, differ arbitrarily. An ordered list of the numbers accorded to the center's neighbors, as required in such a scheme, could represent one stereoisomer in one case and its mirror image in the other case.

The fact that the configuration designations derived by this scheme are defined only in terms of a given numbered structure (and do not have any absolute significance) is irrelevant, as the naming scheme is completely unambiguous. There are well-defined procedures that can recompute the appropriate changes to configuration designations necessary for accommodation of any arbitrary renumbering of a structure. These procedures can be applied when necessary, such as when the configurational forms of two differently numbered structures have to be compared. These manipulations of stereochemistry are discussed briefly in subsequent chapters (e.g., see Chapter VIII under the discussion of stereochemical graph matching).

Examples from **DENDRAL** can serve to illustrate some of these points. The **DENDRAL** method for representation of configurational stereochemical data is, like the method due to Wipke and Dyott, based on a connection table augmented with configurational designations. Configuration is represented explicitly as an atom property rather than by the order of entries in a connection table. Implicit in the scheme, however, is an ordering that represents the sense of the configuration.[39-41] The method is applicable to substructures as well as structures and can, in consequence, be utilized in the many applications where configurational stereochemistry of substructures must be represented.

The programs can process chiral centers involving tetrahedral stereocenters including tetravalent atoms, e.g. carbon, and noninverting, trivalent centers, such as noninverting nitrogen atoms.[41] Configuration at a chiral center is represented by either the value 0 or 1.[39,40] The programs can produce *Fisher projections* of the environment of a stereocenter in either configuration; these displays specify how the computer's internal 0/1 configuration designation relates to the form of a physical model. Figure VII-1 illustrates the display of the two possible configurations at C(1) of 1-chloro-ethanol (**VII-14**, numbered as

```
       4                         4
       :                         :
       :                         :
  2>>>1<<<3                 3>>>1<<<2
       :                         :
       :                         :
       H                         H

CONFIGURATION ONE        CONFIGURATION ZERO
```

FIGURE VII-1. **DENDRAL** configuration designations 1 and 0 for C(1) of **VII-14**.

shown).† The chemist–user never needs to remember the convention because the program can always present its internal definition in the form of conventional stereochemical diagrams.

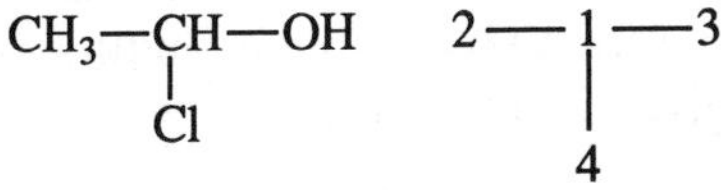

```
CH3—CH—OH      2—1—3
     |           |
     Cl          4
```

VII-14

Displays of the different structural forms corresponding to alternative designated configurations can be exploited in an interactive procedure for augmenting a topological structure description with stereochemical data. For example, to complete the definition of 2-chloro-cyclohexanol (**VII-15**) (with a connection table as shown in Table **VII-11**) it is necessary to specify configurations at

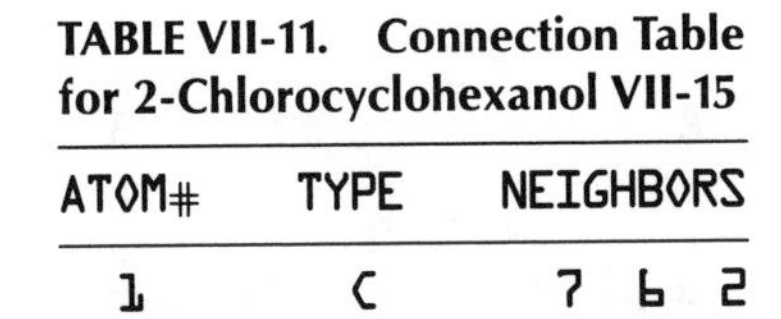

```
O      C—C        7      6—5
 \   /    |         \   /   |
  C       |           1     |
  |       C           |     4
  C—C  /              2—3  /
 /                   /
Cl                  8
```

VII-15

TABLE VII-11. Connection Table for 2-Chlorocyclohexanol VII-15

ATOM#	TYPE	NEIGHBORS
1	C	7 6 2
2	C	8 3 1
3	C	4 2
4	C	5 3
5	C	6 4
6	C	1 5
7	O	1
8	CL	2

† In these displays, hydrogen atoms are added for clarity. (Hydrogens are assigned the number 0, thereby establishing their priority in the numbering scheme).

stereocenters C(1) and C(2). The desired stereochemistry could be as illustrated in structure **VII-16**.

VII-16

A recording of a terminal session is presented in Figure VII-2; in this session the chemist requests projections of the stereocenters C(1) and C(2) and selects the correct configuration based on these projections. In this instance the configurations are both 0 on the basis of atom numbering of **VII-16**. The resulting connection table for **VII-16**, now including the configuration designations just defined, is given in Table VII-12.

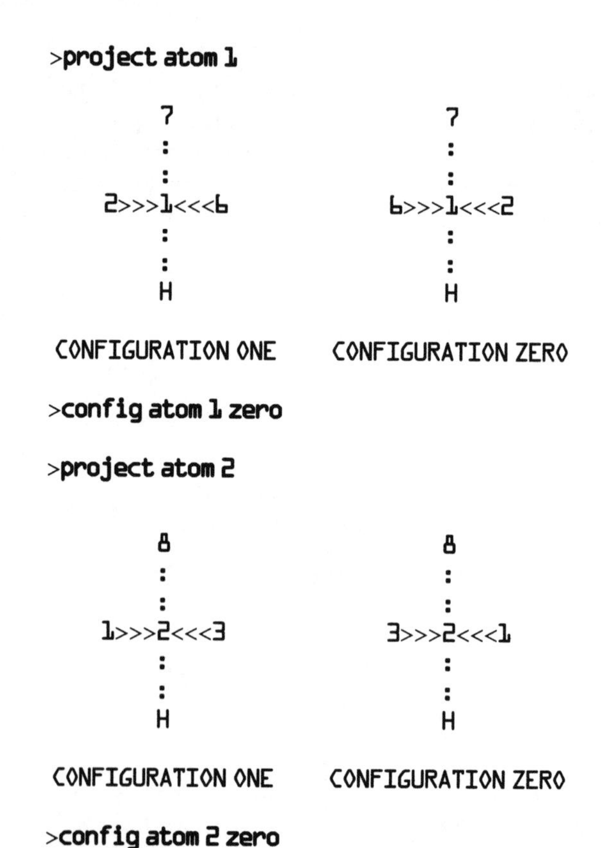

FIGURE VII-2. Summary of steps required in specification of stereochemistry at C(1) and C(2) of **VII-16**.

TABLE VII-12. Connection Table for VII-16, Including Configuration Designations

ATOM#	TYPE	NEIGHBORS	CONFIG
1	C	7 6 2	'0'
2	C	8 3 1	'0'
3	C	4 2	
4	C	5 3	
5	C	6 4	
6	C	1 5	
7	O	1	
8	CL	2	

The **DENDRAL** programs also include mechanisms for presenting the stereochemistry of structures or substructures on standard computer terminals.[41] The interaction with **DENDRAL** and the resulting drawing are shown in Figure VII-3. This drawing, derived from the connection table in Table VII-12, displays the desired, defined stereochemistry.

As an example of the dependence of the 1, 0 configuration designations on structure numbering, assume that **VII-15** had been given the numbering illustrated by **VII-17**.

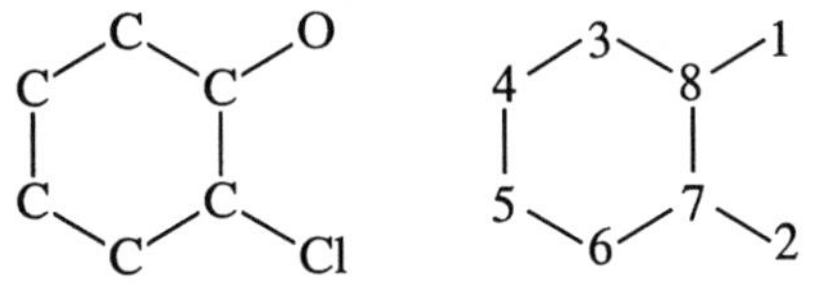

```
    C     O            3     1
 C     C            4     8
 |     |            |     |
 C     C            5     7
    C     Cl           6     2
```

VII-17

```
>sdraw
RINGSIZE:
IDENTIFYING ATOM: 1
IDENTIFYING ATOM:
DO YOU WANT A CROSSRING BOND? (YES OR NO) n
TYPE OF DRAWING: pretty
TYPE OF DRAWING:

8  3
: / \
2    4
|    |
|    |
1    5
! \ /
7  6

CONFIGURATIONS DEFINED:
1->0; 2->0;
```

FIGURE VII-3. "Teletype" drawing of **VII-16** with stereochemistry.

TABLE VII-13. Connection Table for VII-17 with Configuration Designations

ATOM#	TYPE	NEIGHBORS	CONFIG
1	O	8	
2	CL	7	
3	C	8 4	
4	C	5 3	
5	C	6 4	
6	C	7 5	
7	C	2 8 6	'1'
8	C	1 3 7	'1'

This alternative numbering of the structure results in assignment of the configuration designation 1 to the two stereocenters, C(8) and C(7) of **VII-17**. This result is illustrated in Figure VII-4, where the procedures for assigning configuration are applied to structure **VII-17**.

The connection table resulting from the assignment of configurations in Figure VII-4 is shown in Table VII-13.

The example presented in Table VII-13 illustrates that the actual 1, 0 configuration designations are completely dependent on the numbering of the structure. In Table VII-12 the configurations of the two stereocenters are both 0; in Table VII-13 they are both 1. Yet, both connection tables represent the same structure; their identity is revealed in Figure VII-5, where the connection table in Table VII-13 is drawn with stereochemistry and the resulting structure is exactly that of **VII-16** shown in Figure VII-3, except, of course, for the differences in numbering.

Chemists can exploit mental models of the forms of structures that allow differences in numbering to be ignored and thus simplify the perception of the equivalence of structures **VII-16** and **VII-17**. Such tasks are not as simple for computers. The automatic perception of **VII-16** and **VII-17** as the same structure requires special algorithms that are discussed further in Chapter VIII.

The **DENDRAL** programs treat double-bond configurations by assigning configuration designations to both the atoms constituting the double bond. Again, there is a convention for the configuration designations corresponding to *cis* and *trans*. Consider as an example substructure **VII-18**, whose connection table is given in Table VII-14.

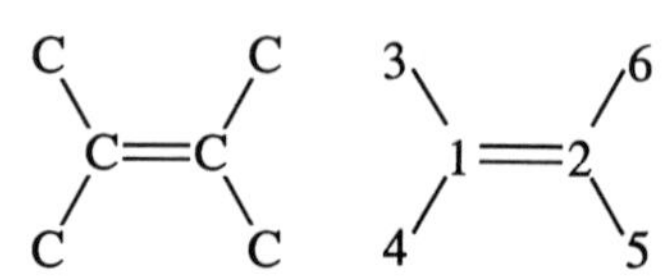

```
C       C      3       6
 \     /        \     /
  C===C          1===2
 /     \        /     \
C       C      4       5
```

VII-18

```
>project atom 8

       1                     1
       :                     :
       :                     :
   3>>>8<<<7             7>>>8<<<3
       :                     :
       :                     :
       H                     H

CONFIGURATION ONE     CONFIGURATION ZERO

>config atom 8 one

>project atom 7
```

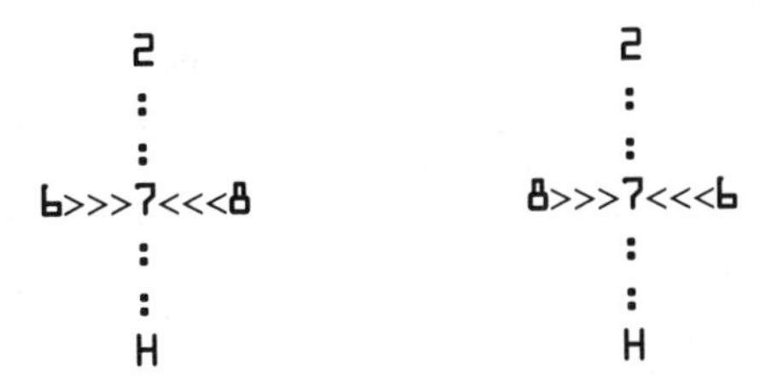


```
CONFIGURATION ONE     CONFIGURATION ZERO

>config atom 7 one
```

FIGURE VII-4. Assignment of stereochemistry of **VII-16** based on the numbering given by **VII-17**.

```
>sdraw
RINGSIZE: 6
IDENTIFYING ATOM: 8
IDENTIFYING ATOM:
DO YOU WANT A CROSSRING BOND? (YES OR NO) no
TYPE OF DRAWING: pretty
TYPE OF DRAWING:

2
   6
: /  \
7     5
|     |
|     |
8     4
! \  /
   3
1

CONFIGURATIONS DEFINED:
7→1; 8→1;
```

FIGURE VII-5. "Teletype" drawing of **VII-17** with stereochemistry.

TABLE VII-14. Connection Table Representation for Substructure VII-18

ATOM#	TYPE	NEIGHBORS			
1	C	4	3	2	2
2	C	6	5	1	1
3	C	1			
4	C	1			
5	C	2			
6	C	2			

The definition of double-bond stereochemistry in **DENDRAL** is illustrated through the data presented in Figure VII-6. The two connection tables, shown in Table VII-15, represent the *cis* and *trans* configurations of structure **VII-18**. These tables are differentiated by the different pairs of configuration designations assigned to the doubly bonded carbon atoms C(1) and C(2).

Configurations on double bonds can be specified easily and naturally merely by selecting the atom numbers of any substituents, one on each end of the double bond, and designating whether they are *cis* or *trans* to one another. From that point on, the program determines, using its internal conventions, the appropriate configuration designations to assign, and how the stereochemistry is to be manipulated in subsequent steps.

These examples dealing with both chiral centers and double bonds illustrate that completely general systems for defining and representing stereochemistry can be developed on the basis of connection table representations of structure.[37,38,39,40,41] The fact that the order of neighbors in the connection table or the configuration designation is dependent on the numbering of the substructure is irrelevant because conventions internal to the computer programs are effectively isolated from the chemists using the programs. As long as the programs to manipulate and display the stereochemistry are cognizant of the scheme for assigning such designations and the convention for representing the alternative stereochemistries, there is no possibility for ambiguity.

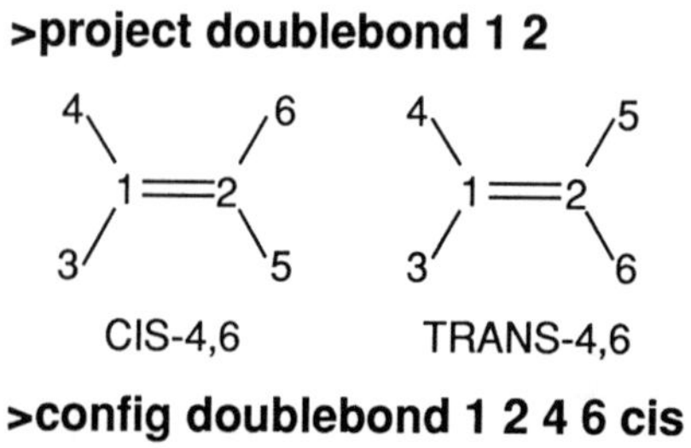

```
>project doublebond 1 2

4\       /6      4\       /5
  \     /          \     /
   1===2            1===2
  /     \          /     \
3/       \5      3/       \6

   CIS-4,6          TRANS-4,6

>config doublebond 1 2 4 6 cis
```

FIGURE VII-6. Conventions for presentation and designation of double bond stereochemistry.

TABLE VII-15. Connection Table Representations for VII-18 in both *cis* and *trans* Configurations

cis-4,6

ATOM#	TYPE	NEIGHBORS				CONFIG
1	C	4	3	2	2	'1'
2	C	6	5	1	1	'0'
3	C	1				
4	C	1				
5	C	2				
6	C	2				

trans-4,6

ATOM#	TYPE	NEIGHBORS				CONFIG
1	C	4	3	2	2	'0'
2	C	6	5	1	1	'0'
3	C	1				
4	C	1				
5	C	2				
6	C	2				

VII.D. REPRESENTATION OF CONFORMATIONAL STEREOCHEMISTRY

The next level of detail in structural representation involves description of the conformation of structures and substructures. It is mainly in the context of studies of structure–activity relationships that a requirement arises for a manipulable conformational description.

There are no suitable representational schemes for conformation based on linear notations. Of course, there are many textual descriptions of conformations, such as the boat, chair, and twist-boat conformations of six-membered rings, but these descriptors are not presently included in any computer-based linear notational system. Such detailed structural information is seldom available. There is no real purpose in designing an information retrieval system on the basis of a linear notation to capture conformational descriptions if the vast majority of structures cannot be so described.

A method for describing conformations can be based on a *connection table* representation of structure, *augmented with labels on bonds that designate torsion angles or ranges of torsion angle*.[42,43] This approach has also been used to define conformations of substructures.[44] These developments are currently research projects and have not yet become incorporated into working systems of practical utility to structural chemistry.

A variation of these methods is a hybrid between connection tables and a graphical representation. This method, due to Gordon and Pople,[45] uses the ordering of the neighbors list in a connection table representation of a structure to define the rotameric state of each bond. Given this information and standard bond lengths and angles, one has a representation of the geometry of a structure.

Representations of conformation can use the set of *X, Y, Z* coordinates of all atoms in a structure. When these data are combined with a connection table (that indicates which atoms are directly bonded), one has a complete description of a conformation. This coordinate information can be obtained by X-ray crystallographic analysis of a structure or from a variety of computer programs designed to compute molecular geometries based on empirical or quantum mechanical methods. A recent book contains several chapters that describe and compare these methods for obtaining and displaying structures using these three-dimensional representations.[46] A major advantage of this type of conformation representation is that subsequent manipulation and display become easy tasks.

VII.E. INPUT AND DISPLAY OF STRUCTURES AND SUBSTRUCTURES

All computer programs for structural analysis have methods for defining structures to the computer and producing output for the chemist. These methods tend to be closely tied to specific types of computer systems and especially the capabilities of the computer terminals available. Some of the methods used for input and display are briefly discussed in this section; the focus is on generic approaches with the illustrative examples drawn from the **DENDRAL** programs. Methods using connection tables as the basis of the computer's internal representation of (sub)structures are accorded most emphasis because these are the most appropriate for structure elucidation systems.

VII.E.1. Linear Notations

The simplest methods for input and output of *linear notations* require manual encoding and decoding. This has the disadvantage of imposing the burden of learning the notational system before one can use a program. Alternatively, linear notations can be generated and used internally in systems where the actual input and output is performed using a graphical representation of structures carried with the name (as exemplified by the CAS registry system),[47] or by using programs that translate between linear notations and connection tables.[48,49]

VII.E.2. Connection Tables

Connection tables themselves can be literally "typed" into the computer. This approach provides a virtual guarantee that few chemists will ever use any program requiring connection tables input in this way. Feldmann[32] pioneered a more satisfactory approach in which a special program provides the chemist with a repertoire of *structure building commands*. Individual commands can, for example, create rings, add acyclic substituent chains, or define the chemical type of particular atoms. The program interprets these commands and builds or modifies its internal representation of a connection table accordingly. The system developed by Feldmann is in routine use as the structure input language for the NIH–EPA CIS.[33] The **DENDRAL** system incorporates a related structure input routine, called ***EDITSTRUC***,[50] which offers a repertoire of structure building commands similar to those used by Feldmann.[32] The commands in these two systems differ somewhat because of the differences in end use of the structures and substructures defined by the two approaches.

A major advantage of Feldmann's approach is that it can be used with any type of computer terminal, thus greatly expanding the accessibility of programs to chemists. It does require that the chemist learn the set of commands comprising the input language. But because the commands are closely related to terms with which chemists are familiar, this approach is simpler to master than one based on a notational system. Some examples of the use of ***EDITSTRUC*** were given in Chapter II in the illustrative analysis of warburganal.

A more complex example is shown in Figure VII-7. Here, the ***EDITSTRUC*** commands required to define the complicated substructure **VII-13** are presented. The effect of these commands is to build up the connection table

COMMAND	RESULT
>ring 6	*Builds a six-membered ring of carbon atoms, singly bonded.*
>branch 3 1 5 2	*Places a 1 carbon branch on atom 3 and a 2 carbon branch on atom 5, singly bonded.*
>bord 1 2 any, 2 3 2, 3 4 any, 4 5 any, 5 6 any, 6 1 any	*Sets the bond order for the indicated pairs of bonded atoms; 1 2, 3 4, 4 5, 5 6, and 6 1 are set to "any" bond order, while 2 3 is designated specifically 2, or double bond.*
>atname 1 (n o), 4 x, 5 x, 6 x, 7 (f cl br i)	*Sets the atom names for noncarbon atoms; 1 and 7 are "polyatoms" of type N or O, and F or Cl or Br or I, respectively, while atoms 4, 5, and 6 are of type X, or any type.*
>artype 1 non, 2 non, 3 non, 4 non, 5 non, 6 non	*Sets the aromatic type of atoms 1–6 to be nonaromatic.*
>hrange 2 1 1, 9 3 3	*Sets the hydrogen range on atom 2 to exactly 1, and on atom 9 to exactly 3*
>lnode 8 1 3	*Establishes atom 8 to be a link node, of length one to three atoms*

[a]Lines of input have been broken and put into columns for clarity.

FIGURE VII-7. EDITSTRUC commands for definition of substructure **VII-13**.[a]

shown in Table VII-10.[†] The commands are annotated in the right-hand column of Figure VII-7. These annotations, together with the structure and connection table for **VII-13** and the accompanying textual description, should clarify the function of the commands. Previous examples in this chapter have illustrated some of the other ***EDITSTRUC*** commands [viz., the **PROJECT** and **CONFIG** commands used to display and assign configurations to chiral centers (substructure **VII-17**, Figure VII-4, Table VII-13) and double bonds (substructure **VII-18**, Figure VII-6, Table VII-15)].

Perhaps the most elegant method for input of connection table information is the use of a graphics terminal in conjunction with a pick device such as stylus and tablet[51] or a light pen[37]. The latter devices are used to draw a structural diagram, with the diagram presented on a graphics terminal as it is being drawn. These are also menu-based systems in that a menu of commands is available along the side of the screen[37] or tablet.[51] By pointing to a particular command, the chemist can initiate or terminate structural input, designate stereochemistry, modify atom and bond properties, and so forth. The program that controls the graphical entry system builds a connection table representation in the computer as the structure is defined. Similar systems have been developed in many laboratories, including CAS, as a more natural way to define structural information to computers. Although there have been no systematic studies, anecdotal information reveals a much higher level of acceptance for such input systems by chemists who are not familiar with computers.

Once connection tables are available in the computer, one is faced with the problem of presenting the structures they represent to chemists. The approach taken depends on the amount of additional information that was collected and stored with the connection table on input.

It is of little utility to chemists to have as primary structural output the connection tables themselves (e.g., Table VII-10). This form of output is useful if no interpretable structural diagram can be obtained from the connection table because then the structure can be reconstructed manually from the table. It is also useful in that considerable detail on atom and bond properties can be conveyed in the connection table, information that is difficult to include in an actual drawing. But the most useful form of output is a structural diagram.

Systems that accept graphical input, by means of a light pen and other devices, have already a copy of a satisfactory representation of a structural diagram. The two-dimensional coordinates of every atom can be recorded as a part of the input process and are then available when required to reconstruct a graphical display. In graphical systems that accept stereochemical configurations in the form of bonds "up" or "down" with respect to the plane of the screen, some three-dimensional information is available as well.[37,38] If this

[†] In this example the long-form names of the commands are used for clarity, but multiple arguments are used for the commands to save space. The short forms and multiple arguments are described in detail elsewhere.[31]

information is saved with the connection table, it is trivial to redraw the structure at any time exactly as it was input.

The intermediate case, where interpretable structural diagrams are to be presented when *only* the connection table is available, presents the most difficulty. There are two limitations here, the lack of geometric information in the connection table and the limitations of output devices. Until low-cost graphics terminals become widely available in the chemical community, most chemists will have to interact with programs using standard teletype, or character-based terminals. It is essential that programs be able to provide structural drawings even on such limited terminals.

Feldmann, exploiting concepts developed by Zimmerman[52], was the first to provide structural drawings on character-based terminals, as part of the structural input system mentioned earlier.[32] This system uses features (e.g., rings and chains) found by analysis of the connection table as keys for selection of standard templates from a built-in library. Once selected, these templates can be used to lay out a structural diagram in the rectilinear grid of a character-based terminal. This system provides pleasing and interpretable drawings for many structures[32], but often fails for more complicated systems, especially bridged ring systems.

Carhart[53] has presented a more general approach to drawing structures on character-based terminals. His method relies on construction in the computer of a simplified three-dimensional model of a structure derived from the connection table itself. The model is flattened into two-dimensions, and an iterative procedure adjusts atom and bond positions to correspond to positions in the rectilinear grid of the terminal. Although this approach still fails to draw an interpretable structure in certain pathalogical cases, the fact that it is not a template-based system allows drawings of most complicated chemical structures including such difficult cases as adamantane and cubane.[53] Simple ring systems often are not drawn as effectively as those by Feldmann's program because of the absence of templates that would dictate a "pleasing" orientation and shape to a ring. It is possible to produce much more elegant drawings from connection tables representing only topology, but such drawings rely on graphical output.[54]

As discussed earlier, configurational stereochemistry can be included as part of a connection table. There are systems that can produce interpretable, but inelegant, structural drawings with stereochemical designations on character-based terminals from such connection tables.[37,41] Such systems rely on special characters to indicate "up" and "down" with respect to the plane of the paper and can convey stereochemistry adequately for structures of moderate complexity, such as steroids.[41]

For drawings that convey both topology and configurational stereochemistry for complex structures based only on connection tables, the problem is computationally much more complicated and high-quality output demands the use of graphics terminals. One recent extension to the **DENDRAL** system is a program that produces three-dimensional representations of structure from connection tables, preserving the configurational stereochemistry of chiral centers and double bonds.[55] This method uses tables of standard bond lengths and angles and distance geometry[56] to obtain *X, Y, Z* coordinates for all atoms in a selected stereoisomer. Structural diagrams produced by this program convey much of the geometry of a structure essential for the chemist's understanding of the molecule, but in the absence of conformational constraints, are of arbitrary conformation.[55]

REFERENCES

1. *Survey of Chemical Notation Systems*, National Academy of Sciences, Committee on Chemical (Publication No. 1150), National Academy of Sciences, Washington, DC, 1964.
2. *Chemical Structure Information Handling*, National Academy of Sciences, Committee on Chemical (Publication No. 1733), National Academy of Sciences, Washington, DC, 1969.
3. P. G. Dittmar, R. E. Stobaugh, and C. E. Watson, "The Chemical Abstracts Service Registry System. 1. General Design," *J. Chem. Inf. Comput. Sci.*, **16** (1976), 111.
4. M. Gielen, "Applications of Graph Theory to Organometallic Chemistry," in *Chemical Applications of Graph Theory*, A. T. Balaban, ed., Academic Press, New York, 1976, 261, Chapter 9.
5. L. Spialter, "The Atom Connectivity Matrix (ACM) and Its Characteristic Polynomial (ACMCP)," *J. Chem. Doc.*, **4** (1964), 261.
6. G. Schill, *Catenanes, Rotaxanes, and Knots*, Academic Press, New York, 1971.
7. *Survey of European Non-Conventional Chemical Notation Systems*, National Academy of Sciences, Committee on Chemical (Publication No. 1278), National Academy of Sciences, Washington, DC, 1965.
8. M. F. Lynch, J. M. Harrison, W. G. Town, and J. E. Ash, *Computer Handling of Chemical Structure Information*, American Elsevier, New York, 1971.
9. R. E. Carhart, D. H. Smith, N. A. B. Gray, J. G. Nourse, and C. Djerassi, "GENOA: A Computer Program for Structure Elucidation Utilizing Overlapping and Alternative Substructures," *J. Org. Chem.*, **46** (1981), 1708.
10. A. T. Balaban, (ed.), *Chemical Applications of Graph Theory*, Academic Press, New York, 1976.
11. A. L. Goodson, "Graph-Based Chemical Nomenclature. 1. Historical Background and Discussion," *J. Chem. Inf. Comput. Sci.*, **20** (1980), 167.
12. J. Mockus, A. C. Isenberg, and G. G. Vander Stouw, "Algorithmic Generation of Chemical Abstracts Index Names. 1. General Design," *J. Chem. Inf. Comput. Sci.*, **21** (1981), 183.
13. E. Garfield, *An Algorithm for Translating Chemical Names to Molecular Formulas*, Institute for Scientific Information, 1961.
14. G. G. Vander Stouw, I. Naznitsky, and J. E. Rush, "Procedures for Converting Systematic Names of Organic Compounds into Atom-Bond Connection Tables," *J. Chem. Doc.*, **7** (1967), 165.
15. G. G. Vander Stouw, P. M. Elliott, and A. C. Isenberg, "Automatic Conversion of Chemical Substance Names to Atom-Bond Connection Tables," *J. Chem. Doc.*, **14** (1974), 185.
16. E. G. Smith, *The Wiswesser Line-Formula Chemical Notation*, McGraw-Hill, New York, 1968.
17. H. W. Hayward, *A New Sequential Enumeration and Line Formula Notation System for Organic Compounds* (Report No. 21), Patent Office, Washington, DC, 1961.

18. H. W. Hayward, H. M. S. Sneed, J. H. Turnipseed, and S. J. Tauber, "Some Experience with the Hayward Linear Notation System," *J. Chem. Doc.*, **5** (1965), 183.
19. G. M. Dyson, M. F. Lynch, and H. L. Morgan, "A Modified IUPAC/Dyson Notation System for Chemical Structures," *Inf. Storage Retr.*, **4** (1968), 27.
20. R. G. Dunn, "A Chemical Substructure Search System Based on Chemical Abstracts Index Nomenclature," *J. Chem. Inf. Comput. Sci.*, **17** (1977), 212.
21. A. V. Tomea and P. F. Sorter, "On-Line Substructure Searching Utilizing Wiswesser Line Notations," *J. Chem. Inf. Comput. Sci.*, **16** (1976), 223.
22. H. Gelernter, N. S. Sridharan, A. J. Hart, F. W. Fowler and H. Shue, "The Discovery of Organic Synthetic Routes by Computer," *Topics Curr. Chem.*, **41** (1973), 113.
23. H. L. Gelernter, A. F. Sanders, D. L. Larsen, K. K. Agarwal, R. H. Boivie, G. A. Spritzer, and J. E. Searleman, "Empirical Explorations of SYNCHEM," *Science*, **197** (1977), 1041.
24. C. N. Mooers, *Ciphering Structural Formulas - the Zatopleg System*, Zator Co., Cambridge MA, 1951.
25. L. C. Ray and R. A. Kirsch, "Finding Chemical Records by Digital Computers," *Science*, **126** (1957), 814.
26. L. Spialter, "The Atom Connectivity Matrix (ACM) and Its Characteristic Polynomial (ACMCP): A New Computer-Oriented Chemical Nomenclature," *J. Am. Chem. Soc.*, **85** (1963), 2012.
27. I. Ugi and P. Gillespie, *Angew. Chem.*, **83** (1971), 980.
28. J. Blair, J. Gasteiger, C. Gillespie, P. D. Gillespie, and I. Ugi, "CICLOPS - A Computer Program for the Design of Syntheses on the Basis of a Mathematical Model," in *Computer Representation and Manipulation of Chemical Information*, W. T. Wipke, S. Heller, R. Feldmann, and E. Hyde, eds, Wiley, New York, 1974, 129, Chapter 6.
29. J. Dugundji, P. Gillespie, D. Marquarding, I. Ugi, and F. Ramirez, "Metric Spaces and Graphs Representing the Logical Structure of Chemistry," in *Chemical Applications of Graph Theory*, A. T. Balaban, ed., Academic Press, New York, 1976, 107, Chapter 6.
30. D. H. Rouvray, "The Topological Matrix in Quantum Chemistry," in *Chemical Applications of Graph Theory*, A. T. Balaban, ed., Academic Press, New York, 1976, 175, Chapter 7.
31. D. H. Smith, *GENOA User's Manual*, The DENDRAL Project, Department of Chemistry, Stanford University, Stanford, CA 94305, 1981.
32. R. J. Feldmann, "Interactive Graphic Chemical Structure Searching," in *Computer Representation and Manipulation of Chemical Information*, W. T. Wipke, S. Heller, R. Feldmann, and E. Hyde, eds., Wiley, New York, 1974, 55, Chapter 3.
33. G. W. A. Milne and S. R. Heller, "NIH/EPA Chemical Information System," *J. Chem. Inf. Comput. Sci.*, **20** (1980), 204.
34. M. Milne, D. Lefkovitz, H. Hill, and R. Powers, "Search of CA Registry (1. 25 Million Compounds) with the Topological Screens System," *J. Chem. Doc.*, **12** (1972), 183.
35. J. E. Blackwood, P. M. Elliott, R. E. Stobaugh, and C. E. Watson, "The Chemical Abstracts Service Chemical Registry System. III. Stereochemistry," *J. Chem. Inf. Comput. Sci.*, **17** (1977), 3.
36. R. S. Kahn, C. K. Ingold, and V. Prelog, "The Specification of Assymetric Configuration in Organic Chemistry," *Experientia*, **12** (1956), 81.
37. W. T. Wipke, "Computer-Assisted Three-Dimensional Synthetic Analysis" in *Computer Representation and Manipulation of Chemical Information*, W. T. Wipke, S. Heller, R. Feldmann, and E. Hyde, eds., Wiley, New York, 1974, 147, Chapter 7.
38. W. T. Wipke and T. M. Dyott, "Simulation and Evaluation of Chemical Synthesis. Computer Representation and Manipulation of Stereochemistry," *J. Am. Chem. Soc.*, **96** (1974), 4825.

39. J. G. Nourse, "The Configuration Symmetry Group and Its Application to Stereoisomer Generation, Specification, and Enumeration," *J. Am. Chem. Soc.*, **101** (1979), 1210.
40. C. Djerassi, D. H. Smith, and T. H. Varkony, "A Novel Role of Computers in the Natural Products Field," *Naturwissenschaften*, **66** (1979), 9.
41. J. G. Nourse, D. H. Smith, and C. Djerassi, "Computer-Assisted Elucidation of Molecular Structure with Stereochemistry," *J. Am. Chem. Soc.*, **102** (1980), 6289.
42. W. T. Wipke and T. M. Dyott, "Stereochemically Unique Naming Algorithm," *J. Am. Chem. Soc.*, **96** (1974), 4834.
43. J. G. Nourse, "Specification and Enumeration of Conformations of Chemical Structures for Computer-Assisted Structure Elucidation," *J. Chem. Inf. Comput. Sci.*, **21** (1981), 168.
44. A. L. Fella, J. G. Nourse, and D. H. Smith, "Conformation Specification of Chemical Structures in Computer Programs," *J. Chem. Inf. Comput. Sci.*, **23** (1983), 43.
45. M. S. Gordon and J. A. Pople, *Program MBLD*, (Program No. 135), Quantum Chemistry Program Exchange, 1975.
46. *Computer-Assisted Drug Design*, E. C. Olson and R. E. Christoffersen, eds. (ACS Symposium Series 112), American Chemical Society, Washington, DC, 1979.
47. A. Zamora and D. L. Dayton, "The Chemical Abstract Service Chemical Registry System. V. Structure Input and Editing," *J. Chem. Inf. Comput. Sci.*, **16** (1976), 219.
48. C. D. Farrell, A. R. Chauvenet, and D. A. Koniver, "Computer Generation of Wiswesser Line Notation," *J. Chem. Doc.*, **11** (1971), 52.
49. G. A. Miller, "Encoding and Decoding WLN," *J. Chem. Doc.*, **12**, (1972), 60.
50. R. E. Carhart, D. H. Smith, H. Brown, and C. Djerassi, "Applications of Artificial Intelligence for Chemical Inference. XVII. An Approach to Computer-Assisted Elucidation of Molecular Structure," *J. Am. Chem. Soc.*, **97** (1975), 5755.
51. E. J. Corey and W. T. Wipke, "Computer-Assisted Design of Complex Organic Syntheses," *Science*, **166** (1969), 178.
52. B. L. Zimmerman, *Computer-Generated Chemical Structural Formulas with Standard Ring Orientations*, Ph. D. Thesis, University of Pennsylvania, 1971.
53. R. E. Carhart, "A Model-Based Approach to the Teletype Drawing of Chemical Structures," *J. Chem. Inf. Comput. Sci.*, **16** (1976), 82.
54. C. A. Shelley, "A Heuristic Approach for Translating Connection Tables to Structural Diagrams," in *Book of Abstracts, 182nd ACS National Meeting, New York*, American Chemical Society, 1981.
55. J. C. Wenger and D. H. Smith, "Deriving Three-Dimensional Representations of Molecular Structure from Connection Tables Augmented with Configuration Designations Using Distance Geometry," *J. Chem. Inf. Comput. Sci.*, **22** (1982), 29.
56. G. Crippen, *Distance Geometry and Conformation Calculations*, Research Studies Press, 1981.

VIII

SUBSTRUCTURE SEARCH AND RELATED SYSTEMS

VIII.A. INTRODUCTION

The connection table representations of structures and substructures, discussed in Chapter VII, allow structural information to be *manipulated* by computer programs. Once structure manipulations are possible, computer programs are no longer limited to simple information retrieval tasks. One major benefit of a program-manipulable structure representation is that it allows for tests to be made on a structure that can identify constituent atoms and bonds with particular properties. A substructure definition, which is simply a specification of a set of required atoms and bonds, can be manipulated to yield, in effect, a sequence of tests that may then be applied to a structure. The application of these tests to some specific structure determines whether that structure contains an instance of the substructure. This ability to identify the occurrence(s) of some specified substructure in a molecule provides a basis for more elaborate computer programs such as those used in studies of substructure–activity relationships or for the prediction of spectral properties or analysis of how molecules react. An appropriate starting point for consideration of structure manipulation programs is, therefore, *substructure search*.

The most common application of *substructure search* routines is in systems for searching computer files of structures for those that contain a specified substructure (or set of substructures). These search systems can be used to identify sets, or families, of structures that are related by their substructural content. The reasons for these identifications are many and include at least the following:

1. Location of known compounds of a given class, where compound class is interpreted very generally to mean any group of compounds possessing particular functionalities or ring systems.
2. Identification of compounds that may possess similar pharmacological or toxicological behavior on the basis of common shared substructures.
3. Location of close relatives of new compounds as part of patent development or search.

Occasionally, it may be necessary to count the number of occurrences of the substructure rather than simply determine whether it occurs. A substructure search system that can find all distinct occurrences of a substructure in some structure allows more complex search requests to be specified and used.

A general-purpose substructure search system must allow for a rich language of substructural description, including a wide selection of atom and bond properties. This aspect was introduced in Chapter VII in the discussions of substructure representations; there, some examples were given to illustrate various atom and bond properties, including variable atom and bond types.

Stereochemical detail can also be incorporated into both substructure and structure specifications. If stereochemical detail is available, a search can be constrained to retrieve only those reference structures with appropriate stereochemistry. Currently, relatively few systems allow for stereochemical constraints. The extensions to standard substructure search systems that are necessary to accommodate stereochemistry are discussed later.

It is also useful for a substructure search system to allow the use of multiple substructures and to permit search requests that specify particular combinations of these substructures. For example, one might wish to search for all structures possessing as substructures a perhydrophenanthrene skeleton (**VIII-1**) *and* a 5–6-ring fused system (**VIII-2**) that were *not* also steroids represented by the skeleton **VIII-3**.

VIII-1 **VIII-2** **VIII-3**

Given appropriate definitions of **VIII-1**–**VIII-3**, one needs to perform the search with the logical expression:

(VIII-1 and VIII-2) and (not VIII-3)

The *SURVEY* program, illustrated in Chapter II, is just one of many substructure search programs that incorporates these capabilities.

A basic substructure search system first reads in definitions of the required substructure(s) and then works through the reference file of structure definitions, processing these one at a time. The processing of each reference structure entails calling the subroutine for attempting to match a substructure onto the current structure. If multiple substructures are used, the substructure matching routine must be called once for each substructure and records must be kept to identify the presence or absence of each. When all the substructures have been matched against the current structure, the main program must analyze any logical expressions to determine the overall success or failure of the required matching process.

With large reference files, the computational costs of matching the substructure(s) onto every reference structure can become excessive. The problem is analogous to that considered in Chapter IV, where search systems for matching observed spectra with reference data in large files were discussed. The solution is also analogous. The spectral file search systems used various "prefilters" that served to eliminate the vast majority of reference spectra prior to any attempt at computing a detailed spectrum matching coefficient. Similarly, in substructure searches, various preliminary screening steps can be employed to limit the number of reference structures onto which the substructure(s) must be matched in detail.

Search systems for retrieving structures containing particular substructures entail a rather passive and restricted version of the concept of *substructure search*. The basic concept of substructure search can be elaborated; the **GENOA** structure elucidation system[1] is one such elaboration. The **GENOA** system has a rather "procrustean" approach to substructure search—specifically, if a structure does not already contain a requested substructure, it is forced to (by the construction of such additional bonds and atom properties as may be needed).

As illustrated in the example analysis of warburganal presented in Chapter II, **GENOA**'s task is to start with the set of constituent atoms of an unknown structure and build up partial structures, or *cases*, that incorporate those substructures specified as constraints. The first substructural constraint given to **GENOA** is interpreted simply as a request to build the specified substructure from the available atoms; often, such a constraint will define a presumed molecular skeleton. Subsequent constraints request that if the cases do not already contain the desired substructure, they should be elaborated through new bond building steps or by assignment of more detailed attributes for constituent atoms.

Sometimes, a constraint may already be satisfied in the existing constructed case(s). The **GENOA** system can determine whether this is true by using a conventional substructure search subroutine to find all distinct occurrences of the substructure in the case. The analysis of warburganal in Chapter II includes one rather simple instance where previously specified constraints resulted in cases that already satisfied a subsequently specified constraint. There, the requirements for an enal system and $>C{-}CH{=}O$ group meant that the cases already contained the four C_{sp^2} carbons requested. More typically, a substructure specified as a subsequent constraint may be partially matched in several places. Completion of any of these partial matchings requires either construction of new bonds or, sometimes, simply a specification of atom properties. In a structure generator, all distinct ways of completing each matching must be found.

For example, suppose that initial constraints in some structural problem have established two reasonably large fragments: the cyclopropane system **VIII-4** and the branched oxirane **VIII-5**, leaving the bonding of three carbons and nine hydrogens still to be specified. (The free valences shown on **VIII-4** and **VIII-5** are to be permitted to bond to hydrogens.) The next constraint applied in this structural problem is a requirement for an instance of substructure **VIII-6**.

VIII-4 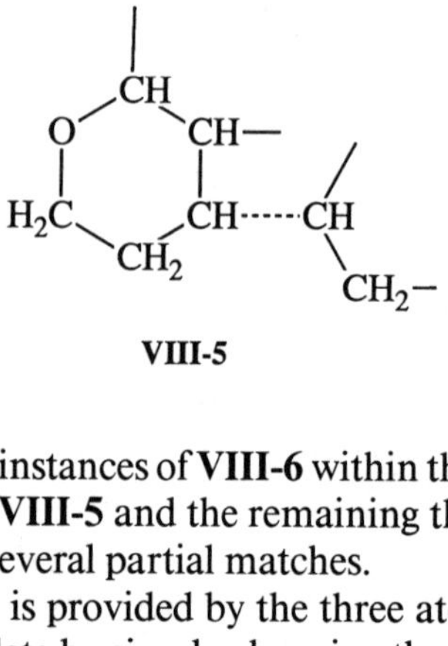 **VIII-5** —CH—CH_2—CH— **VIII-6**

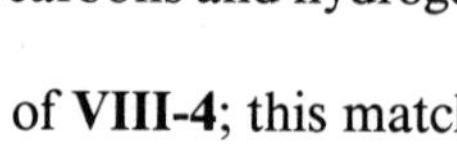

There are no existing instances of **VIII-6** within the existing fragments defining the case (i.e., **VIII-4**, **VIII-5** and the remaining three carbons and hydrogens). There are, however, several partial matches.

One partial match is provided by the three atoms of **VIII-4**; this matching could be made complete by simply changing the hydrogen range of one of the cyclopropane carbons from being (1 or 2) to (2). This yields **VIII-7**, one of the many more elaborate cases that **GENOA** would create when performing this *constructive substructure search*.

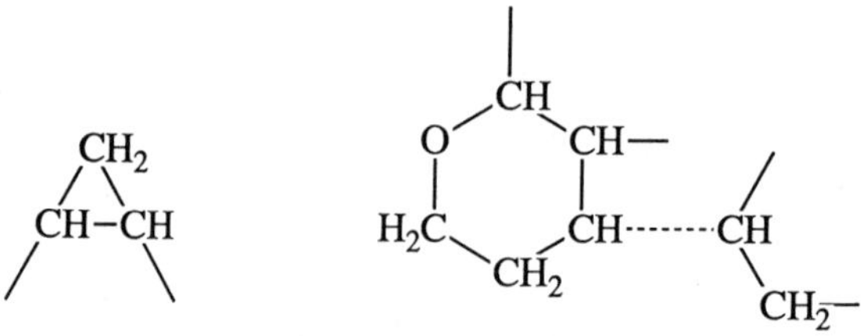

3C + 9H

VIII-7

A rather similar matching of **VIII-6** into **VIII-5** would again require only changes to attributes on atoms; this constructive matching forces another methylene group into the oxirane ring. Other matchings onto atoms from **VIII-5** necessitate bond building steps. The three atoms of **VIII-6** could be matched to the two atoms in the substituent chain of **VIII-5** and to either methine in the ring. Completion of such partial matchings requires a bond building step from the methylene to the chosen methine. One of the two cases that would result from these constructive matchings is shown as **VIII-8**.

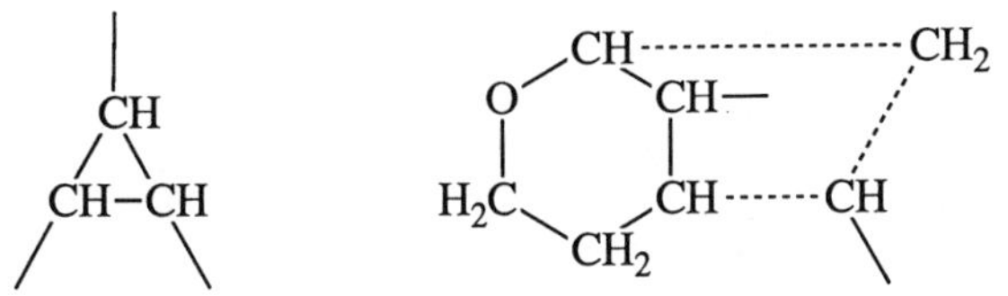

```
       |              CH·················CH2
      CH           O      CH—           :
     / \           |      |            :
   CH—CH         H2C      CH·······CH              3C + 9H
   /    \            CH2             \
```

VIII-8

Alternative matchings of **VIII-6** can take atoms from both **VIII-4** and **VIII-5**, yielding structures like **VIII-9**, or can use some of the previously unattached carbon and hydrogen atoms to produce structures such as **VIII-10**. Still another constructive matching builds the required instance of **VIII-6** solely from the previously unused carbons and results in **VIII-11**.

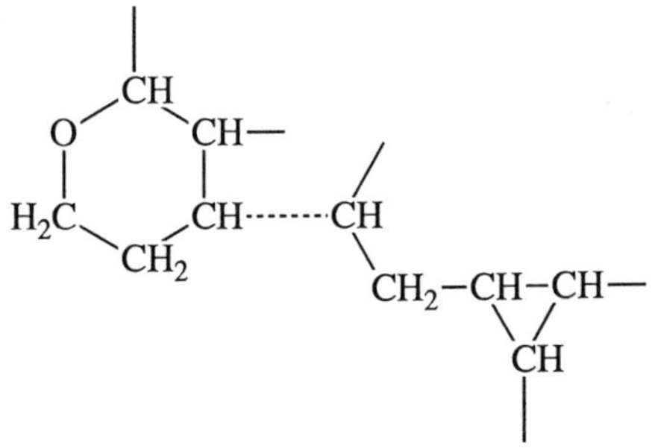

```
           |
           CH
        O      CH—
        |      |       /
      H2C      CH·······CH                         3C + 9H
          CH2            \
                          CH2—CH—CH—
                              \ /
                              CH
                              |
```

VIII-9

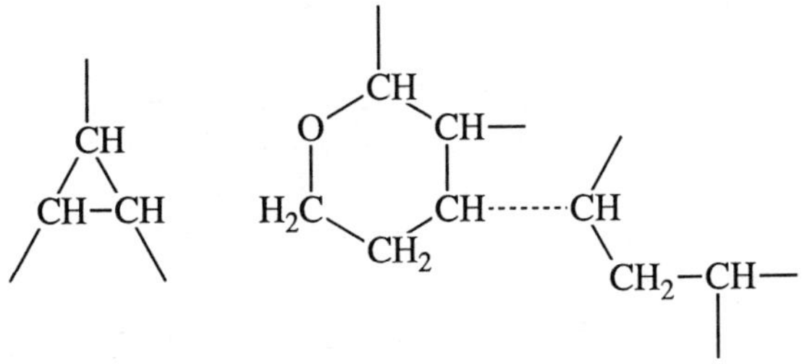

```
                 |
     |           CH
    CH        O      CH—
   / \        |      |       /
 CH—CH      H2C      CH·······CH                   2C + 8H
 /    \         CH2            \
                                CH2—CH—
                                     |
```

VIII-10

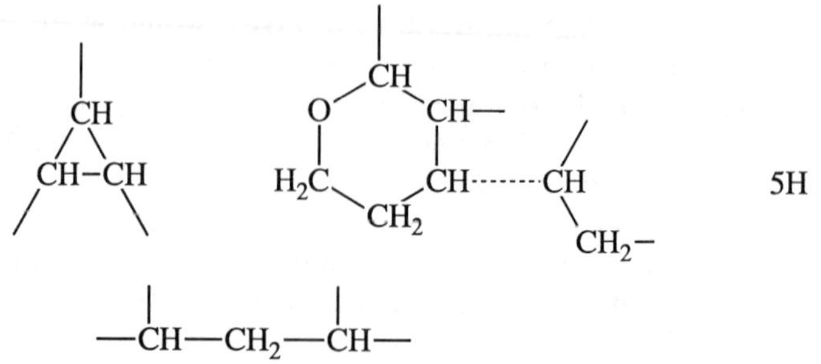

CH
CH–CH

CH
O CH–
H_2C CH·······CH 5H
CH_2 CH_2–

–CH–CH_2–CH–

VIII-11

In summary, conventional substructure searches analyze structures to determine the number of occurrences of a substructure; *constructive* substructure searches transform a "structure" so that it incorporates a specified number of occurrences of a substructure. Although the final outcomes of these procedures are quite different, the underlying analysis is very similar. The substructure definition becomes a specification of tests that prescribe how a search for atoms is to be performed. This search is performed exhaustively, with every possible matching of similar structural and substructural atoms being considered. In ordinary substructure search, partial matchings that cannot be extended to perfect matches are abandoned; in the constructive search, one proceeds by exploring all possible transformations of the structure that would allow the matching to be completed.

The following sections present details of (1) conventional searches for finding instances of topological defined substructures within structures, (2) extensions to accommodate stereochemical detail in structures and substructures, (3) methods developed to optimize searches against large files of reference structures, and (4) of the *constructive* searches used in **GENOA**. The discussion of **GENOA**'s algorithms introduces problems of recognizing and using symmetries in structures; solutions for these new problems are presented in Chapter IX.

VIII.B. SUBSTRUCTURE SEARCH USING CONSTITUTIONAL REPRESENTATIONS OF STRUCTURE

There are two main methods for searching for correspondences between the atoms of a substructure and those of a structure: (1) the set reduction technique proposed by Sussenguth,[2] and (2) the node-by-node technique proposed by Ray and Kirsch.[3] Both fall under the generic term *graph matching*, a term that results from the view of topological representations of structure as graphs. In this view, substructure search is equivalent to comparison of graphs. In both procedures a substructure is said to be found, or contained within, a structure if the graph represented by a substructure matches a portion of the graph of a structure exactly, considering the connectivity of both entities and all defined atom and bond properties, that is all relevant descriptions of constitution.

VIII.B.1. Graph Matching by Set Reduction

The general procedure of set reduction involves an initial selection of descriptors used to characterize atoms and bonds in structures. Sussenguth proposed atom type, branch value (the type of bond efferent from an atom, single, double, etc.), degree (number of neighbors), order (smallest ring membership), and connectivity (identity of nearest neighbors).[2] These are all properties found explicitly in the connection table or determined by other routines that examine the connection table. Obviously this list could be expanded to include other atom and bond properties, each inclusion serving to reduce the size of sets used in the procedure itself outlined in the following paragraph. Here the issue of efficiency arises, because although the inclusion of additional properties serves to make the subsequent computations more rapid, each property requires additional time to compute. There are no firm rules for deciding on a useful set of properties, because it is always possible to design pathological examples that make any given choice inefficient or incomplete.

The procedure itself consists of the following basic steps:[2]

1. Establish set membership for atoms in both the substructure and structure, for each selected property, except connectivity.
2. Partition the sets using the logical operation of set intersection to reduce the membership of each set.
3. If a unique assignment of atoms in the substructure to atoms in the structure has not been made, use the property of connectivity to obtain new sets.
4. Partition the resulting sets as in step 2.
5. If a unique assignment has not been made, use an assignment procedure to resolve ambiguities.

This procedure is best illustrated with an example. This example illustrates steps 1–4 (above) and is used in subsequent sections of this chapter to illustrate other methods of substructure search. Additional complexities of the algorithm are treated in some detail in the original paper.[2]

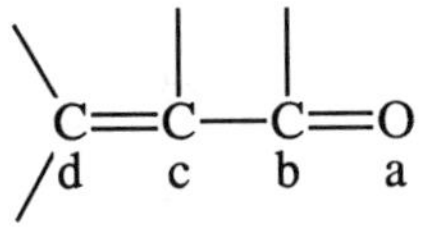

VIII-12 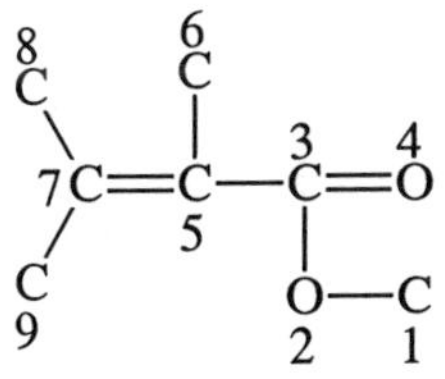 **VIII-13**

This example illustrates how the set reduction procedure would apply to the problem of determining whether substructure **VIII-12** can be found within structure **VIII-13**.† In general, a substructure may be described with any number of atom and bond properties, and a successful match can occur only if the substructure and structure are in agreement with respect to these properties. For this example the free valences of **VIII-12** are assumed to be to nonhydrogen atoms, and hydrogen atoms are ignored in the procedure. In addition, substructure **VIII-12** is characterized by specific atom names and bond types.

Of course, the result is obvious by visual inspection. Remember, however, that a program has no direct means of visualizing **VIII-12** and **VIII-13** in the way that a chemist would. A program must work with its internal representation, here a connection table, using a well-defined procedure. Set reduction is such a procedure, one that succeeds in this simple case and also in extremely complex cases, such as bridged polycyclic systems, that would defeat any attempt at visual inspection.

TABLE VIII-1. Set Reduction Procedure Applied to Location of Substructure VIII-12 in Structure VIII-13

	Initial Subsets		
	Subset of **VIII-12**	Subset of **VIII-13**	Line
Node value			
C	{b,c,d}	{1,3,5,6,7,8,9}	1
O	{a}	{2,4}	2
Branch value			
Single	{b,c,d}	{1,2,3,5,6,7,8,9}	3
Double	{a,b,c,d}	{3,4,5,7}	4
Degree			
1	{a}	{1,4,6,8,9}	5
2		{2}	6
3	{b,c,d}	{3,5,7}	7
Order			
oo	{a,b,c,d}	{1,2,3,4,5,6,7,8,9}	8
Partition			
Lines 1–8	{a}	{4}	9
	{b,c,d}	{3,5,7}	10
Connectivity			
From line 9	{b}	{3}	11
From line 10	{a,b,c,d}	{2,3,4,5,6,7,8,9}	12
From line 11	{a,c}	{4,5}	13
From line 13	{b,d}	{3,6,7}	14
Partition			
Lines 9–14	{a}	{4}	15
	{b}	{3}	16
	{c}	{5}	17
	{d}	{7}	18

†The (arbitrary) labelings a–d for atoms in substructure **VIII-12**, and 1–9 for atoms in the structure **VIII-13**, are employed to allow for reference to specific atoms in the following discussions.

Step 1 of the procedure involves description of the set membership of the atoms of the substructure and structure. As shown in lines 1–8 in Table VIII-1, this step involves the formation of a series of subsets that differentiate the atoms of **VIII-12** and **VIII-13** as follows:

1. *By Node Value*. Atoms b,c,d of **VIII-12** are carbon atoms that must then correspond to three of carbon atoms 1,3,5,6,7,8,9 of **VIII-13** (line 1). Atom a of **VIII-12** is an oxygen and must correspond to either oxygen atom 2 or 4 of **VIII-13** (line 2).
2. *By Branch Value*. Atoms b,c,d have efferent single bonds, so this subset must correspond to three of atoms 1,2,3,5,6,7,8,9 (line 3). All atoms of **VIII-12** have efferent double bonds, whereas only atoms 3,4,5,7 of **VIII-13** possess this property (line 4).
3. *By Degree*. Only atom a of **VIII-12** possesses a single neighbor, degree 1, whereas atoms 1,4,6,8,9 of **VIII-13** meet this criterion (line 5). There are no atoms of **VIII-12** of degree 2 (line 6). Atoms b,c,d and the corresponding subset 3,5,7 are of degree three (line 7).
4. *By Order*. No atoms of either **VIII-12** or **VIII-13** are in a ring, so all atoms of each structure are in one set.

In the second step of the procedure, *partitioning*, the data obtained from the individual classifications of step-1 are combined. As illustrated in line 9 in Table VIII-1, atom a of **VIII-12** must correspond to atom 4 of **VIII-13** because this is the only atom common to the sets 2,4 (atoms of same node value, line 2) and 3,4,5,7 (same branch value, line 4) already identified as possible matches for a. The best that can be said about the remaining atoms b,c,d is that they must correspond to 3,5,7, although the correspondence cannot yet be determined (line 10).

The results of step 3 of the procedure, application of the criterion of connectivity, are presented in lines 11–14 in Table VIII-1. The one-to-one correspondence of atoms b and 3 (line 11) is established by consideration of the neighbor(s) of atom a (i.e., atom b) and the corresponding neighbor(s) of atom 4 to which a must match (as in line 9). The subsets shown in line 12 result by taking the neighbors of the atoms in the sets b,c,d and 3,5,7 as defined in line 10. Then, by utilizing the results in line 11, the required correspondence of the neighbors of b and the neighbors of 3 gives the sets of line 13. These sets are then used to establish the correspondences of the neighbors of a,c (b,d) and 4,5 (3,6,7), shown in line 14.

Step 4 of the procedure applies another partitioning step, this time including lines 9–14. This results in establishment of a unique correspondence of the atoms of substructure **VIII-12** with atoms of structure **VIII-13** (Table VIII-1).

At this point in the procedure, success is noted and the algorithm terminates with a positive result and the correspondences of the atoms. Another obvious terminating condition, failure, is obtained when the formation of subsets yields no atoms of the structure to correspond with an atom of the substructure or when the partitioning procedure can find no consistent way to establish

TABLE VIII-2. Four 6-Membered Rings of Adamantane (VIII-14), by Atom Number

Ring No.	Comprised of Atoms
1	{1,2,3,4,5,6}
2	{1,2,7,10,8,6}
3	{4,5,6,8,10,9}
4	{2,3,4,9,10,7}

correspondences. At this point the algorithm returns the result that the substructure is *not* included within the structure.

Step 5 of the procedure is required when correspondences exist but a unique correspondence has not been found on the basis of the preceding descriptors. This happens, for example, when there is more than one mapping of the substructural atoms to the atoms of the structure. The *assignment* procedure to discover these mappings is described by Sussenguth.[2] This procedure will eventually return a successful match and, if properly written, can return *all* the matchings. The assignment procedure would be required, for example, to locate the four 6-membered rings in adamantane (**VIII-14**), summarized in Table VIII-2.

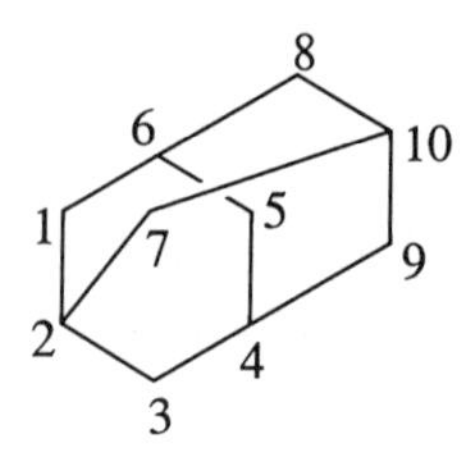

VIII-14

Such problems require a *backtracking* procedure because an assignment of a substructural atom to a structural atom is made at each step. If this assignment does not lead to a legal matching of the complete substructure, alternative assignments at this and previous steps must be considered. The assignment procedure begins to acquire the appearance of the second technique for graph matching in substructure search, the atom-by-atom procedure described in the next section.

One attractive feature of set reduction is that it is relatively straightforward to program in any language with provision for logical operations for set manipulation.[2] As long as the assignment procedure with backtracking is not required, conventional, iterative programming techniques suffice. When provision for handling backtracking is required, the problem becomes more complicated, requiring recursive programming techniques. Because the atom-by-atom search technique inherently requires backtracking, this issue is discussed in more detail in the next section.

VIII.B.2. Graph Matching by Atom-by-Atom Search

The atom-by-atom technique for graph matching in substructure search is simpler to describe than set reduction. It bears more resemblance to manual methods used by chemists for comparison of substructures and structures. The basic strategy given a substructure and a structure (e.g., **VIII-12** and **VIII-13**) is to pick a starting atom common to both and then attempt to "superimpose" the representations further, beginning with the atoms connected to the starting atom. Eventually a terminating condition is reached where the matching either succeeds or fails because additional neighbors cannot be matched. This method can be characterized as a *neighbor-first* procedure in that at each step the only candidates for extending a partial match are those atoms that are neighbors of the previous atom considered, information that is explicit in a connection table representation.

Before presenting an example, there is an important strategy to consider. Atom-by-atom search is most efficient when it is programmed to *fail as quickly as possible*. Although this may sound illogical, it is very important to detect when a substructure cannot be found in a structure as soon as possible to avoid time-consuming computations that eventually lead to this failure condition. Whenever a choice must be made for the starting atom in the substructure, or a neighboring atom in subsequent steps, the most "unique" atom should be chosen with the hope that this atom cannot be found in the structure.

There are many ways to define "unique," but uniqueness does reflect chemical intuition. A good definition would consider, for example, heteroatoms before carbon atoms, the least abundant heteroatom before other heteroatoms, and the heteroatom with the most specific properties (hybridization, hydrogen range, etc.) before other heteroatoms of the same type. Some preprocessing of the substructure is required if this information is not already available, but the preprocessing and ordering of next selection can save considerable time.

The atom-by-atom search method is illustrated here, using the same example problem as in the previous section, matching of substructure **VIII-12** to structure **VIII-13**. A more detailed analysis of this same example has been presented by Lynch et al.[4]

For this example it is assumed that the representation of structure includes only the atom property of atom type and the bond property of multiplicity. The procedure is illustrated in Figure VIII-1.

Through application of a criterion of uniqueness based on atom type, the procedure would begin with the oxygen atom a of **VIII-12** because there are fewer oxygen atoms than carbon atoms in structure **VIII-13**. There are two oxygen atoms of **VIII-13** that must be considered, atoms 2 and 4 (step 1, Figure VIII-1). The substructure next has a double bond from atom a to its only

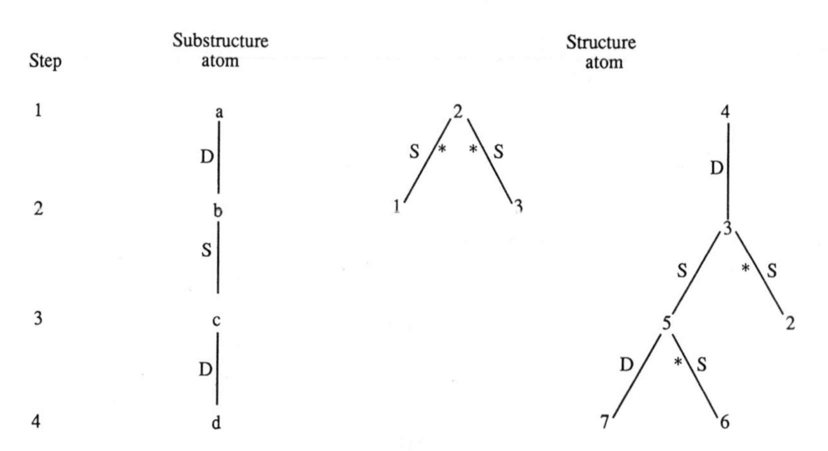

FIGURE VIII-1. Illustration of graph matching procedure of atom-by-atom search, applied to matching substructure **VIII-12** to structure **VIII-13**.

neighbor, carbon atom b. In Figure VIII-1 the bond multiplicities are indicated by S and D for single and double bonds, respectively, placed next to the lines interconnecting the atoms. In **VIII-13**, atom 2 has two neighbors, atoms 1 and 3, both of which are carbon atoms, but both are connected to 2 by single bonds. Graph matching along this path thus terminates, as indicated in Figure VIII-1 by the asterisk character placed next to the connection. However, atom 4 has a single neighbor, carbon atom 3, connected to atom 4 with a double bond, so this path represents a partial match and is retained for the next step.

In step 3, **VIII-12** possesses a single bond to the remaining, unused neighbor of atom b, carbon atom c. There are two remaining neighbors of the corresponding atom 3 in **VIII-13**, atoms 5 and 2. Atom 2, however, is an oxygen atom, so this path is rejected. Atom 5 is a carbon atom connected to 3 with a single bond, so the partial matching is successful and continues to the next step. The final step requires the matching to continue from atom c to atom d through a double bond. In **VIII-13**, there are two neighbors of atom 5, atoms 6 and 7. The former is rejected because its connection to atom 5 is through a single bond. The latter is accepted because it is a carbon atom connected to 5 by a double bond. Thus the matching is successful and the procedure terminates.

It is obvious even from this simple example that the inclusion of additional atom properties can increase the efficiency of the procedure. If, for example, the atom properties of number of neighbors or hybridization were used, atom 2 of **VIII-13** would never be considered at step 1, Figure VIII-1.

This example requires no backtracking to find a successful match. However, the capability for backtracking is essential in atom-by-atom search for the general case. Often partial matchings are extended through several atoms before a disagreement is found. The procedure must then back up to the last point at which there were two or more choices and continue with the exploration of alternative extensions of partial matchings. This problem is treated most conveniently by algorithmic computer languages such as ALGOL, SAIL, Pascal, C and BCPL, or list-processing languages such as LISP. These languages allow recursive procedures to be written that make backtracking simple.

VIII.C. GRAPH MATCHING OF CONFIGURATIONAL REPRESENTATIONS OF STRUCTURE

As discussed in Chapter VII, it is both meaningful and practical to define configurational stereochemistry at the chiral centers and double bonds within substructures. Given the importance of stereochemistry in determining molecular reactivities and properties, it is desirable (if not essential) for substructure searches to utilize all available stereochemical detail. Fortunately, the graph matching methods for finding "constitutional" substructures can be extended readily to handle substructures with defined configurational stereochemistry. In fact, stereochemical substructure matching comprises two quite distinct phases; one must first find the "constitutional" substructure and then verify a match of stereochemical detail.

As previously noted, the main available chemical data bases do not contain computer-manipulable records of molecular stereochemistry. Consequently, despite the chemical importance of stereochemistry, there are currently only limited applications for configurational substructure matching systems. Proprietary systems for constructing and searching private data bases of stereochemical structures do now exist (one example is Molecular Design's **MACCS** system[5]).

Thus far, however, the main uses of configurational graph matching have not been in data base searching but rather in synthesis planning and structure elucidation. Stereochemical data has been extensively exploited in both **SECS**[6,7] and **LHASA**[8]; other synthesis planning programs have also been extended to include stereochemical detail.[9,10] Some of the uses of stereochemical data in synthesis planning require checks on the relative stereochemistry of various atoms in some reaction or transform site, that is, graph matching a stereochemically defined substructure onto a molecule with specified stereochemistry. The **DENDRAL** structure elucidation system includes its **STEREO** module. The **STEREO** module can be used to generate either all or some chosen subset of the stereoisomers of a structure.[11] Many of the constraints that can be used within **STEREO** also involve graph matching of substructures with defined configurations.

A method for assigning configuration designations to chiral centers and double bonds was illustrated for structures and substructures in Chapter VII, where details were also given of the 0,1 convention for specifying the chirality or *cis,trans* substitution based on the numbering of atoms attached to the center or double bond. Given the results of graph matching, it is necessary only to determine whether the *sense of the configurations in substructure and structure are the same*. Now, because the substructure and structure are numbered arbitrarily and the actual configuration designations depend on the numbering, one cannot simply compare the 0s and 1s for correspondence. For example, consider the structure of (*R*)-1-aminoethanol (**VIII-15**).

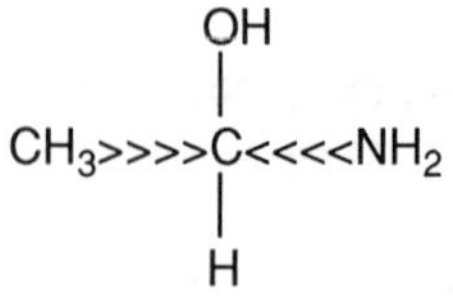

```
         OH
         |
CH3>>>>C<<<<NH2
         |
         H
```

VIII-15

If this structure were numbered as shown in **VIII-16**, the configuration designation corresponding to the *R* configuration is "0," as shown in **VIII-17**.[12,13]

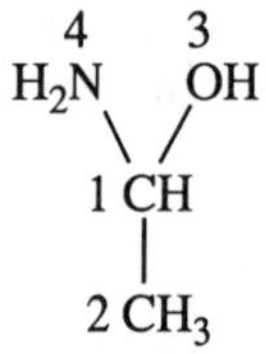

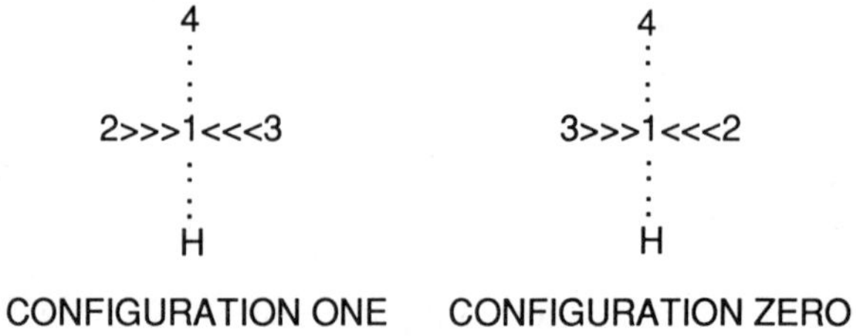

```
  4    3
 H2N   OH
    \ /
   1 CH
     |
   2 CH3
```

VIII-16

```
       4                        4
       :                        :
   2>>>1<<<3                3>>>1<<<2
       :                        :
       H                        H

CONFIGURATION ONE       CONFIGURATION ZERO
```

VIII-17

If, however, the structure had been numbered as shown in **VIII-18**, then by the convention in **STEREO**, the configuration designation corresponding to *R* is "1," as shown in **VIII-19**.

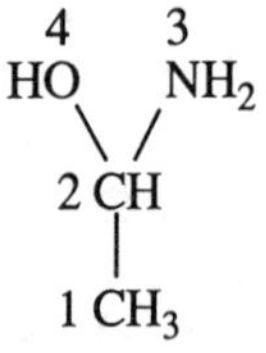

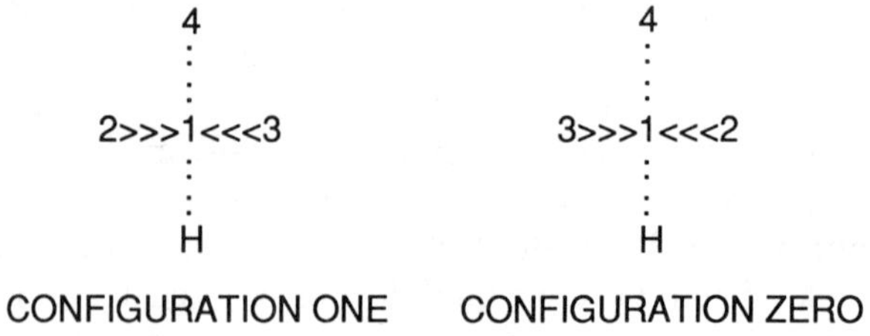

```
  4    3
 HO    NH2
    \ /
   2 CH
     |
   1 CH3
```

VIII-18

```
       4                        4
       :                        :
   2>>>1<<<3                3>>>1<<<2
       :                        :
       H                        H

CONFIGURATION ONE       CONFIGURATION ZERO
```

VIII-19

This result is expected intuitively, because by renumbering of the oxygen and nitrogen atoms from 3,4 in **VIII-16** to 4,3 in **VIII-18**, respectively, the configuration has, in effect, been inverted. The configuration at a chiral center is inverted by the interchange of any two substituents. This simple principle underlies the method for comparison of configuration designations for equivalence of substructures.

The essential problem is to determine whether the sense of the configuration imposed by the numbering of the substructure is the same as the sense imposed by the numbering of the relevant atoms in the structure. There is a simple algorithm for testing the equivalence of configuration designations to establish the overall matching of a substructure to a structure.[11] This algorithm is based on formalizing the rule describing stereocenter inversion mentioned above and applying it to comparison of two fixed configurations rather than inversion of a single configuration. Consider a single stereocenter in a substructure, such as substructure **VIII-20**, which corresponds closely to the structures presented above for 1-aminoethanol. Furthermore, assume that **VIII-20** is numbered as shown and that C(2) is assigned the configuration designation "0" as shown in the accompanying Fischer projection (this is, of course, the same configuration designation "0" given for **VIII-16**).†

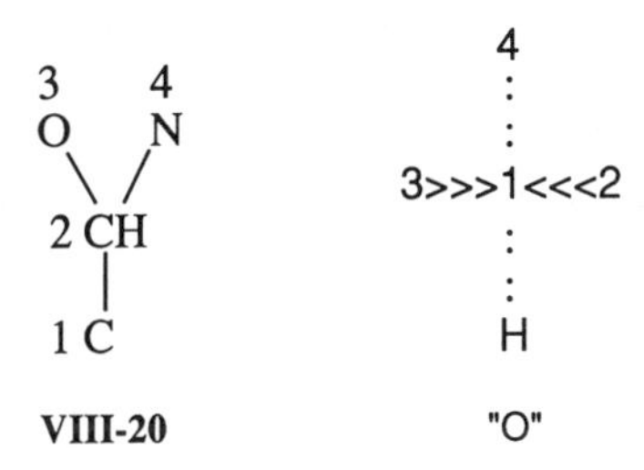

A test to determine determine whether the substructure as defined in **VIII-20** could be found in the structure of (*R*)-1-aminoethanol (**VIII-15**), numbered as shown in **VIII-18**, would first involve conventional graph matching. A standard graph matcher could readily determine that the constitution of **VIII-20** was instantiated within the structure of **VIII-18**. Table VIII-3 summarizes the results of such a graph matching.

TABLE VIII-3. Results Obtained by Graph Matching of Substructure VIII-20 to Structure VIII-18

Atom in **VIII-20**	Matches	Atom in **VIII-18**
H(0)		H(0)
C(1)		C(1)
C(2)		C(2)
N(4)		N(3)
O(3)		O(4)

†By convention, the hydrogen atom is numbered zero to assign its priority with respect to other atom numbers as needed to establish configuration designations.

The next step in the analysis entails determination of whether the configuration designations of the stereocenter in the structure and the substructure correspond to the same configuration. This is accomplished by determining the parity (*odd* or *even*) of the permutation relating the numbers of the neighbors of the chiral atom in the substructure and the structure (Table VIII-3). In this example, comparison of numbers of neighbors of the stereocenters C(2) reveals that atom 0 corresponds to atom 0, atom 1 to atom 1, atom 4 to atom 3, and atom 3 to atom 4. These relationships are merely a *permutation* of the numbers of the neighbors, and there is an established formalism for expressing such permutations.[12] Here, the permutation can be written as

$$(0)\,(1)\,(34)$$

which is shorthand for the preceding textual description of the correspondence of neighbors. The elements within parentheses correspond to *cycles*;[12] thus, cycles (0) and (1) mean that atoms 0 and 1 are in correspondence. Cycle (34) means atom 3 goes to atom 4 and atom 4 goes to atom 3, that is, it is read as a circular list. The parity of a permutation is determined by the following rules:

1. Determine the parity of each cycle. A cycle is *even* if its length is *odd*, *odd* if its length is *even*.
2. Take the sum of the odd and even parities of the cycles using the rules:
   a. odd + odd = even
   b. even + even = even
   c. odd + even = odd
3. The parity of the permutation is the result of step 2.

Therefore, the parity of the permutation for this example is

$$(0)\,(1)\,(34) \rightarrow \text{even} + \text{even} + \text{odd} = \text{odd}$$

Finally, the correspondence of configurations is determined by the following rule. If the parity of the permutation is *odd*, the configurations are in correspondence if they are *different*. If the parity is *even*, the configuration designations must be the *same* for the substructure to match the structure at the given chiral center. In the preceding example, the configuration designations for the respective C(2) chiral centers are different, and the parity of the permutation of substituents' numbers is odd, so the configurations must be the same and substructure **VIII-20** matches structure **VIII-18** in both constitution and configuration.

This procedure generalizes readily to more complex problems involving large substructures and many chiral centers. In addition, because double bonds are treated with a similar formalism, the correspondence of *cis,trans* configurations is readily established by the same procedure.[12,13]

VIII.D. SEARCHING LARGE DATA BASES

The methods of atom-by-atom search or set reduction are, in principle, completely adequate for solving the problem of substructure search for both topological and configurational representations of structure. In practice, however, the methods consume significant amounts of computer time per structure. These computation times are not so excessive as to inhibit the use of these methods in interactive structure elucidation or synthesis planning programs (because in these applications it is rarely necessary to analyze more than a few hundred structures). However, when the number of structures increases to the thousands or millions, as may be present in a large data base, computationally more economic methods must be employed.

One standard approach is to characterize the structures comprising a data base of structural information with a set of descriptors called *fragment codes*, or *screens*. These descriptors capture various features of the structures, such as ring sizes and systems, substructures of various types, molecular formulas, and so forth. Conceptually, when a substructure search is to be performed, the substructural query is characterized by use of the same vocabulary of descriptors and the subset of the total structural file that possesses the appropriate descriptors is retrieved. One method for graph matching described in Section VIII.B is then used to determine actual matching of the substructure against the retrieved structures. Clearly, the goal of this method is to perform the time-consuming graph matching against as few of the structures as possible.

Fragment codes tend to be descriptions of chemically-relevant portions of structures, such as chemical class, ring system, and functionality. Early systems relied on *ad hoc* definitions of what terms were to be included in the codes. This is an obvious limitation in that the addition of new terms forces a reevaluation of the structural file and a reconstruction of the ancillary files used in the search procedures. Later systems, for example, the GREMAS system,[14,15] possess rules for deriving fragment codes from structures in such a way that the codes are open-ended and can easily accommodate new fragments found in new structures.

The term "screens" is a very general one, implying only that structures are characterized by a set of properties, the screens. Normally, each screen is a descriptor of a property in binary form; that is, a structure either does or does not possses the property. Several examples were cited above, including ring size. A single screen may represent a range of a property, for example, 5–10 carbon atoms. Because each screen has a binary value of 0 or 1, the set of n screens for a structure can be represented in the computer as a string of n bits, where n is usually chosen to correspond to multiples of the word length for a given computer. In this way each structure is "represented" by a small number of computer words, and comparisons of the screens can be done with extremely fast, bitwise logical operations.

Several problems are involved in selection of the terms to be used as screens, as discussed in detail by Lynch et al.[4] There are essentially two methods for selection, one tied to chemically "meaningful" terms and the other algorithmic. A selection based on chemically meaningful terms leads to the same problems mentioned above for fragment codes. In addition, the concept is difficult to express simply in any computer program designed to compute the screens for the structure file itself and for substructures defined for searching the structure file.

Several methods for algorithmic generation of screens have been presented. Feldmann has proposed a hierarchical scheme[16] that divides the screens into several categories, each of which represents a separate method for searching a structure file. Three methods, searching based on CAS registry number, molecular formula, and molecular properties, are somewhat apart from the focus of this chapter, but do represent alternative methods for focusing subsequent substructure search on a subset of the original file. Two other methods, searching based on either ring systems or structural fragments, correspond to actual substructure search functions. The fragment search uses as screens *atom-centered fragments*. An atom-centered fragment is simply an atom in a structure or substructure together with its nearest bonded neighbors. This system is open-ended in that any new, unique atom-centered fragment is captured automatically by programs that process structural information, and added to the list of screens.

A somewhat different approach has been taken by Milne et al.[17] Screens are constructed on the basis of acyclic and cyclic portions of a structure. Acyclic screens describe central atoms and emanating branches, and in addition include subscreens that describe partial branching patterns about a central atom. Cyclic screens describe ring systems and molecular formulas of ring systems. Interactive substructure search is performed by using all screens in one search request. Again, structures retrieved on the basis of screens must be subjected to graph matching to establish precise correspondence between retrieved structures and the substructure. The use of screens that capture more of acyclic features than a simple atom-centered code makes the screens more precise at the cost of a large number of screens. This system, too, is open-ended. A number of other approaches to algorithmic screen generation have been described.[18]

Algorithmic approaches to selection of structural features to be used as screens have the disadvantage that many selections will either be overrepresented in the structural file (i.e., the screen will be inefficient because too many structures possess the feature) or underrepresented (i.e., the screen will be so detailed that only a very few structures possess the feature). The latter condition implies that the number of screens will proliferate rapidly, each screen referencing only a few structures. Algorithmic generation of screens can, however, be controlled by tying screen generation to the actual contents of the structure file so that a more uniform distribution of structures possessing each screen is obtained. A method for choosing an optimum set of screens given a data base of structures has been presented by Feldman and Hodes.[19] This method results in a more even distribution of structures among the screens.

This results in more efficient use of storage, because the number of screens is reduced significantly and more efficient searching because the screens are designed so that large numbers of structures are never retrieved (unless, of course, the substructure specified in the search request was a very general one, such as a six-membered ring).

VIII.E. CONSTRUCTIVE SUBSTRUCTURE SEARCH: AN APPROACH TO STRUCTURE GENERATION

The procedure of *constructive substructure search* (CSS) can be illustrated by taking a generalization of the earlier problem of matching **VIII-12** onto **VIII-13**. In the generalized problem, an α,β-unsaturated ester forms a component of a partially assembled C_{15} structure (**VIII-21**) that is to be further constrained by a requirement for at least two instances of α,β-unsaturated carbonyl groups (**VIII-22**). The data shown in Figure VIII-2 summarize this initial situation; connection tables for the partially assembled structure and the substructural constraint are shown in Tables VIII-4 and VIII-5, respectively.

In this example, fixed hydrogen ranges have been assigned for certain atoms in case **VIII-21** where the valence allows potential new bonding sites, for example, at atoms C(9)-C(12) of **VIII-21**. The hybridization of other atoms, which might have more than one possibility, has also been fixed [e.g., C(11), C(12)]. Other hydrogen ranges and hybridizations are set automatically by **GENOA**. In the constraint (**VIII-22**) all hybridizations are fixed to be sp^2.

Partially assembled structure (case):

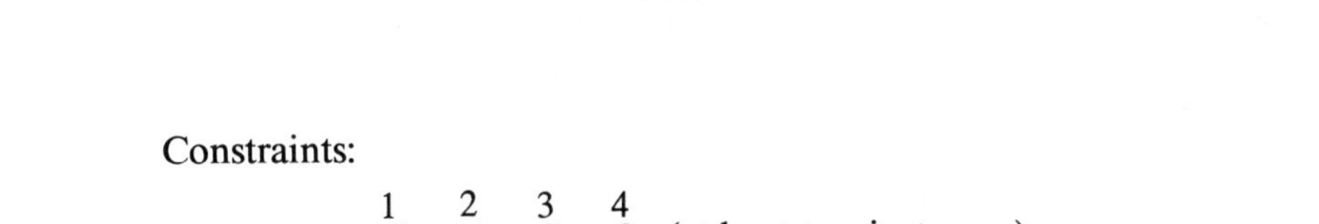

```
             4
             O
1       2    ||   5    7    8       10      11         12       13  14  15  16  17  18
CH3 ——— O ——— C ——— C ═ C ——— CH3    CH3—    —CH<     >C<      C   C   C   C   C   O
                   /    |
                 6 C   9 CH—
                        |
```

VIII-21

Constraints:

```
1   2   3   4
C ═ C — C ═ O   (at least two instances)
```

VIII-22

FIGURE VIII-2. Initial state in a constructive substructure search for two instances of an α-β-unsaturated carbonyl group in a C_{15} structure.

TABLE VIII-4. Connection Table for Case VIII-21 Prior to Specification of Constraint VIII-22

ATOM#	TYPE	NEIGHBORS[a]	HRANGE[b]	HYBRID[c]
1	C	2	3-3	SP3
2	O	1 3	0-0	SP3
3	C	2 4 4 5	0-0	SP2
4	O	3 3	0-0	SP2
5	C	3 6 7 7	0-0	SP2
6	C	5 FV FV FV	0-3	SP3 SP2 SP1A
7	C	5 5 8 9	0-0	SP2
8	C	7	3-3	SP3
9	C	7 FV FV	1-1	SP3 SP2
10	C	FV	3-3	SP3
11	C	FV FV FV	1-1	SP2
12	C	FV FV FV FV	0-0	SP2
13	C	FV FV FV FV	0-3	
14	C	FV FV FV FV	0-3	
15	C	FV FV FV FV	0-3	
16	C	FV FV FV FV	0-3	
17	C	FV FV FV FV	0-3	
18	O	FV FV	0-1	SP3 SP2

[a]A list of the atom numbers to which the given atom is connected. The entry FV in the neighbors list indicates a free valence, an available bonding site. Whether the free valence can be bonded to a hydrogen atom depends on the hydrogen range for that atom.
[b]The allowed hydrogen range for the given atom.
[c]The allowed hybridization of the given atom. The absence of an entry in this column indicates any hybridization possible.

The examples given earlier have illustrated how graph matching may be used to determine that atoms 1,2,3,4 of **VIII-22** match atoms 7,5,3,4 of **VIII-21**, respectively, thereby determining that one occurrence of **VIII-22** already exists. Thus one additional copy of substructure **VIII-22** must be constructed from other elements of **VIII-21**. There are numerous ways to incorporate the second copy of **VIII-22**, each involving different overlaps of atoms in **VIII-21**. Four examples have been chosen to illustrate various types of overlap. These examples show how **GENOA** creates new bonds and adjusts atom properties during the constructive procedures.

TABLE VIII-5. Connection Table for Constraint VIII-22

ATOM#	TYPE	NEIGHBORS	HRANGE	HYBRID
1	C	2 2 FV FV	0-2	SP2
2	C	1 1 3 FV	0-1	SP2
3	C	2 4 4 FV	0-1	SP2
4	O	3 3	0-0	SP2

For this illustrative example, **GENOA** can be assumed to choose C(1) of **VIII-22** as the starting point for matching. The construction procedure is recursive and depth-first. In other words, once a candidate match is found in **VIII-21** for C(1), the construction procedure follows an atom-by-atom search for candidates for its neighbor, continuing to neighbors of neighbors until either the substructural constraint is incorporated or it is impossible to continue. At either one of these terminating conditions, the procedure backs up to the previous neighbor and explores any other candidates at that level. Eventually the procedure will explore all possibilities beginning with the first candidate match for C(1) at which point **GENOA** searches for another candidate for C(1), repeats the depth-first search, and continues until all candidates for C(1) have been explored.

These procedures are illustrated by the constructive steps summarized in Table VIII-6, leading to the four new cases **VIII-23–VIII-26** chosen as exam-

TABLE VIII-6. Steps in Constructive Graph-Matching Procedure Leading to Construction of New Cases VIII-23–VIII-26 from VIII-21 Under the Constraint of at Least Two VIII-22s

Atom in **VIII-22**	Matched to Atom in **VIII-21**	Bond Formation[a]	New Atom Properties	Final Hydrogen Range
		*Construction of **VIII-23***		
C(1)	C(12)	–	–	C(12), 0–0
C(2)	C(13)	DB C(12), C(13)	C(13) → sp^2	C(13), 0–1
C(3)	C(14)	SB C(13), C(14)	C(14) → sp^2	C(14), 0–1
O(4)	O(18)	DB C(14), O(18)	O(18) → sp^2	O(18), 0–0
		*Construction of **VIII-24***		
C(1)	(Same as Case **VIII-23**)			
C(2)	(Same as Case **VIII-23**)			
C(3)	C(6)	SB C(6), C(13)	C(6) → sp^2	C(6), 0–0
O(4)	O(18)	DB C(6), O(18)	O(18) → sp^2	O(18), 0–0
		*Construction of **VIII-25***		
C(1)	C(5)	–	–	C(5), 0–0
C(2)	C(7)	–	–	C(7), 0–0
C(3)	C(9)	–	C(9) → sp^2	C(9), 1–1
O(4)	O(18)	DB C(9), O(18)	O(18) → sp^2	O(18), 0–0
		*Construction of **VIII-26***		
C(1)	C(9)	–	C(9) → sp^2	C(9), 1–1
C(2)	C(12)	DB C(9), C(12)	–	C(12), 0–0
C(3)	C(6)	SB C(6), C(12)	C(6) → sp^2	C(6), 0–0
O(4)	O(18)	DB C(6), O(18)	O(18) → sp^2	O(18), 0–0

[a] DB, stands for double bond; SB, for single bond.

ples. Each row in the table is a step in the constructive graph matching within the CSS procedure. Because search paths that fail to construct a new case are omitted for simplicity, each example proceeds successfully to incorporation of the second copy of **VIII-22**. The first two columns of Table VIII-6 simply summarize the atoms in **VIII-22** that were matched to atoms in **VIII-21**. Where it was necessary to form new bonds, such bonds are indicated in the third column. The fourth column summarizes the modifications necessary to atom properties in **VIII-21**. The assignment of hydrogen ranges actually takes place in steps throughout the procedure. Once a new case is constructed, **GENOA** makes a final pass through the connection table to determine any ramifications of previous assignments. Rather than attempt to trace the complex procedure, the fifth column simply summarizes the final result.

Construction of case **VIII-23** is as follows:

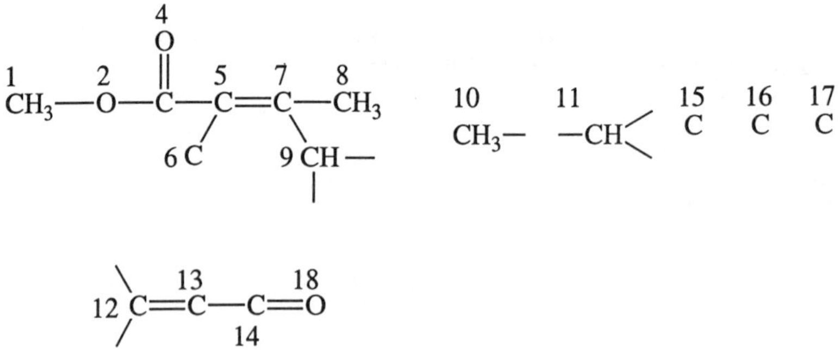

```
              4
              O
  1      2    ||   5    7     8
 CH3 — O — C — C = C — CH3        10          11            15   16   17
                   /      \                 CH3—     —CH<          C    C    C
               6 C     9 CH —
                           |

  \    13           18
12 C = C — C = O
  /           14
```

VIII-23

The construction procedure is best visualized by comparison of the steps in Table VIII-6 with the atom numbers for **VIII-21** and **VIII-22** in Figure VIII-2. The first case constructed by **GENOA** begins by matching C(1) of **VIII-22** to C(12) of **VIII-21**. No bonds are formed, and no adjustment is necessary to hybridization or hydrogen ranges because C(12) already has those properties specified. The first candidate chosen for a match to C(2) of **VIII-22** is C(13) of **VIII-21**, one of the remaining carbon atoms of the case whose properties are unspecified (Table VIII-6). The bond order of the C(12)–C(13) bond is set to double to correspond with the required C(1)–C(2) double bond of **VIII-22**. The hybridization of C(13) is set to sp^2 and its hydrogen range reduced from 0–3 to 0–1 to correspond with the properties of C(2) in **VIII-22**. The next atom of **VIII-22**, C(3), is matched to C(14) of **VIII-21**. A C(13)–C(14) single bond is formed and the properties of C(14) adjusted accordingly.

Finally, O(4) of **VIII-22** is matched to the remaining oxygen of **VIII-21**, O(18); the double bond is established; and the necessary hybridization and hydrogen ranges are set. The result is case **VIII-23**; this case is constructed with **no** overlap with the multiatom substructure of **VIII-21**. The connection table

TABLE VIII-7. Final Connection Table for New Case VIII-23

ATOM#	TYPE	NEIGHBORS	HRANGE	HYBRID
1	C	2	3-3	SP3
2	O	1 3	0-0	SP3
3	C	2 4 4 5	0-0	SP2
4	O	3 3	0-0	SP2
5	C	3 6 7 7	0-0	SP2
6	C	5 FV FV FV	0-3	SP3 SP2 SP1A
7	C	5 5 8 9	0-0	SP2
8	C	7	3-3	SP3
9	C	7 FV FV	1-1	SP3 SP2
10	C	FV	3-3	SP3
11	C	FV FV FV	1-1	SP2
12	C	13 13 FV FV	0-0	SP2
13	C	12 12 14 FV	0-1	SP2
14	C	13 18 18 FV	0-1	SP2
15	C	FV FV FV FV	0-3	
16	C	FV FV FV FV	0-3	
17	C	FV FV FV FV	0-3	
18	O	14 14	0-0	SP2

for **VIII-23** is presented in Table VIII-7 to indicate the changes represented by the steps in Table VIII-6 (compare Table VIII-4 with Table VIII-7).

Construction of case **VIII-24** is as follows:

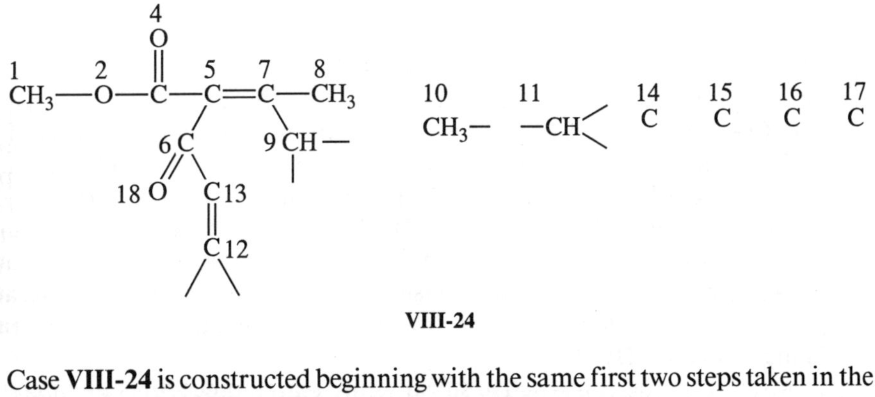

4
O
1 2 5 7 8
CH_3—O—C—C=C—CH_3
6 C 9 CH—
18 O C13
C12

10 11 14 15 16 17
CH_3— —CH< C C C C

VIII-24

Case **VIII-24** is constructed beginning with the same first two steps taken in the construction of **VIII-23** (Table VIII-6). For C(3) in **VIII-22**, however, another candidate in **VIII-21** is chosen, C(6), which is one of the atoms already built into the multiatom substructure in **VIII-21**. A single bond is formed [C(6)–C(13)] and the properties of C(6) adjusted accordingly. The required oxygen atom of the carbonyl is then affixed to C(6) with a double bond, resulting in case **VIII-24**. This new case incorporates the second copy of **VIII-22** utilizing a one atom overlap with the previous substructure.

Construction of case **VIII-25** is as follows:

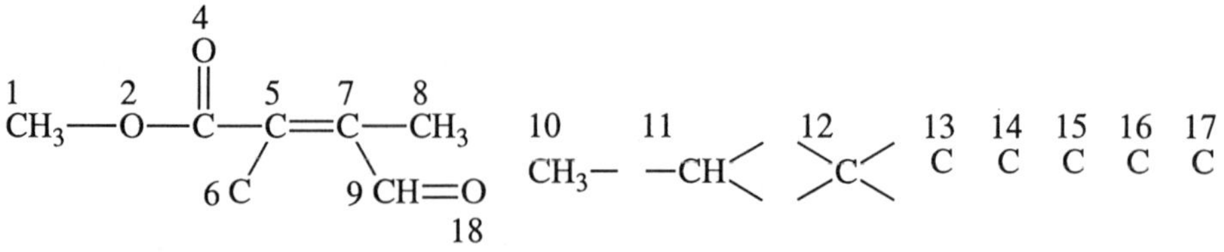

```
              4
              O
              ‖
1      2      ‖   5   7   8
CH3—O—C—C=C—CH3     10        11           12         13  14  15  16  17
                /        \                  CH3–   –CH<   >C<      C   C   C   C   C
            6 C        9 CH=O
                              18
```

VIII-25

Case **VIII-25** is constructed with a three-atom overlap with the existing substructure in **VIII-21** (Table VIII-6). Atoms C(1)–C(3) of **VIII-22** are matched to the existing substituted double bond comprised of C(5), C(7), and C(9). No new bonds are formed. Only the hybridization of C(9) is changed. The required carbonyl is constructed by joining O(18) to C(9) with a double bond and adjusting the properties of O(18).

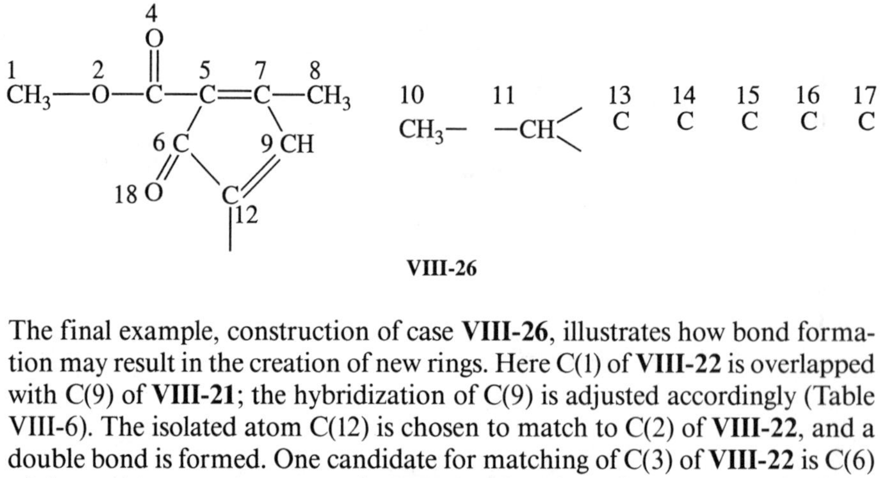

```
              4
              O
              ‖
1      2      ‖   5   7   8
CH3—O—C—C=C—CH3     10        11        13   14   15   16   17
                /        \                  CH3–   –CH<     C    C    C    C    C
            6 C        9 CH
           //   \      //
       18 O      C
                 |12
                 |
```

VIII-26

The final example, construction of case **VIII-26**, illustrates how bond formation may result in the creation of new rings. Here C(1) of **VIII-22** is overlapped with C(9) of **VIII-21**; the hybridization of C(9) is adjusted accordingly (Table VIII-6). The isolated atom C(12) is chosen to match to C(2) of **VIII-22**, and a double bond is formed. One candidate for matching of C(3) of **VIII-22** is C(6) of the multiatom substructure in **VIII-21**. Creation of the C(6)–C(12) single bond closes the ring. The construction is finished by attaching the oxygen atom for the carbonyl to C(6) and adjusting hybridization and hydrogen range properties. This results in construction of the cyclopentadienone functionality in the new case **VIII-26**.

Apart from the constructive substructure search procedures themselves, a structure generator such as **GENOA** must perform other analyses of the constraints, of the cases to which these constraints are applied, and the new elaborated cases that are generated. Some of these additional analysis procedures are concerned with the recognition and exploitation of symmetries in substructures. Other analysis routines are needed to determine whether two structures, or cases, are identical with the exception of some arbitrary numbering of their constituent atoms. Such problems could, of course, be solved by graph matching the two structures, but frequently rather different, more efficient procedures are required.

The need for some analyses of symmetries is apparent in an example problem involving the determination of the structure of the reduced aglycon portion from lemnalialoside.[20] (This structural problem was the original inspiration for **GENOA**'s constructive substructure approach to structure generation.)[21] This $C_{20}H_{34}O$ compound was known to incorporate a cembranolide skeleton, **VIII-27**. (Structure **VIII-27** is shown with single bonds and free valences at possible bonding sites; however, the bond orders are not in fact restricted and the free valences shown are permitted to be used in bonds to hydrogens.) This skeletal structure includes all the nonhydrogen atoms of this molecule; the structural problem concerns the placing of some olefinic bonds and an additional single bond between atoms in this skeleton.

VIII-27

a b c d

$CH_3—C{=}CH—CH_2—$

VIII-28

Data obtained by ^{1}H nmr have established the presence of the substructure shown as **VIII-28**. Substructure **VIII-28** can be matched into **VIII-27** in the vicinity of C(3), C(9), or C(13). The constructive matching process entails simply establishing a extra (π) bond between two carbons and the fixing of a few hydrogen ranges. It is possible to obtain six different matchings, shown as **VIII-29–VIII-34**; however, because of symmetries, there are only three topologically distinct structures: **VIII-29**, **VIII-31**, and **VIII-33**.

VIII-29

VIII-30

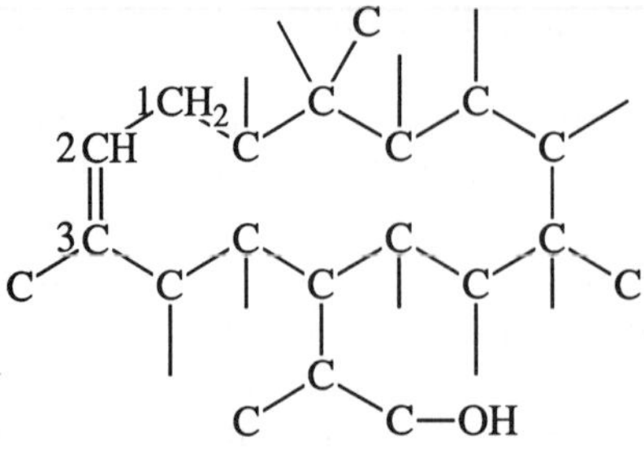

VIII-31

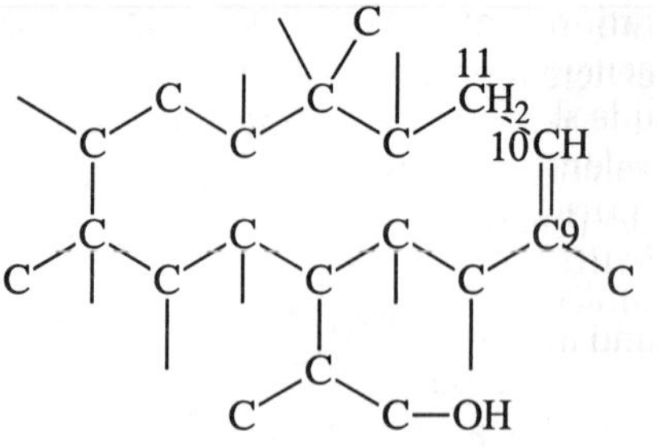

VIII-32

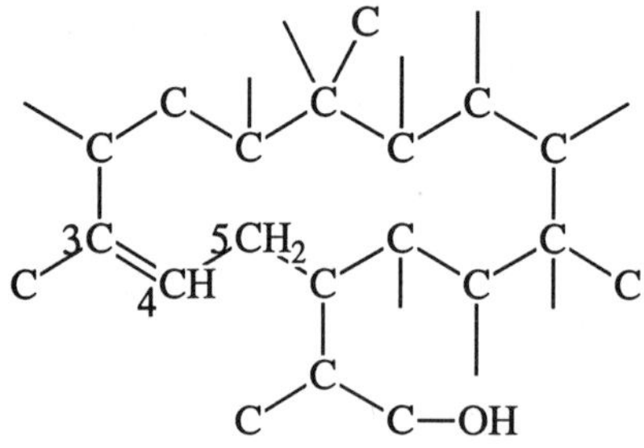

VIII-33

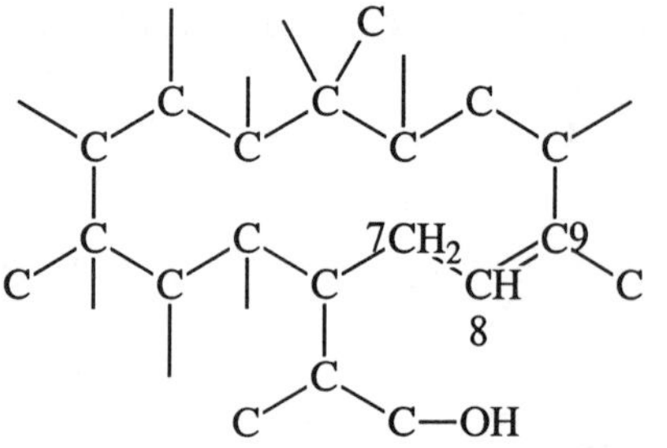

VIII-34

Any potential for duplication can often, as in this example, be revealed by an analysis of symmetries prior to the constructive graph-matching process. In **VIII-27** there is a plane of symmetry; the pairs of atoms [C(1),C(11); C(2),C(10); C(3),C(9); C(4),C(8);. . .; C(14),C(12)] are in topologically identical environments. Inevitably, any new structure generated by forming, say, an additional bond between C(9) and C(10) will be identical to a structure created by building the double bond between the equivalent atoms C(3) and C(2). A preliminary symmetry analysis of the case can provide data that identify these atom equivalences in a form that can be used to restrict the constructive graph-matching process.

At each level of recursion in a constructive graph matching it is necessary to search for all possible candidate atoms in the case onto which the next atom of the substructural constraint can be overlapped. The possible target atoms in the case are limited by restrictions on atom type, existing bonding, hydrogen ranges, and so forth. Data defining symmetries can also be used. If several atoms in the case are known to be equivalent, one may be identified as a representative for its equivalence class. Only the representative atom in any equivalence class can constitute a target for overlap of an atom from the substructural constraint.

Thus in this example there are only two targets for C(b), the vinylic >C= atom, of **VIII-28**, namely, C(13) and C(3) of **VIII-27**, where C(3) has been taken as the representative for the equivalence class C(3),C(9). With the use of this symmetry restriction on C(b), there are only three targets for overlapping C(c) of **VIII-28**: C(14), C(2), and C(4). In this selection of targets for overlap, symmetry restrictions have again been used. After overlapping of C(b) with C(13), both C(14) and C(12) might have constituted potential targets for C(c). However, C(14) and C(12) again form an equivalence class of **VIII-27**; furthermore, in these cases their symmetry equivalences are not altered by the process expressing the overlapping of C(b) and C(13). Consequently, once again it is necessary to select only the representative member of the equivalence class, and only C(14) becomes a candidate for overlap of C(c). By use of data on equivalences of atoms, it is possible to control constructive substructure search so that only the unique solutions, **VIII-29**, **VIII-31**, and **VIII-33** result.

Of course, a matching step that is constructive will normally change the symmetries of a case. For example, one might have a case containing a six-membered ring (where additional bridging bonds are to be permitted); initially, all six atoms would be equivalent. A constraint requiring a CH_3–C–C–OH moiety should yield new cases with the C–C carbons of the substructure overlapped onto, say, C(1)–C(2), C(1)–C(3), and C(1)–C(4) of the ring [without duplicates like C(1)–C(6), C(2)–C(3) etc.]. The original symmetry of the ring can, of course, be used to restrict the matching of the first atom of the substructure [and consequently avoid generation of equivalent structures involving C(2)–C(3), C(2)–C(4) etc.]. However, this first matching changes the remaining symmetry of the case being worked on. The remaining atoms of the ring are neither all still equivalent, nor are they all dissimilar. The symmetry analysis and symmetry data representation must suffice to reveal that, following the overlap of C(1) of the case with a carbon of the substructure, the remaining equivalence classes of case atoms are [C(2), C(6)], [C(3), C(5)] and [C(4)].

As each individual constructive matching step is completed and an atom from the case selected, some form of further analysis must be performed to determine the remaining elements of symmetry present in the case. Only those symmetries that remain can be used in restricting the subsequent searches for those target atoms that can overlap with the other atoms comprising a substructural constraint. Such a reanalysis does not require reidentification of symmetries after each step. It is, indeed, possible to represent the symmetries present in the original case so as to both identify initially equivalent atoms and allow the symmetries remaining after any change to be readily identified. The **GENOA** methods for analyzing symmetries in cases, and for the slightly simpler analysis of symmetries in substructural constraints, have been described in some detail;[1] these methods depend on algorithms outlined in Chapter IX.

Some duplication of cases produced by constructive graph matching can also result from the interplay of successive constraints. A simple example is provided by considering the application of two constructive substructure matches to an initial case consisting of a five-atom carbon chain. First, under the constraint of at least two C–OH groups, nine cases, **VIII-35–VIII-43**, would be obtained.

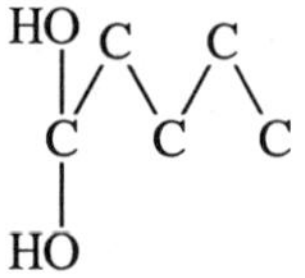

```
HO C   C
 |/ \ / \
 C   C   C
 |
HO
```

VIII-35

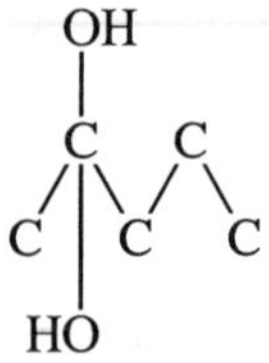

```
   OH
   |
   C   C
  /|\ / \
 C | C   C
   |
  HO
```

VIII-36

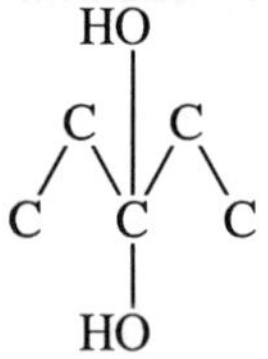

```
     HO
     |
   C | C
  / \|/ \
 C   C   C
     |
    HO
```

VIII-37

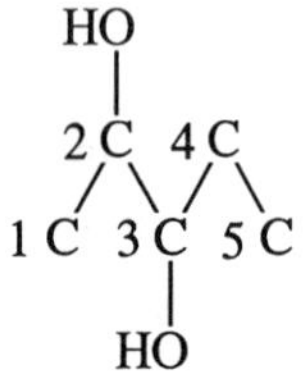

```
     HO
     |
    2C  4C
   /  \/  \
 1C   3C   5C
       |
      HO
```

VIII-38

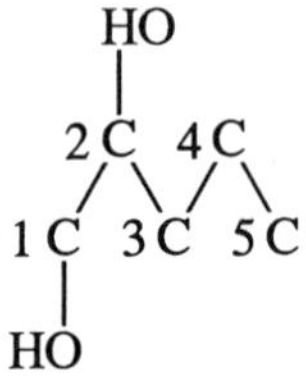

```
     HO
     |
    2C  4C
   /  \/  \
 1C   3C   5C
  |
 HO
```

VIII-39

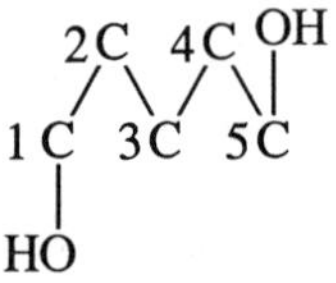

```
    2C  4C OH
   /  \/  \|
 1C   3C   5C
  |
 HO
```

VIII-40

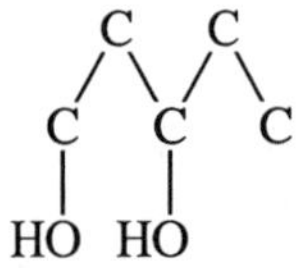

```
   C   C
  / \ / \
 C   C   C
 |   |
HO  HO
```

VIII-41

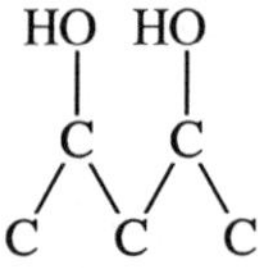

```
  HO  HO
  |   |
  C   C
 / \ / \
C   C   C
```

VIII-42

```
       HO
       |
   C   C
  / \ / \
 C   C   C
 |
HO
```

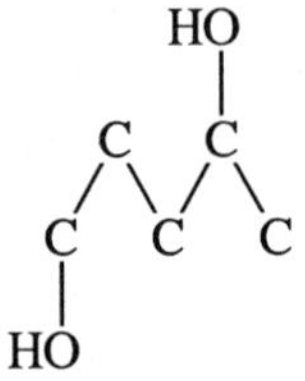

VIII-43

A subsequent constraint requiring a five-membered ring would reduce these nine to only three final, unique cases, **VIII-44–VIII-46**:

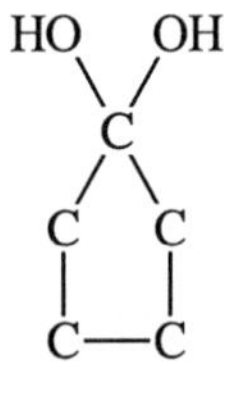

```
HO   OH
  \ /
   C
  / \
 C   C
 |   |
 C---C
```

VIII-44

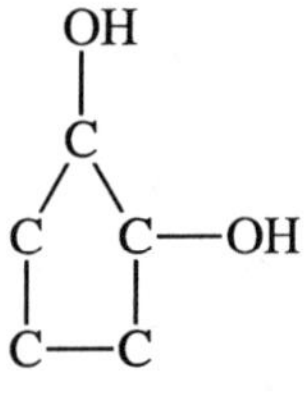

```
   OH
   |
   C
  / \
 C   C---OH
 |   |
 C---C
```

VIII-45

```
   OH
   |
   C
  / \
 C   C
 |   |
 C---C---OH
```

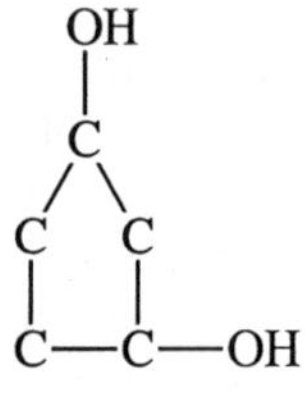

VIII-46

Duplicates of these would necessarily be obtained. Thus from **VIII-38** one would obtain the form **VIII-47** whereas **VIII-39** would yield **VIII-48** and **VIII-40** would result in **VIII-49**. Of course, **VIII-47**, **VIII-48**, and **VIII-49** are topologically identical and represent different ways of numbering case **VIII-45**. Their identity could be revealed by standard substructure matching. It is, however, more efficient to make the constructive procedures convert each case that is built into some standardized form. This standard form would have to be defined by encoded rules that determine a standard, or *"canonical,"* numbering for a structure or case. Recognizing the identity of two structures is much simpler if they are numbered identically. No search for corresponding atoms is then necessary, one simply compares the environments of atoms that have been accorded the same numbers.

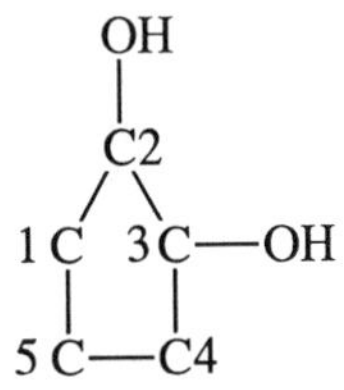

```
     OH               OH                OH
     |                |                 |
     C2              1C                1C
    /  \            /  \              /  \
 1 C   3C—OH     5 C   2C—OH       2 C   5C—OH
   |    |          |    |            |    |
 5 C—— C4        4 C—— C3          3 C—— C4
```

VIII-47 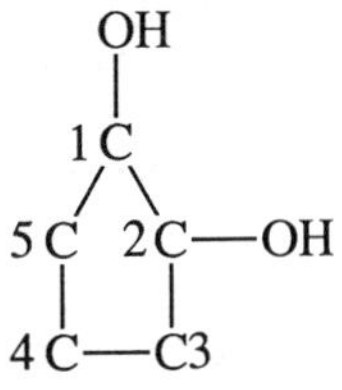 **VIII-48** 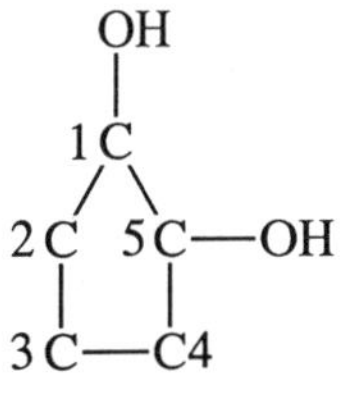 **VIII-49**

A simplified version of a canonicalization rule might require that the substituent–OH groups be placed on the lower numbered ring atoms. This rule establishes **VIII-48** as the canonical numbering of **VIII-45**. The constructive substructure search would proceed by first assembling **VIII-47** from **VIII-38** and renumbering it in accord with the canonicalization rule to obtain the form shown as **VIII-48**. This new case, represented as **VIII-48**, would be found to be distinct from the previously generated **VIII-44** already derived from **VIII-35**, **VIII-36**, and **VIII-37**. Consequently, since it would represent a distinct new structural form, the case would be kept; it would be saved in some list of generated structures *in its canonically numbered form*. Then **VIII-39** would be analyzed by the constructive procedures and **VIII-48** would be created, and the canonicalization rules would confirm that it was already in its standard numbering. However, it would then be recognized (by searching the list of previously created cases) as having already been generated and thus would be discarded. Similarly, **VIII-49** would be created from **VIII-40**, renumbered to again yield the form shown as **VIII-48**, recognized as a duplicate and discarded.

Canonicalization procedures form an essential part of any structure generator. They are also useful for applications involving structure data bases. If the structures in the data base are held in canonical form, it becomes much simpler to identify any duplicates or to perform searches for query structures (also represented in canonical form).

The procedures for identifying symmetries and converting structures to canonical forms are very closely interrelated. These procedures form the main topic in Chapter IX.

REFERENCES

1. R. E. Carhart, D. H. Smith, N. A.B. Gray, J. G. Nourse, and C. Djerassi, "GENOA: A Computer Program for Structure Elucidation Utilizing Overlapping and Alternative Substructures," *J. Org. Chem.*, **46** (1981), 1708.

2. E. H. Sussenguth, Jr., "A Graph-Theoretic Algorithm for Matching Chemical Structures," *J. Chem. Doc.*, **5** (1965), 36.
3. L. C. Ray and R. A. Kirsch, "Finding Chemical Records by Digital Computers," *Science*, **126** (1957), 814.
4. M. F. Lynch, J. M. Harrison, W. G. Town, and J. E. Ash, *Computer Handling of Chemical Structure Information*, American Elsevier, New York, 1971.
5. Molecular Design, Ltd., 1122 B St., Hayward, CA 94541.
6. W. T. Wipke, "Computer-Assisted Three-Dimensional Synthetic Analysis," in *Computer Representation and Manipulation of Chemical Information*, W. T. Wipke, S. Heller, R. Feldmann, and E. Hyde, eds, Wiley, New York, 1974, 147, Chapter 7.
7. W. T. Wipke and T. M. Dyott, "Simulation and Evaluation of Chemical Synthesis. Computer Representation and Manipulation of Stereochemistry," *J. Am. Chem. Soc.*, **96** (1974), 4825.
8. A. K. Long, S. D. Rubenstein, and L. J. Joncas, "A Computer Program for Organic Synthesis," *Chem. & Eng. News*, (May 1983), 22.
9. A. Esack and M. Bersohn, "Computer Manipulation of Central Chirality," *J. Chem. Soc. Perkin. Trans. I*, **1975** 1124.
10. H. W. Davis, *Computer Representation of the Stereochemistry of Organic Molecules*, Birkhauser, Basel, 1976.
11. J. G. Nourse, D. H. Smith, and C. Djerassi, "Computer-Assisted Elucidation of Molecular Structure with Stereochemistry," *J. Am. Chem. Soc.*, **102** (1980), 6289.
12. J. G. Nourse, "The Configuration Symmetry Group and Its Application to Stereoisomer Generation, Specification, and Enumeration," *J. Am. Chem. Soc.*, **101** (1979), 1210.
13. J. G. Nourse, R. E. Carhart, D. H. Smith, and C. Djerassi, "Exhaustive Generation of Stereoisomers for Structure Elucidation," *J. Am. Chem. Soc.*, **101** (1979), 1216.
14. R. Fugmann, W. Braun, and W. Vaupel, "Zur Dokumentation chemischer Forschungsergebnisse," *Angew. Chem.*, **73** (1961), 745.
15. E. Meyer, "Topological Search for Classes of Compounds in Large Files—Even of Markush Formulas—at Reasonable Machine Cost," in *Computer Representation and Manipulation of Chemical Information*, W. T. Wipke, S. Heller, R. Feldmann, and E. Hyde, eds., Wiley, New York, 1974, 105, Chapter 5.
16. R. J. Feldmann, "Interactive Graphic Chemical Structure Searching," in *Computer Representation and Manipulation of Chemical Information*, W. T. Wipke, S. Heller, R. Feldmann, and E. Hyde, eds., Wiley, New York, 1974, 55, Chapter 3.
17. M. Milne, D. Lefkovitz, H. Hill, and R. Powers, "Search of CA Registry (1.25 Million Compounds) with the Topological Screens System," *J. Chem. Doc.*, **12** (1972), 183.
18. G. W. Adamson, J. Cowell, M. F. Lynch, A. H.W. McLure, W. G. Town, and A. M. Yapp, "Strategic Considerations in the Design of a Screening System for Substructure Searches on Chemical Structure Files," *J. Chem. Doc.*, **13** (1973), 153.
19. A. Feldman and L. Hodes, "An Efficient Design for Chemical Structure Searching. I. The Screens," *J. Chem. Inf. Comput. Sci.*, **15** (1975), 147.
20. C. Charles, *License of Chemical Science Thesis*, Faculty of Sciences, Free University of Brussels, 1974.
21. R. E. Carhart, T. H. Varkony, and D. H. Smith, "Computer Assistance for the Structural Chemist" in *Computer-Assisted Structure Elucidation*, D. H. Smith, ed., American Chemical Society, Washington, D. C., 1977, 126, Chapter 9.

IX

CANONICALIZATION AND TOPOLOGICAL SYMMETRY

IX.A. INTRODUCTION

In the preceding chapters we have illustrated how one (1) can define computer-compatible representations of molecular structures, (2) search for the occurrence of a small substructure within some larger structure, and (3) might generate structures comprising a number of (possibly overlapping) substructures. Now, a new problem arises: how to determine whether some library of standard compounds or previously generated candidate structures contains an instance of a particular hypothesized structure as derived by a program (or directly inferred by a chemist).

The basic problem is to take a complete molecular structure and "compare" it with each member in some library of known structures. This problem pervades much of computer-aided chemistry. Frequently one needs to access some data base of structures, such as the Chemical Abstracts Service data base, to find information on the known properties of a given structure. Or, as when planning chemical syntheses, it is useful to know when a particular postulated intermediate is available and registered in some manufacturer's catalog; it is essential to identify cases where the same intermediate is engendered on different postulated synthetic paths (so that one can avoid the work of rediscovering ways of synthesizing it).

As illustrated graphically in Figure IX-1, it is not always simple to perceive the equivalence or nonequivalence of two slightly differently presented

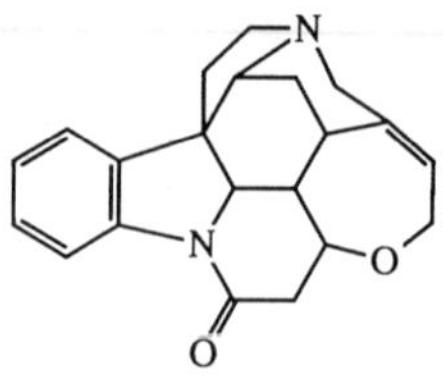

IX-1 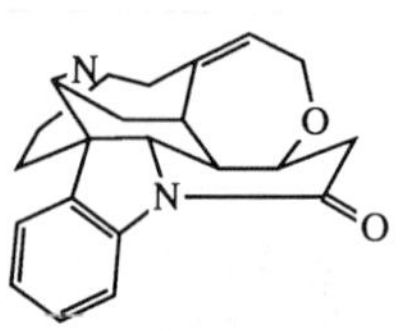 IX-2

FIGURE IX-1. Two structures, or two representations of the same structure?

molecules. For the structures in Figure IX-1, one would most probably proceed by first noting that both molecules contain nitrogens bonded to aromatic rings. One could then intercompare the neighboring atoms of these nitrogens, establishing their equivalence or nonequivalence and thus search through all the atoms of the molecules. This process involves simply graph matching as discussed in Chapter VIII. Such an analysis might be feasible for structures **IX-1** and **IX-2**, where unique atoms, suitable as starting points for the comparison process, are readily apparent.

In other cases, such as the arbitrarily numbered structures illustrated in Figure IX-2, one lacks even a starting point for comparison. All possible initial matchings of, say, node A in **IX-3** to a, b. . ., j in structure **IX-4**, 1, 2. . ., in **IX-5**, and so on, would have to be considered and the matchings explored. A graph-matching approach would constitute an intrinsically very costly method of searching for a given structure, namely **IX-3**, in a file of reference structures, specifically **IX-4–IX-6**.

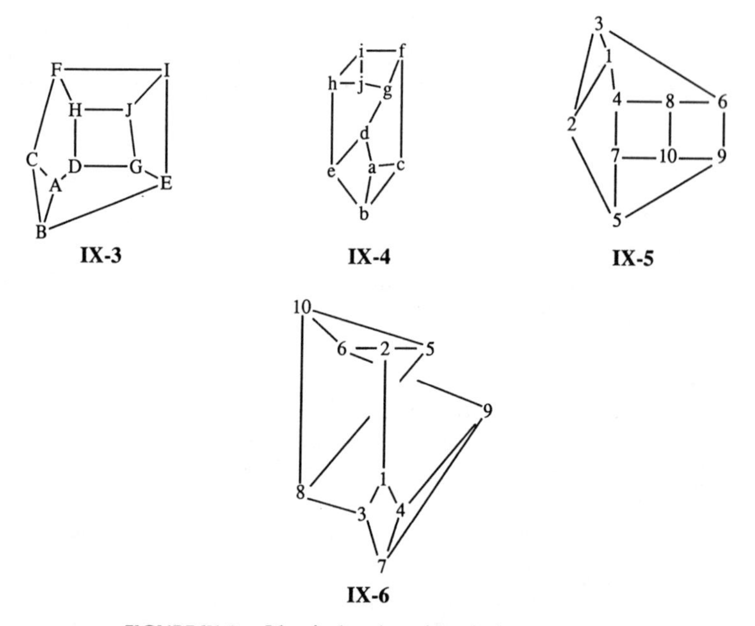

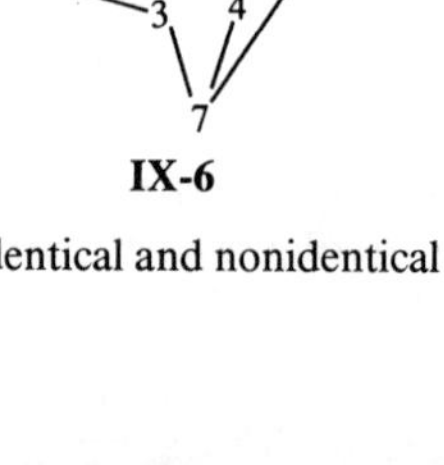

FIGURE IX-2. Identical and nonidentical structures.

The problem with structures such as those in Figure IX-2 is that there are no distinctive atom properties to simplify intercomparison. Identification of the equivalence or nonequivalence of the structures must be based solely on an analysis of the atom connectivities. One must verify that every atom in one structure has a corresponding atom, with identical connectivity, in the other structure.

Atom connectivities are (somewhat) more clearly revealed in the "connection," or "adjacency," matrix of the atoms. Binary adjacency matrices for the structures (with ones indicating bonds and zeros where there are no bonds) in their given "arbitrary" numberings are shown in Figure IX-3. (The adjacency matrix representations can, of course, be easily generalized to allow for multiple bonds between atoms.) Comparison of the adjacency matrices for structures **IX-3** and **IX-5** reveals that these are identical and distinct from both **IX-4** and **IX-6**.

```
  ABCDEFGHIJ
 +-------------
A!0111000000
B!1010100000
C!1100010000
D!1000001100
E!0100001010
F!0010000110
G!0001100001
H!0001010001
I!0000110001
J!0000001110
```

for IX-3

```
  abcdefgjij
 +------------
a!0111000000
b!1010100000
c!1100010000
d!1000101000
e!0101000100
f!0010001010
g!0001010001
h!0000100011
i!0000010101
j!0000001110
```

for IX-4

```
            1
   1234567890
  +------------
 1!0111000000
 2!1010100000
 3!1100010000
 4!1000001100
 5!0100001010
 6!0010000110
 7!0001100001
 8!0001010001
 9!0000110001
10!0000001110
```

for IX-5

```
            1
   1234567890
  +------------
 1!0111000000
 2!1000110000
 3!1000001100
 4!1000001010
 5!0100000101
 6!0100000011
 7!0011000010
 8!0010100001
 9!0001011000
10!0000110100
```

for IX-6

FIGURE IX-3. Adjacency matrices for the structures shown in Figure IX-2.

Unfortunately, such a simple-minded comparison of adjacency matrices might suggest that structures **IX-4** and **IX-6** also differed. In fact, **IX-4** and **IX-6** are identical structures, merely presented from different viewpoints and with their atoms numbered in accord with differing arbitrary numbering schemes. Their equivalence can be revealed if the atoms of **IX-6** are relabeled by (a,4), (b,7), (c,9), (d,1), (e,3), (f,6), (g,2), (h,8), (i,10), and (j,5). Any comparison of adjacency matrices of two arbitrarily numbered structures must include consideration of possible renumberings of the atoms in one structure, and hence the different resulting adjacency matrices, before it can be established that the structures differ.

If there were a method for deriving a *standard numbering* for a structure, comparison of adjacency matrices could serve as the basis of a method for searching for identical structures in large files. The structures in the reference file would be recorded with their standard, or "*canonical*" numbering. The test structure, for which the search was being made, would first be converted into its standard or canonical form. Then the comparison of the canonical test structure with each canonical reference structure would be simple. Formally, it would require checking for correspondence of the adjacency matrices, or at least the $n*(n - 1)/2$ elements in the top half of the adjacency matrices. Other representations of the canonical adjacency matrix encode these data as strings of character symbols, thus allowing checks for identity to be performed by string comparisons, possibly implemented directly in the computer's hardware. (Where there are distinct atom types, the procedure for comparing two structures must, of course, verify the correspondence of atom types of each canonically numbered atom as well as check the connectivity defined through the adjacency matrix)

To make such a structure identification method feasible, one requires a set of rules that define a standard canonical numbering for a structure and an algorithm for renumbering a structure in accord with these rules.

Conventional chemical nomenclature is similarly based on rules that are supposed to allow a unique name to be derived for a given structure. For example, structure **IX-7** might be named *4-ethyl-5-ethenyl-octane*, *4-ethenyl-5-propyl-heptane*, *3-propyl-4-propyl-1-hexene*, *4-ethyl-5-propyl-6-heptene*, *3-propyl-4-ethyl-1-heptene*, *3-ethenyl-4-ethyl-octane*, or similar.

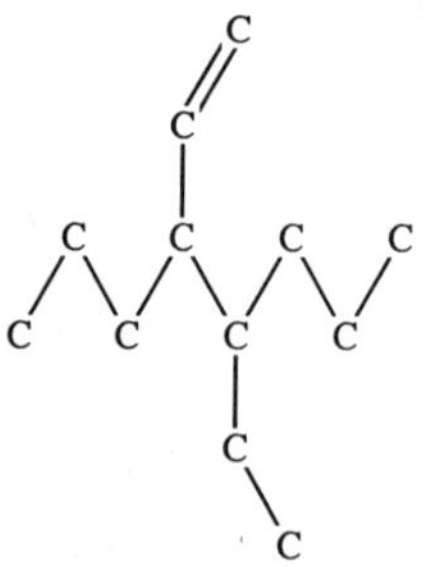

```
            C
           //
          C
          |
    C     C     C     C
   / \   / \   / \   /
  C   C     C     C
            |
            C
             \
              C
```

IX-7

The rules of chemical nomenclature, as applicable to this molecule, require that the foundation for the compound's name be derived from the name of the longest chain bearing unsaturation. This chain must be numbered so that it gives the lowest number to the multiple bond. One might find the correct name by generating various possibilities and finding which best satisfied these rules; names such as *3-propyl-4-propyl-1-hexene* would thus be found but rejected in favor of a subsequently derived name such as *4-ethyl-5-propyl-6-heptene* (because the second name involves the longer basic chain). More effectively, the rules might be used to guide the name generation procedure. Regardless of the method used, it should be possible to establish that the correct name for the compound is *3-propyl-4-ethyl-1-heptene*.

The naming rules of conventional chemical nomenclature are to some degree arbitrary but, provided that they are adhered to, will permit a unique name to be derived for each structure. One can define an analogous set of rules that permit the derivation of a unique numbering, and hence unique adjacency matrix, for a structure being manipulated by a computer program. In general, one requires various criteria to both define appropriate numberings (akin to the rules that require the conventional name to have the unsaturation-functional groups in the basic chain) and rank different numberings satisfying these requirements (like the rules that favor the *1-heptene* name over *1-hexene*).

One criterion that might be used is to require that the canonical numbering be "greatest" (or "least") as measured according to some scheme for ranking adjacency matrices. Some alternative numberings for a simple structure, and their corresponding binary adjacency matrices, are shown in Figure IX-4. (The atoms of structure **IX-8** are supposedly identical in all respects except their connectivity; they are labeled for subsequent reference.)

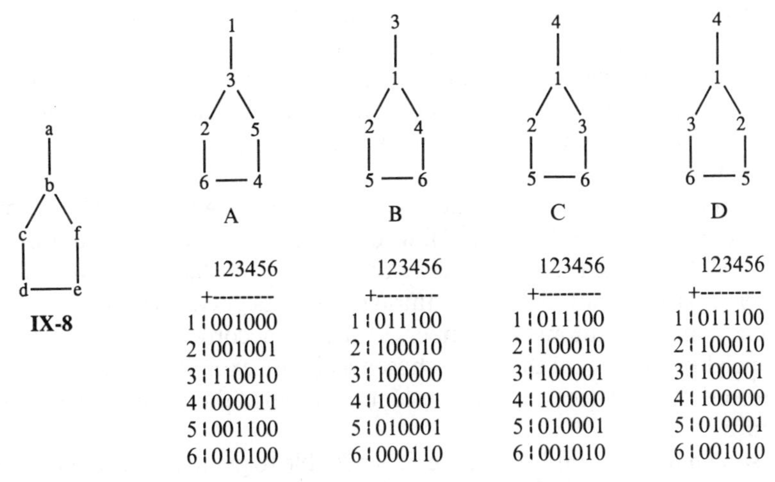

FIGURE IX-4. A simple structure and some of its possible numberings and their corresponding adjacency matrices.

One definition of "greatest," and thus "best," matrix might accord a higher ranking to 1s in the lowest numbered rows and columns of the matrix. Two adjacency matrices, differing in their patterns of nonzero entries, can be compared row by row until some difference is found. When a difference is revealed, the adjacency matrix with a 1 in the lowest numbered column is favored. In the example in Figure IX-4, numbering B would be favored over numbering A, because the adjacency matrix corresponding to numbering B has a 1 in row=1, column=2, where A has a zero. Numbering B would then replace A as the currently identified 'best" and putative canonical numbering. When the numbering C was subsequently generated, a similar matrix comparison would reveal it as "better" (because of the difference at row=3, column=6).

Sometimes, differing numberings of the atoms of a structure will yield identical adjacency matrices; such is the case with the last two examples (C and D) in Figure IX-4. Equivalences of adjacency matrices are a reflection of *symmetries* in the structure. As is further developed later, the process of finding a canonical form for a molecule will also reveal symmetry. (Furthermore, it is possible to exploit any previously determined symmetry in a molecule to reduce the computation necessary to find its canonical numbering.)

An n atom structure allows $n!$ possible numberings. It is impractical to rely on a scheme whereby one generates each possible numbering and then tests it by comparing its corresponding adjacency matrix with the best matrix found thus far. A canonicalization procedure must exploit those rules that define the best numbering to guide its search through the possible renumberings of a given structure. Only those numberings that might reasonably be expected to constitute best numberings should be generated and tested.

The definition of best adjacency matrix used in relation to the structure in Figure IX-4 requires as many possible "1s" (i.e., bonds) in the low-numbered columns of the first row of the matrix. Obviously, to comply with this requirement, the canonical numbering 1 should be accorded to the atom of highest degree in the structure; its neighbors should then receive canonical numberings 2, 3, and so on. In structure **IX-8** there is a unique atom of highest degree, and consequently the canonical numbering 1 can be unequivocally assigned to atom b.

Similar analysis shows that canonical numberings 2 and 3 must be accorded to b's two bivalent neighbors c and f, whereas its monovalent neighbor must be numbered 4. Once either c or f has been chosen to be labeled with canonical number 2, label 5 can also be unambiguously assigned. (To obtain the best adjacency matrix, canonical number 5 must be accorded to the as yet unlabeled neighbor, d or e, of the atom chosen as 2.) By exploiting such an analysis of connectivity, it is necessary to consider only two of the 5! numberings of this structure.

As shown by this simple example, it is possible to greatly reduce the number of permutations of canonical numbers among atoms that must be considered when trying to find the best numbering. More complex structures require more elaborate analyses to achieve corresponding reductions in the number of candidate canonical numberings that need to be generated and compared. Such analyses can be based purely on topological properties. However, atom–node properties are usually also available and can be used to help restrict the allocation of canonical numbers and thus reduce the computational effort.

If the structures incorporate atoms of different types, the atom-type property can be used. Structure **IX-9** represents a compound topologically similar to **IX-8** but comprising one **P**, two **Q**, and three **R** atoms. A canonicalization scheme could require that the lowest canonical numbers be ascribed to the least common atoms. Here, canonical number 1 would be reserved for the P atom, and canonical numbers 2 and 3 would be assigned to the two Qs and the Rs numbered 4–6. Even if no other data were used, this partitioning of atoms into equivalent groups reduces the problem from one requiring the generation of 5! numberings to one requiring 1!2!3! numberings. (There is only one choice for the first canonical number, two ways of assigning numbers 2 and 3 to the two Q atoms, and six (3!) ways of permuting the numbers 4–6 among the R atoms).

```
    P
    |
    R
   / \
  R   Q
  |   |
  Q — R
```

IX-9

Twelve different adjacency matrices would be generated and ranked by the matrix comparison function. (Use of data on atom connectivities would reduce the number adjacency matrices generated, possibly allowing only the final canonical form ever to be created.) The canonical numbering, with the best adjacency matrix would be:

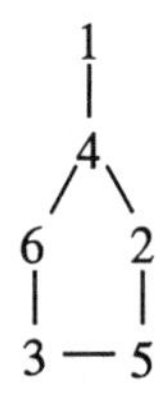

Of course, this canonical adjacency matrix is no longer "best" in the previous sense. It is simply the best given the additional requirement on the distribution of canonical numbers according to atom type.

As with conventional nomenclature, a change in priorities accorded to different facets of the naming procedure results in a differing, but equally well defined and unique, name. If conventional nomenclature schemes accorded highest priority to the longest chain, structure **IX-7** would be a substituted *octane*; the rules, however, accord greater priority to defining the position of unsaturation, and thus it is a *heptene*.

There are applications for which it is argued that any canonicalization scheme should be based solely on topological properties;[1] the resulting canonical numberings of atoms then have possible interpretations in terms of structural properties and may be of value in speculative investigations of structure–activity relationships. More typically, the major concern is speed. Automatic synthesis planning programs, and structure generators for structure elucidation programs, may need to process very large numbers of structures. Some of these structures may well prove to be duplicates. Such duplicates must be eliminated. It may also be appropriate to check whether any of the unique structures derived are in some standard library. All such checks require the structures to be in some canonical form, but the particular distribution of canonical numbers to atoms is of no consequence. It is then appropriate to use all such data as are available and whose use reduces the number of computational steps and speeds up the overall calculation procedure.

IX.B. CANONICALIZATION

IX.B.1. The Basic Canonicalization Algorithm

Most canonicalization algorithms presented in the chemical literature are elaborations of, and variations on a method originally outlined by Morgan.[2] It is useful to distinguish three basic procedures that together comprise a typical canonicalization algorithm. These basic procedures are implicit, if not explicitly coded, in all conventional canonicalization algorithms.

In the first procedure, the atoms constituting a molecule are partitioned according to various atom properties or on the basis of some limited analysis of their topological environments. The partitioning will yield, say, i different classes of atoms, with n_1 atoms in class 1, n_2 atoms in class 2, and so on. This partitioning changes the numbering problem for an n atom molecule from being $O(n!)$ to one where the computational cost is proportional to $n_1!n_2!...n_i!$ $(n=n_1+ n_2+...+n_i)$. Second, a numbering scheme is applied that explores allowed permutations of canonical numbers among the various atoms, using the partitioning already obtained to limit the number of permutations considered. In some schemes the assignment of each successive canonical number may permit a further partitioning of the as yet unnumbered atoms. This additional partitioning provides further constraints on the permutations that must be subsequently processed.

Finally, in the third procedure of the canonicalizer, candidate numberings are analyzed by some function that identifies the best numbering. Usually, the best numbering will be defined through the comparison of adjacency matrices (or an equivalent operation on code strings).

Morgan's original algorithm used a simple analysis of connectivity as the basis for its atom partitioning scheme. Exact details of the connectivity analysis vary;[2–4] the following scheme is typical. Initial connectivity values assigned to atoms measure the number of nonhydrogen atom neighbors that they possess. The number of different connectivity values is determined; for typical organic molecules, there will be at most four different initial connectivity values. "Extended connectivities" are then computed by means of an iterative procedure.

At each iterative cycle, a new connectivity value is determined for each atom in a structure by summing the current connectivities of its neighbors. This process discriminates between atoms according to a crude measure of their larger structural environment. For example, initially all the methyl carbons in a hydrocarbon molecule would have a connectivity value of 1 and would appear equivalent. After the first iterative cycle, the methyls would be split into groups according to whether they were bonded to secondary methylene groups (obtaining an extended connectivity value of 2), tertiary methines (and, consequently, obtaining an extended connectivity of 3), or quaternary carbons.

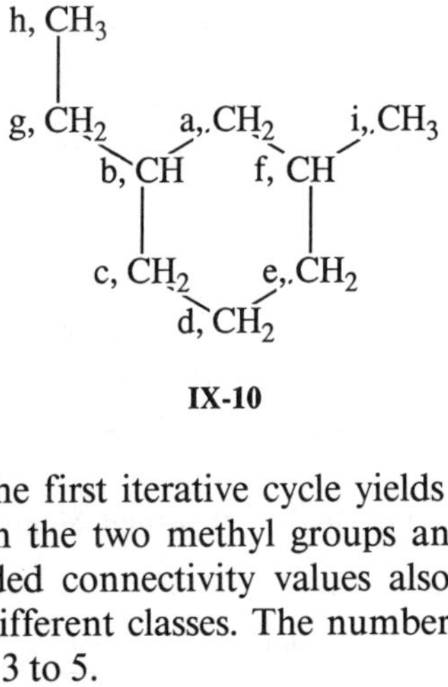

h, CH_3

g, CH_2 a, CH_2 i, CH_3

b, CH f, CH

c, CH_2 e, CH_2

d, CH_2

IX-10

In structure **IX-10**, the first iterative cycle yields extended connectivities that distinguish between the two methyl groups and also between the two methines. These extended connectivity values also serve to break up the methylenes into three different classes. The number of distinct connectivity values is increased from 3 to 5.

This iterative cycle is repeated for so long as an increase in the number of distinct connectivity values is obtained. Reapplication of the Morgan extended connectivity summation algorithm to structure **IX-10** increases the number of distinct extended connectivity values from 5 to 8. The new extended connectivities are given in Table IX-1. These new values distinguish among all atoms except the two methylenes c and d.

Another application of the iterative cycle **reduces** the number of distinct connectivity values. As shown by the data in Table IX-1, numerical coincidences result in similar extended connectivities being obtained for a number of pairs of atoms, such as d, g or a, b. Because the reapplication of the summation procedure failed to increase the number of distinct connectivity values and hence the number of distinguished classes of atoms, the iterative algorithm terminates. The extended connectivity values used to finally characterize the atoms are those obtained in the last successful iterative step. For structure **IX-10**, the values are those shown in Table IX-1 as resulting from the second iterative cycle of the summation procedure.

When the iterative procedure terminates, the atoms will have been partitioned among a number of classes. Each class comprises those atoms characterized by the same extended connectivity value that, it has been suggested, is a measure of how "centrally involved an atom is in a structure."[4]

The partitioning of atoms may be imperfect. Quite different atoms may, fortuitously, possess the same extended connectivity value at the point where the iterative cycle terminates. Atoms c and d in **IX-10** provide an example of this effect.

Atoms with the same final extended connectivities may well have been distinguished at some earlier stage in the analysis. In the example of structure **IX-10**, atoms c and d did possess different connectivities at the end of the first iteration. However, this distinction between them was lost on the next iteration. With reliance on a single numeric value to characterize connectivity, the basic Morgan algorithm can lose some of the structural information that it has in fact derived.

In the second stage of the analysis tentative canonical numbers are placed on each atom. The atom of highest extended connectivity is selected to be labeled with canonical number 1. Canonical labels 2, 3, and so on are assigned to its neighbors. Essentially, this corresponds to an *additional partitioning* step; unlabeled atoms are being partitioned according to their distance from the canonically labeled atom. In structure **IX-10**, atom b has the greatest extended connectivity and consequently is accorded canonical number 1. Canonical numbers 2–4 must be distributed over atoms g, a, and c.

TABLE IX-1. "Extended" Connectivities at Various Stages in Analysis of Structure IX-10

		Initial	First	Second	Third
	a	2	6	11	29
	b	3	6	15	29
	c	2	5	10	25
	d	2	4	10	19
	e	2	5	9	24
	f	3	5	14	25
	g	2	4	8	19
	h	1	2	4	8
	i	1	3	5	14
Number of connectivity values		3	5	8	6

The extended connectivities of neighboring atoms define their priorities in this next step of the numbering process. The next, lowest, canonical number is accorded to the neighboring atom with the highest extended connectivity value. There may be several neighbors with equal extended connectivity; then, if possible, the choice of neighboring atom to receive the next canonical number is based on some other property, such as atom type. If there is no such property that resolves between neighbors, each possible numbering for the neighboring atoms is explored.

Once the atom of highest connectivity and its immediate bonded neighbors have had canonical numbers assigned, the numbering algorithm proceeds with atoms at a greater bond radius. At each step the program identifies the atom that has both the lowest already assigned canonical number and some as yet unnumbered neighbors. The next canonical numbers are assigned to its neighbors, again using their extended connectivities and, if necessary, additional atom properties to determine their priorities.

With structure **IX-10**, this numbering procedure generates only the one, canonical, numbering. Atom b has the highest extended connectivity and so is accorded canonical number 1; numbers 2, 3 and 4 are assigned to atoms a, c and g, and their differing extended connectivity values uniquely determine their priorities. After canonical numbers 2–4 are assigned, atom a is the atom with lowest canonical number and unnumbered neighbor(s) f; canonical number 5 can thus be assigned to f. The final numbering derived is as shown in Table IX-2.

In some cases (e.g., structure **IX-10**) the scheme results in the generation of a single numbering. It can happen that the partitioning of atoms is incomplete (or symmetries may mean that some atoms cannot be distinguished); it is then possible for several alternative numberings to be generated and checked by the matrix comparison function. Structure **IX-11** provides an example of these effects.

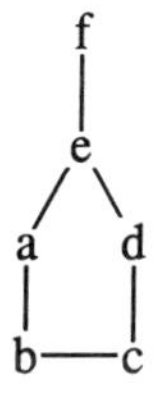

IX-11

The extended connectivities, as derived by the Morgan algorithm, are summarized in Table IX-3. At the point where the algorithm terminates, numerical coincidences have resulted in atoms a, d and e all possessing the same maximal value for their extended connectivities. (Again, the use of a single numeric value to characterize an atom's topological environment is losing information, such as the difference between trivalent atom e and bivalents a and d). If we assume that there are no differentiating atom

TABLE IX-2. Numbering of Structure IX-10

Atom	Extended Connectivity	Step-1 Largest Connectivity (i.e., b)	Step-2 Neighbors a,c,g	Step-3 Neighbors of a	Step-4 Neighbors of c	Step-5 Neighbors of g	Step-6 Neighbors of f
a	11		#2	#2	#2	#2	#2
b	15	#1	#1	#1	#1	#1	#1
c	10		#3	#3	#3	#3	#3
d	10				#6	#6	#6
e	9						#8
f	14			#5	#5	#5	#5
g	8		#4	#4	#4	#4	#4
h	4					#7	#7
i	5						#9
	123456789						
b	1 011100000						
a	2 100010000						
c	3 100001000						
g	4 100000100						
f	5 010000011						
d	6 001000010						
h	7 000100000						
e	8 000011000						
i	9 000010000						

TABLE IX-3. Extended Connectivity Computations for Structure IX-11

Atom	Connectivity	Extended Connectivities Step 1	Extended Connectivities Step 2
a	2	5	9
b	2	4	9
c	2	4	9
d	2	5	9
e	3	5	13
f	1	3	5
		"Final" values	

properties, atoms a, d and e are equally good candidates for receiving canonical number 1.

If the numbering scheme commenced by trying to assign canonical number 1 to atom a, adjacency matrix A in Figure IX-5 would result. Adjacency matrix B, with identically the same pattern of zero and one entries, results if d is labeled with canonical number 1. When e is chosen to receive number 1, the algorithm must explore alternative assignments of canonical numbers 2 and 3 to atoms a and d. These two alternatives yield adjacency matrices C and D.

Adjacency matrix A would be determined first and saved as the best adjacency matrix found thus far; the numbering [(a,1) (e,2) (b,3) (d,4) (f,5) (c,6)] would be saved as the best candidate canonical numbering. The second numbering [(d,1) (e,2) (c,3) (a,4) (f,5) (b,6)] gives an equal connectivity matrix and hence does not represent any improvement. However, the third numbering [(e,1) (a,2) (d,3) (f,4) (b,5) (c,6)] yields a better adjacency matrix. Our comparison function rates matrix C as better than matrix A because of the difference at row = 1, column = 3, where C has a 1 but A has a 0. Matrix C would replace A as the best adjacency matrix and the corresponding numberings for atoms would replace the previous best candidate canonical numbering. The final numbering yields adjacency matrix D, equal to matrix C; consequently,

```
       123456            123456            123456            123456
      +----------       +----------       +----------       +----------
a  1 |011000      d  1 |011000      e  1 |011100      e  1 |011100
e  2 |100110      e  2 |100110      a  2 |100010      d  2 |100010
b  3 |100001      c  3 |100001      d  3 |100001      a  3 |100001
d  4 |010001      a  4 |010001      f  4 |100000      f  4 |100000
f  5 |010000      f  5 |010000      b  5 |010001      c  5 |010001
c  6 |001100      b  6 |001100      c  6 |001010      b  6 |001010
        A                 B                 C                 D
```

FIGURE IX-5. Adjacency matrices resulting from differing numberings attempted for structure **IX-11**.

matrix C and numbering [(e,1) (a,2) (d,3) (f,4) (b,5) (c,6)] represent the final canonical form found for the structure. (Matrix C is, of course, the same as that derived from an initially differently labeled version of structure **IX-11** in the example given in Figure IX-4.)

Sometimes, the initial atom partitioning may fail to find *any* differences among the atoms. This may be due to symmetries, as with prismane-type structure **IX-12**, or because the extended connectivity sums obscure real differences amongst atoms, as with structure **IX-13**. Such structures are unfortunate but not intractable; the repartitioning process, effected as the Morgan algorithm explores possible candidate canonical numberings, still greatly reduces the number of adjacency matrices derived and tested.

With structure **IX-12**, of the possible 720 (6!) adjacency matrices, only some 48 need to be generated and tested. With **IX-12**, the trial assignment of, say, canonical number 1 to atom b, restricts the placements of 2–4 to atoms a, c, and e. Of the six possible arrangements for labels 2–4, four would have 2 assigned either to atom a or c; for these, there is only one possible allocation for the remaining labels 5 and 6. If label 2 is placed on atom e, choices remain for placements of both label pairs 3,4 and 5,6. Thus there are total of eight adjacency matrices generated for each initial arbitrary placing of label 1.

With structure **IX-13** some 90-odd adjacency matrices might have to be analyzed, but this is still only a small fraction of the 40,000 or so potentially possible. As illustrated by these examples, the repartitioning of unlabeled atoms, as candidate canonical numbers are assigned, is a very effective method for limiting computations. (Examples given in the literature might be taken to suggest that an extremal adjacency matrix, for a structure with n nodes each of degree d, can be found by exploring just $(n) \cdot (d!)$ possible numberings; this is true only in certain special cases, such as in a hexane ring, where the assignment of the first $d + 1$ canonical numbers implicitly defines placements of all subsequent canonical numbers).

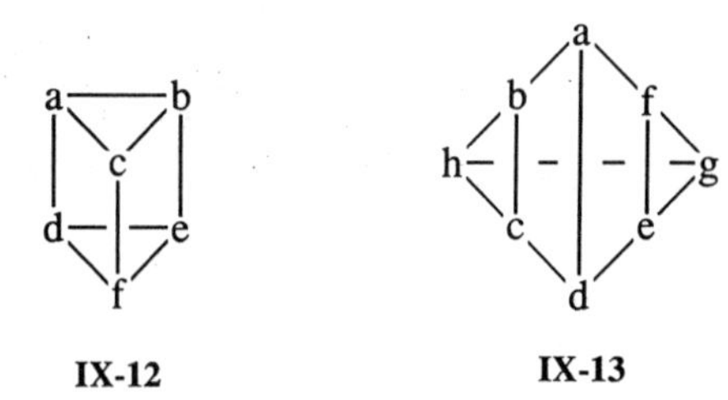

IX-12 **IX-13**

These examples have been chosen to illustrate the various procedures that must be executed in order to derive a canonical numbering for the structure. The structures were deliberately selected to provide simple cases where different adjacency matrices would be generated and would have to be compared to find the best.

The extended connectivity atom partitioning scheme has not been too effective in these examples. With most routine asymmetric organic structures, even the very simple extended connectivity analysis will partition the atoms completely. With something such as a sterane, **IX-14**, at most five iterations of the extended connectivity algorithm are necessary to obtain distinct extended connectivities for all atoms (see Table IX-4). (As is almost always the case, however, discriminatory data are lost at various intermediate stages; examples are atoms 5 and 13 appearing equivalent at the first extended connectivity step, with the distinction between atoms 1 and 7 becoming lost, temporarily, at the second step and that between atoms 2 and 3, at the third step).

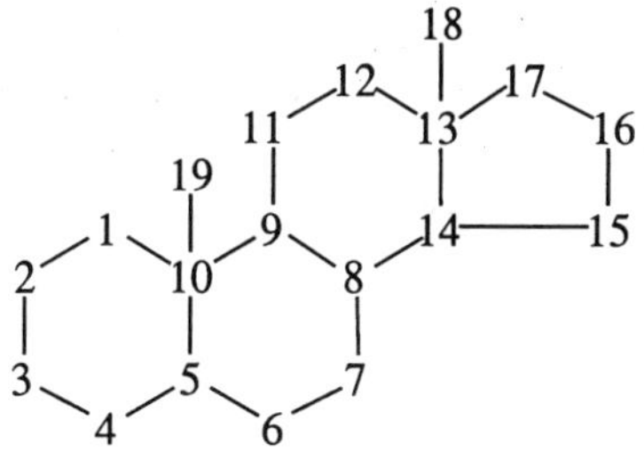

IX-14

TABLE IX-4. Extended Connectivity Calculations for Sterane IX-14

Atom	Initial Connectivity	Extended Connectivities at Successive Iterations				
		1	2	3	4	5
1	2	6	13	37	85	240
2	2	4	10	22	59	135
3	2	4	9	22	50	133
4	2	5	12	28	74	173
5	3	8	19	52	123	343
6	2	5	13	32	88	211
7	2	5	13	36	88	250
8	3	8	23	56	162	384
9	3	9	22	65	154	448
10	4	9	27	63	181	425
11	2	5	15	35	105	243
12	2	6	13	40	89	267
13	4	8	25	54	162	364
14	3	9	21	61	142	410
15	2	5	13	32	86	210
16	2	4	11	25	68	165
17	2	6	12	36	79	230
18	1	4	8	25	54	162
19	1	4	9	27	63	181
Number of distinct labels	4	5	13	15	17	19

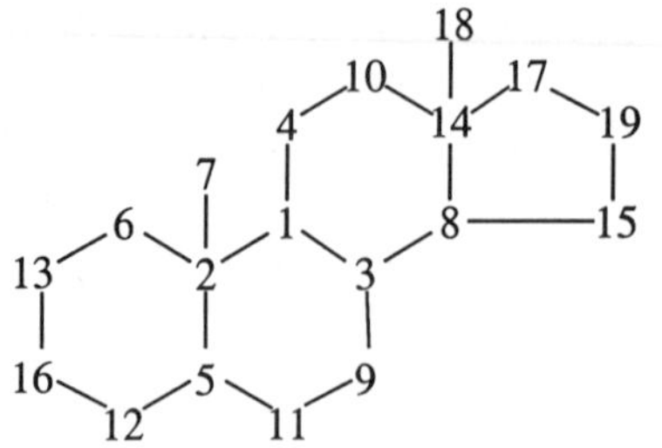

IX-15 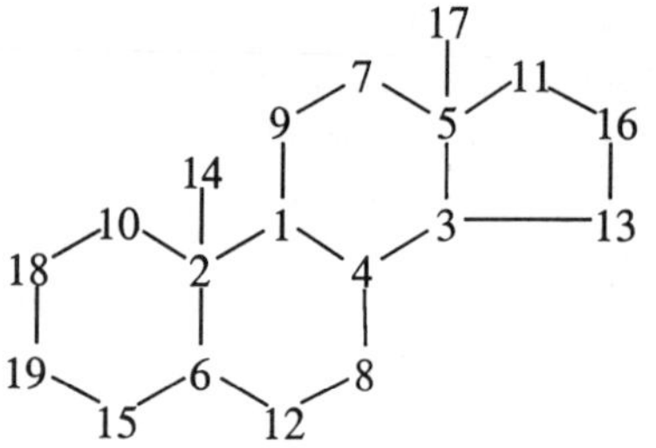 IX-16

FIGURE IX-6. Canonical numberings for the sterane skeleton obtained by using a standard Morgan style algorithm and considering solely extended connectivities.

As soon as the number of distinct extended connectivities equals the number of atoms, the iterative cycle may be terminated. These final extended connectivities provide a unique ordering of the atoms dependent only on topological properties of the structure. Consequently, these extended connectivities can be used directly as the basis for a canonical numbering scheme. Atoms can simply be listed in order of their extended connectivities, with canonical number 1 being accorded to the atom of highest extended connectivity, canonical number n going to the atom of lowest extended connectivity. The distribution of canonical numbers produced by this scheme differs from that derived by the complete Morgan scheme (where higher priority is given to the subsequent partitioning of atoms by distance from the atom accorded canonical number 1). The two different numberings for the skeleton would be as shown in Figure IX-6.

The full three-phase canonicalization procedure, with its exploration of different candidate canonical numberings and comparisons of adjacency matrices, is required only when the atom partitioning process does not completely separate the atoms into individual classes. When atoms do receive equivalent extended connectivity values, all their alternative numberings must be explored. Of course, as already illustrated, atoms related by symmetry will inevitably possess identical connectivity scores, and the alternative numberings of these atoms will simply result in identical adjacency matrices. However, only if identical values for extended connectivities of the atoms were **known** to result from symmetry would it possible to avoid exploration of the alternative numberings of the atoms.

IX.B.2. Improvements on the Basic Morgan Algorithm

Much attention has been devoted to supposed defects of the Morgan algorithm. If one examines the different extended connectivities accorded to atoms at each successive cycle, it is apparent that there is no uniform convergence to a final ordering. At each iteration, different sets of nonequivalent atoms may obtain the same extended connectivity values. The relative ordering of atoms, according to their extended connectivity values, may change from iteration to iteration. The equivalence of final extended connectivity values does not necessarily imply equivalence of atoms.

These "defects" are of little practical consequence. The Morgan extended connectivities generally limit the number of choices for canonical atom 1 to a very small subset of the constituent atoms in a structure. The extended connectivities of remaining atoms, together with the repartitioning of atoms as canonical numbers are assigned (as realized by the requirement that the next canonical number must be accorded to a neighbor of an already canonically numbered atom), provide effective constraints on the number of permutations of canonical number labels subsequently considered. Occasionally, the algorithm will generate a few redundant, nonoptimal, adjacency matrices, but it will always find a unique numbering for a structure. There are no structures that cannot be handled. Data on the use of the Morgan algorithm by the Chemical Abstracts Service show that it is efficient for almost all real molecules.[5]

In fact, there are only a couple of problems with the basic Morgan algorithm. The analysis does not exploit all the data that could be used to partition the atoms. Often, additional data are available to characterize atoms; these additional data can be exploited to speed up the partitioning process by reducing the number of iterative cycles. More importantly, the representation of connectivity data, as a single numeric index, does lose information. This loss of information can either result in incomplete partitioning or require extra iterative cycles.

There are many ways of improving on the atom partitioning stage of the Morgan canonicalization algorithm. Some are purely *ad hoc* extensions. These may work on a limited variety of test examples but do not represent any fundamental improvement. For example, the algorithm for computation of extended connectivities can be modified by initializing an atom's *new* extended connectivity to its *current* value before adding in the *current* connectivities of its neighbors. Often, this modification will lead to some small improvement in the number of distinct extended connectivity values found after each iterative cycle and thus may reduce the number of iterations necessary. Various schemes for weighting different contributions to an extended connectivity sum have also been noted.[6] Monovalent atoms may be excluded from the process whereby connectivities are updated. The terminating criterion may be modified so that iterations stop only after there have been *two* failures to improve on the number of distinct extended connectivity values. (If that modification is used, the atoms of structure **IX-11** are completely partitioned).

Most proposed extensions simply make more complete use of available data characterizing atoms. For example, atoms may first be combined into functional groups;[7] this not only increases the number of different atom types, but also decreases the effective size of the structure that must be processed. If the atom partitioning and numbering schemes give precedence to atoms in functional groups so that these are accorded lower canonical numbers, then as a side effect of the canonicalization process the structure has been reordered so that its functionality is easier to find.[8] Such a reordering of atoms would have no significance in systems concerned solely with the testing for duplicate structures but could be of considerable value in a synthesis planning program.

Any graph invariant can be used to help in the partitioning process. A *graph invariant* is a property of the structure (or "graph") that does not depend on its initial arbitrary numbering. All atom properties are graph invariants, so it would be appropriate to partition the atoms according to any (or all) of the properties *atom type, hybridization, number of hydrogens, and number of π-electrons*. If the rings in a structure are already known, each atom can be characterized by data defining the sizes of the various rings that include it. These data are again graph invariants and suitable as extra features whereby the initial atom partitioning might sometimes be improved. In one typical canonicalization algorithm,[9] the initial partitioning of atoms is based on an ordered list of properties for each atom that includes the element type, the number of covalent bonds to nonhydrogen atoms and the number of rings of various sizes that incorporate the atom.

Another property that has been suggested as a useful discriminant is the "Number of the Outermost Occupied Neighbor sphere" (NOON) value.[10] The NOON value of an atom is a measure of the length of the shortest path from it to the most remote atom(s) in the structure. Atoms with the least NOON values are the most centrally placed in the structure. The NOON values for the atoms in *n*-propyl-cyclopentane are shown in structure **IX-17**. The NOON value can be determined by "walking" through the graph or, possibly, by mathematical manipulations of the adjacency matrix. (The entries in each row of the product of a binary adjacency matrix with itself identify all atoms that can be reached by paths of 2 or less, paths of length 3 are determined by another matrix multiplication step, and so on).

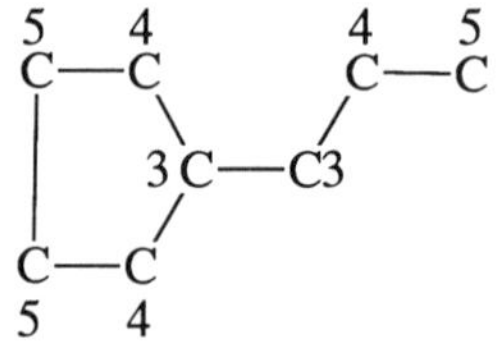

```
5   4          4   5
C———C          C———C
|    \        /
|     3C———C3
|     /
C———C
5   4
```

IX-17

Read and Corneil[11] include a brief discussion of suitable graph invariants in their general review of work on graph isomorphism. It should be noted that the use of some graph invariants, as aids to atom partitioning, may not be worthwhile except in special cases. For example, the number of rings in a structure can grow exponentially with the number of atoms; it would normally not be computationally effective to find all these rings simply to achieve some slight improvement in the partitioning process.

Irrespective of the number and variety of different atomic attributes used, any attempt to combine these values into a single numeric index is liable to suffer from chance numerical coincidences that result in distinct atoms fortuitously obtaining identical index scores. This problem of loss of discriminating information can be fairly readily overcome by a simple change in the data structures used to represent the sets of possibly equivalent atoms.

The following example illustrates more effective approaches to atom partitioning using the same data as given in Table IX-4 for the extended connectivity analysis for the sterane molecule (**IX-14**). Initially, all the atoms would be assigned to the same "class." The first partitioning step would be essentially identical to the first step in the previous analysis; the atoms would be split up into four classes according to degree. Computation of extended connectivities, as before, provides new "scores" that allow further partitioning of these initial classes. Although atoms 5, 8, and 13 obtain the same score at this stage (as do atoms 9, 14, and 10), the existing distinction between the tri- and tetravalents is now preserved. Each class of atoms as obtained in a previous partitioning step either remains unchanged, as does class 1 with atoms 18 and 19, or is split into a number of new classes as is class 2, which is split into three new classes. The initial partitioning steps are illustrated by the data in Table IX-5.

These data manipulations require a vector for the atom scores and two vectors in which atoms are grouped by their current classes, and in which new groupings are constructed. After each stage of computing new "scores" for the atoms, a new class vector is constructed. Singleton classes, those comprising only one atom, can be simply copied from old to new class vectors. Hopefully, the new scores will discriminate among members of multielement classes; for these, new classes can be constructed by what is, essentially, a "selection sort"

TABLE IX-5. Using Extended Connectivity Scores to Achieve a More Effective Partitioning of Atoms of IX-14

One initial class comprising all atoms:

Class:	1
Members:	1 2 3 4 5 6 7 8 9 10 11 12 13 14 15 16 17 18 19

Initial Connectivity Scores:

"Scores":	2 2 2 2 3 2 2 3 3 4 2 2 4 3 2 2 2 1 1

Yields 4 classes:

Class:	1	2	3	4
Members:	18 19	1 2 3 4 6 7 11 12 15 16 17	5 8 9 14	10 13

Morgan extended connectivity scores (First iteration):

"Scores":	4 4	6 4 4 5 5 5 5 6 5 4 6	8 8 9 9	9 8

Yields 8 classes:

Class:	1	2	3	4	5	6	7	8
Members:	18 19	2 3 16	4 6 7 11 15	1 12 17	5 8	9 14	13	10

algorithm. Thus when the "extended connectivity data" shown in Table IX-5 are used, the 11 bivalent atoms of class 2 are split into three classes. For each current class, the minimum atom score is determined (here, score 4 for atoms 2, 3, and 16), and all atoms with that score are collected together to form the next class in the new class vector. The process is repeated until all members of the current class have been entered into the new class vector.

The subsequent stages in this approach to using extended connectivity scores for structure **IX-14** are summarized in Table IX-6.

Separation of the atoms into classes, and application of subsequently derived scores to further partition these classes reduces the frequency of incomplete partitioning (e.g., the atoms of structures **IX-10** and **IX-11** would

TABLE IX-6. Completion of the Partitioning Process for IX-14

Eight Classes:

Class:	1	2	3	4	5	6	7	[illegible]
Members:	18 19	2 3 16	4 6 7 11 15	1 12 17	5 8	9 14	13	1[illegible]

Morgan extended connectivity scores (second iteration):

"Scores":	8 9	10 9 11	12 13 13 15 13	13 13 12	19 23	22 21	25	2[illegible]

Yields 16 classes:

Class:	1	2	3	4	5	6	7	8	9
Members:	18	19	3	2	16	4	6 7 15	11	17

Class (contd.):	10	11	12	13	14	15	16
Members (contd.):	1 12	5	8	14	9	13	10

Morgan extended connectivity scores (third iteration):

"Scores":	25	27	22	22	25	28	32 36 32	35	36
"Scores" (contd):	37 40	52	56	61	65	54	63		

Yields 18 classes:

Class:	1	2	3	4	5	6	7	8	9	10
Members:	18	19	3	2	16	4	6 15	7	11	17

Class (contd.):	11	12	13	14	15	16	17	18
Members (contd.):	1	12	5	8	14	9	13	10

Morgan extended connectivity scores (fourth iteration):

"Scores":	54	63	50	59	68	74	88 86	88	105	79
"Scores"(contd.):	85	89	123	162	142	154	162	181		

Yields 19 classes:

Members:	18	19	3	2	16	4	15	6	7	11	17
Members (contd.):	1	12	5	8	14	9	13	10			

be properly partitioned) and can reduce the amount of computation. In the sterane example only one iteration of the cycle computing extended connectivities was avoided. Such small savings in the number of iterations might be offset by the slightly more complex data manipulations. However, it is possible to further reduce computations by not recomputing scores for atoms already known to be in singleton classes and by slight improvements to the form of the computed discriminant property.

For example, one reasonable discriminant function takes a sum of the squares of the current class numbers of its neighbors. In the example in Table IX-6 at the stage where eight classes of atoms have been identified, the new discriminant score computed for node 6 would be $5^2 + 3^2 = 34$ (neighbor 5 is in class 5, neighbor 7 in class 3), wheras for node 15 the corresponding score would be $6^2 + 2^2 = 40$. The atoms of structure **IX-14** can be completely partitioned in only two steps, using the sum of squares of neighbor classes, following an initial partitioning by degree. Taking squares reduces the frequency of numerical coincidences; if a simple sum of neighbor classes or extended connectivities is used, atoms 6 and 15 would not be differentiated at this stage. A similar but slightly more elaborate approach uses as its "score" an ordered list of the class numbers of an atom's neighbors.[12]

In some structures, such as **IX-13**, local connectivity analysis (summing neighbor connectivities, class-numbers, etc.) will not reveal differences between atoms. Typically, some partitioning of the atoms can be achieved by calculating some more global measure of the each atom's environment. Essentially, the molecule is "viewed" from each atom in turn and a score is derived that expresses some simple aspect of each such view of the molecule. This viewing of the molecule involves some walk to neighbors, next nearest neighbors, and so on. NOON values represent one global property that could be calculated. For **IX-13**, NOON values would discriminate among the atoms as shown in Figure IX-7.

Viewing an atom's molecular environment is computationally rather expensive. In general, the determination simply of the NOON value, the

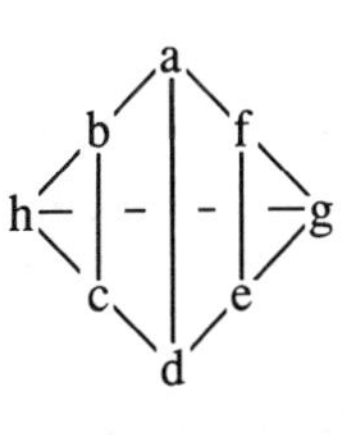

IX-13

Atom	NOON-values	"Neighbors"	"Parities"
a	2	3,4,0	1
b	3	3,3,1	7
c	3	3,3,1	7
d	2	3,4,0	1
e	3	3,3,1	7
f	3	3,3,1	7
g	2	3,4,0	1
h	2	3,4,0	1

Partitioning by NOON values:
Class #1: a d g h
Class #2: b c e f

FIGURE IX-7. Viewing of a molecule from each atom can provide discriminant scores dependent on global molecular properties.

minimal distance to the most remote atom(s), would not make most effective use of such data as can be derived when walking through the molecule. One more complete "score" would consist of a vector whose elements define the number of atoms at various bond radii; these vector scores are shown as "Neighbors" in Figure IX-7. Since manipulation and comparison of such vector scores can be costly, a less complete but more economic score could be based on a bit pattern, 1 bit for each bond radius. The 0/1 value for each bit could be determined according to whether there were an even or an odd number of neighbors at each bond radius. Such scores are shown as "Parities" in Figure IX-7. For instance, atom b has an odd number of neighbors at bond radii 1, 2, and 3 and thus has each bit set and a Parities score of 7. Although the more complete Neighbors or Parities scores do not improve on the partitioning obtained by NOON values for structure **IX-13**, they typically do provide better discrimination.

A partitioning of atoms, more complete than that obtained from the standard Morgan algorithm, can be achieved through devices such as the viewing of the molecule from each atom to derive discriminant scores, the separation of atoms into classes that are prevented from recombining and the iterative reclassification schemes for further breaking up classes. However, no atom partitioning scheme can achieve better than the *automorphism partitioning* in which atoms have been split into groups related by the symmetries of the structure. No partitioning scheme can guarantee that it will have achieved this automorphism partitioning on termination.

When more work is performed in the initial partitioning phase, subsequent numbering and adjacency matrix testing phases are less onerous. However, the real objective is, or should be, to minimize *overall* computational costs. There do not appear to be any data defining the relative costs of atom partitioning and other steps in the overall canonicalization process. The optimal analysis procedures for chemical structures have not really been defined.

It is possible to modify other steps in the canonicalization procedure and thus obtain increased performance. In some algorithms, terminal atoms are treated as special cases.[10] Rather than being analyzed by the main numbering algorithm, the terminal atoms receive canonical numbers in a subsequent processing step. A unique numbering of the terminal atoms can be defined through rules specifying precedences by atom type and by the previously derived canonical numbering of the atoms to which the terminals are attached. Identical terminals attached to a common atom (e.g., *gem*-dimethyls) may be recognized and only one numbering sequence derived for them;[4] this represents a special case of the more general form of symmetry analysis outlined later.

IX.B.3. Canonicalization of Structures with Defined Stereochemistry, Disconnected Structures, and Substructures

The analyses presented so far have presumed that only constitutional, that is, topological, properties of structures need be defined and manipulated. Often, this simplification is justified. In structure elucidation, the constitution of a molecule is generally determined prior to any consideration of stereochemistry. Most data-retrieval systems work in terms of constitutional structures, with any stereochemical data held in supplementary text records.

Planning of chemical syntheses is one area of computer assisted analysis where stereochemistry must be considered from the outset. The products of reactions are often determined by steric factors, and, not infrequently, the correct manipulation of stereochemistry is really the essence of the entire synthesis plan. Configurational stereochemical details must be represented and manipulated as synthesis plans are developed. Consequently, the ability to generate canonical forms for structures incorporating defined stereochemistry is essential. The first stereochemically extended structure canonicalizers were developed for various synthesis planning program.[4,13]

As discussed in preceding chapters, configurational stereochemistry can be specified by either an explicit ordering imposed on neighbor lists[4] or a "parity tag" and an implicit ordering.[14] Either way, the stereochemical designation is specified *relative to the given numbering of the neighbors of the stereocenter*. On canonicalization, the numbering of the neighbors will be changed; consequently, it may be necessary to change the value of the stereochemical tag. For example, if stereochemistry is defined in terms of some explicit ascending ordering of an atom's neighbors, any parity change can be determined from the number of pairwise exchanges of the canonically numbered neighbors that would be necessary to again have the neighbors listed in ascending order.

Assignment of a unique canonical stereochemical name can be complex in structures with topological symmetries. It is necessary to take configurational data into account for canonicalization of a symmetrical, arbitrarily numbered structure given stereochemical designators specified relative to the original arbitrary numbering.[4] Procedures for generation of all stereoisomers, or some constrained set of stereoisomers, of a given structure are based on a comprehensive analysis of symmetry and inherently yield "canonical" forms.[15]

Stereochemical data are normally in a form that can be appended to the canonical constitutional representation of a structure to give a canonical stereochemical representation. Thus in the **SECS** system the normal Morgan style representation of a canonical topological structure is supplemented by two extra records, one designating configurations in double bonds and the other detailing tetravalent stereocenters.[4]

Occasionally, it is necessary to process "structures" that are comprised of a number of disconnected parts. For example, **GENOA** requires canonical representations of its "cases" while it is combining substructural constraints. The **GENOA** cases will normally consist of a few multiatom fragments, such as two each of $(CH_3)_2C<$ and $>CH-OH$ groups, together with a collection of left-over atoms. Canonicalization of such multicomponent structures presents no particular difficulties. Each connected component is canonicalized by standard procedures as already outlined. Once in their canonical forms, identical constituent components can be recognized and grouped together. Ordering relationships for the canonical components can be defined and used to derive a final canonical form for the overall multicomponent structure.

Canonical substructures are frequently required in systems for analyzing (sub)structure–activity relationships. The most common such use has been in work on various atom properties that can be rationalized in terms of the structural environment of an atom. Such canonical substructures are typically defined in terms of *atom-centered codes*.[16,17]

These codes are generated by procedures similar to those used in a structure canonicalizer. The central atom for the code is, necessarily, accorded an effective "canonical number 1." Its immediate neighbor(s) is(are) assigned the canonical numbers 2, and so on; the ordering of neighbors will be determined by various priority rules, usually based on some form of extended atom types (including Hs, hybridization, aromaticity, etc.). Neighbors at two bonds radius are then incorporated; again, the ordering of these beta neighbors would be determined by some priority rules.

There are usually requirements that these "codes" be canonical at each shell level. This requirement means that priorities used to determine the numbering of neighbors can depend only on properties of these atoms or details of their connections either within a shell level or back to lower shell levels. All various possible numberings of an atom's neighbors at each shell level, out to the maximum represented by a code, must be generated and some preferred canonical form selected for the code.

Configurational stereochemical descriptors can be incorporated into such substructure codes.[17] Configurationally extended substructure codes are important in, for example, a system intended to correlate ^{13}C shifts with atom environments. Shifts of substituents on double bonds, or rings, are markedly dependent on the groups to which they are *cis* or *trans*. If only topology is represented, such structural details cannot be captured. The indistinguishability of purely topological codes for different stereoisomers would, necessarily, obscure certain spectral–structural relationships.

It is possible to represent conformational data in canonical substructures.[18] An obvious application for substructures with conformational descriptors would be the prediction of ^{1}H NMR spectra, where shifts and coupling patterns are determined by local geometry. Many other biochemical properties can be interpreted in terms of the nature and shape of particular substructures. Consequently, it is probable that there will be further development of canonical representations of substructures with defined configurational and conformational stereochemistry.

IX.C. SYMMETRY

IX.C.1. Introduction

The objective of canonicalization is to recognize the identity of two structures from the equivalence of their canonical forms. Atoms of the two structures that are accorded the same final canonical numbers are in identical molecular environments. Bond by bond, atom by atom, their equivalent molecular environment can be read back from the common canonical adjacency matrix.

If two different candidate canonical numberings of a single structure yield identical adjacency matrices, equivalences of atoms are again revealed. Atoms accorded the same canonical number are in symmetrically equivalent environments within the single structure. A representation of molecular symmetry, suitable for subsequent computer manipulation, can be built up as candidate canonical adjacency matrices are generated and compared.

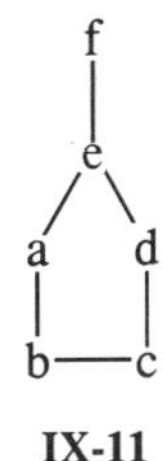

IX-11

Whenever comparison shows that the adjacency matrices are equal, the original atom identifiers, as given in the initial structure, need to be recorded for each canonical number. With structure **IX-11** used in earlier examples, a numbering [(e,1) (a,2) (d,3) (f,4) (b,4) (c,6)] gave the "best" adjacency matrix (C in Figure IX-5) found at a particular stage in the processing; this numbering would be saved to form the first element in a representation of the molecular symmetry. The subsequent numbering [(e,1) (d,2) (a,3) (f,4) (c,5) (d,6)] gave an equal adjacency matrix and thus would also be saved. These data would be recorded in some array similar to that illustrated in Table IX-7.

This process groups symmetrically related atoms into classes; these classes are known as *orbits*. Structure **IX-11** has four orbits: (e), (f), (a d), and (b c). The complete data array, built in this manner, is the *permutation symmetry group of the atoms of the structure*.

In structures without symmetry, each atom forms its own unique orbit and the permutation symmetry group comprises only the *identity* element (which records the correspondence of canonical and original atom numbers). Structures possessing elaborate symmetries may have rather large permutation symmetry groups. Thus even little tetrahedrane (**IX-18**) has 24 permutations in

TABLE IX-7. Data Identifying Equivalences Among Atoms as Discovered During the Canonicalization Process Applied to IX-11

Canonical Number	:	1	2	3	4	5	6
accorded to atom ---							
Adjacency matrix C	:	e	a	d	f	b	c
Adjacency matrix D	:	e	d	a	f	c	b

its symmetry group (which would require some 96 words of storage in a computer). A relatively simple structure such as **IX-19** has a symmetry group of order 128, (i.e., there are 128 different permutations of atom numbers over canonical labels); the complete permutation group for this structure requires more than 2500 words of storage. Sometimes, a specification of the orbits in the permutation symmetry group will suffice, but many uses of symmetry are best served by some representation of the complete group.

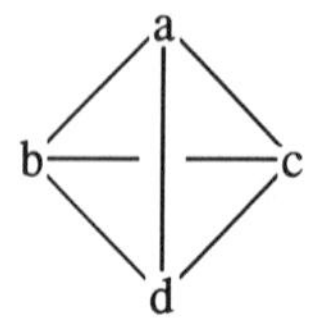

IX-18 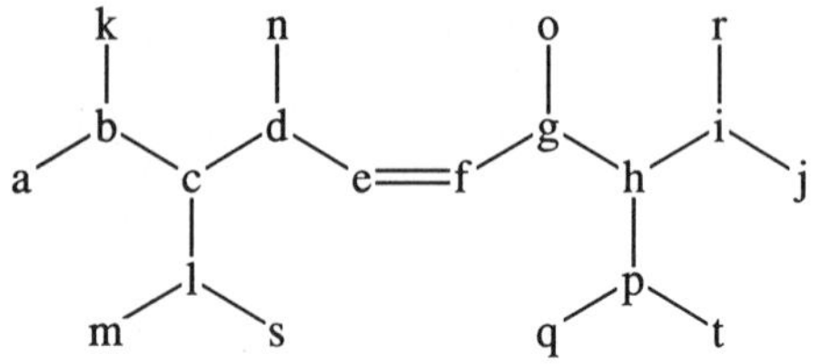 **IX-19**

In general, one is interested in the permutation symmetries of the atoms and the atom orbits. The symmetries of a structure can also be described in terms of a permutation symmetry group of its bonds (or "edges"). The edge permutation symmetry group of the dimethyl-cyclobutene structure (**IX-20**) are shown in Table **IX-8**. Double bonds in structures should be represented explicitly as two distinct "edges." Some of the permutations of the edge symmetry group involve only exchanges of pairs of edges in double bonds.

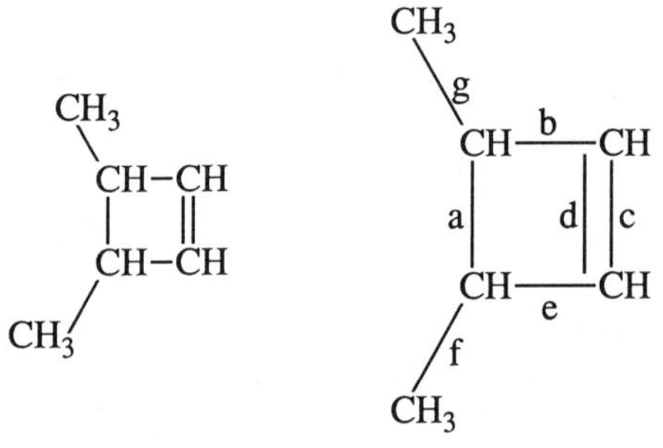

IX-20

TABLE IX-8. Edge Permutation Symmetry Group for IX-20

a	b	c	d	e	f	g
a	b	d	c	e	f	g
a	e	c	d	b	g	f
a	e	d	c	b	g	f

There are alternative ways of presenting the permutations in both atom and edge symmetry groups. The permutations shown in Table IX-8 might be given as [(a) (b) (c) (d) (e) (f) (g)], [(a) (b) (c d) (e) (f) (g)], [(a) (b e) (c) (d) (f g)], and so on. Such alternative representations focus attention on the atoms or edges being interchanged by the permutation (the *orbits of the permutation*) and the number of interchanges performed (and hence the *parity* of the permutation).

Although generally less used, permutation groups of edges are required in the analysis of the stereochemistry of structures with double bonds. The symmetry of a structure with specified configurations at double bonds and tetrahedral centers can be described in terms of the *configurational symmetry group*.[14] The configuration symmetry group combines information concerning double bonds (from the edge symmetry group) with information about the tetrahedral stereocenters (from the atom symmetry group).

The analysis of the symmetry of structures is of considerable importance in programs for structure elucidation and synthesis planning. For example, consider a structural problem wherein it has been (1) inferred that the unknown structure incorporates a decalin ring and (2) subsequently determined that this ring bears at least one substituent hydroxy group. If a structure assembly program had no model of the symmetries of the ring system, it would be necessary to try to build up larger structures by placing the hydroxy group on each of the ring atoms. Two of the 10 possible placings are shown in Figure IX-8.

After assembly, these structures would have to be converted into their canonical forms, at which point the second arrangement, as shown in Figure IX-8, would be recognized as a duplicate and discarded.

Without a symmetry analysis, a structure generator may spend an exorbitant amount of time on unprofitable assembling and canonicalization of duplicate structures. Similarly, unless it exploits symmetries in target

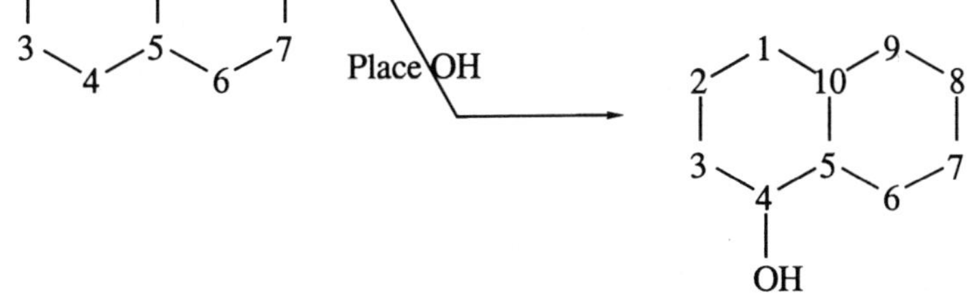

FIGURE IX-8. Substitution of a hydroxy group on a decalin ring without regard for symmetry.

molecules or reactants, a synthesis planner program may waste effort on the generation of duplicate synthetic paths.

The task of a symmetry analysis routine is basically to determine the number of distinct ways in which "labels" may be placed on a structure's atoms. This labeling task may be a simple one, similar to finding the distinct ways of labeling a decalin ring with a single substituent hydroxy group. Another simple example of labeling arises when one uses known symmetries to aid in a canonicalization procedure; after all, canonicalization entails simply finding the distinct ways of placing n different labels, *the integers* $1 \cdots n$, onto the atoms of a structure subject to various additional constraints. More complex labeling tasks arise when one has several different sets of labels, each set comprising one or several identical labels. An example of such a problem would be to find all isomers incorporating a decalin ring substituted with three methyls and two hydroxy groups.

IX.C.2. Use of Known Symmetries to Aid the Canonicalization Procedure

Sometimes the symmetry group of a structure may have been determined prior to the invoking of a structure canonicalization procedure. In some synthesis planning schemes, for example, the symmetry of a hypothesized precursor might be determined as a part of the analysis of the transform by which it is converted into a desired target compound of known symmetry. Knowledge of the symmetries of the precursor structure could reduce the work necessary to convert it into canonical form.

A canonicalization procedure can be presumed to have partitioned the structure's atoms into various classes and use this partitioning to determine the allocation of canonical numbers. If the structure has no symmetry and the partitioning process is effective, the canonical numbering of atoms is uniquely determined. Otherwise, at some stage in the numbering procedure, the next canonical label must be assigned to one of a set of atoms that could not be differentiated by the partitioning process.

Usually, it is necessary to try each possible atom in turn and derive different adjacency matrices from among which the best matrix, and numbering, is chosen. If the atoms in the set representing alternative choices are **known to be symmetrically related**, it is also known that different candidate canonical numberings for these atoms will ultimately yield identical adjacency matrices. Consequently, any one member of the set can be arbitrarily selected to receive the next canonical number, without any possibility of such arbitrary selection resulting in an invalid final numbering.

Of course, once an atom in a symmetrical structure has been labeled with some specific canonical number, the symmetry of the structure has been changed. For example, consider structure **IX-21**, with the permutation symmetry group given in Table IX-9.

TABLE IX-9. A Representation of Symmetries of Structure IX-21 Showing Complete Permutation Symmetry Group and Orbits of Atoms

j	c	d	i	a	b	e	h	f	g
j	c	i	d	b	a	h	e	g	f
j	d	c	i	e	f	a	g	b	h
j	d	i	c	f	e	g	a	h	b
j	i	c	d	h	g	b	f	a	e
j	i	d	c	g	h	f	b	e	a

IX-21

The first canonical number could be given to the unique atom j. There could then be three choices for atoms to be labeled as canonical number 2; these are atoms c, d, and i of structure **IX-21**. Given the symmetry group, it is apparent that c, d, and i are in the same orbit and thus any one, say, c, can be selected arbitrarily.

This assignment of a specific label, canonical number 2, to c changes the symmetry. As shown in Table IX-10, only two of the six original symmetry permutations are still valid. The other four permutations would exchange c with some other atom bearing a different label. The only symmetry elements remaining are those that correspond to permutations that do not involve c. (These are said to be the permutations that "stabilize" c.)

TABLE IX-10. Permutation Symmetry Group After Arbitrary Allocation of Label 2 to Atom c, and Permutations that Stabilize c

j-1	c-2	d-?	i-?	a-?	b-?	e-?	h-?	f-?	g-?
j-1	c-2	i-?	d-?	b-?	a-?	h-?	e-?	g-?	f-?
j-1	d-?	c-2	i-?	e-?	f-?	a-?	g-?	b-?	h-?
j-1	d-?	i-?	c-2	f-?	e-?	g-?	a-?	h-?	b-?
j-1	i-?	c-2	d-?	h-?	g-?	b-?	f-?	a-?	e-?
j-1	i-?	d-?	c-2	g-?	h-?	f-?	b-?	e-?	a-?
				Permutations Stabilizing j and c					
j-1	c-2	d-?	i-?	a-?	b-?	e-?	h-?	f-?	g-?
j-1	c-2	i-?	d-?	b-?	a-?	h-?	e-?	g-?	f-?

The partially labeled structure still possesses some elements of symmetry. The next canonical label, 3, might have to be assigned to a neighbor of c-2, for example. Examination of the reduced symmetry group of the partially labeled structure, as shown in Table IX-10, shows that c's unlabeled neighbors a and b are in the same orbit and so, again, an arbitrary choice can be made. Once label 3 has been assigned, as shown in Table IX-11, no elements of symmetry remain. (The analysis is similar if the label 3 had to be assigned to one or other of d or i).

If, at some subsequent stage in the numbering process, it were again necessary to choose between atoms, all alternative choices would have to be considered in order to guarantee that the best adjacency matrix would indeed be discovered.

This analysis of symmetry can be performed, as outlined, using some representation of the permutation symmetry group that is updated as canonical numbers are assigned and symmetries change. Alternative approaches use simply information identifying the orbit of each atom in the original symmetry group. (Usually, this is represented as a "class number" with all atoms in the same orbit accorded the same class number.) If the representation of symmetry is not updated as canonical numbers are assigned to atoms, other methods must be used to guarantee that arbitrary choices of numbering are made only for those atoms related by symmetries still remaining in the partially labeled structure.

The problem with any such use of symmetry is that the symmetries must be known prior to commencement of the canonicalization procedure. If the complete permutation symmetry group is not automatically available as a result of the processes leading to a particular structure, it will have to be computed and saved, or at the very least the orbits of the symmetry group (i.e., the *automorphism partitioning* of the atoms) will need to be found. There are many algorithms that give excellent **approximations** to the automorphism partitioning of atoms of a given structure (using simply standard atom partitioning tricks and a sufficient repertoire of atom properties and other graph invariants). However, there is no known graph invariant that guarantees finding the automorphism partitioning of atoms.[11]

Establishment of the equivalence of two atoms assigned to the same class by some atom partitioning process requires essentially a graph-matching procedure. The overall canonicalization process will have provided many data characterizing atoms; these data can be used to reduce the number of atom

TABLE IX-11. Permutation Symmetry Group After Two Labeling Steps

j-1	c-2	d-?	i-?	a-3	b-?	e-?	h-?	f-?	g-?
j-1	c-2	i-?	d-?	b-?	a-3	h-?	e-?	g-?	f-?
Permutations Stabilizing atoms j, c and a									
j-1	c-2	d-?	i-?	a-3	b-?	e-?	h-?	f-?	g-?

correspondences that must be considered by a graph-matcher. Nevertheless, it will still be necessary to perform a complete graph matching of the structure against itself for every pair of atoms whose equivalence must be proved. The two atoms must be placed in correspondence, their neighbors and bonds to and between neighbors checked, then next-neighbors, and so forth until the entire structures have been compared (or a difference is found).[19]

The cost of the necessary graph-matching procedures might be expected to make expensive an analysis that requires the automorphism partitioning of nodes. There are few comparative figures of performance of different canonicalization algorithms, but such figures as do exist suggest that algorithms that do first derive an automorphism partitioning of atoms are more costly than those that proceed more directly.[20]

Symmetry can, however, still play some role in reducing computations. As adjacency matrices are derived for different candidate canonical numberings, equalities can be detected and implied symmetries extracted. These symmetries can then be used to limit the number of permutations of canonical number labels that must be explored in subsequent steps. Such symmetry analyses have been incorporated into at least two of the canonicalization algorithms reported in the literature.[4,12]

Although relatively few chemical structures are symmetrical, there are practical reasons for the interest in algorithms that can efficiently handle symmetrical structures. Taking a structure elucidation problem as an example, one might seek to derive all isomers compatible with some chemical and spectral data characterizing a C_{20} structure. The final molecules obtained might well exhibit no symmetries, or at most trivial symmetries in their (topological) structures (e.g., *gem*-dimethyls). At intermediate stages in the analysis, however, symmetries may well be present. The first constraint in the structural analysis might be the requirement of a skeleton such as **IX-19** or maybe for a 14-membered ring system, as is found in a number of natural compounds. At this stage, then, there could be considerable symmetry to the partially assembled structures. Even with algorithms that repartition atoms as canonical numbers are assigned, in some cases many hundreds of adjacency matrices will be unnecessarily created if symmetries are not perceived and exploited. It is unfortunate for a program, even if generally efficient, to sometimes become preoccupied with some essentially $n!$ numbering problem.

IX.D. LABELING THE ATOMS OF A STRUCTURE

IX.D.1. Introduction: Simple Labeling Problems

As previously noted, *labeling* procedures are required whenever it is necessary to consider how many distinct ways there are of reacting, or substituting, some structure of known symmetry. Simple labeling tasks can be solved directly by considering the *orbits*, the sets of symmetrically related atoms, in the permutation symmetry group of the structure.

1	2	3	4	5	6	7	8	9	10
9	8	7	6	5	4	3	2	1	10
4	3	2	1	10	9	8	7	6	5
6	7	8	9	10	1	2	3	4	5

FIGURE IX-9. Atom Permutation Symmetry Group for the Decalin Skeleton.

Given the symmetry group of the decalin skeleton, the problem of finding the distinct placements of a single hydroxy group is trivial. The permutations of the symmetry group of the skeleton show that there are three orbits [viz., (5 10), (1 4 6 9), and (2 3 7 8)] (see Figure IX-9).

The three distinct monohydroxy decalins can be derived by selecting, as the point of attachment, a representative atom from each of the three orbits in turn. In the original structure numbering, this would give 1-hydroxy, 2-hydroxy, and 5-hydroxy decalins.

The analysis of a subsequent labeling step can be performed by using the updated representation of the permutation symmetry group. Consider, for example, the task of finding all decalin structures with both a hydroxy and a methyl substituent. (For simplicity, these groups are not permitted to be geminal.) First, the initial labeling with a hydroxy group could be performed, the new structure assembled, and the original symmetry group amended.

The amendments to the symmetry group are readily realized. When a specific label has been assigned to an atom, one can check through all the permutations in the symmetry group and mark as "disabled" those that involve the exchange of that (labeled) atom with some other atom. The three hydroxy decalins and their permutation symmetry groups are shown in Figure IX-10; permutations that would exchange the substituted ring atom with some other atom are marked as *disabled*.

For the first two monohydroxy decalins, only the *identity* element remains in the permutation symmetry group. Each atom in the decalin skeleton is in its own individual orbit. Each can be attached to the methyl substituent, and each such attachment of the methyl gives a unique structure.

For the third case, there are several orbits in the reduced permutation symmetry group. These orbits are (5), (10), (1 9), (2 8), (3 7), and (4 6). Placement of the methyl on node 5 is excluded, so there are five possible placements of the methyl that will yield unique structures. As before, only one representative node need be selected from each orbit and five structures are derived. Four of these are devoid of symmetry, but the fifth still possesses some symmetries that would be relevant to any subsequent labeling step (see Figure IX-11).

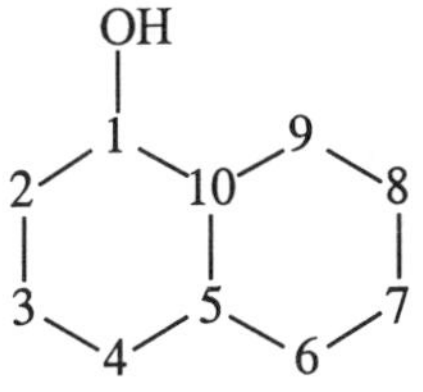

IX-10A

1*	2	3	4	5	6	7	8	9	10	
9x	8	7	6	5	4	3	2	1	10	(disabled)
4x	3	2	1	10	9	8	7	6	5	(disabled)
6x	7	8	9	10	1	2	3	4	5	(disabled)

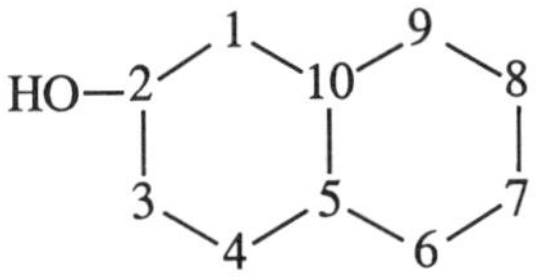

IX-10B

1	2*	3	4	5	6	7	8	9	10	
9	8x	7	6	5	4	3	2	1	10	(disabled)
4	3x	2	1	10	9	8	7	6	5	(disabled)
6	7x	8	9	10	1	2	3	4	5	(disabled)

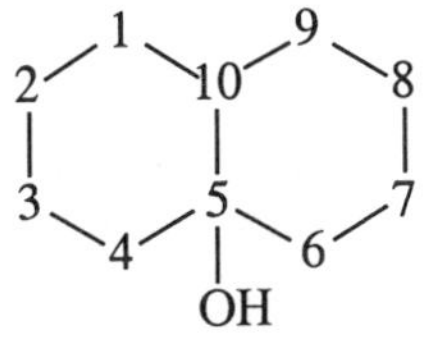

IX-10C

1	2	3	4	5*	6	7	8	9	10	
9	8	7	6	5*	4	3	2	1	10	
4	3	2	1	10x	9	8	7	6	5	(disabled)
6	7	8	9	10x	1	2	3	4	5	(disabled)

FIGURE IX-10. The three monohydroxy decalins and their corresponding reduced permutation symmetry groups

The analysis of symmetries used by the **GENOA** structure generating program is based on such considerations of orbits of atoms.[21] The intermediate "cases" in a **GENOA** problem often have some symmetries. The symmetries may be trivial; for instance, we have some "symmetry" simply because we have four identical unbonded quaternary carbons >C<, but even these symmetries can still be useful. (Given the requirement for a $\rightarrow$C—OH, it is necessary to consider only one of the >C<s as a potential bonding site for the–OH.[21]) Most of the constructive steps in a **GENOA** problem involve either taking a set of

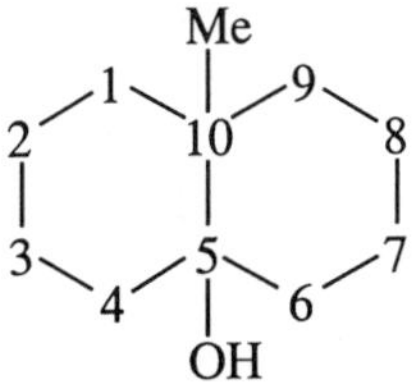

1	2	3	4	5*	6	7	8	9	10+	
9	8	7	6	5*	4	3	2	1	10+	
4	3	2	1	10+	9	8	7	6	5*	(disabled)
6	7	8	9	10+	1	2	3	4	5*	(disabled)

FIGURE IX-11. The only resulting structure with symmetries after two labeling steps.

unused atoms and building another discrete substructure or finding the distinct ways of extending an existing substructure by bonding on another atom. The constructive steps that extend substructures do involve a labeling process, but it is usually of the simplest kind where a single label (the new bond to another atom) has to be placed on a component of known symmetry.

IX.D.2. The General Labeling Problem

Analyses of symmetry such as those just outlined will suffice for most practical applications. However, such analyses are oversimplified. Sometimes, these simple analyses will suggest that a partially labeled structure has no remaining symmetry and that, consequently, any further labeling step will automatically yield a distinct result. However, when one is labeling a structure that has some symmetry with several different sets of identical labels, a partially labeled form may possess symmetries not apparent to the simple analysis. If these symmetries in the partially labeled form are ignored, subsequent labeling steps will inevitably lead to duplication.

In a structure generation procedure, for example, one might want to place two methyl groups, along with some other substituents as yet undefined, on a predefined prismane skeleton (**IX-22**) with the atom permutation symmetry group shown in Table IX-12. Here, one has a skeleton that is to be labeled with two CMe labels and four c labels.

One could start by placing a c label on node 1. Since node 1 of the structure has now been made unique by its c label, the symmetry permutations allowed are only those that do not interchange node 1 with any other node. These are shown in Table IX-13.

If node 2 is now labeled, also with a "c", there are no apparent elements of symmetry. Each remaining placement of the various labels supposedly yields a different structure. Some of these structures are shown as **IX-23–IX-26**.

IX-22

TABLE IX-12. Symmetry Permutations of Structure IX-22

1	2	3	4	5	6
1	3	2	4	6	5
2	1	3	5	4	6
2	3	1	5	6	4
3	1	2	6	4	5
3	2	1	6	5	4
4	5	6	1	2	3
4	6	5	1	3	2
5	4	6	2	1	3
5	6	4	2	3	1
6	4	5	3	1	2
6	5	4	3	2	1

TABLE IX-13. Reduced Set of Symmetry Permutations of Structure IX-22 After Labeling Node 1 with a "c"

1	2	3	4	5	6	
1	3	2	4	6	5	
2	1	3	5	4	6	(disabled)
2	3	1	5	6	4	(disabled)
			.			
			.			
			.			
6	5	4	3	2	1	(disabled)

Of course, because the essential equivalence of each of the c labels has been ignored, duplicates have been created. Structure **IX-25** is the same as **IX-24**. When structure **IX-25** is canonicalized it will be renumbered [*viz*., (1–2), (2–1), (3–3), (4–5), (5–4), and (6–6)]. In its renumbered form, **IX-25** will be recognizably identical to the previously created **IX-24** and will thus have to be discarded.

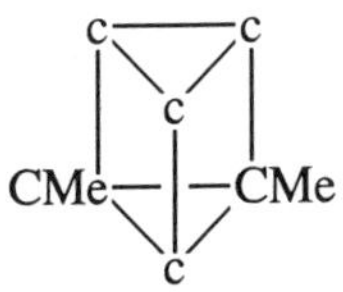

IX-23 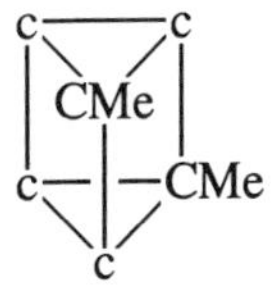 **IX-24** 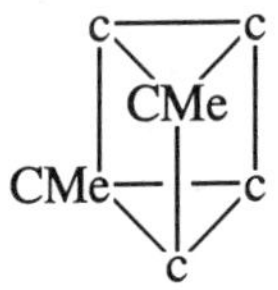 **IX-25**  **IX-26**

A labeling procedure should check its intermediate results rather than pass structures through to a full canonicalization procedure for duplicate elimination. In general, it is possible to make such checks well before a labeling has been completed and, thereby, eliminate invalid partial labelings and avoid unnecessary computations. Such checks are related to the requirements that the final canonical form be "greatest" (or "least") according to some suitable measure. Partial labelings can be tested to determine whether it is feasible that they might yield a "greatest" valued form.

Such tests on partial labelings are performed by considering the symmetries of the specific orbit currently being labeled. Each permutation of atoms in the orbit is applied to a given partial labeling and the resulting permuted partial labeling analyzed. This analysis is based on comparison of the given and permuted labelings to determine which is "better"; the comparison function is akin to that used in comparisons of candidate canonical adjacency matrices. If such a test reveals that the *permuted* labeling would be better, then the current partial labeling can be discarded. The better labeling either will have already been, or will subsequently be generated by the procedure that explores all possible labelings.

The comparison of given partial labeling and its permuted form has to cater for atoms to which no specific label can yet be assigned. For these atoms, the comparison process must assume the result most favorable to the current partial labeling. Partial labelings should be discarded only if they can be shown, unequivocally, to be inferior to symmetrically related labelings.

As an example of this analysis, consider the case where a labeling procedure has generated the partially labeled structure **IX-27** and is considering the extension of this partial labeling to give **IX-28**:

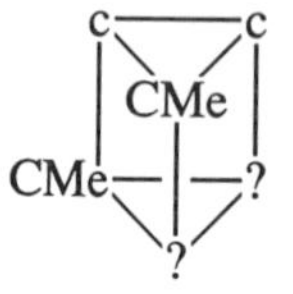

IX-27 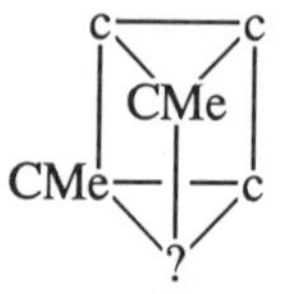 **IX-28**

At this stage the partial labeling is (c c CMe CMe c ?), where "?" indicates an unknown label. In the original unlabeled structure, all the atoms were in the same orbit, and thus all the permutations of the symmetry group, shown in Table IX-12, must be analyzed. The effects of these permutations on the partial labeling are summarized in Table IX-14.

The comparison function is presumed to rate cs as preferable to CMes and to work by comparing successive entries in the given labeling. The second permutation would be revealed as "worse" when comparing the second entries, no further comparisons would be necessary. As far as can be judged from these data, the current partial labeling is worth developing.

The third permutation reveals that a better labeling *must* exist and that consequently the partial labeling (c c CMe CMe c ?) is really not worth pursuing. The difference in the labelings is apparent when we compare the fourth entry where, assuming that the comparison function rates cs as better than CMes, the given partial labeling is inferior to its permuted image (irrespective of how any unlabeled atoms, ?s, may become labeled).

TABLE IX-14. Effects of Symmetry Permutations of Structure IX-22 on Partial Labeling c c CMe CMe c ?

Permutation	Labeling	Comparison
1 2 3 4 5 6	c c CMe CMe c ?	Identity
1 3 2 4 6 5	c CMe c CMe ? c	Worse
2 1 3 5 4 6	c c CMe c CMe ?	Better
⋮		
6 5 4 3 2 1	? c CMe CMe c c	?

Sometimes the result of the comparison would depend on the label eventually accorded to an unlabeled atom, as with the final permutation in Table IX-14; here, the "indeterminate" result is interpreted as meaning that the current partial labeling is still viable.

In the example in Table IX-14, the checks on the partial labeling did not effect much reduction in the number of final labelings generated. However, a similar analysis applied at earlier stages in the development of other possible labelings would eliminate many invalid permutations of labels. Thus the partial labeling (c CMe c ? ? ?), shown as structure **IX-29**, can be rejected on the basis of an analysis of the effect of the (1 3 2 4 6 5) permutation.

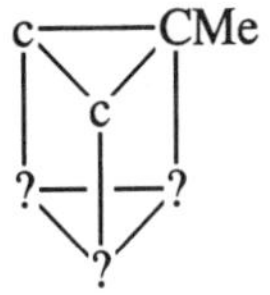

c——CMe
c
?—|—?
?

IX-29

This analysis of symmetry requires one orbit to be considered at a time. With the prismane structure, all atoms were in the same orbit. The labeling process could proceed directly by using the full permutation symmetry group of the unlabeled structure. If the original symmetry is such that there are several distinct orbits of atoms, the analysis becomes slightly more complex.

In order to make use of symmetry, it is necessary to focus on one orbit at a time. Before the labeling can begin, it is necessary to determine how many labels of each type are to be accorded to the atoms in each different orbit. The problem of labeling a decalin skeleton with three CMe and seven c labels provides examples of most of the steps necessary in these more elaborate labeling tasks.

As shown earlier, in Figure IX-9, there are three distinct orbits in decalin: (5 10), (1 4 6 9), and (2 3 7 8). The three 'CMe's must be partitioned among these three orbits; orbit (1 4 6 9) may be labeled with [(0,CMe) (4,c)], [(1,CMe) (3,c)], [(2,CMe) (2,c)], or [(3,CMe) (1,c)]. These various possible labelings of the (1 4 6 9) orbit would have to be analyzed. Then the remaining symmetries of the resulting structures, with the atoms in one orbit all labeled, would have to be determined, and then the entire procedure would have to be called recursively to analyze the partitioning and allocation of remaining labels among the atoms (2 3 5 7 8 10). (This process of partitioning the various label sets among orbits is known as *orbit recursion*).[22]

Three of the four possible allocations of CMe and c labels to the (1 4 6 9) orbit represent simple special cases. As shown in Table IX-15, if all four atoms in this orbit are labeled with cs, the symmetries of the skeleton have not been altered. In contrast, as shown in Table IX-16, both of the cases involving three labels of one type and one of the other result in partly labeled structures with no remaining symmetries. These two labelings involving only one instance of one

TABLE IX-15. Labeling the (1 4 6 9) Orbit with [(0,CMe) (4,c)] Labels Does Not Affect Symmetries of Decalin Ring System

	1c	2–	3–	4c	5–	6c	7–	8–	9c	10–
OK	9c	8–	7–	6c	5–	4c	3–	2–	1c	10–
OK	4c	3–	2–	1c	10–	9c	8–	7–	6c	5–
OK	6c	7–	8–	9c	10–	1c	2–	3–	4c	5–

TABLE IX-16. Labelings of the (1 4 6 9) Orbit with [(1,CMe) (3,c)], or [(3,CMe) (1,c)], Eliminate All Symmetries of Ring System

[(1,CMe) (3,c)]

	1CMe	2–	3–	4c	5–	6c	7–	8–	9c	10–
X	9c	8–	7–	6c	5–	4c	3–	2–	1CMe	10–
X	4c	3–	2–	1CMe	10–	9c	8–	7–	6c	5–
X	6c	7–	8–	9c	10–	1CMe	2–	3–	4c	5–

[(3,CMe) (1,c)]

	1c	2–	3–	4CMe	5–	6CMe	7–	8–	9CMe	10–
X	9CMe	8–	7–	6CMe	5–	4CMe	3–	2–	1c	10–
X	4CMe	3–	2–	1c	10–	9CMe	8–	7–	6CMe	5–
X	6CMe	7–	8–	9CMe	10–	1c	2–	3–	4CMe	5–

TABLE IX-17. Symmetries Remaining After the (1 4 6 9) Orbit Has Been Labeled with [(2,CMe) (2,c)]

	1c	2–	3–	4CMe	5–	6CMe	7–	8–	9c	10–
OK	9c	8–	7–	6CMe	5–	4CMe	3–	2–	1c	10–
X	4CMe	3–	2–	1c	10–	9c	8–	7–	6CMe	5–
X	6CMe	7–	8–	9c	10–	1c	2–	3–	4CMe	5–
	1c	2–	3–	4c	5–	6CMe	7–	8–	9CMe	10–
X	9CMe	8–	7–	6CMe	5–	4c	3–	2–	1c	10–
OK	4c	3–	2–	1c	10–	9CMe	8–	7–	6CMe	5–
X	6CMe	7–	8–	9CMe	10–	1c	2–	3–	4c	5–
	1c	2–	3–	4CMe	5–	6c	7–	8–	9CMe	10–
X	9CMe	8–	7–	6c	5–	4CMe	3–	2–	1c	10–
X	4CMe	3–	2–	1c	10–	9CMe	8–	7–	6c	5–
OK	6c	7–	8–	9CMe	10–	1c	2–	3–	4CMe	5–

type of label are, of course, solved by simply assigning that unique label to the first atom in the orbit.

There are three different ways of placing two labels of each type on the atoms of the (1 4 6 9) orbit. As shown by the data in Table IX-17, each such labeling leaves some symmetries in the partially labeled structure. When labeling an orbit, checks such as those previously illustrated for prismane would be needed. A partial labeling CMe→1, c→4, ?→5, and ?→9, would be rejected by considering the effect of the various permutations as is illustrated in Table IX-18. (This particular invalid labeling can be avoided altogether by the requirement that the first label allocated be of the type preferred by the comparison function.)

Each structure corresponding to these various labelings of the (1 4 6 9) orbit with from zero to three CMes, would have to be analyzed to determine the possible further labelings with the remaining members of the label set. The case where three CMes were used for the (1 4 6 9) orbit is trivial, and the labeling can be completed by assigning c labels to all remaining atoms. If no CMes were used for the (1 4 6 9) orbit, the situation would be almost identical to the original labeling problem; one to three CMes would now have to be placed on atoms of the (2 3 7 8) orbit.

The structures resulting from placement of two CMes in the (1 4 6 9) orbit are easy to label with the one remaining CMe and the five c's. In each of these structures the atoms still to be labeled form three orbits. [For example, in the last case shown in Table IX-17, these are (2 7) (3 8) (5 10).] Fully labeled structures can be obtained by associating the last remaining CMe label with one representative node taken in turn from each of these orbits.

The final case is where one CMe and three c labels have been placed on the (1 4 6 9) orbit leaving a structure, without symmetry, whose six remaining atoms must be labeled with two CMe and four c labels. This is a simple

TABLE IX-18. Testing, and Rejecting, a Partial Labeling of the (1 4 6 9) Orbit

Permutations				Effect on Partial Labeling				Comparison
				CMe	c	?	?	
1	4	6	9	CMe	c	?	?	Identity
9	6	4	1	?	?	c	CMe	? → partial labeling ok
4	1	9	6	c	CMe	?	?	Better! → reject partial labeling
6	9	1	4	(don't need to test, partial labeling already rejected)				

combinatorial problem, equivalent to finding all distinct ways of selecting two out of a set of six distinct objects. Standard combinatorial methods can handle this case.[22]

Labeling problems come in yet more elaborate forms. One may have several sets of distinct labels to place on some structure with complex symmetry. For instance, one might want to generate all 5–10-bridged, 10-membered heterocyclic rings with two nitrogens, three oxygens, and five carbons. Here, one has the label set [(2,N), (3,O), (5,C)] with these labels to be applied to the nodes of the decalin-type skeleton.

There are techniques whereby such complex labeling tasks can be solved. It is always possible to reduce the original problem involving several labels, $[(N_1,L_1), (N_2,L_2), \cdots, (N_i,L_i)]$, to a recursive sequence of problems each involving only two labels, starting with $((N_1,L_1), (N_2+\cdots N_i,?))$.[21,22] In the example, one could start by labeling the decalin-type skeleton with 2 Ns and 8 ?s. Each such labeling would then effectively define a new skeleton, with some reduced symmetry. Each such "new skeleton" and the remaining labels would be used in a recursive call to the same labeling procedure. This overall procedure, based on the analysis of just two label types at each recursive step, is known as *label recursion*.

Various simplifying special cases can arise in the recursive labeling process, particularly at the deeper levels in the recursion. For example, the partially labeled structure may have no symmetry left, or there may be only a single instance of one of the two label types. When there are several labels of each type, and the structure has symmetry, it is necessary to analyze the various orbits of nodes in the structure and apply the procedures outlined previously.

These general labeling algorithms have direct applications to such problems as finding all substitutional isomers of a ring system.[24] These algorithms are also necessary for some structure generators. For example, the **CONGEN** structure generation programs use *superatom* components when generating intermediate structures.[25] Superatoms are multiatom substructures with many potential bonding sites; one could, for example, have a decalin system as a superatom. The intermediate structures produced by **CONGEN** are defined in terms of bonding between its constituent superatoms and residual atoms. Final structures have to be derived by embedding the superatoms, that is, expanding them out to reveal their internal structure. When a superatom is thus embedded, bonds from other structural components have to be allocated specifically to its various constituent atoms; this allocation process is really just a special case of the standard problem of finding substitutional isomers.

IX.E. FINDING THE RINGS IN A MOLECULE

The canonical representation of a compound and its atom permutation symmetry group provide a complete specification of its structure. Data characterizing just the *ring systems of a molecule* are often needed to supplement the complete structure definition so that attention may be more readily focused on particular aspects of its form. For example, the planning of good synthetic routes to a compound depends on knowledge of its cyclic structure. The overall importance of the ring systems in chemistry is reflected in conventional chemical nomenclature and in the various "ring indices" detailing known structures.

Ring analysis tends to be of less importance in a structure elucidation system than in a synthesis planner or program for data base manipulations. On the whole, experiments do not yield data simply specifying the existence of a ring of given size. Data *requiring* the presence of some ring system will further characterize it, at least, as carbocyclic or heterocyclic. Ring systems so inferred from spectral and chemical data can be defined as substructures and then used as constraints in a structure assembly process as outlined previously.

Spectral evidence *excluding* small rings, for example, the absence of appropriate ^{13}C and ^{1}H NMR resonances for a cyclopropyl system, can be readily accommodated in even simple structure generators. Such structural constraints can be expressed either by defining certain ring systems as *prohibited* substructures or through the use of special *atom tags* that inhibit the construction of any bonds that would result in undesired rings.

Ring analysis does have some role in the structure evaluation stage of the overall interpret-generate-evaluate scheme. For example, aspects of the analysis of conformational stereoisomerism and the prediction of plausible mass spectral fragmentations both depend on information characterizing ring systems.

The essentials of ring-finding algorithms have been clearly summarized in the work of Wipke and Dyott.[26] As detailed in their work, there are two major stages in a procedure for finding all rings in a given structure. A *fundamental* set of rings must first be found. The number of rings in this fundamental set is determined by the number of atoms and bonds in the given structure as follows: (number of fundamental rings = 1 + number of bonds − number of atoms).

The total number of rings in the structure can be much larger (as large as $2^{\text{number of fundamental rings}}$ -1). Once a fundamental set of rings has been found (and characterized by the sets of bonds making up each ring), the other rings in the structure can be identified in the second major stage of the process. These other rings can be found by taking appropriate combinations of the bonds from different fundamental rings.

Although the number of fundamental rings is well defined, for most cyclic structures there will be many alternative possible sets of rings that could be taken as the fundamental set. An example is provided by the kaurene-type skeleton shown in Figure IX-12. This structure has 16 atoms and 19 bonds (or edges) and thus has four fundamental rings. A particular set of fundamental rings (as found by the ring-analysis routines from **GENOA-STRCHK**) is also shown in Figure IX-12.

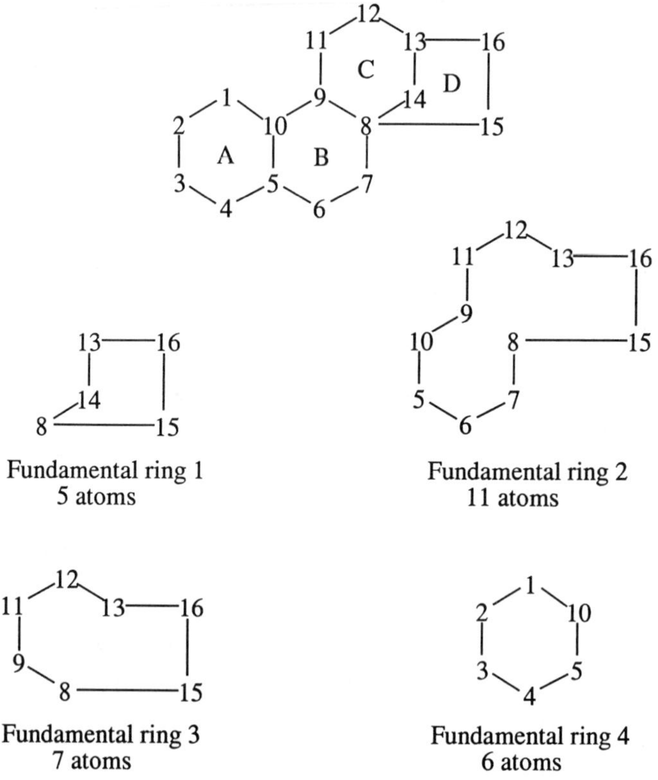

Fundamental ring 1
5 atoms

Fundamental ring 2
11 atoms

Fundamental ring 3
7 atoms

Fundamental ring 4
6 atoms

FIGURE IX-12. A kaurene like structure and a set of fundamental rings for this structure.

A set of three 6- and one 5-membered rings (i.e., the A, B, C, and D rings of the structure) would constitute an alternative set of fundamental rings. Such an alternative set would be much more relevant to a description of the chemical nature of this structure. Considering topology only, however, either of these sets of fundamental rings (or any of the others possible for this structure) is equally good.

The particular set of fundamental rings derived by most ring-finding algorithms will be arbitrary in nature, dictated by the order in which paths through the structure are developed. The fundamental rings will normally be determined by first growing a *spanning tree* for the structure.

A spanning tree can be found by a recursive algorithm that explores out along each alternative bond from a given atom to as yet unvisited neighbors. (Iterative algorithms are also possible.) The development of a spanning tree can be illustrated for the kaurene structure shown in Figure IX-12. Starting from atom 1, the sequence of bonds traversed might be 1: 1–10, 2: 10–5, 3: 5–6,..., 9: 13–14. As each atom is reached, it should be marked as having been visited and characterized by data defining the path connecting it back to the root node (here, atom 1).

Traversing the skeleton:

The spanning tree derived:

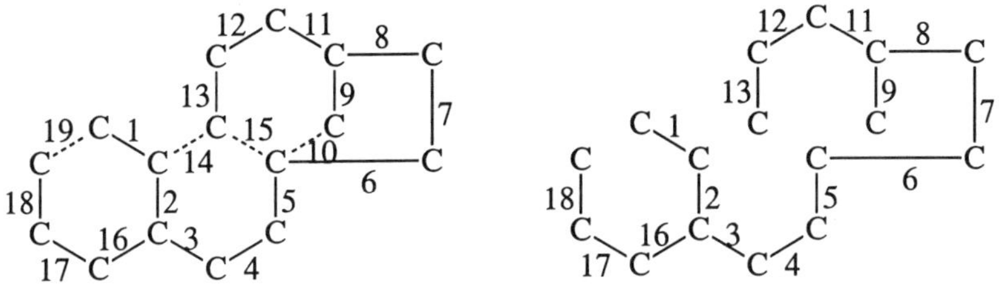

The ring closure bonds:

Bond 10, joining C(8) - C(14)
Bond 14, joining C(9) - C(10)
Bond 15, joining C(9) - C(8)
Bond 19, joining C(2) - C(1)

FIGURE IX-13. Development of a spanning tree for the kaurene skeleton.

On reaching atom 14 of the kaurene skeleton, it would be apparent that the only bond (apart from that by which this atom was reached) leads back to an already visited neighbor. This bond, bond 10 in Figure IX-13, represents a *ring closure bond*. The first fundamental ring could be identified at this stage. This fundamental ring comprises just the ring closure bond, bond 10, and bonds 9, 8, 7, and 6 leading back from atom 14 to atom 8.

After generation of the first fundamental ring, the recursive sequence of function calls would unwind to the point where there was an atom with as yet unvisited neighbors. Recursion would then proceed with exploration of these alternative paths. For the example in Figure IX-13, bonds 11: 13–12, 12: 12–11, and 13: 11–9 would thus be traversed. On reaching atom 9, it would again be found that the atom's neighbors had previously been visited. Bonds from atom 9 to atom 10 and from atom 9 to atom 8 are, again, ring closure bonds. The second fundamental cycle shown in Figure IX-12 is obtained by taking bond 14 (14: 9–10) and the sequence of bonds 13, 12, 11, 8, 7, 6, 5, 4, 3, and 2 along the path from atom 9 back to the root. Similarly, the third fundamental cycle in Figure IX-12 is based on the ring closure bond 15 (15: 9–8) and bonds that are on the path from atom 9 to the root and lie between atoms 9 and 8 on this path.

Again, recursion would be unwound. All atom 8's neighbors would have been marked as visited. (There could be no redundant exploration back along bonds from atom 8 to atom 14 or 9.) The unwinding of recursion would proceed as far as atom 5, which still has an as yet unvisited neighbor. Exploration out along the bond from atom 5 to atom 4 would reveal the last of the fundamental rings.

Given any set of fundamental rings, the other rings of a structure can be derived. These other rings will be comprised of bonds taken from two, or more,

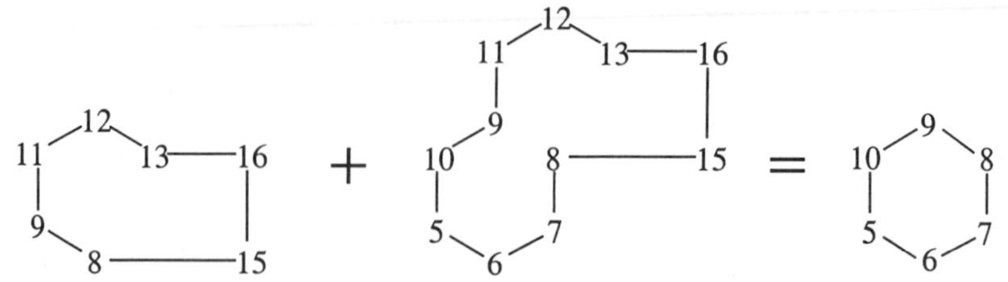

Bond sets:

Fundamental ring 2:	2,	3,	4,	5,	6,	7,	8,	11,	12,	13	14,	15
Fundamental ring 3:					6,	7,	8,	11,	12,	13,		
New ring :	2,	3,	4,	5,							14,	15

FIGURE IX-14. Combination of a pair of fundamental cycles to reveal an additional ring.

of the fundamental rings. For example, as shown in Figure IX-14, the B ring of the kaurene structure is revealed if one selects those bonds in either one, but not both, of fundamental rings 2 and 3 from Figure IX-12.

Other rings in this structure can be revealed by taking combinations of three fundamental rings. Thus the basic 10-membered ring can be found by combining fundamental cycles 2, 3, and 4 as shown in Figure IX-15. All possible combinations of the fundamental cycles have to be explored in this manner.

Not all such combinations of fundamental cycles will yield connected rings. Obviously, for the example in Figure IX-12, no new rings would be revealed by taking the combination of fundamental rings 1 and 4. There are no edges common to this pair of rings; the combination consists, of course, of the two separate rings themselves. Another unsatisfactory combination would be obtained by taking fundamental rings 1, 2, and 3; this combination again yields

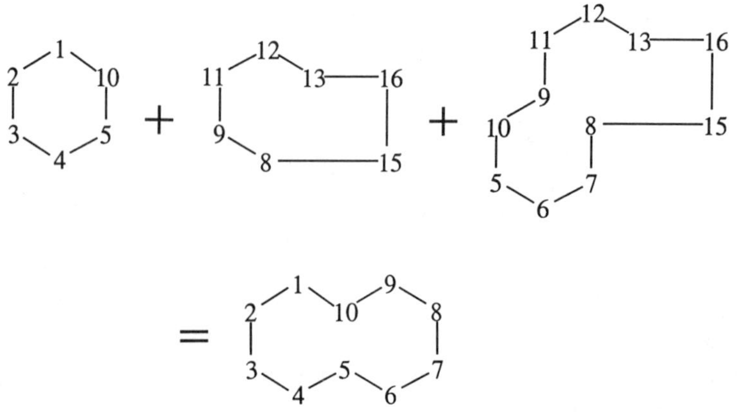

FIGURE IX-15. Combination of three fundamental cycles for identification of 10-membered A-B ring system.

two rings, this time spiro fused, rather than a single new ring. As combinations of fundamental rings are generated, it is necessary to check that the set of bonds thus obtained does represent a single connected ring comprising all the bonds in the set. For this structure, six of the combinations of the fundamental cycles yield new rings.

The process of combining fundamental rings can be performed very simply on a computer. The requirement for bonds that are in one ring, but not two, corresponds to an EXCLUSIVE OR operation on sets of bonds. "Sets" can be conveniently represented by bit strings. A bond's index number (possibly determined as the spanning tree is grown) corresponds to a specific bit in the bit string; a bond's presence or absence in a particular ring is defined by the 1/0 value of its corresponding bit.

A simple recursive algorithm for finding all rings from a given set of fundamental rings would be:

```
Combine (depth)
   if (depth > number_fundamental_cycles) then test current_bond_set
  else begin
       {try including the bonds from fundamental cycle [depth]}
       From Exclusive Or of current_bond_set and bond_set_cycle [depth]
       Combine (depth + 1)
       {Take these bonds back out of current bond set}
       Form EXCLUSIVE OR of current_bond_set and bond_set_cycle [depth]
       {and now try those further combinations that do not}
       {involve any bonds from this fundamental cycle}
       Combine (depth + 1)
   end
```

Initially, the *current bond set* would be empty. Sets of bonds from the different fundamental cycles would be combined with the current bond set as the procedure recurses in.† When the recursion depth exceeds the number of fundamental cycles, the current bond set represents another particular combination of fundamental cycles. The "test" procedure must verify that the set of bonds so obtained does represent a true ring.

A complete FORTRAN program for growing a spanning tree and finding a fundamental set of rings, and for further generating all rings by combining these fundamental rings, has been published by Wipke and Dyott.[26] This program exploits bit strings to allow the efficient representation and manipulation of sets of bonds. The published program contains a further refinement that results in considerable gains in efficiency. Quite commonly, there will be several distinct "assemblies" of rings in a polycyclic molecule. A simple example is provided by the clerodane **IX-30**.

†The bond sets for the different fundamental rings are assumed to be held is some array of sets: bond set cycle = array [$1 \cdots n$] of set.

IX-30

Here there are five fundamental cycles, the epoxide ring, the furan ring, the lactone, and two rings (either 6,6 or 6,10 rings) in the decalin system. The furan ring is obviously completely disjoint from the rest of the cyclic system, and there is no possibility of finding any additional rings by combining bonds from the furan ring with bonds taken from other rings. Although the epoxy and lactone rings both incorporate atoms that are also members of the decalin system, they again share no bonds with any other ring. Like the furan, the two fundamental rings represented by the epoxide and lactone can not be combined with any other fundamental ring to yield additional rings.

Structure **IX-30** can be said to incorporate four distinct ring assemblies.[26] A ring assembly is found by collecting all fundamental rings that have bonds in common. This collection operation can be performed after selection of a set of fundamental rings and prior to the step wherein additional rings are sought. In that subsequent step, where fundamental rings are combined to derive additional rings, it is possible to restrict attention to only one ring assembly at one time. Quite dramatic time savings have been reported for complex organic molecules such as large alkaloids.[26]

As rings are generated, they can be passed to routines for analysis or selection of those rings of particular interest in some specific application. For example, some structure generation programs do provide for ring-size constraints.[25] Such constraints allow the user to define minimum and maximum numbers of rings of various sizes in a structure. In such programs, the ring analysis will involve simply updating some counter determined by the size of each ring discovered in the structure.

Synthesis planning programs utilize ring information in a number of ways. The presence of rings of certain specific sizes may be used to suggest the use of transformations particularly suited to generating these rings. Evaluation of the effectiveness of a possible transform will generally require data on other rings in the structures; a proposed transform might, for example, be contraindicated by steric constraints implied by other bridging or fused rings. Rather than saving details of all the rings in each intermediate structure, most synthesis planning programs take a subset chosen to represent the rings of synthetic interest.

There are a number of varied but basically fairly similar definitions of *synthetically significant rings*. For example, in the **SECS** program sets of synthetically significant rings are defined for each ring assembly in a structure. For **SECS**, the chemically interesting rings in a ring assembly are those forming a reduced basis set together with any additional rings with up to eight bonds. The reduced basis set is developed from the set of fundamental rings characterizing a particular ring assembly. The reduced basis set is an approximation to the *smallest set of smallest rings* , that is, a set of rings consisting of only the smallest rings that can be used together to make up a fundamental set of rings. The reduced basis set is found by a simple iterative procedure that tries pairwise combinations rings starting from the particular (arbitrary) set of fundamental rings found for a ring assembly; if some combination yields a smaller ring, this replaces the larger of the pair of rings that were combined.

Selection of a set of "chemically interesting" rings is, obviously, a somewhat empirical process. The rings chosen should probably incorporate some fundamental set of rings (so that all other rings can again be derived easily if they should be subsequently required). Any additional rings are redundant, but this redundancy may be useful at later stages in the processing of a structure in that required data can be simply "looked up" rather than recomputed. There are obvious advantages to a system, such as the Wipke-Dyott system, which separates the basically *ad hoc* ring selection procedures from the well defined algorithmic procedures for ring generation. The ring generators can then be relied on in all circumstances while the ring selectors can be modified as may be necessary for particular applications.

However, programs for planning syntheses are primarily concerned with small rings. The strain energy of a structure can be analyzed in terms of the small ring systems that it incorporates. The relative reactivities of different functional groups and double bonds are related to the smallest rings in which they occur. Although there are many reactions designed to create small rings, there are few reactions for generating specific large ring systems, or for inserting additional bridging bonds into existing large ring systems. Since only small rings are apparently useful, approaches based on finding all rings and selecting a subset are frequently held to be overly indirect.

Similarly, when ring systems are used to classify structures for storage in some data base, the rings of interest are, again, the small rings. Data base systems generally use the *smallest set of smallest rings* (SSSR). The SSSR can be found from a complete set of rings in a structure. As rings are generated, they can be sorted by size and stored. Then, starting with the smallest ring, a fundamental set of rings can be collected by taking the next smallest ring that is linearly independent of those already chosen. This chosen set would be the smallest set of smallest rings. Although these procedures are fairly simple, their perceived costs have justified the search for alternative, more direct methods of obtaining this particular fundamental set of rings.

Alternative algorithms for finding only small ring systems have been frequently proposed. A number of algorithms have been given that involve following paths through a structure until these either return to their starting points[28,29] or are otherwise closed.[30] These methods have a number of disadvantages. The same rings may be "discovered" repeatedly. To discover rings of up to some maximum size it is necessary either to thoroughly explore the entire structure (and thereby find all rings, albeit inefficiently), or to perform repeated searches from, say, all branched atoms.

A more efficient approach entails first the development of a spanning tree.[31] This identifies a set of ring closure bonds and one set of fundamental rings. A "reduced basis set" of rings can be found by iterative pairwise combinations of these rings. Additional rings, up to a defined maximum size, can then be developed by searching paths out from each end of each known ring closure bond. The ring closure bonds, identified while growing the spanning tree, define all the necessary starting points involved in exploration for additional rings.

These algorithms[28–31] were devised to find "chemically interesting" rings for various synthesis programs. Programs designed to identify directly only the *smallest set of smallest rings* have also been reported.[32,33] These programs encounter problems with unusual structures, such as those that incorporate a ring that is a necessary member of the smallest set of smallest rings but all of whose edges are incorporated in other still smaller rings.

Arguments have been presented justifying the use of solely the smallest set of smallest rings for characterization of structures in a chemical synthesis program.[34] Another algorithm for identification of the set of smallest rings was also given. The discovery of rings is based on the initial development of a spanning tree. The method suggested for finding a spanning tree and initial set of ring closure bonds is elaborate requiring prior computation of the bond radius from an arbitrarily chosen root atom to all other atoms in the structure. The ring closure bonds are ordered by distance from the arbitrarily chosen root atom. Then each ring closure bond is used in a process that grows paths, back down through the spanning tree from each end of the bond, until a complete ring containing that ring closure bond is revealed. For certain structures, it is necessary to repeat the entire procedure of growing a spanning tree from some alternative starting point, and then again developing rings in order to guarantee that the smallest set of smallest rings is indeed derived.

Yet another algorithm for finding the smallest set of smallest rings has been developed for the **CAMEO** program for mechanistic evaluation of organic reactions.[35] Structures are first pruned by the elimination of acyclic appendages, and branched atoms are located. These branched atoms serve as initiating points from where paths are grown by traversing the bonds of the structure. As paths are developed from each initiating point, they are checked pairwise to determine whether any rings have been revealed. Duplicate rings may be generated; therefore, the program must check for and eliminate any duplicates. It may be necessary to explore paths from several initiating atoms to find all rings. Special checks are applied to reveal "buried" rings whose edges are all incorporated in other ring systems. Large circumscribing rings, such as the outer framework ring of a porphyrin, are analyzed separately.

Most of these algorithms for discovering small rings are complex. Some have limitations that prevent the correct processing of certain (rare) complex ring systems; other algorithms need to make provision for various, already discovered, special cases; still other algorithms may have to repeatedly reanalyze a structure. A complete analysis of the comparative performance of the various algorithms is not possible on the basis of published data. Generally, these ring-finding procedures form a part of some more elaborate program, and such computation times as are given sometimes include the calculation of other structural attributes (e.g., aromaticity) that are required by the particular application. Computation times are, of course, markedly dependent on machine and programming language. From such data as are available, however, it does not appear as if any of these alternatives algorithms even approaches the efficiency of the Wipke-Dyott method for selecting rings of interest from a complete set as generated by an efficient algorithm.

IX.F. SUMMARY

Procedures for canonicalization, identification of symmetry, and finding rings are fundamental to any computer-assisted structure manipulation system. The basic algorithms are now well established. The published literature includes complete algorithms,[4] even program listings,[26] of some of the better schemes. Briefer details of many subsequent developments, making more use of symmetry in canonicalization, and so on, are also available. It should not be necessary to develop further variations of existing algorithms for any of these basic procedures.

A particular application may dictate the use of an algorithm that, in some more general context, would not be optimum. For example, if one is interested only in small rings (and does not require a fundamental set of rings), it may well be sufficient to use some simple graph-exploring algorithm that starts from each atom and, if necessary, includes some mechanism for discarding duplicate rings. Most applications are likely to be better served by an efficient ring-finding algorithm generating the set of all rings, with the ring systems of interest being selected from this complete set.

As a further example, it is quite appropriate to use all available atom properties, including exotic properties such as ring memberships, NOON values, distance from nearest unsaturation, and so on, to improve the initial partitioning of atoms for a canonicalizer. However, it is usually more effective to employ a sophisticated canonicalization algorithm rather than invest substantial effort in computing extra atom properties that, in some few selected test cases, happen to have provided some small improvement in the partitioning process.

The appropriate detail in a symmetry analysis depends on the application. Generally, chemical structures have little symmetry and the manipulations are limited to the placement of either a single "label" or perhaps two distinct labels on the structure. For such applications, the simplified symmetry analysis outlined above will suffice. However, these simple schemes can lead to many duplicates in tasks such as finding all the isomers of some large ring system bearing several different groups of identical substituents. For these applications, the full symmetry analysis, with its elaborate "orbit" and "label" recursion procedures, is necessary. Published descriptive[22] and mathematical[23] treatments of the general labeling problem are in sufficient detail for the coding of labeling programs to be feasible.

The whole area of graph isomorphism, canonicalization, and labeling has recently again become of some interest to mathematicians. Graph isomorphism constitutes a convenient example of a class of related computational problems; mathematicians are primarily interested in characterizing the inherent complexity of these computations. Several significant advances have recently been made, largely as a result of work in group theory from where algorithms for determination of properties of permutation groups have been taken and adapted to be tests for graph isomorphism. For certain restricted classes of graphs, there are now isomorphism algorithms that require "polynomial" computation time. [For instance, for graphs comprising n trivalent -nodes there are isomorphism tests requiring $O(n^4)$ computational steps.] These mathematical analyses are reviewed by Hoffmann.[36]

REFERENCES

1. M. Randic, "On Unique Numbering of Atoms and Unique Codes for Molecular Graphs," *J. Chem. Inf. Comput. Sci.*, **15** (1975), 105.
2. H. L. Morgan, "Generation of Unique Machine Description for Chemical Structures, a Technique Developed at Chemical Abstracts Service," *J. Chem. Doc.*, **5** (1965), 107.
3. J. E. Ash, "Connection Tables and their Role in a System," in *Chemical Information Systems*, J. E. Ash and E. Hyde, eds., Ellis Horwood, Chichester, UK, 1975, 156, Chapter 11.
4. W. T. Wipke and T. M. Dyott, "Stereochemically Unique Naming Algorithm," *J. Am. Chem. Soc.*, **96** (1974), 4834.
5. L. J. O'Korn, "Algorithms in the Computer Handling of Chemical Information," in *Algorithms for Chemical Computations*, R. E. Christoffersen, ed., American Chemical Society, New York, 1977, 122, Chapter 6.
6. C. A. Shelley and M. E. Munk, "Computer Perception of Topological Symmetry," *J. Chem. Inf. Comput. Sci.*, **17** (1977), 110.
7. M. Uchino, "Algorithms for Unique and Unambiguous Coding and Symmetry Perception of Molecular Structure Diagrams. II. Basic Algorithm for Unique Coding and Computation of Group," *J. Chem. Inf. Comput. Sci.*, **20** (1980), 116.

8. M. Bersohn and A. Esack, "A. Canonical Connection Table Representation of Molecular Structure," *Chimica Scripta*, **6** (1974), 122.
9. C. A. Shelley, M. E. Munk, and R. V. Roman, "A Unique Computer Representation for Molecular Structures," *Anal. Chim. Acta*, **103** (1978), 245.
10. C. Jochum and J. Gasteiger, "Canonical Numbering and Constitutional Symmetry," *J. Chem. Inf. Comput. Sci.*, **17** (1977), 113.
11. R. C. Read and D. G. Corneil, "The Graph Isomorphism Disease," *J. Graph Theory*, **1** (1977), 339.
12. C. A. Shelley and M. E. Munk, "An Approach to the Assignment of Canonical Connection Tables and Topological Symmetry Perception," *J. Chem. Inf. Comput. Sci.*, **19** (1979), 247.
13. A. Esack and M. Bersohn, "Computer Manipulation of Central Chirality," *J. Chem. Soc. Perkin Trans. I*, **1975**, 1124.
14. J. G. Nourse, "The Configuration Symmetry Group and its Applications to Stereoisomer Generation, Specification and Enumeration," *J. Am. Chem. Soc.*, **101** (1979), 1210.
15. J. G. Nourse, R. E. Carhart, D. H. Smith, and C. Djerassi, "Exhaustive Generation of Stereoisomers for Structure Elucidation," *J. Am. Chem. Soc.*, **101** (1979), 1216.
16. W. Bremser, "HOSE—a Novel Substructure Code," *Anal. Chim. Acta*, **103** (1978), 355.
17. N. A. B. Gray, J. G. Nourse, C. W. Crandell, D. H. Smith, and C. Djerassi, "Stereochemical Substructure Codes for 13-C Spectral Analysis," *Org. Magn. Reson.*, **15** (1981), 375.
18. A. L. Fella, J. G. Nourse, and D. H. Smith, "Conformational Specification of Chemical Structures in Computer Programs," *J. Chem. Inf. Comput. Sci.*, **23** (1983), 43.
19. C. Jochum and J. Gasteiger, "On the Misinterpretation of our Algorithm for the Perception of Constitutional Symmetry," *J. Chem. Inf. Comput. Sci.*, **19** (1979), 43.
20. M. Bersohn, "A. Sum Algorithm for Numbering the Atoms of a Molecule," *Comput. Chem.*, **2**, (1978), 112.
21. R. E. Carhart, D. H. Smith, N. A. B. Gray, J. G. Nourse, and C. Djerassi, "GENOA: A Computer Program for Structure Elucidation Based on Overlapping and Alternative Substructures," *J. Org. Chem.* , **46** (1981), 1708.
22. L. M. Masinter, N. S. Sridharan, R. E. Carhart, and D. H. Smith, "Applications of Artificial Intelligence for Chemical Inference. XIII. Labeling of Objects Having Symmetry," *J. Am. Chem. Soc.*, **96** (1974), 7714.
23. H. Brown, L. Hjelmeland, and L. Masinter, "Constructive Graph Labeling Using Double Cosets," *Discrete Math.*, **7** (1974), 1.
24. D. H. Smith, "Applications of Artificial Intelligence for Chemical Inference. Constructive Graph Labeling Applied to Chemical Problems. Chlorinated Hydrocarbons," *Anal. Chem.*, **47** (1975), 1176.
25. R. E. Carhart, D. H. Smith, H. Brown, and C. Djerassi, "Applications of Artificial Intelligence for Chemical Inference. XVII. An Approach to Computer Assisted Elucidation of Molecular Structure," *J. Am. Chem. Soc.*, **97** (1975), 5755.
26. W. T. Wipke and T. M. Dyott, "Use of Ring Assemblies in Ring Perception Algorithms," *J. Chem. Inf. Comput. Sci.*, **15** (1975), 105.
27. M. Plotkin, "Mathematical Basis of Ring-Finding Algorithms at CIDS," *J. Chem. Doc.*, **11** (1971), 60.
28. M. Bersohn, "An Algorithm for Finding the Synthetically Important Rings of a Molecule," *J. Chem. Soc. Perkin Trans. I*, 1973, 1239.
29. A. Esack, "A. Procedure for Rapid Recognition of the Rings of a Molecule," *J. Chem. Soc. Perkin Trans. I*, 1975, 1120.
30. E. J. Corey and W. T. Wipke, "Computer-Assisted Design of Complex Organic Syntheses," *Science*, **166** (1969), 178.

31. E. J. Corey and G. A. Petersson, "An Algorithm for Machine Perception of Synthetically Significant Rings in Complex Organic Structures," *J. Am. Chem. Soc.*, **94** (1972), 460.
32. A. Zamora, "An Algorithm for Finding the Smallest Set of Smallest Rings," *J. Chem. Inf. Comput. Sci.*, **16** (1976), 40.
33. B. Schmidt and J. Fleischauer, "A Fortran-IV Program for Finding the Smallest Set of Smallest Rings in a Graph," *J. Chem. Inf. Comput. Sci.*, **18** (1978), 204.
34. J. Gasteiger and C. Jochum, "An Algorithm for the Perception of Synthetically Important Rings," *J. Chem. Inf. Comput. Sci.*, **19** (1979), 43.
35. B. L. Roos-Kozel and W. L. Jorgensen, "Computer-Assisted Mechanistic Evaluation of Organic Reactions. 2. Perception of Rings, Aromaticity and Tautomers," *J. Chem. Inf. Comput. Sci.*, **21** (1981), 101.
36. C. M. Hoffmann, *Group-Theoretic Algorithms and Graph Isomorphism*, Springer-Verlag, Berlin, 1982.

X

STRUCTURE GENERATION

X.A. INTRODUCTION

The primary role of the computer in a structure elucidation system is the generation of an exhaustive list of candidate structures. Although computer programs can assist the chemist in the interpretation of data and the final evaluation of candidates, the chemist retains control (and responsibility for the results) in these stages. The chemist must select, or at least approve, the constraints that apply during structure generation and determine appropriate criteria for evaluating the generated candidates. The task of combining substructures, subject to given constraints, is a purely combinatorial one, needing neither judgment nor intuition. Consequently, it is a task well suited to a computer.

The ability to delegate the structure generation task to a computer is not merely a convenience for the structural chemist; it is often a necessity. Attempts to use manual methods to develop possible candidates will often fail to find all structures. Inevitably, there are biases toward structures similar to those previously encountered; radically different forms that are also consistent with available data may be overlooked. Examples discussed previously in the literature[1,2] have shown that many alternative structures can sometimes remain when a structure has been "identified" on the basis of limited spectral data and analogies to known compounds. Often, the alternatives incorporate highly strained ring systems or comprise unusual arrangements of several distinct fused or bridged ring systems joined by acyclic chains. Such alternative hypothesized structures may be easily refuted on the basis of small amounts of additional chemical, biochemical, or spectral data. However, the process of identifying an unknown must take cognizance of all possible structural forms and involve sufficient experiments to provide the definitive structural data.

Computerized structure generation will normally be performed in some stepwise manner dealing in turn with constitution, configurational stereochemistry and conformational form. First, all constitutional structures, compatible with already specified constraints, must be developed. Typically, a review of these structures will reveal some possessing features incompatible with already known data. Constraints eliminating these structures could have been specified prior to generation. However, once having been generated, any such structures can always be eliminated simply by the subsequent application of additional constraints. The first generation step will normally be followed by a "pruning" step wherein these inappropriate constitutional isomers are eliminated. Most current structure elucidation systems stop at this point.

Any more complete analysis of the candidate structures is likely to require definition of details of the configurational stereochemistry of these structures. A knowledge of stereochemistry is necessary for any precise interpretation of magnetic resonance data or other spectral properties; any search for analogous known compounds will normally need to take stereochemistry into account. Consequently, there will have to be a second generation step in which the configurational stereoisomers are developed for each surviving constitutional structure. Some constraints can usually be applied either during this generation process or in some subsequent pruning step. For example, if an unknown compound is optically active, all achiral candidate stereostructures can be eliminated. Sometimes, such constraints will eliminate all the stereoisomers of one of the candidate constitutional structures and, consequently, that constitutional structure.

Finally, it may be appropriate to explore the conformational forms of some few remaining stereoisomers. Such an analysis would not normally be required in a simple structure elucidation problem. However, information on conformational forms could be important in more general, long-term studies of structure–activity relationships among a set of related compounds.

The analyses of conformations, configurations, and constitutions all involve some exhaustive, combinatorial generation process together with a mechanism for defining and applying constraints. Major requirements in the design of any generation algorithm include:

1. *A Guarantee of Exhaustiveness*. The algorithms, which will typically be recursive in nature, must guarantee the generation of all distinct solutions.
2. *Reasonable Efficiency*. Many of the algorithms involve a generation scheme that can be viewed as some form of depth-first search through a tree structure; different branches of the tree corresponding to different choices for the generation process. Some of these branches are "dead ends" whose exploration cannot possibly yield any solutions to the

generation problem. If the algorithm cannot completely avoid the exploration of such dead ends, it should perceive as soon as possible the futility of exploring further any such branches of its search tree. On detecting that a branch is not worth exploration, the algorithm should backtrack and explore alternative branches.

3. *Irredundancy.* Ideally, the algorithm should never result in the creation of duplicates; that is, the generation process should be irredundant. Sometimes, this can not be achieved, but every effort should be made to minimize duplication.
4. *Duplicate Elimination.* If it is not possible to completely avoid the generation of duplicates, procedures for duplicate elimination must be included. If necessary, these must involve comparison of a canonical form of each generated structure with the canonical forms of all previously generated distinct structures.
5. *Constraints.* The generation process should admit the use of constraints expressed in a form convenient to the structural chemist.
6. *Prospective Use of Constraints.* Rather than generate structures and then, retrospectively, check for consistency with constraints, the generation procedure should use the constraint data to limit its search for possible solutions.

Effective generation algorithms are closely related to those interrelated algorithms discussed in Chapter IX for canonicalization, finding the permutation symmetry groups of atoms and of bonds, and labeling structures.

X.B. GENERATION OF CONSTITUTIONAL ISOMERS

X.B.1. Some Approaches to the Isomer Generation Problem

The typical isomer generation problem involves finding all ways of bonding together "fragments" that will yield connected structures. The fragments can be standard chemical atoms, or small assemblies of atoms with some "free valences" indicating potential bonding sites, or larger "superatoms" with elaborate internal structure and many distinct bonding sites. (A different type of generation problem involves characterization of all structures that can be derived, from a given structure, by some sequence of bond formation and deletion steps. Such problems arise when one considers reactions that could be used to degrade an unknown as a part of a structure elucidation task, or when planning chemical syntheses. Methods for these structure interconversion problems are considered briefly in Chapter XII.)

The type of fragments manipulated by a generation program and the structure representation produced can both vary widely. One may have a connection matrix of distinctly labeled individual chemical atoms, sets of possibly equivalent atomic fragments, or collections of multiatom substructures. The generator may be invoked just once to create all candidate structures. Alternatively, previous steps of the overall analysis may have broken down the structural problem into several different possible cases. Each of these cases would then entail a separate call to the generator with some different set of fragments; the complete set of structural candidates would be obtained by merging or appending the lists of structures produced in each separate call to the generator.

X.B.1.a. Structure Assembly from a Set of "Chemical Atoms"

At the simplest level, one might have a program for processing $^{13}C-^{13}C$ coupling data. Such a program would use, as its fragments, individual carbon atoms labeled with specific shifts and multiplicities. Structural constraints used by such a program would, primarily, define bonds between specific labeled atoms. The output might take the form of alternative connection matrices for these atoms.

The **CHEMICS** structure generator works with somewhat more complex problems; but still involves the combination of essentially atomic fragments. The interpretation systems of **CHEMICS** identify *sets of standard fragments* that satisfy the requirements of molecular composition and are consistent with observed data. Each such set of fragments constitutes a separate problem for the **CHEMICS** structure generator. The total set of structural candidates can be obtained by appending together the lists of structures produced from each set of fragments.

The standard fragments of **CHEMICS** include both individual atoms and

Fragment	Standard fragment index number	Atoms	Type	Additional restrictions on bonding
1	188	1	>C<	Bonds only to saturated carbons
2	12	2, 3	Pair of geminal-methyls	Must bond to a quaternary alkyl carbon, further restriction that this carbon bond only to other saturated carbons
3	38	4, 5, & 6	CH3–CO–	Should bond to an olefinic carbon
4	107	7	–CH2–	Should bond to one alkyl and one olefinic carbon
5	107	8	–CH2–	Should bond to one alkyl and one olefinic carbon
6	118	9	–CH=	Must bond to another olefinic carbon
7	143	10	>C=	Must bond to another olefinic carbon

FIGURE X-1. Fragments for structure assembly as in **CHEMICS** system.

	a	b	c	d	e	...
a		1	2	4	7	...
b			3	5	...	
c				6		
d						
.						
.						
.						

FIGURE X-2. Ordering of elements of an adjacency matrix that corresponds to entries in the Kudo–Sasaki connectivity stack (and may be used in adjacency matrix comparison operations for generation of canonical structures).

small assemblies of atoms in which only one atom bears any free valences for use in bonding.[3–5] The free valences on these fragments have, in effect, tags that define restrictions on allowed types of neighbor. A typical set of fragments as derived for an illustrative analysis of a $C_9H_{14}O$ compound is shown in Figure X-1.

The **CHEMICS** program generates "connectivity stacks" whose elements identify the interconnections of these fragments. A connectivity stack defines the elements of the top half of an adjacency matrix of fragments as listed in a standard order. This standard ordering is shown in Figure X-2.

The process of building up alternative connectivity stacks is equivalent to exploring, in a systematic manner, how the elements of the adjacency matrix might be filled in. Taking the fragments in the order shown in Figure X-1, a connectivity stack, adjacency matrix and partial structure are illustrated in Figure X-3. (In this example, bonding is indicated by a 0/1 value even though,

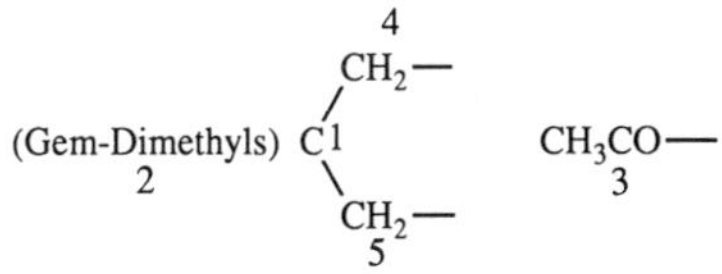

```
                       4
                      CH2—
                     /
(Gem-Dimethyls) C1              CH3CO—
        2            \                3
                      CH2—
                       5
```

	1	2	3	4	5	...
1		1	0	1	1	
2	1		0	0	0	
3	0	0		0	0	
4	1	0	0		0	
5	1	0	0	0		

Stack: 1 0 0 1 0 0 1 0 0 0 ...

FIGURE X-3. A partially assembled structure with adjacency matrix and connectivity stack.

in cases such as the "bond" to the *gem*-dimethyl fragment, more than one chemical bond may really be defined). In this example, the bonding and composition restrictions are such that this is the only acceptable way of building up an initial stack with the first five fragments. No olefinic carbon has yet been incorporated; thus fragment 3, though formally included in the partially assembled structure, does not as yet have any bonds defined.

There are two valid developments from this partial structure. In one, both the –CH2–s are bonded to the olefinic >C= carbon (fragment 7); in the other development, the –CH2– (identified as fragment 4) can be bonded to the –CH= olefinic carbon (fragment 6) while the other end of the olefin is bonded to the other –CH2–(i.e., a bond between fragments 5 and 7).

It might seem that other structures could be created by the alternative bonding of the –CH2–s to the –CH= and >C=. In fact, this would only create duplicate structures; one of these structures represents the canonical form with any others corresponding to arbitrary renumberings of the atoms in the canonical form. As is discussed later, the **CHEMICS** structure generator can test the "canonicity" of intermediate, partially filled connectivity stacks. If a stack is not canonical, development of that partially assembled structure can be abandoned and, thus, duplicates prospectively avoided.

With small molecules, the number of separate fragments in any set used for structure assembly will be limited, and most will be of different types (since the bonding restrictions are incorporated in the standard atom types). In such cases, the structure building procedures of **CHEMICS** can be reasonably effective. However, **CHEMICS**'s need to work with these very small standard fragments is a considerable disadvantage when more complex chemical structures must be analyzed. Extensions of **CHEMICS** allow user-defined substructures to be taken into account in the structure building process;[6] but this analysis has, ultimately, to convert these user substructures back into standard fragments and bonding restrictions on these fragments.

X.B.1.b. Structure Assembly as in GENOA

The **GENOA** program[7] is somewhat unusual in that the majority of the work of creating structures will have been performed in the various constructive graph matching steps (Section VIII.D) prior to any request for structure generation. One major consequent advantage of **GENOA**'s incremental approach is that a review of possible chemical structural forms does not necessitate generation of all candidates, and the final structural forms can usually be inferred from cases based on a reasonable amount of data.

When working with **GENOA**, structure generation is typically deferred until sufficient evidence has been acquired to allow most, if not all, of a molecule's constituent atoms to have been incorporated into multiatom fragments. Usually, this will have been accomplished in quite a number of ways. These different ways correspond to **GENOA**'s final cases, each one of which is comprised of some set of arbitrarily complex fragments. These various final cases can be taken to constitute separate problems for the **GENOA** structure generator. The data on cases are represented as connection tables in which the atoms still possessing free valences have unfilled entries. The generation procedure finds all ways of completing the connection tables that result in satisfactory structures.

Typical example data are shown in Figure X-4. (The numbers, as shown associated with each atom in Figure X-4, correspond to internal canonical numbers assigned by **GENOA**.) These data correspond to the first of the final cases resulting for warburganal. Free valences remain on six of the atoms; final structure generation requires all distinct interconnections of these atoms to be developed, subject to previously specified constraints. [The only one of the constraints given for warburganal (Section II.A.1) that is still relevant is that prohibiting the formation of three-membered rings.]

Structure generation involves a systematic search for valid interconnections of the free valences. In this example the generator could start by considering a bond incorporating the free valence on atom 15. There are four possible atoms bearing free valences that might be used to complete this first bond: atoms 12, 9, 6, and 2. (The program can detect the equivalence of atoms 1 and 2 and circumvent the need to consider atom 1 as a possible neighbor for 15.) Each of these would be tried in turn.

Interconnection of 15 and 12 would result in the formation of a five-membered ring and leave atom 9 as the next site for bond building. The immediately following steps would involve interconnection of atoms 9 and 6 and then joining 6 to both 1 and 2 to yield **X-1** (the first of the structures listed among the 42 solutions proposed for warburganal).

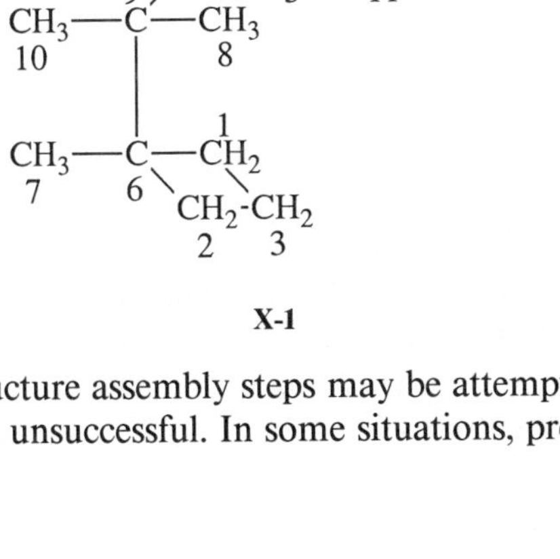

```
         18        14   17
         HO        CH=O
           \      /        12       13   16
           15 C ------------ C ---- CH=O
              |              ||
              |              ||
            4 CH - CH2 - CH
           9 /      5     11
  CH3 --- C --- CH3
  10      |      8
          |           1
  CH3 --- C --- CH2
   7     6 \       \
            CH2 - CH2
             2     3
```

X-1

Some partial structure assembly steps may be attempted only to be abandoned when proved unsuccessful. In some situations, prospective tests avoid

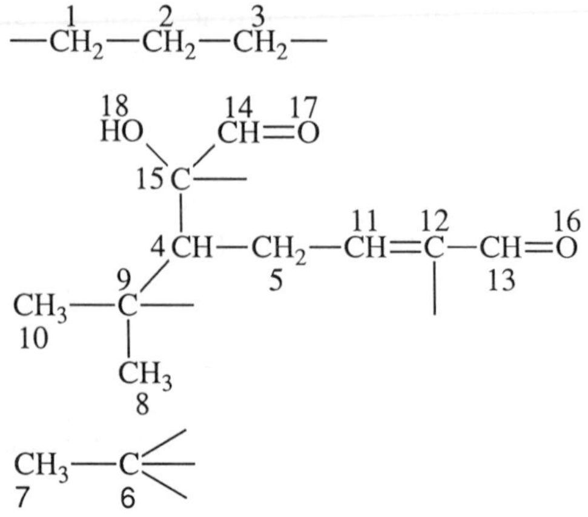

```
  1       2       3
—CH2—CH2—CH2—

        18        14  17
        HO        CH=O
           \     /
          15 C—
             |                 11   12          16
           4 CH—CH2—CH=C—CH=O
            9 /       5          |        13
CH3—C—                           |
10        |
         CH3
          8

CH3—C<
7       6
```

FIGURE X-4. Typical fragments comprising one of *GENOA*'s cases and forming the input to its structure generator.

excessive fruitless exploration. For instance, in the example just given, **GENOA**'s structure generator would determine, in advance, that it would not be worth considering the alternative of a 9–2 bond (rather than a 9–6 bond), because any 9–2 bonding step would inevitably leave unsatisfiable free valences on atom 6. The **GENOA** program would try the 15–9 interconnection and build up candidates, only to reject these during final tests to eliminate three-membered rings.

The final structure generator procedure in the **GENOA** system is relatively simple. Usually, it will be working with only a very small number of large, asymmetric substructural fragments requiring only a few additional interconnections. It can afford to use limited analyses of remaining symmetries at this final stage; all the more complex situations will normally have been analyzed, and resolved, during early constructive substructure searches.

Final canonicalization tests are always necessary in **GENOA**. It is not unusual for the same final structure to be created from two, or more, of **GENOA**'s cases. The examples given previously to illustrate the constructive substructure search procedure included the situation where formation of a ring resulted in the creation of the same structure from several previously distinct cases; of course, bond formation steps performed during final assembly can similarly result in the derivation of identical structures from different input cases. Since the cases are analyzed individually by the generator, this duplication must be dealt with by retrospective testing. These final canonicalization tests will also remove any duplicates resulting from **GENOA**'s failure to perceive induced equivalences of atoms while generating structures for any single case. The extent of generation of duplicates in **GENOA** depends on the data for the particular structural problem being analyzed. In some examples, such as the warburganal problem, there is no duplication; each structure that is created is found to be unique by the final canonicalizer.

X.B.1.c. Structure Generation in ASSEMBLE and Similar Programs

Whereas **GENOA**'s constructive substructure search procedure allows structural data to be utilized as available, without thought of possible overlaps, other structure elucidation systems require all data to be acquired and fully analyzed prior to any structure generation. Furthermore, since structure generation is the process of putting together discrete fragments, these alternatives to **GENOA** require spectral and chemical data to be interpreted in terms of discrete, nonoverlapping fragments. There will normally be more than one way of analyzing the data in terms of nonoverlapping substructures, and it is quite possible for different analyses to result in structural problems of widely varying complexity.

For example, given all the data on warburganal, one could identify a number of nonoverlapping substructures and some restrictions on the bonding among these substructures. One possible way of interpreting the data on warburganal is shown in Figure X-5. Here, an attempt has been made to create the largest possible substructures. The data characterizing the enal fragment has been used together with the results of the decoupling experiment to yield the "ENal-&-Decoupling" (END) substructure. The chain of three–CH2–s, ("CHaiN"), is the same substructure as used by **GENOA**.

The other data available for this structure included requirements for three methyls bonded to quaternary alkyl carbons, a $\gt$C–OH, and a $\gt$C–CH=O substructure. There were also prohibitions on CH3–C–OH substructures, *t*-butyl groups, and cyclopropyl rings. These other required substructural features can not be defined directly for structure generators such as **CONGEN**[1]

(Name, number)	Substructure	Possible bonding constraints:
(CHN,1)	$—CH_2—CH_2—CH_2—$	Prohibition on bonds to METs (methyls), to ALD (aldehyde), or OH (hydroxy)
(END,1)	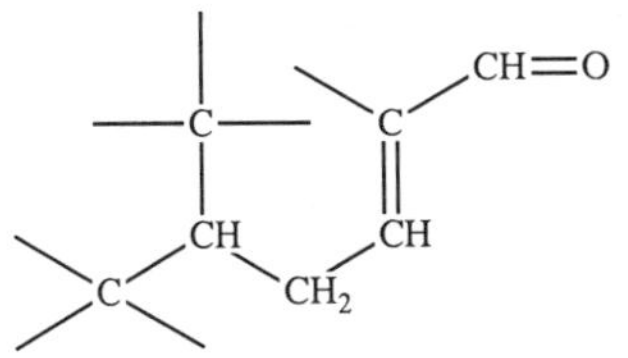 C, C, CH=O, CH, CH, CH_2, C	Quaternary alkyl carbons might have tags limiting number of substituent methylis to at most 2, and prohibiting membership of three-membered rings
(ALD,1)	—CH=O	Restriction to bond to a quaternary alkyl carbon
(OH,1)	—OH	Restriction to bond to a quaternary alkyl carbon
(MET,3)	$—CH_3$	Restriction to bond to a quaternary alkyl carbon
(C,1)	$\gt C \lt$	

FIGURE X-5. One possible set of nonoverlapping substructures that might be developed for data concerning warburganal.

or **ASSEMBLE**.[8–10] They all involve quaternary alkyl carbons, but it is not known whether these are the carbons already incorporated into substructure END, or the residual quaternary alkyl carbon necessary to make up the molecular composition of $C_{15}H_{22}O_3$. These other constituents of the molecule have to be defined as separate fragments, namely ALD, OH, 3 x MET, and C.

The structure generators in **CONGEN** and **ASSEMBLE** normally will have rather more fragments to manipulate than the structure generator used in **GENOA**. There are more likely to be sets of several equivalent fragments and sets of equivalent bonding sites within individual fragments. The data in Figure X-5 provide examples of both of these effects.

The generator in **GENOA** need only be concerned with prohibited substructures, because the individual cases used as input must necessarily satisfy all constraints specifying the presence of particular substructures. In contrast, the generators of **CONGEN** and **ASSEMBLE** must be concerned with requirements for particular bonding patterns and with prohibitions on substructures. These two programs approach these problems in rather different ways.

In **ASSEMBLE**, final structures are directly generated from fragments such as those shown in Figure X-5. Again the process is basically one of taking a partially completed connection table and filling in the remaining bonds in all ways that yield distinct structures that satisfy constraints. Constraints in **ASSEMBLE** can take varying forms. Some are defined in terms of prohibited substructures. Many other constraints for **ASSEMBLE** are expressed through "tags" on those atoms of known constituent fragments that still bear free valences. The **ASSEMBLE** program has a more extensive repertoire of atom tags than does **CHEMICS**; tags can be used to place restrictions on an atom's membership in ring systems or define requirements on the atom type, hybridization, or number of substituent hydrogens on an atom.

Some examples of the sort of data used by the **ASSEMBLE** program are summarized in Figure X-5. For example, the quaternary alkyl carbons in fragment END could bear tags prohibiting incorporation into a three-membered ring and restricting the maximum number of methyl neighbors to two (thus avoiding *t*-butyl substructures). Similar restrictions could be placed on the other quaternary alkyl carbon and corresponding restrictions devised for the other constituent fragments. Some structural requirements, such as the prohibition of CH3–C–OH substructures, are not readily expressed in terms of bonding tags on atoms. These have to be specified as required or prohibited substructures.

The **ASSEMBLE** program uses known equivalences of fragments, and bonding sites within fragments, to limit the generation of duplicate structures. At the beginning of the structure assembly process, **ASSEMBLE** would recognize the equivalences of the two quaternary atoms in END, the three methyls, and the two ends of the chain CHN. Just as in the structure generator of **GENOA**, the use of such equivalences can limit the search for valid structures. The **ASSEMBLE** program can nominate one of the two quater-

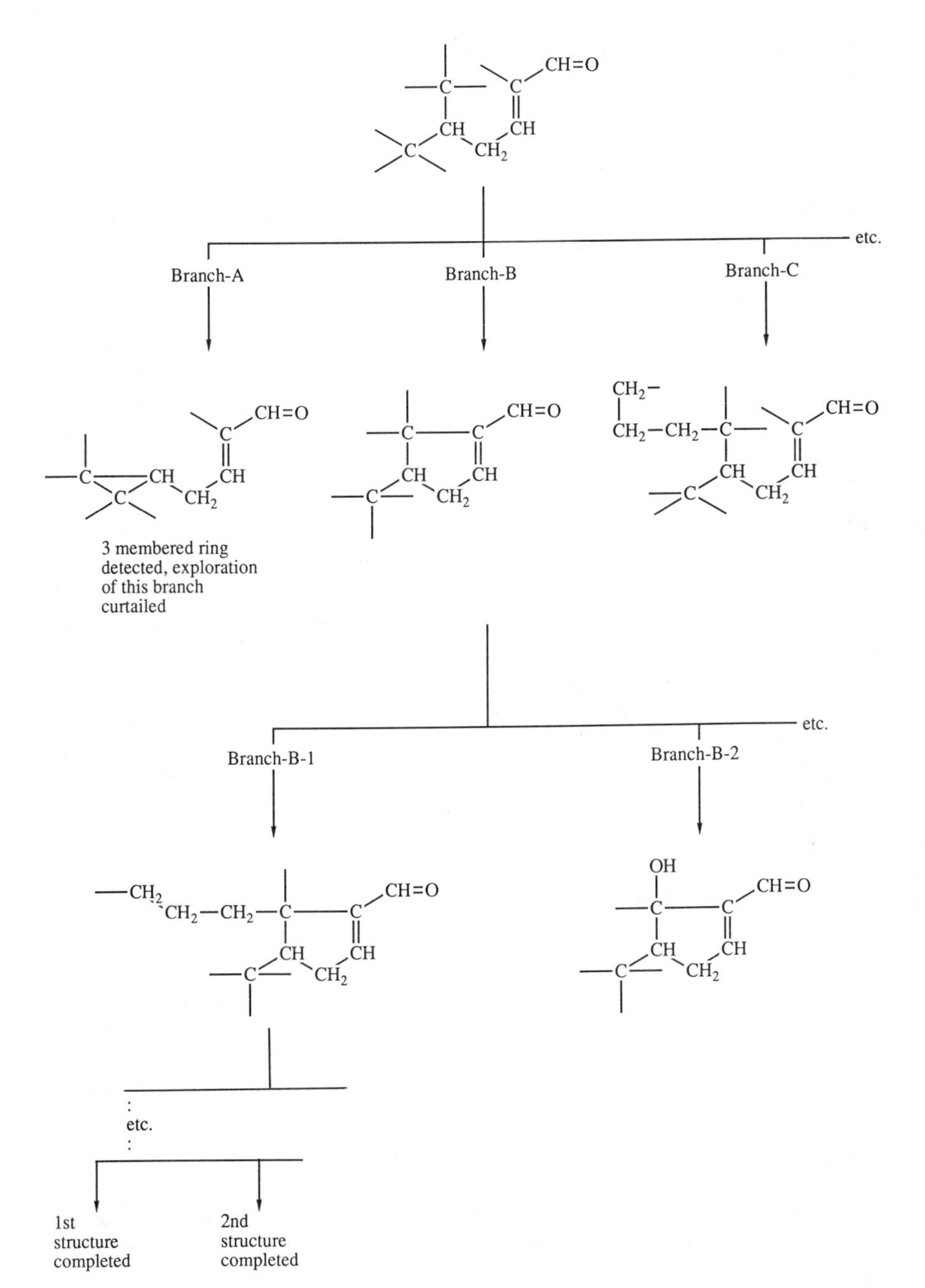

FIGURE X-6. Part of a structure generation search tree that might be developed by the ASSEMBLE program.

nary carbons in END as the starting point for a search for bonds, recognizing that a similar analysis starting from the other atom would yield identical structures.

The **ASSEMBLE** program makes more use of this analysis of equivalences than does **GENOA**'s generator. As structures are built up, **ASSEMBLE** reanalyses them in an attempt to determine whether any new elements of symmetry have been created which could be used to limit exploration down particular paths in its search tree.

The example shown in Figure X-6 illustrates how a generation algorithm, similar to that of **ASSEMBLE**, would process the data in Figure X-5. One of the given substructural fragments, typically the largest or most unique, is selected to form the root of the generation tree. Each way of satisfying the valence requirements of its atoms is then explored; each choice starts a different branch of the search tree.

Fragment END would represent the most logical choice of root for the generation tree for warburganal. The first bonding choice might be that shown as "branch A" in Figure X-6; here, the two quaternary carbons of END have been joined by a single bond. The **ASSEMBLE** program maintains data identifying the fragment containing each individual atom; the two atoms joined by this bond are already in the same fragment; thus, obviously, a ring is being formed. The **ASSEMBLE** program can determine the size of ring (by "walking" through the fragment) and test against any ring constraints on the atoms. In this example, the bond would be seen to result in a prohibited ring and thus exploration of this branch of the search tree would be curtailed. The possibility of a bond joining the two quaternary carbons of END would never subsequently be considered.

The generator in **ASSEMBLE** would then proceed to explore other branches of the tree that are defined by alternative bonding selections for the first quaternary carbon atom in END. Exploration along branch B would lead to all structures incorporating the olefinic bond into a five-membered ring. As the tree is explored depth first, all these structures will be created prior to the generation of alternatives reached by means of branch C and other branches.

Of course, as with any other structure generator that takes as input both defined substructures and residual atoms, there is always a possibility that the residual atoms may be combined to form a copy of of a substructure, or at least a part of one of the substructures. This can result in duplication among the structures produced. Further duplication can arise because the symmetry heuristics employed cannot avoid all duplication. Like **GENOA**, **ASSEMBLE** must use a retrospective canonicalization check that determines whether a generated structure is new or merely a duplicate of one already in the list of unique generated candidates.

X.B.1.d. CONGEN's Multistep GENERATE–EMBED Approach to Structure Generation

The inputs to the generation phase of **CONGEN** are somewhat similar to those of **ASSEMBLE**. Like **ASSEMBLE**, **CONGEN** works with some given set of fragments including defined substructures and residual atoms; constraints on required and prohibited substructures and on rings can be specified. In **CONGEN**, however, the role of the generation step is in some ways more limited than in the other programs already described. *Final structures are not produced in a single "generation" step*.

At the generation stage, **CONGEN** suppresses detail of the internal form of any constituent substructures; they are treated, in effect, as atoms with unusual names and valences. (In **CONGEN** terminology, the pseudoatoms that really correspond to complex substructures are *superatoms*.) Rather than final representations of chemical structures, **CONGEN** produces *intermediate structures* defined in terms of the superatoms, and any residual chemical atoms, making up the molecular composition.

For example, **CONGEN** would view the problem of structure generation shown in Figure X-5 as involving the identification of all structures comprising three monovalent MET atoms together with one each of a heptavalent END superatom, a bivalent CHN superatom, a monovalent ALD superatom, a monovalent OH atom, and a tetravalent C atom.

For **CONGEN**, structural constraints have to be expressed in terms of the atoms and superatoms. Constraints are mainly specified in terms of required and prohibited substructures. Initial constraints that might be appropriate for use in **CONGEN**'s generation step would include [–CHN–(MET,OH,ALD), None], [MET–C–OH,None], [$(MET)_3$C–, None]. The first of these constraints prevents any attempt to construct an acyclic chain terminated by a methyl, hydroxy, or aldehyde group; the other two prevent formation of *t*butyl or CH_3–C–OH substructures at this stage.

Different versions of **CONGEN**[1,11] have used very different algorithms in the actual generation process. The original version was based on an analysis of the various cyclic forms that can be obtained given a set of atoms of known valence.[1,12] A subsequent version of **CONGEN**[11] used a method similar in concept to that of Kudo and Sasaki's **CHEMICS** in which canonical connection matrices are developed. These algorithms differ in their efficiency for different types of structure generation problem; the canonical connection matrix approach is more effective for routine chemical structure problems. Although the generation algorithms differed, details of inputs and the type of intermediate structures output were identical.

Some of the 15 or so intermediate structures that would be produced by **CONGEN** for the given warburganal data are shown in Figure X-7. Interpretation of diagrams of **CONGEN**'s intermediate structures requires some skill. For example, in the intermediate structure CONGEN-i, the residual quaternary carbon atom has been bonded to the aldehyde group, the hydroxy group, and two of the bonding sites of END. Although the connection between END and C, and between END and CHN, are shown formally as double bonds, these bonds are most likely to be allocated eventually to different atoms in END and thus would not necessarily correspond to olefinic bonds in final structures. Similarly, the "three-membered ring" of END-CHN-C in CONGEN-iii is not really a three-membered ring; in fact, it will correspond to a ring of at least five atoms in any final structures (CHN must contribute three atoms to this ring, C adds one more and END must contribute at least one atom).

It is possible for some of the free valences of END not to be used for bonds to other fragments, instead being reserved to make further interconnections among the constituent atoms of that substructure. Such an "internal" bond in END is obviously needed if **CONGEN** is to generate all those final chemical structures with a bond from the olefinic carbon of END to one of its quaternary carbons. (Structure **X-1** is an example of one of these final structures.) Two of the example intermediate structures shown in Figure X-7 have "internal" bonds; in CONGEN-iii two of END's free valences are used to form a single internal bond, and in CONGEN-iv four of END's free valence are used to form two such bonds.

Development of the final structures, represented as connection tables of chemical atoms, requires the superatom parts to be "embedded". This "embedding" process is simply an application of the graph labeling procedures discussed in Chapter IX. In this example, there is only one complex superatom: END. The main embedding step will be that involving END.

The embedding process essentially involves analysis of the permutation symmetries of the free valences on a superatom such as "END" and the symmetries of the various sets of labels, i.e. the groups substituted on superatoms like END. There are usually several kinds of symmetry to be considered. In END, two of the atoms bearing free valences are symmetrically equivalent (as can be determined by examining the permutation symmetry group of its nodes). Since stereochemical properties are not considered at this stage, the three free valences on each one of END's quaternary carbons are again equivalent. The CHN superatom provides two identical labels for END's free valences, and the various MET labels are, of course, also identical. The original article[13] on graph labeling shows how the fairly complex resulting labeling problems can be solved by using extensions of the simple labeling procedures discussed in Chapter IX.

When END is embedded, it will again be necessary to specify the constraints prohibiting MET–C–OH and $(MET)_3$–C–substructures. Obviously, all intermediate structures with the form MET–END–OH have the potential to yield partially elaborated structures containing a MET–C–OH system. Constraints

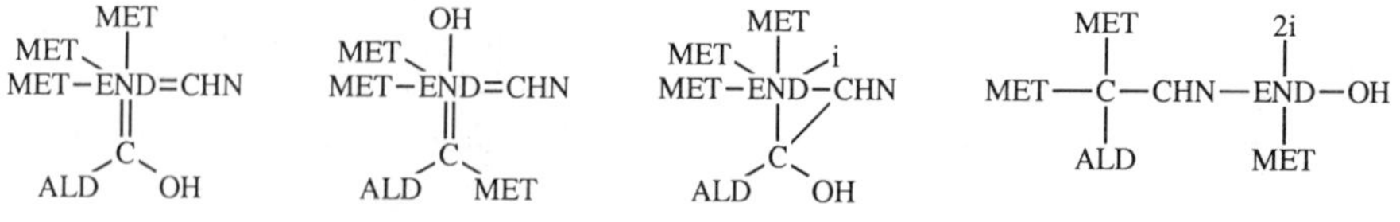

MET, MET, MET–END=CHN, C, ALD, OH — CONGEN i

OH, MET, MET–END=CHN, C, ALD, MET — CONGEN ii

MET, MET, i, MET–END–CHN, C, ALD, OH — CONGEN iii

MET, 2i, MET—C—CHN—END—OH, ALD, MET — CONGEN iv

FIGURE X-7. Some of the intermediate structures that would be produced for the warburganal example at the generation stage in **CONGEN**.

prohibiting cyclopropyl rings should also be specified during this embedding process, for it is at this stage that such substructures might be created. The requirements for final structures with >C–CH=O, >C–OH, and >C–CH3 groups also need to be stated (otherwise, structures with vinyl methyls, enols, etc. might possibly be formed). The simplest expression of these constraints for this embedding step might be a prohibition of structures of the form [–CH=C–(ALD,OH,MET)].

Figure X-8 shows examples of the intermediate structures that would be obtained by embedding END into the various intermediate structures shown in Figure X-7. In some cases, all embeddings of a superatom in some intermediate structure contain features that violate given constraints; that intermedi-

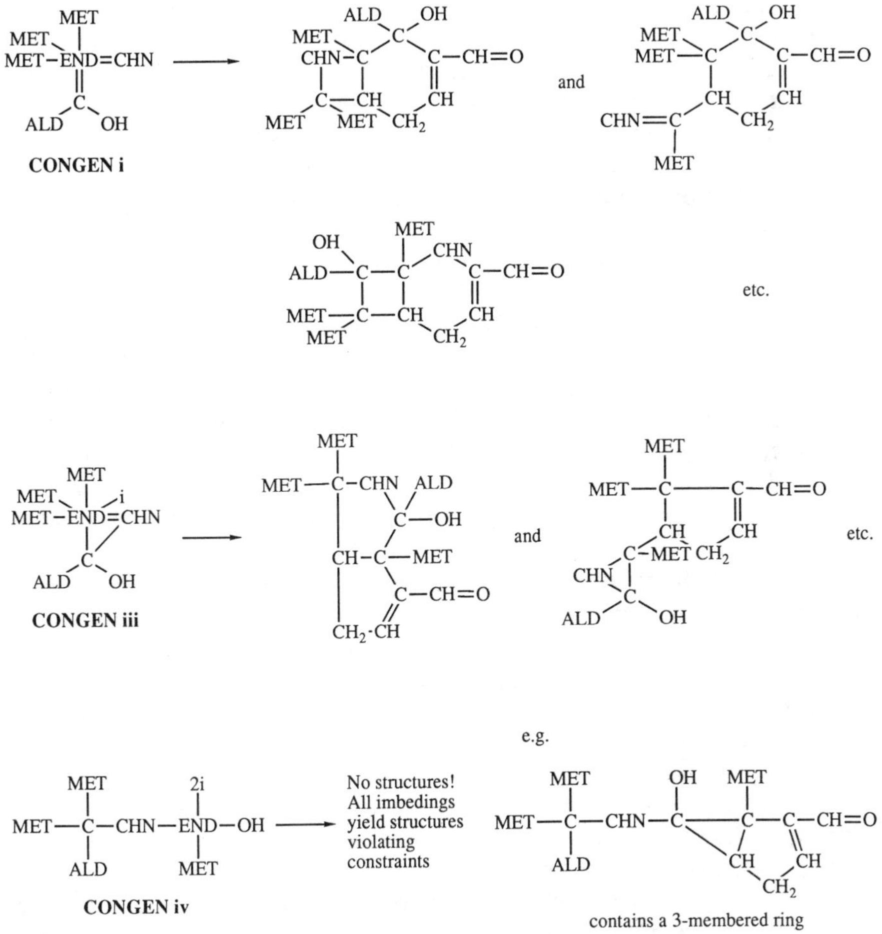

CONGEN i → ... and ... etc.

CONGEN iii → ... and ... etc.

CONGEN iv → No structures! All imbedings yield structures violating constraints

e.g. ... contains a 3-membered ring

FIGURE X-8. Some intermediate structures that would result from the embedding of superatom END.

ate structure is effectively removed from further consideration. An instance of this effect is provided by intermediate structure CONGEN-iv; attempts to form the two rings internal to superatom END inevitably lead to three-membered rings.

The different embeddings of a superatom can result in quite different structural forms. This effect can be seen for both CONGEN-i and CONGEN-iii. The intermediate forms produced from CONGEN-i will ultimately yield a 6,6-ring fused (decalin) system, a 4,8-ring fused system, and a structure with separate four- and six-membered rings. Of course, before final structures are derived it is necessary to embed the other superatoms such as CHN.

During embedding, **CONGEN** must consider the possibility of duplication. It is again the duplication that can arise when one assembles structures from fragments that comprise some mixture of defined superatom substructures and residual atoms. Sometimes it is possible to take residual atoms and small superatoms and use them to build copies of complete large superatoms, or perhaps only relevant parts of superatoms. This can result in duplication when the superatoms are embedded. This effect, which is not observed with warburganal, is illustrated by the simplified example in Figure X-9.

The data in Figure X-9 represent a situation where there is a large superatom with some symmetrically related bonding sites (R6ME), two substituent superatoms (MET, and OH) and some implicit hydrogens. Generation results in a single intermediate structure with the two substituents attached to the R6ME superatom.

In such a structural problem, the embedding of the R6ME superatom would result in a number of isomers with the remaining OH and MET superatoms substituted at different positions on the ring. When the MET superatom subsequently becomes embedded, some of the previously distinct intermediate structures will yield the same final canonical structure. As with the other structure generators, a list of all previously generated distinct structures (in their canonical forms) must be maintained and each newly derived structure must be checked against this list.

There are both advantages and disadvantages to **CONGEN**'s "GENERATE then EMBED" approach to developing complete structures. On the merit side, there are efficient algorithms for generating unique (intermediate) structures from a given set of atoms and superatoms, and for then subsequently embedding superatoms. The alternative generation algorithms, which must consider the internal form of and different bonding sites within component substructures, are generally less efficient.

Furthermore, in **CONGEN** it is possible to inspect intermediate structures, both those resulting from the initial generation step and those produced by subsequent embeddings of some subset of the superatoms. Each such intermediate structure represents an entire class of molecules. Sometimes, new constraints can be identified that can be used to prune the sets of intermediate structures; the pruning away of a single intermediate structure may effectively eliminate scores, or hundreds of final structures.

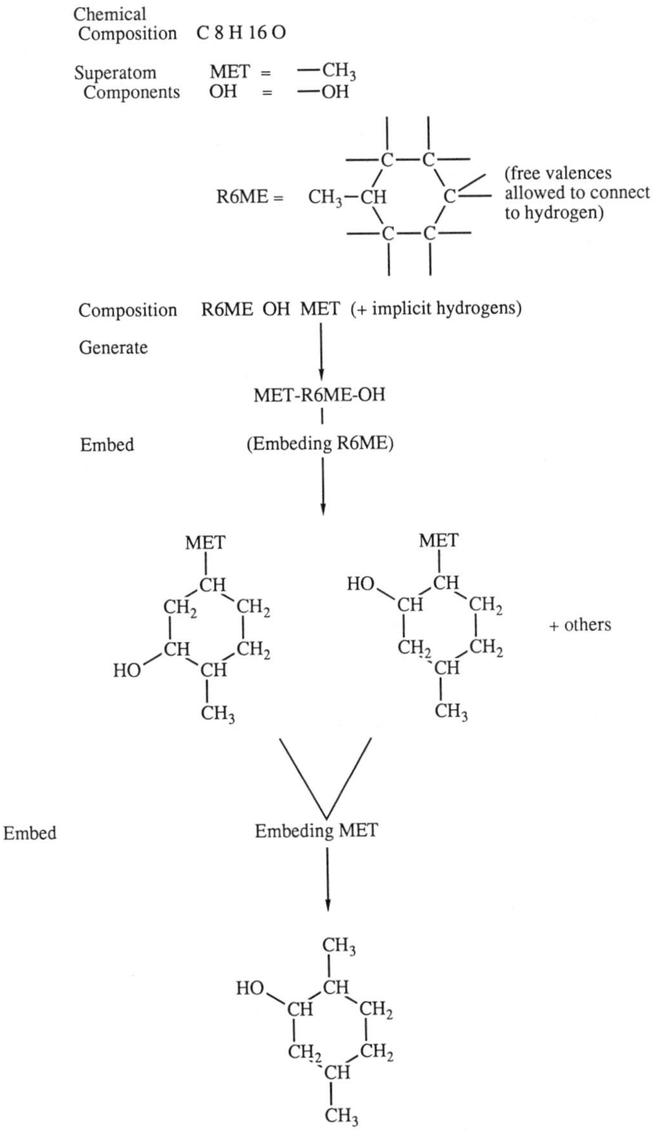

Chemical Composition C 8 H 16 O

Superatom Components MET = $-CH_3$
OH = $-OH$

R6ME = CH_3-CH ring with free valences on the ring carbons (free valences allowed to connect to hydrogen)

Composition R6ME OH MET (+ implicit hydrogens)

Generate

MET-R6ME-OH

Embed (Embeding R6ME)

+ others

Embed Embeding MET

FIGURE X-9. The embedding process may result in structure duplication.

The increased complexity of the overall approach to structure generation must however be set against these advantages. Specification of structural constraints becomes more involved; careful consideration is necessary to determine which constraints are appropriate at the generate stage and which are appropriate in each subsequent embedding step. Although the inspection of intermediate structures, with unembedded superatoms, can be useful and can suggest additional constraints, this process is not always simple. As shown in the examples given above, the same intermediate structure may correspond to a number of quite different final structural forms. The intermediate structures may be so different from normal structures that it can be difficult to relate them to normal chemical–spectral interpretation processes. The need to allow for bonds internal to a superatom may seem to be a rather esoteric feature. In some applications, however, it is appropriate to define constraints on permitted numbers of internal bonds. This, again, adds to the complexity of the constraint definition process. Furthermore, the allocation of internal bonds requires additional analysis procedures in the generator program.

X.B.2. Generation of Unique Structures

Although the nature of the fragments and the representation of the resulting structures vary, the nature of the processing in the generation step is essentially the same. The basic task of all these constitutional structure generators is simple—through some defined combinatorial procedure, all unique structures that can be derived by interconnecting a given set of fragments are to be produced.

It is not difficult to devise simple recursive algorithms that explore all possible ways of interconnecting a set of fragments of defined valence. The simplest of such algorithms are inefficient. Inefficiencies result because these algorithms may entail extensive searches for ways of completing a connected structure following selection of some set of bonds that must inevitably lead solely to disconnected or other invalid structures. More refined combinatorial algorithms can prospectively avoid such fruitless searches. It is possible to develop provably correct algorithms that are exhaustive and irredundant in the sense that they will produce every possible arrangement of bonds among fragments, each possible arrangement of bonds being produced just once.[11]

The major source of complexity in generation algorithms is the requirement for *uniqueness of the final structures*. It is normal for distinct allocations of bonds among fragments to result merely in alternative representations of identical structures.

Given an exhaustive generation algorithm, the requirement for structural uniqueness could always be satisfied by a subsequent canonicalization process. Each generated "structure" would have to be converted to a canonical form and then compared with all previously created distinct canonical structures. As already noted, such a step is in any case necessary for generators such as those in **GENOA** and **ASSEMBLE**, and a similar process is also necessary during the embedding steps of **CONGEN**. Except in the simplest of problems, however, reliance solely on retrospective canonicalization tests is liable to prove prohibitively expensive. A structure generation procedure should try to avoid, prospectively, such potential duplication; as far as possible, it should produce *unique structures* rather than *unique distributions of bonds between fragments*.

Obviously, there will be trade-offs between the analysis performed during the bond arrangement process and the work in subsequent canonicalization and duplicate checking processes. A more complete, and costly, analysis prior to assembling a structure might, by revealing symmetries, indicate that some possible arrangements of bonds would result in duplicate structures. In such cases, only one representative example of the set of duplicates should be created and passed on for canonicalization. The canonicalizer and duplicate removal process would then be saved the work that would otherwise have been expended on the symmetrically equivalent, duplicate structures.

The appropriate detail of the analysis in the generation procedure depends on the nature of the fragments that it must manipulate. In the simplest case, each fragment is essentially unique. Consequently, each distinct arrangement of bonds among the various fragments will yield a unique structure. A simple generation algorithm, incorporating only checks to prevent creation of disconnected structures, will suffice.

In slightly more complex cases, one may have small numbers of fragments, some few of which are equivalent, and only some very limited number of additional bonds to be created. Here, an analysis of the symmetries of the initial problem will usually suffice. Although some duplicates may be produced, they should be sufficiently limited in number that retrospective canonicalization tests are adequate.

More difficult are those structural problems involving several sets of identical fragments. One might, for example, have three or four distinct polyatomic substructures together with a dozen methyls, methylenes, methines, and ethereal oxygens that interconnect or are substituted on the larger known substructures. Here, there is considerable potential for duplication. A simple analysis of equivalences among the initial fragments will not prove sufficient. Such problems require either a reanalysis of symmetries at each bonding step or some scheme checking whether a given partially assembled structure could conceivably be extended to yield a complete structure in canonical form.

Of course, structural problems become much more difficult as the available data become less specific. In particular, if the complete degree sequence of the carbons is not known (no ^{13}C NMR data, or at least no resonance multiplicities), really many distinct structural problems must be solved. These distinct problems result from differing distributions of hydrogens among other atoms of the structure. A variety of approaches have been devised to deal with such problems.

The **CHEMICS** program considers, separately, each possible set of fragments that are consistent with composition and spectral constraints; these different sets of fragments should cover for all alternative allocations of hydrogens. In **ASSEMBLE**, free valences on atoms can be reserved for bonds to residual hydrogens. Hydrogens are not represented explicitly in the **ASSEMBLE** system; special case analyses of symmetries must be invoked when free valences of atoms are being reserved for hydrogens.[8]

Different versions of **CONGEN** have used distinct approaches to the hydrogen allocation problem. In the original version of **CONGEN**[1,12] hydrogens were left implicit, simply saturating the free valences of atoms in final structures. The distributions of hydrogens were thus determined by the allocations of free valences. These free valence allocations were performed at several distinct stages in the overall structure generation process. Initially, free valences were partitioned among various groups of atoms used subsequently to build different cyclic and acyclic parts of the molecule. Subsequently, the free valences accorded to each group of atoms were partitioned out among the individual atoms. In the later version of **CONGEN**,[11] the allocation of monovalents (not necessarily just hydrogens) to higher valent fragments is performed as a preliminary step in the structure generation procedure. The allocation algorithms are based on mathematically proven routines that find all ways of partitioning n objects (hydrogens, etc.) among N slots (the higher valent atoms) subject to minimum and maximum constraints for each individual allocation.[11] Each such distribution of monovalents defines a distinct problem for the main structure generation routines in **CONGEN**.

The most complex structural problems are those involving the assembly of cyclic structures comprising large numbers of identical fragments. In such problems, methods for exploring the interconnections of the fragments, testing for symmetries and canonical forms at intermediate stages of assembly, and so on, are of limited efficacy. For these problems, a rather different approach to structure generation is appropriate. This alternative approach is based on an analysis of those fundamental cyclic forms, or *vertex graphs*, that can incorporate a given set of nodes (atoms) of known valence. The structure generation process consists of a series of steps wherein these *vertex graphs* are elaborated into representations of complete structures. At each stage of elaboration, symmetries can be fully exploited and duplication of structures is avoided. Most chemical structure elucidation problems are amenable to the simpler approaches; the analysis based on vertex graphs does have esoteric applications such as determination of the scope and limits for isomerization.[14]

X.B.3. Structure Generation Algorithms

X.B.3.a. A Simple Generator

Some general understanding of standard structure generation procedures can be obtained by considering a simple algorithm that finds all ways of completing a partially specified adjacency matrix. This adjacency matrix identifies bonds between "unique" atoms. Some of the matrix entries can be presumed to have been initialized, either by an automated spectral interpretation system or some "manual" scheme for substructure specification. These initial entries will specify both known bonds and indicate where bonds are prohibited.

An example of such an adjacency matrix is shown in Table X-1. These data are supposed to represent the structural inferences that might have been made given the spectral information (shown in Figure X-10) for a $C_{12}H_{22}O_2$ com-

TABLE X-1. An Initial, Partially Defined Adjacency Matrix that Could Form Input to a Simple Structure Generator

"Unique Atom Labels"

	CH_3 /18	CH_3 /20	CH_3 /21	CH_3 /26	CH_2 /26	CH_2 /36	CH_2 /37	CH_2 /63	CH/ 30	CH/ 125	C/ 131	C/ 171	O	O	Remaining Free Valences
	1	2	3	4	5	6	7	8	9	10	11	12	13	14	
1	.	0	0	0	0	0	0	0	?	0	?	0	0	0	1
2	0	.	0	0	0	0	0	0	?	0	?	0	0	0	1
3	0	0	.	0	0	0	0	0	0	0	0	1	0	0	0
4	0	0	0	.	0	0	0	0	?	0	?	0	0	0	1
5	0	0	0	0	.	?	?	?	?	?	?	0	0	0	2
6	0	0	0	0	?	.	?	?	?	?	?	0	0	0	2
7	0	0	0	0	?	?	.	?	?	?	?	0	0	0	2
8	0	0	0	0	?	?	?	.	0	0	0	0	1	0	1
9	?	?	0	?	?	?	?	0	.	?	?	0	0	0	3
10	0	0	0	0	?	?	?	0	?	.	2	0	0	0	1
11	?	?	0	?	?	?	?	0	?	2	.	0	0	0	2
12	0	0	1	0	0	0	0	0	0	0	0	.	1	2	0
13	0	0	0	0	0	0	0	1	0	0	0	1	.	0	0
14	0	0	0	0	0	0	0	0	0	0	0	2	0	.	0

pound. Zero entries correspond to cases where bonds have been eliminated. Interconnections between methyls are prohibited (by the requirement for one connected molecule), connections from methyls to methylenes violate constraints from ^{1}H NMR data, and bonds between the –O– and methylenes C(5)–C(7) would be incompatible with their ^{13}C shifts. The bond order of known bonds is shown explicitly. The other matrix entries indicate where bonds might exist. The number of free valences remaining on each atom can be determined from the atom's intrinsic valence and count of bonds already made.

At this stage, the different ^{13}C shifts are taken to uniquely label the individual carbon atoms. Carbon C(3) is shown as bonded to the carbonyl carbon (for it is the only methyl carbon with an appropriate ^{13}C shift). ^{1}H NMR data suggests that one methyl must bond to the >CH– methine, but in the absence of data interrelating ^{1}H and ^{13}C resonances, it is not possible to establish which of the distinct methyls is so bonded.

All alternative arrangements for the remaining bonds, necessary to complete a structure and satisfy all atoms' free valences, must be found. These bond arrangements can be derived by considering the possible interconnections

Atom #	Type	^{13}C	1-H NMR
1	–CH3	17.6 q	)
2	–CH3	19.5 q	) Three of the methyls are
3	–CH3	20.8 q	) singlets, one is doublet and,
4	–CH3	25.7 q	) from its shift, is part of a >CH–CH3 group
5	–CH2–	25.5 t	)
6	–CH2–	35.7 t	) Proton resonances all
7	–CH2–	37.2 t	) multiplets
8	–CH2–	62.9 t	) Proton resonance is triplet
9	>CH–	29.6 d	) Multiplet
10	–CH=	124.8 d	
11	>C=	131.1 s	
12	>C=	170.9 s	Acetoxy group
13	–O–		
14	=O		

FIGURE X-10. Spectral data and inferred substructures for a $C_{12}H_{22}O_2$ compound used to illustrate aspects of a simple structure generation procedure.

among atoms still possessing free valences. A search for bonds can start by taking the first atom with free valences as an "initiating" atom and trying, in turn, bonds to all those higher-numbered atoms to which it is permitted to bond. Once a particular bond has been made, the search can continue either by analyzing further free valences on the same initiating atom or, if valence requirements of that atom are now saturated, by proceeding to the next atom with free valences.

Calls for each successive bond-formation check are made *recursively*, with the recursion terminating when n (the index number of the initiating atom) equals the number of atoms in the structure. When the recursion terminates, a candidate structure can be specified by listing all predefined bonds together with the particular current selection of generated bonds. A basic structure for a recursive bond forming routine is as follows:

```
Formbonds(n)
   if (n = number_of_atoms) then Print_structure()
   else
   if (number_free_valences[n]=0) then Formbonds(n+1)
   else
   for j=(n+1) to number_of_atoms do {for_loop
      if (number_free_valences[j] > 0) then {if_fv
       if Ok_to_bond(n,j) then {if_okbond
```

```
            decrement free valence of n
            decrement free valence of j
            increment count of bonds made
            record ends of this bond as n and j
            Formbonds(n)
            decrement count of bonds made
            increment (i.e. restore) free valence of j
            increment free valence of n
            }if_okbond
    }if_fv
  }for_loop
```

In the example algorithm the first check is for the terminating condition of the recursion. When recursion is terminating, a call is made to a "Print_structure( )" routine; in other cases a check is made on bonding for the current atom n. If this current atom has no remaining free valences, it cannot participate in further bonds; the "Formbonds( )" routine can be called, recursively, to begin analyzing the next atom in sequence.

Possible choices of bonds for a given initiating atom with free valences can be determined by iterating through all higher numbered atoms. Each of these is checked in turn to determine whether it still possesses free valences and, if so, whether it can bond to the current initiating atom. The routine "Ok_to_bond( )" might involve no more than a check for a "?-possible bond" entry in the given adjacency matrix, although, as noted later, it is appropriate to include more elaborate checks in this routine.

If a bond can be made, then before the next recursive call to the "Formbonds( )" procedure, the number of free valences for both initiating and terminating atoms are decremented (to allow for formation of this bond). Details of the initiating and terminating atoms of this bond are stacked in some array structure for subsequent reference [e.g., in "Print_structure( )"]. (Alternatively, the entry in the adjacency matrix could be changed to indicate a generated bond.) The recursive call on "Formbonds( )" is made with the same initiating atom number because this atom may possess further free valences. On return from recursion, the bond formation steps are reversed by restoring previous values of free valences. The very first call on the recursive routine is to form bonds for atom 1; with the count of bonds formed thus far set to zero.

For the example in Table X-1, the initial call to the "Formbonds( )" routine for atom C(1) would result in tests for bonds from C(1) to C(2)...O(14). Although C(2), C(4), C(5), and so on, all still possess free valences, bonds to these atoms are not permitted. The first possible terminating atom for a bond from C(1) is C(9); this bond would be recorded and a recursive call to "Formbonds( )" made. On this subsequent call, the routine would determine that the valence of C(1) was now satisfied and recursion would again occur, and the routine would begin the examination of the bonding options of C(2).

Again, C(9) could be identified as the first possible terminating atom for a bond starting at C(2). Formation of such a bond would violate the restriction on the number of methyls bonded to methine carbons. Some such constraint violations are fairly readily detected and can be tested by the "Ok_to_bond( )" routine. In a simple case such as this, the constraint could have been expressed in terms of bonding restriction tags *on the methine*. The "Ok_to_bond( )" routine could then detect that, given an already generated C(1)–C(9) bond (recorded in the generated bonds' stack), the C(2)–C(9) connection was no longer permitted.

Some limited analyses of this form are appropriate at each bond formation step. However, these checks will for the most part be restricted to simple atom properties such as can be expressed in terms of atom tags. Bonding tags specifying limits on neighbors of particular types are readily checked, as would be other similar tags defining restrictions on hybridization and participation in multiple bonds.

More general structural constraints cannot be so readily tested. Most structural constraints would require graph matching of a substructure onto the partially assembled structure, the number of matchings found being checked against any specified limits. Such a matching process would normally be too complex to be invoked at each bond building step. Checks on the number of occurrences of complex substructures have to be applied to finished structures at the completion of the recursive bond building process prior to printing of results.

For the purposes of example, it is useful to presume that some form of atom tag restriction on the methine, C(9), inhibits further bonds to methyl atoms following the formation of the C(1)–C(9) bond. Then, the next two bond formation steps would connect C(2)–C(11) and C(4)–C(11), thereby saturating C(11) as well as C(2) and C(4). Having made such an initial selection of possible bonds for the methyls, the program would continue by seeking possible bonds for the methylene C(5). A C(5)–C(6) bond would be the first generated.

A recursively called "Formbonds()" routine would again find C(5) with free valences and so try a further bond to C(6). This would result in a C(5)–C(6) double bond that would, presumably, violate constraints on the hybridization predefined for these atoms. A second C(5)–C(6) bond should be rejected by the "Ok_to_bond( )" routine. Instead, a C(5)–C(7) bond would be generated, thus saturating the valence of C(5) and causing further recursion leading to consideration of further bonding for C(6).

A C(6)–C(7) bond would apparently be acceptable, for it would violate neither bonding nor hybridization constraints. However, such a C(6)–C(7) bond would saturate the valences of C(6) and C(7) and create a cyclopropane *molecule*. Further recursion to consider the bonding for C(8) and all subsequent bond building steps would be futile because only disjoint structures would be assembled.

The simple recursive algorithm outlined above *must* be elaborated so that it can avoid generation of such disconnected structures. (Bond formation steps that would yield disconnected structures are common and would result in much wastage of computer time.) The simplest approach for avoiding such structures uses some extra data records (really just integer vectors) to hold a "component number," and component total free valence, for each atom. These data are adjusted as bonds are generated and removed.

Initially, each atom can be assigned to a separate "component" (so its component number will be the same as its index number, and its component valence will equal its free valence). When a bond is created, either in the initial interpretation of spectral derived constraints or by the subsequently applied recursive "Formbonds( )" routine, the atoms joined can be allocated a common component number. (For simplicity, this can be the smaller of their original component numbers.) The new total free valence of the component formed can be determined from the free valences of the components joined. If separate components are joined by a bond, the free valence of a combined component is 2 less than the sum of their original free valences. However, if the atoms being connected are already assigned to the same component (i.e., a ring or multiple bond is being formed), its new valence is simply reduced by two.

For the example in Figure X-10, the initial data on "components" using results of spectral interpretation and data following some bond formation steps in "Formbonds()" are summarized in Table X-2. The initial constraints define a couple of multiatom fragments; thus the two olefinic carbons, C(10) and C(11), are part of the same component that has a total of three remaining free valences. The bond building steps (1–9), (2–11), (4–11), (5–6), (5–7) further reduce the number of separate components. Atoms C(6) and C(7) are recorded as being in the same component with a total free valence of 2. If a bond is formed between them, the resulting component would possess zero free valences—that is, it must represent a separate molecule.

Given data on component number and component total free valence, an "Ok_to_bond( )" routine can readily detect, and reject, cases where a particular possible bond would yield a component with zero free valence. Such a bond formation step is allowed only if it is the last step in molecular assembly (which is easily determined from counts of the total number of bonds needed and made, or counts of number of distinct components remaining). The simple algorithm, as outlined earlier, thus requires some extra testing code in the "Ok_to__bond( )" routine and additional code, in the bond formation step, that will allocate a new component vector and fill this with an appropriately updated component number for each atom. The component number and free valence vectors must be passed as arguments in the recursive calls on "Formbonds( )."

If disconnected structures are avoided in this manner, the next few bond formation steps create the bonds C(6)–C(8), C(7)–C(9), and C(9)–C(10) thus leading to the first complete candidate structure, **X-2**. Recursion would then unwind. The C(9)–C(10) bond would be deleted; since there are no alternative possible terminating atoms for a bond from C(9), the unwinding of the recursion continues back to the level where bonds for C(7) are being processed. A C(7)–C(10) bond could be rejected for it would again lead to a disconnected

TABLE X-2. Data Defining "Component Number" and "Component Valence" at Various Stages in the Structure Assembly Process[a]

Initial			After 1–9, 2–11, 4–11 5–6 & 5–7 Bonds Formed			After 6–8 Bond also Formed		
Atom Number	Component Number	Component Valence	Atom Number	Component Number	Component Valence	Atom Number	Component Number	Component Valence
1	1	1	1	1	2	1	1	2
2	2	1	2	2	1	2	2	1
3	3	1	3	3	1	3	3	1
4	4	1	4	2	1	4	2	1
5	5	2	5	5	2	5	3	1
6	6	2	6	5	2	6	3	1
7	7	2	7	5	2	7	3	1
8	3	1	8	3	1	8	3	1
9	9	3	9	1	2	9	1	2
10	10	3	10	2	1	10	2	1
11	10	3	11	2	1	11	2	1
12	3	1	12	3	1	12	3	1
13	3	1	13	3	1	13	3	1
14	3	1	14	3	1	14	3	1

[a]As discussed in the text, such data permit the avoidance of disconnected structures.

final structure and recursion would again unwind to the previous level and resume examination of bonds for C(6).

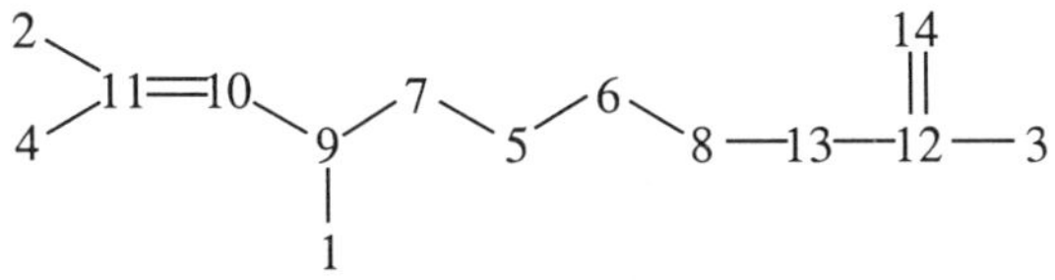

X-2

The alternative choice of bond, C(6)–C(9), would be followed by C(7)–C(8) and C(9)–C(10) leading to structure **X-3**. Structures **X-2** and **X-3** really differ only in the alternative assignments of the two methylene shifts of 35.7 C(6) and 37.2 ppm C(7) to two different structural environments:

Alternative bonding arrangements following from C(6)–C(9) all fail to yield structures; the first, C(7)–C(9), would give a disconnected molecule, and C(7)–C(10) would leave no bonding sites for C(8). Unwinding recursion would delete the C(6)–C(9) bond; C(6)–C(10) would then be tried, but all continuations

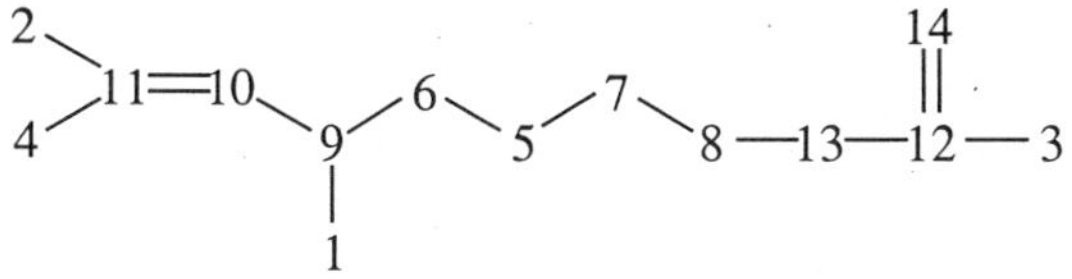

X-3

from this choice of bond fail because these yield only disconnected structures or cases where there would be no bonding possibilities for C(8).

Further unwinding of the recursion would eliminate the C(5)–C(7) bond; a series of structures involving C(5)–C(8) bonds would then be produced. The first two are shown as **X-4** and **X-5**. Structure **X-4** again possesses the same

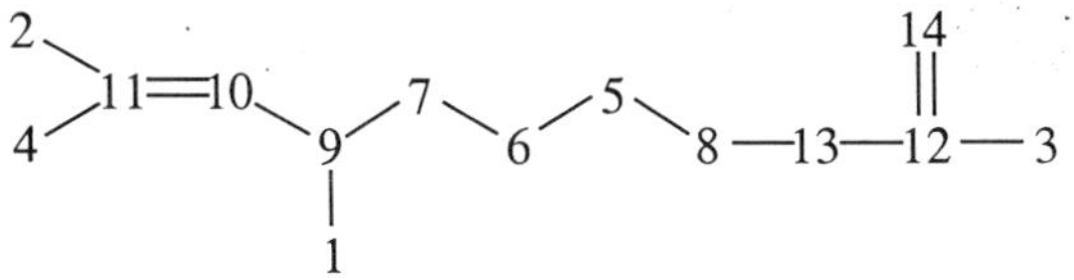

X-4

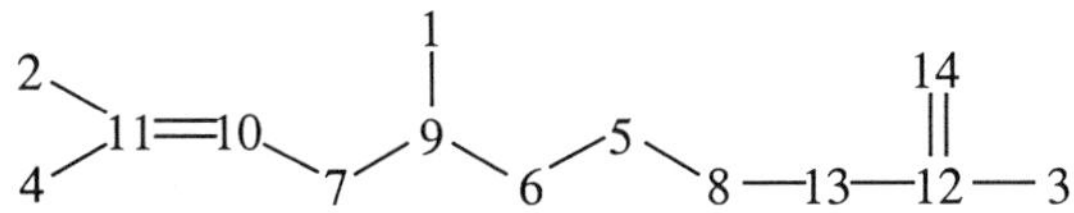

X-5

topology as **X-2** but differs this time in respect to the assignment of the methylene resonances at 25.5 and 35.7 ppm. Structure **X-5**, produced by the bond formation steps C(1)–C(9), C(2)–C(11), C(4)–C(11), C(5)–C(6), C(5)–C(8), C(6)–C(9), C(7)–C(9), and C(7)–C(10), represents an alternative topological form.

After generating structure **X-5**, recursion would be unwound, eliminating the C(7)–C(10) bond; there would then be no remaining alternative terminating atoms for a bond from C(7), so recursion would again unwind. The C(7)–C(9) bond would be deleted, and the iterative cycle would continue to examine alternative placements for a bond using the first of the two free valences available on of a C(7). The possibility C(7)–C(10) bond would be discovered. At this point another shortcoming of the simple algorithm would be revealed; after forming the C(7)–C(10) bond, a further recursive call would find the search for a second bond for C(7) starting once again with $j = 8$. A bond to C(9) would be selected, thus exactly duplicating the previous structure.

This problem can be overcome by using stronger constraints on the search for possible terminating atoms of a bond. If one of the several free valences on some initiating atom n has already been used to form a bond to another atom i ($i >= n+1$), in the following recursive call to "Formbonds( )" the iterative search for possible terminating atoms for bonds for n's remaining free valences should start at i rather than at $n + 1$. It is simplest if the lower limit for the iterative search forms part of the arguments in each call to "Formbonds( )."

An outline for a slightly more elaborate version of "Formbonds( )" follows at the end of this paragraph. The procedure call has been modified to pass in a starting point for the iterative search for terminating atoms, along with data defining "component number" and "total component free valence" for each atom. The data on components are presumed to be checked by the "Ok_to_bond( )" routine. This version of "Formbonds( )" also indicates where a check routine, "Satisfies_constraints( )," should be called to eliminate those generated structures that fail to satisfy any available elaborate structural constraints. The algorithm indicates that new vectors, for example, "new_component_no", must be created at appropriate levels in the recursion. Depending on the implementation language, the processes of creating (and later freeing) vectors may be implicit or may have to be performed explicitly by the programmer (in which case statements to free the new component vectors would need to be included at the end of the "try_n" block of code).

```
Formbonds(n,start_nbrs,component_no,component_valence)
   if (n = number_of_atoms) then {
            if Satisfies_constraints() then Print_structure()
            }
   else
   if (number_free_valences[n]=0) then Formbonds(n+1,n+2,component_
   no,component_valence)
   else {try_n
```

```
(   dynamically allocate vectors to hold new component number )
( and new component valence data                              )
for j=start_nbrs to number_of_atoms do {for_loop
   if (number_free_valences[j] > 0) then {if_fv
    if Ok_to_bond(n,j,component_no,component_valence) then {if_
    okbond
         new_num = minimum(component_no[n],component_no[j])
         if component_no[n] = component_no[j] then
                new_valence = component_valence[n] - 2
         else
                new_valence = component_valence[n] +
                                    component_valence[j] - 2;
         for k=1 to number_of_atoms do {fill_vectors
                if (component_no[k] = component_no[n]) or
                   (component_no[k] = component_no[j]) then
                   {changed
                        new_component_no[k] = new_num
                        new_cv[k] = new_valence
                   }changed
                else {unchanged
                        new_component_no[k] = component_no[k]
                        new_cv[k] = component_valence[k]
                     }unchanged
                }fill_vectors
         decrement free valence of n
         decrement free valence of j
         increment count of bonds made
         record ends of this bond as n and j
         Formbonds(n,j,new_component_no,new_cv)
         decrement count of bonds made
         increment (i.e. restore) free valence of j
         increment free valence of n
         }if_okbond
      }if_fv
   }for_loop
  }try_n
```

The sequences of recursive calls on "Formbonds()" would continue to explore all possible structures incorporating a C(5)–C(6) bond. Structures produced would include **X-6–X-9**. Many other lines would be explored but be abandoned when only disconnected structures resulted.

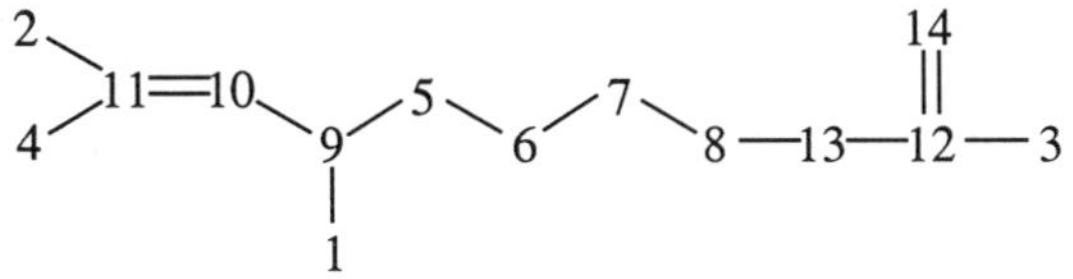

```
2                                                   14
  \                                                 ||
   11=10                                            ||
  /     \    5       7                              ||
4        9/    \ 6/    \ 8 —13—12— 3
         |
         1
```

X-6

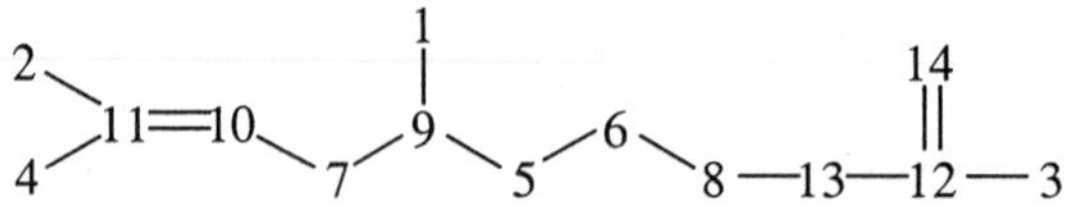

1
2 14
11═10 9 6
4 7 5 8—13—12—3

X-7

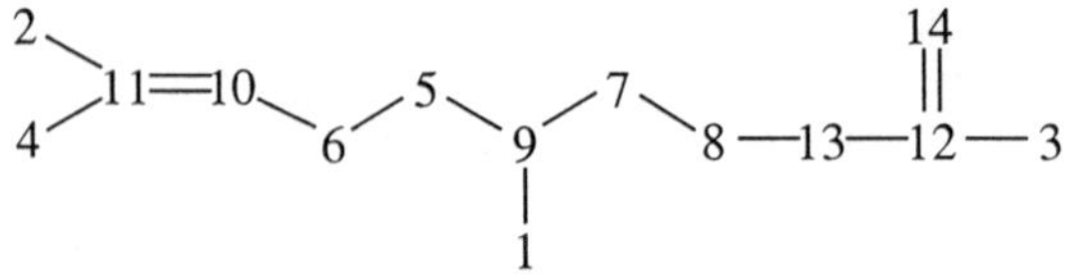

2 14
11═10 5 7
4 6 9 8—13—12—3
1

X-8

2 14
11═10 6 7
4 5 9 8—13—12—3
1

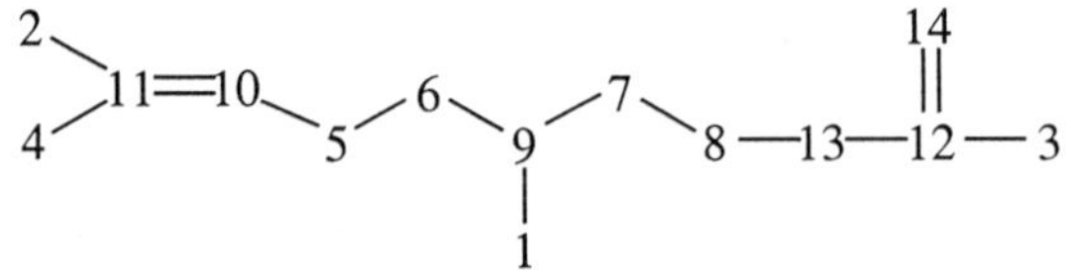

X-9

Once all the structures incorporating a C(5)–C(6) bond had been created, generation would continue with all C(5)–C(7) structures. Another sequence of candidates, including **X-10** and **X-11**, would thus be produced. The entire analysis would then have to be repeated for the two alternative arrangements of the methyls, >CH–and >C= atoms. The same unacceptable combinations of bonds for the–CH_2–s and so on would be rediscovered and again rejected. The

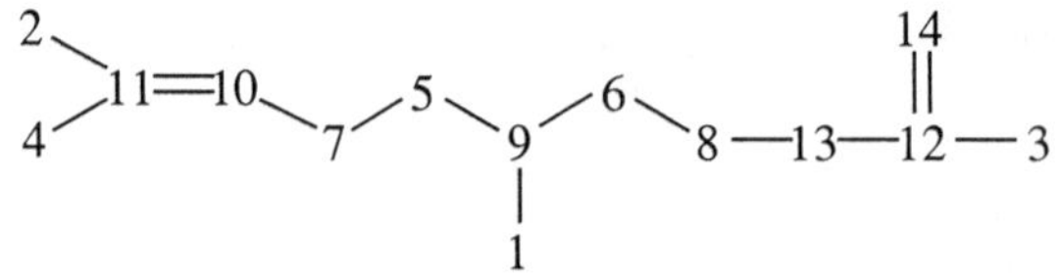

2 14
11═10 5 6
4 7 9 8—13—12—3
1

X-10

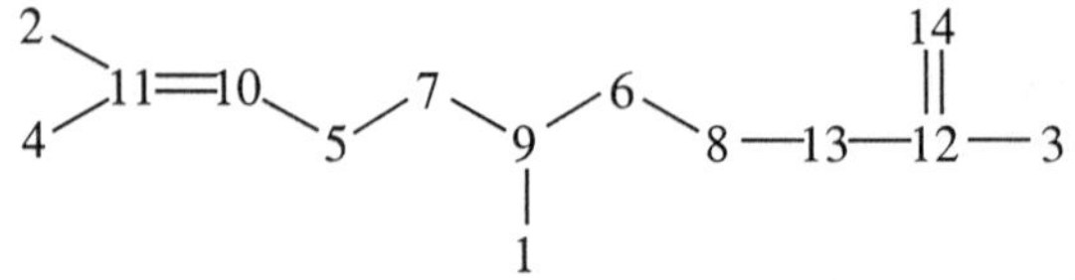

2 14
11═10 7 6
4 5 9 8—13—12—3
1

X-11

1 14
11═10 7 6
4 9 5 8—13—12—3
2

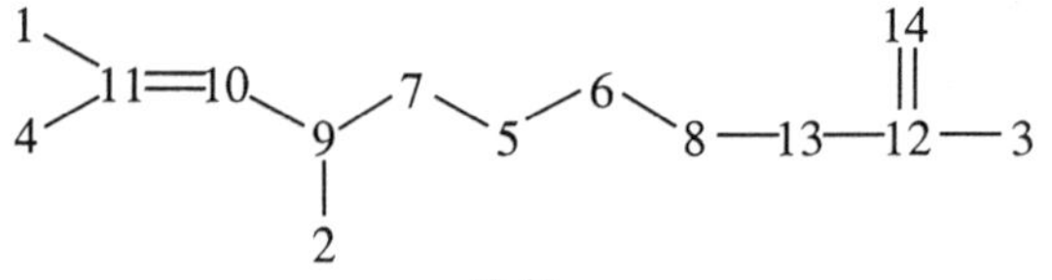

X-12

same valid bonding patterns of the other atoms would be determined and structures like **X-12** would be formed.

Thus, exhaustive structure generation is possible using such an algorithm (although "exhaustive" may come to describe more than just the thoroughness of the process). However, there still remain problems with the "structural" candidates suggested as solutions to the problem.

These are all different solutions, but they represent *solutions to the problem of simultaneously generating structures and assigning* ^{13}C *shifts*. If one were only interested in the form of the final structures then, after generation, it would be necessary to remove the "chemical shift labels" from the various carbons. Of course, the duplication of topological structures would then be revealed. The distinct candidate structures would have to be found by a process of converting each generated candidate to its canonical form and comparing this with entries in a list of previously generated distinct canonical structures.

More appropriately, the chemical shift labels should be removed from atoms once these data have served their purpose of providing structural constraints and prior to the generation process. The generation procedure should be capable of determining that some constituent atoms are identical and that different arrangements of these atoms will not yield distinct topological structures. The generation procedures must be further capable of using data on atom equivalences.

The algorithm outlined does not satisfy these additional requirements. For this algorithm, atoms are intrinsically different—their arbitrary index numbers act as "uniqueness labels." Additional code must be incorporated so that in a search for a terminating atom for a bond, for example, only one of any set of equivalent atoms is considered. Obviously, further code is also required identify any equivalences among atoms.

X.B.3.b. Extending the Simple Algorithm by Analysis of Atom Equivalences

It is simple to identify any equivalences among the constituent fragments of a molecule prior to the start of the generation procedure. If the fragments are essentially atomic, as in **CHEMICS** or the generation step of **CONGEN**, equivalences are obvious from their atom types. All $-CH_2-$s can be gathered into one set, and $(CH_3)_2C$<s may be grouped into another. (In **CHEMICS**, any bonding restriction tags that might differentiate among the atoms can be considered as forming a part of the specification of an extended atom type and used in the process that identifies classes of equivalent atoms.)

In more complex cases, the fragments may include large multiatom substructures with many distinct bonding sites. As discussed in Chapter IX, multicomponent "structures" can be converted to canonical forms, and, as a part of this canonicalization process, both equivalences of different component fragments and equivalences of bonding sites within fragments can be determined. The permutation symmetry group of atoms in a (multicomponent) structure can be determined as part of the canonicalization process, and the sets of equivalent atoms can be determined directly from this symmetry group. Although it is possible to proceed with data defining only the orbits of atoms (i.e., the sets of equivalent atoms), there are advantages in using the complete permutation symmetry groups to identify equivalences among atoms in substructures.

For the example in Table X-1, if the ^{13}C shift tags are suppressed, there are two sets of three equivalent atoms [viz. (1 2 4) and (5 6 7)] along with a number of singleton atoms. Such data can be used in a slightly extended version of the simple "Formbonds( )" routine.

The use of data on equivalences among atoms can be illustrated in relation to the analysis of bonding for C(5). One source of duplication among the generated structures was "Formbonds( )" creating a sequence of structures with C(6)–C(5)–C(8), C(6)–C(5)–C(9), and C(6)–C(5)–C(10) and then, subsequently, producing similar structures with C(7)–C(5)–C(8), and so on. Examples representative of the resulting duplication include structures **X-8** and **X-10**.

Of course, even when C(5) has first been selected to be an initiating atom for a bond, C(6) and C(7) remain equivalent. If C(5) is to form only one bond to atoms in this set of equivalent atoms, only the first, or *orbit representative*, need be considered. Data identifying the initial equivalence of C(5), C(6), and C(7) could be represented in terms of the orbit of nodes (5 6 7). Selection of C(5) as an initiating atom would leave the orbit (6 7). Thus C(6) could be recognized as the orbit representative, and generation of structures with C(7)–C(5)–C(8), and so on, could then be avoided.

Inclusion of some multicomponent structure canonicalization routine would be the major extension to a structure generator necessary to use such analyses of symmetry. This routine, called before "Formbonds( )," would build some data structure recording equivalences of atoms. For the example of Figure X-10, the recorded data would define only the sets [viz. (1 2 4) (5 6 7) (8) (3) etc.]. The extensions to "Formbonds( )" itself are fairly simple. Changes are necessary at the stage where atom n is selected to initiate a bond, at the stage where atom j is tested to determine whether it represents an appropriate terminator for the bond and at the stage where j has been chosen and a further recursive call to "Formbonds( )" is made.

Thus when atom n is selected as an initiating bond, during inward recursion, the data structure recording equivalences should be updated. At the very least, this updating will involve removal the initiating atom n from any set of equivalent atoms. For j to be an appropriate terminating atom, it must be the "orbit representative" of any set of equivalent atoms. If j is selected, it also should be removed from any record of the set of equivalent atoms. All such changes need to be reversed as recursion is unwound again.

Such simple analyses will suffice for sets of equivalent atomic fragments. More elaborate approaches are needed when the initial structural fragments include substructures with some symmetry and several bonding sites with remaining free valences. For example, suppose that fragment **X-13** were among those predefined for a structure generator. (There are no restrictions on hybridization or existence of multiple-bonds in **X-13**.) Analysis of the topological symmetry of **X-13** would determine that, initially, all the atoms were in the same orbit.

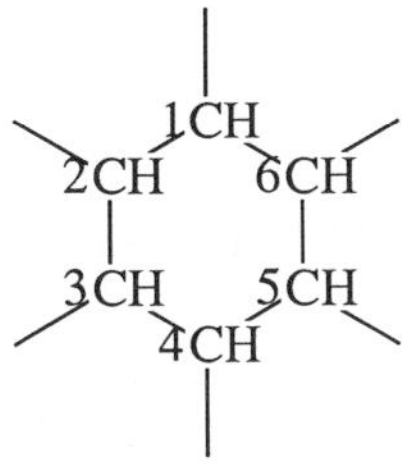

1CH
2CH 6CH
3CH 5CH
4CH

X-13

If atom 1 of **X-13** were to be selected as an initiating atom for a bond, bonds to other atoms in the ring would have to be considered. It would obviously be wrong to analyze this bond formation problem solely in terms of the original orbits of atoms as computed prior to structure generation. If the simple method of analysis were used, the selection of atom 1 would apparently leave atoms (2 3 4 5 6) as "equivalent" and only atom 2, "the orbit representative," would be selected. The structure generator would then not be exhaustive! Only the C(1)–C(2) connection in **X–13** would be tried, and the possibilities for a C(1)–C(3) or C(1)–C(4) bond would be ignored.

It is in situations such as this that the complete atom permutation symmetry group of the original structural fragment proves useful. A part of the atom permutation symmetry group for **X-13** is summarized in Table X-3. The "orbits" of atoms are defined by the vertical columns; initially, all nodes are in the same orbit.

TABLE X-3. A Part of the Permutation Symmetry Group of Atoms of Structural Fragment X-13[a]

1	2	3	4	5	6	
1	6	5	4	3	2	
2	1	6	5	4	3	Disabled
2	3	4	5	6	1	Disabled
3	2	1	6	5	4	Disabled
3	4	.	.			..
.						..
.						
6	.	.	.			..

[a]The permutations marked as "disabled" correspond to those symmetry elements lost when C(1) is selected as a bond initiating atom.

The effect of any changes in symmetry of the substructure, such as would follow from selection of C(1) as a bond initiating atom, are easily derived from the data in the tables. Once C(1) has been made "unique" by such a selection step, permutations that interchange it with other atoms should be "disabled." In this example, only the first two permutations would remain. From these two permutations, it is possible to identify the new orbits of nodes: (1) (4) (2 6) (3 5). Given these data, it is apparent that a structure generator should consider not just the C(1)–C(2) bond but also C(1)–C(3) and C(1)–C(4) bonds. The remaining symmetries are, however, such that it is still appropriate to ignore C(1)–C(6) or C(1)–C(5) bonds. (It is possible to proceed with only data on atoms' orbits instead of the permutation symmetry group, but more complex recomputations of symmetries must then follow each selection of an initiating atom.)

A version of "Formbonds( )" that uses permutation symmetry groups computed for the initial set of fragments, and appropriately updates these data as bonds are made and deleted, can serve as the basis for a reasonable structure generator. (The analysis for the final generation step of **GENOA** is performed at about this level.) However, testing whether terminating atoms are orbit representatives only avoids some of the sources of duplication.

An extended "Formbonds( )" routine, with the orbit representative checks as suggested, would still produce **X-2** and **X-12** as alternative structural candidates given the data of Figure X-10 and Table X-2. The duplication between **X-2** and **X-12** results here because the alternative possible bondings of C(1) and C(2) are considered in isolation. Other similar but more complex cases also arise. For example, while bonding patterns such as C(7)–C(5)–C(6)–C(8) and C(8)–C(5)–C(6)–C(9) are acceptable with the data in Figure X-10, the pattern C(8)–C(5)–C(6)–C(7) is not. (It leads to structure **X-4**, a duplicate of **X-2**.) In this case, the duplication "error" can be taken as resulting from the allocation of the "least" set of neighbors to C(6), rather than to C(5). Allocation of the least set of neighbors to the first member of any set of equivalent atoms could be viewed as the only valid allocation, the only one leading to a standard, or canonical molecular form.

The extensions suggested for "Formbonds()" do not define any restrictions on the bonding of equivalent atoms that are considered, separately, as initiating atoms. It is obviously possible to define simple *ad hoc* restrictions that could prevent the generation of **X-12**, and possibly even avoid the generation of **X-4**. (For example, *subsequent initiating atoms, from an equivalent set, must bond to atoms with the larger index numbers. . . .*) The problem with such patching of the basic algorithm is that it increases the scope for errors; an incorrectly defined restriction might result in a structure generator that avoids duplicates and also, in obscure circumstances, omits a few of the genuinely distinct solutions. Furthermore, even if implemented correctly, such patching of the simple algorithm will still leave open the possibility of a considerable degree of duplication.

In more complex cases, new equivalences among atoms may result from the bond formation steps themselves; these new equivalences can result in duplicates at later bond formation steps. Figure X-11 illustrates one way in which this may occur. The initial fragments, being used to assemble the structure, supposedly include the–CH_2–CO–CH(–CH_2–)–moiety and various additional–CH_2–s and other fragments. Initially, each–CH_2–shown would be in different equivalence classes. The first bond formation step could be bonding the free–CH_2–, atom 6, to the methine (atom 3) of the ketone fragment. This would saturate the valence of atom 3, so the next bond building step would involve atom 4. One possibility for atom 4's bonding would be formation of a four-membered ring. As shown in Figure X-11, there are formally two different ways, bond formation steps 2a and 2b, in which this might be done. Of course, these two ways yield identical structures.

One approach to dealing with such problems is to recompute the equivalence classes of fragments and atoms after each bond building step. This is the approach of **ASSEMBLE**, the structure generator for the **CASE** system.[8,15] The **ASSEMBLE** program basically works in terms of orbits of atoms. Initial symmetries are computed prior to structure assembly and then symmetries are recomputed as bonds are formed. For the data in Figure X-11, the induced

Initial fragments for structure assembly:

```
                    5
         O1        CH2–
         ||        /                 6
 —CH2—C—CH          ,     —CH2—  , etc., etc.
   4       2    3\
```

Bond formation step 1:

```
                    5
         O1        CH2–
         ||        /
 —CH2—C—CH              , etc., etc.
   4       2    3\
                     CH2–
                      6
```

Bond formation step 2a

```
                  5
  O              CH2–
   \\          3/
     C—CH
     |        |
    CH2-CH2
     4      6
```

Bond formation step2b

```
          5
 4CH2-CH2
     |       |
 1C—CH
  //       3\
 O           CH2–
              6
```

FIGURE X-11. Bond formation steps can induce equivalences among previously distinct atoms. Initially, all the–CH_2–s are in distinct equivalence classes. The first bond formation step results in atoms 5 and 6 becoming equivalent. If this new equivalence is not perceived then and exploited, duplicate structures will result as shown by bond formation steps 2a and 2b.

equivalence of C(5) and C(6) would be recognized when the orbits of atoms are recomputed following the formation of the C(3)–C(6) bond. The **ASSEMBLE** program also solves problems related to symmetries in a given substructure, as illustrated earlier for structure **X-13**, by a recomputation of symmetries after selection of a bond initiating atom.

In **ASSEMBLE**, symmetrically equivalent atoms are identified by an elaborate atom partitioning process similar to those used in a structure canonicalizer and described in Chapter IX. Atom partitioning schemes cannot guarantee to identify symmetry orbits.[16] It is possible for any scheme to terminate prematurely with what are really nonequivalent atoms still grouped together in the same class. Then, such nonequivalent atoms will appear equivalent to the program. Because the generation procedure considers building bonds to only a single representative atom from each "equivalence" class, this could result in a failure to generate all possible structures.

The heuristics used for atom partitioning in **ASSEMBLE** should prove adequate for any application in routine chemical structure generation. The atom properties are extensive and include even data on all ring systems known to incorporate each atom. These detailed atom properties, together with the iterative scheme for breaking up classes,[15] are likely to identify correctly equivalences of atoms in any chemical structure. However, the method could not be relied on in some more general application requiring the generation of arbitrary structures or graphs. Furthermore, even in chemical applications where the symmetry heuristics should be adequate, there are limits to the overall efficacy of these approaches. In complex structure generation problems, duplicates will still be created and must still be removed by retrospective canonicalization and testing.

X.B.3.c. Efficient Generation of Solely Canonical Forms

It is, in fact, possible to organize the processes of structure assembly so that only unique canonical structures result. The methods for generating only unique structures differ in several ways from those outlined in the previous sections. One important difference is that these new approaches need to control all allocations of bonds between constituent fragments. The constituent fragments may be simple chemical atoms or may be "superatoms" that represent polyatomic substructures. There must not be any predefined bonds interconnecting these fragments. Equivalences among the constituent fragments are determined solely by their atom types.

In the simple algorithm outlined earlier, initial and generated bonds tend to be "scattered" around the adjacency matrix in an arbitrary sequence. In the example data in Table X-1, some bonds [e.g., C(3)–C(12)] were predefined; the next few generated bonds were C(1)–C(9), C(2)–C(11), and so on. Alternative schemes for generating unique canonical structures fill out the adjacency matrix in a more systematic manner. The order for filling in elements is defined as shown in Figure X-2. This ordering effectively focuses attention on the interconnections of increasing subsets of the set of constituent atoms.

A number of advantages accrue from this systematic approach to building up the adjacency matrix;[11] the following example illustrates at least some of these advantages. The major advantage is that it becomes possible to test the canonicity of partially assembled structures. Essentially, the canonicity tests determine whether any renumbering of constituent atoms of identical type could yield a "better" partially assembled structure. These tests provide a uniform way of dealing with the various sources of duplication illustrated previously. Such tests can recognize that, as in the example given earlier, a particular bond pattern, C(8)–C(5)–C(6)–C(7), might be equivalent to, but not as "good" as, an alternative such as C(8)–C(6)–C(5)–C(7). Development of "not so good" (i.e., noncanonical) partial structures can simply be abandoned.

A secondary advantage of these new schemes is that they make it simpler to define comprehensive tests on bond allocations. These tests extend the previous checks used to avoid creating disconnected submolecules. It is possible to have bond allocations that do not immediately result in disconnected submolecules but will force such results at some subsequent step. Such invalid bond allocations can be identified and discarded, so avoiding fruitless elaboration of unacceptable partial structures.

The example problem for illustrating these new methods is to generate all isomers from the fragment set $(CH_2)_2\,(CH)_4$, where there are no restrictions on atom hybridization and thus multiple bonds are permitted. This example is one where analysis of initial and induced equivalences of atoms will not suffice to avoid all duplicates. The initial stages in developing structural solutions to this problem are illustrated in Figure X-12.

The constituent atoms would be grouped by chemical type with some arbitrary criterion defining an ordering for the different groups. For this example, the two methylenes could be the first atoms incorporated into the structure, with the four methines following. The initial structure assembly step requires the definition of the 1–2 element of the adjacency matrix. Two lines for development follow; in one all structures incorporate a–CH_2–CH_2–bond, the other line develops structures with separated methylenes. The first line of development will serve to illustrate some illegal bond allocation steps that can be avoided given appropriate prospective testing.

The next step, shown only for the first development line in Figure X-12, introduces a methine. There will usually be a number of alternative possible bond allocations. A newly introduced atom of valence i can, in general, form $0 \ldots i$ bonds to previously incorporated atoms, some of these could be multiple bonds. For this example, there are fewer possibilities; the methine may form a bond to one or both of the methylenes. There is no need to consider the possibility of a 2–3 bond but no 1–3 bond (because a "renumbering" of atoms 1 and 2 in such a partially assembled structure results in the preferred–2–1–3< system). Some of the subsequent steps in continued elaboration of the cyclopropyl-type structure are shown in Figure X-13.

There is only one way to introduce a second methine, atom 4, into the

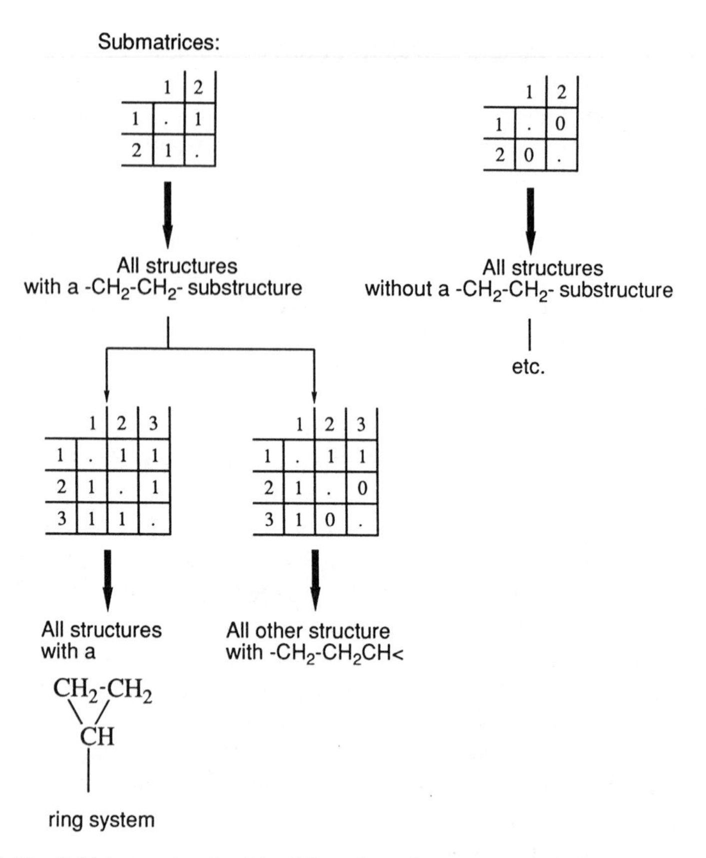

FIGURE X-12. Initial steps involved in elaboration of structures comprising two methylenes, atoms 1 and 2, and four methines, atoms 3–6.

cyclopropyl structure; this is shown as step 4a-i in Figure X-13. A complete structure can then be derived in two more bond building steps. Introduction of atom 5, with only a single bond back to atom 4, gives partial structure 5a–i–1 which can be elaborated as shown by step 6 in Figure X-13.

Following step 4a-i of Figure X-13, an alternative line of development through 5a-i-2 might be explored. No valid final structure can result from this line.

The problem here is similar to that discussed earlier where disconnected structures resulted from particular allocations of bonds. That other problem can be easily avoided by prohibiting the formation of bonds that, at intermediate stages of the assembly process, would result in components with no remaining free valences. (Even in this example, such a check would be necessary to prevent consideration of structures with a double bond connecting atoms 1 and 2).

The partially assembled structure 5a-i-2 of Figure X-13 still possesses

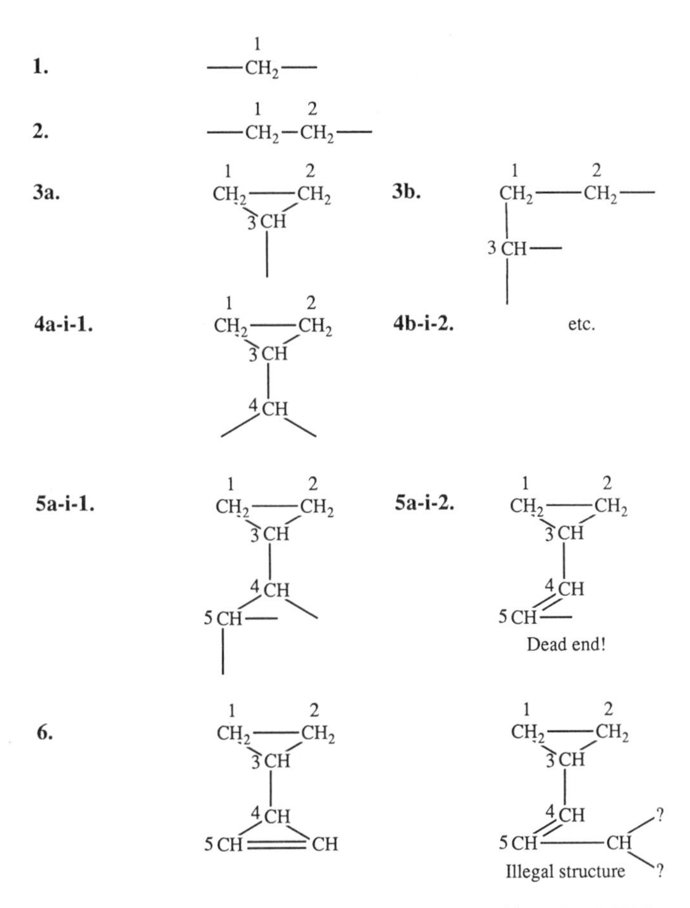

FIGURE X-13. Structure generation for the composition $(CH_2)_2(CH)_4$.

bonding sites, but it is no longer possible to satisfy the valence requirements of those other atoms that remain to be incorporated into the structure.

There are a number of similar types of invalid partial structure. A partial structure incorporating a disjoint submolecule may have been created. Too many bonds may have been formed between the atoms already incorporated into a partial structure, consequently leaving too few free valences for remaining atoms. Too few bonds may have been allocated to already incorporated atoms, thus resulting in a partial structure whose valence requirements cannot be satisfied by subsequent bonds to as yet unincorporated atoms. It is possible to determine whether any given partially assembled structure contains any of these unacceptable bond distributions. The necessary analysis involves consideration of the number and valence of separate fragments either already incorporated or remaining to be added to the partial structure. Furthermore, given such data on numbers and valences of fragments, it is possible to

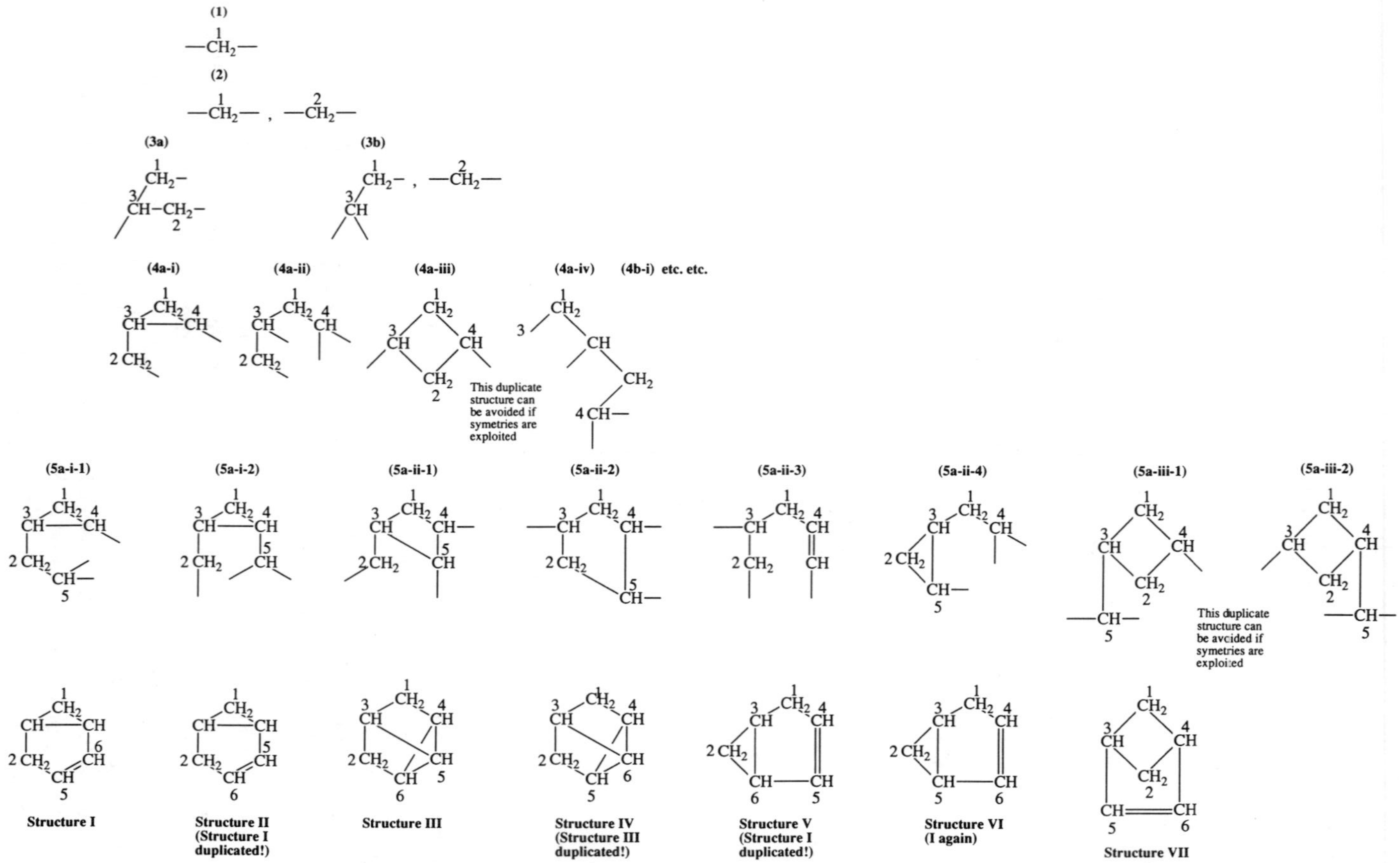

(1)

(2)

(3a) (3b)

(4a-i) (4a-ii) (4a-iii) (4a-iv) (4b-i) etc. etc.

This duplicate structure can be avoided if symetries are exploited

(5a-i-1) (5a-i-2) (5a-ii-1) (5a-ii-2) (5a-ii-3) (5a-ii-4) (5a-iii-1) (5a-iii-2)

This duplicate structure can be avoided if symetries are exploited

Structure I

Structure II (Structure I duplicated!)

Structure III

Structure IV (Structure III duplicated!)

Structure V (Structure I duplicated!)

Structure VI (I again)

Structure VII

FIGURE X-14. Alternative structure generation paths for the $(CH_2)_2(CH)_4$ problem.

determine upper and lower limits on the number of bonds a fragment must make to lower-numbered atoms when it is introduced into a partially assembled structure.[11] The use of such computed restrictions on the numbers of bonds makes it possible to avoid exploration of paths, such as that from 4a-i to 5a-i-2, which cannot yield valid structures. Such restrictions can considerably increase overall computational efficiency.

The major benefits of this approach to filling the adjacency matrix can be seen in relation to the alternative initial line of development for this structural problem. Some of the subsequent possible bonding steps are shown in Figure X-14. These alternative bonding steps lead to many duplicate structures.

Although use of symmetries does cut off a few branches of this search tree, many duplicates are still generated. The duplication among generated structures would, normally, necessitate some retrospective canonicalization step and comparison with all previously created structures. Thus, in a normal structure generator, after Structure-III has been formed it would be passed first to a canonicalizer. Canonicalization of Structure-III would result in a renumbering (atoms 5 and 6 would be accorded canonical numbers 6 and 5 respectively). This canonically numbered form of Structure-III is, of course, Structure-IV. The canonical form would be searched for in the list of previously generated structures; when not found, it would be appended to that list. Subsequently, Structure-IV would be created, canonicalized (in this case experiencing no change), searched for in the list of generated structures, found, and discarded.

However in this approach to structure generation, *it is possible to determine that a noncanonical connection matrix is being created before the structure is complete*. Consider, for example, the development leading to Structure-VI in Figure X-14. The connection matrix for partial structure when the fifth atom has been incorporated, **X-14**, is as shown in Table X-4.

The two $-CH_2-$s were originally equivalent, before structure assembly, and it is possible to rewrite this partial structure with the $-CH_2-$ atoms 1 and 2

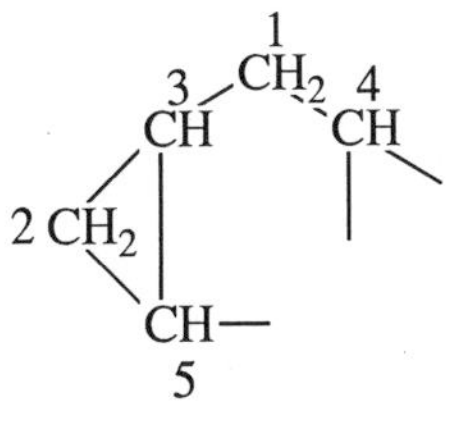

X-14

TABLE X-4. Connection Matrix for Partially Assembled Structure X-14 in the Given Numbering

	1	2	3	4	5
1	·	0	1	1	0
2	0	·	1	0	1
3	1	1	·	0	1
4	1	0	0	·	0
5	0	1	1	0	·
⋮					

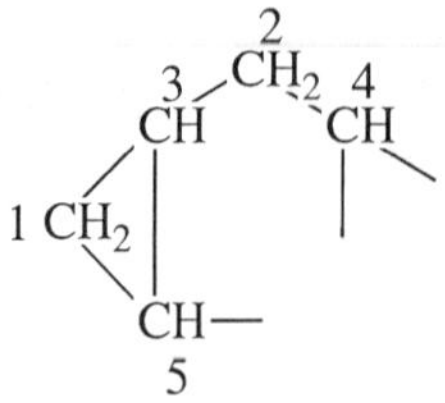

X-15

TABLE X-5. Connection Matrix for the Partially Assembled Structure X-14 After Exchanging $-CH_2-$ Fragments to Give Structure X-15

	1	2	3	4	5
1	·	0	1	0	1
2	0	·	1	1	0
3	1	1	·	0	1
4	0	1	0	·	0
5	1	0	1	0	·

exchanged, as shown as **X-15** with the partial connection matrix shown in Table X-5.

Permutations of the >CH–s are also permitted, as are permutations of the $-CH_2-$s and permutations of the >CH–s taken in combination. One particular permutation of atoms is (2 1 3 5 4–). This permutation rearranges the partial structure **X-14** to give **X-16** with the connection matrix shown in Table X-6.

If the connection tables shown in Tables X-4 and X-6 are compared, in the order specified in Figure X-2, a difference is found at the sixth element compared (row 3, column 4). Here, the matrix shown in Table X-6 has a bond order entry of one, where the other has a zero. If, as in previous examples, the matrix comparison function favors the greater matrix, the connection table for

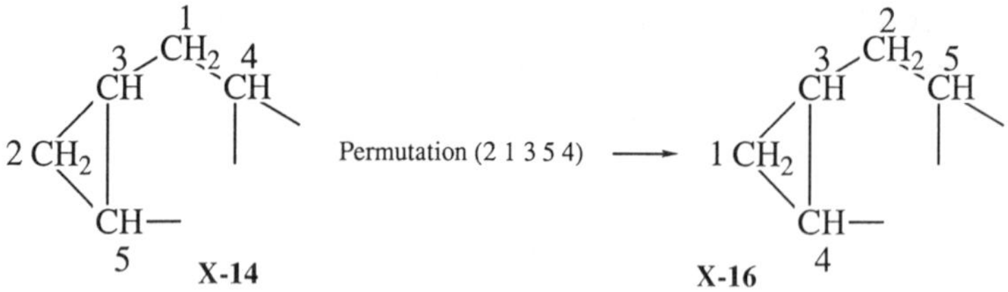

X-14 Permutation (2 1 3 5 4) → X-16

TABLE X-6. Connection Matrix for the Partially Assembled Structure X-14 After Applying the Permutation (2 1 3 5 4) to Give X-16

	1	2	3	4	5
1	·	0	1	1	0
2	0	·	1	0	1
3	1	1	·	1	0
4	1	0	0	·	0
5	0	1	0	0	·

X-16 would be rated "better" than that for **X-14**. (The order of comparison for elements of an adjacency matrix as defined in Figure X-2 makes it possible to compare incomplete adjacency matrices in which the rightmost columns remain to be filled in.)

*At this stage, therefore, it can be determined that the partially assembled structure **X-14** will not be in canonical form*. Since the generation procedure can be shown to yield all possible adjacency matrices, the correct canonical form of any structure incorporating **X-14** will either have already been generated or will be generated at some subsequent stage of the analysis. Processing of the partial structure **X-14** can be abandoned at this stage.

This approach was first exploited by Kudo and Sasaki[3,4] in the structure generation algorithms developed for **CHEMICS**. The method by which structures are developed in **CHEMICS** permits checks for "canonicity" to be performed as each additional atom is combined into the growing partially assembled structure. The inclusion of the final atom involves, naturally, a check on the canonicity of the final structure. "A structure can examine by itself whether it is canonical or not."[4] It is impossible to create duplicate structures through the use of the **CHEMICS** generator algorithm. Unlike the other algorithms outlined above, it is not necessary to keep a list of all structures already created and check each newly generated structure against this list. (There are cases, such as the case of Structure-III and Structure-IV in Figure X-14, where the noncanonical nature of a structure is only apparent when the very last atom is introduced. Although the canonicity checks cannot then save any of the cost of generation, they do permit the noncanonical structure to be recognized and discarded without any requirement for checks in lists of previously generated candidates.)

The method of Kudo and Sasaki does involve considering the permutations of the various sets of initial equivalent fragments.[3,4,11] As illustrated by the preceding example, various combinations of permutations must be considered. For n_1 fragments of type 1, n_2 fragments of type 2, and so on, it is may be necessary to consider all $n_1!n_2! \cdots$ permutations while verifying the canonicity of a partially assembled structure. This process can be computationally costly as is illustrated by the timing data given by Kudo and Sasaki.[3,4]

The canonicity checks of Kudo and Sasaki essentially specify just an ordering by atom type and then a check for a maximal adjacency matrix. The numbers $1, \ldots, n_1$ are accorded to the n_1 fragments of some first type, numbers $n_1+1, \ldots, n_1+n_2$ to the n_2 atoms of the second type, and so forth. The checks for a maximal adjacency matrix are expressed in terms of manipulations of a "connectivity stack" that contains an ordered list of entries from the connection matrix. These requirements are akin to the very simplest of criteria used to define canonical form in the examples given in Chapter IX. As illustrated in the discussions of the Morgan canonicalization algorithm,[17] and some of the various elaborations on that algorithm, other atom properties may be used to enhance the performance of a canonicalization algorithm.

Carhart has developed the basic concepts of the Kudo–Sasaki generation scheme and, in particular, provided a much more effective mechanism for checking canonicity of the partial structures.[11] Essentially, Carhart uses an elaborate set of rules that define a canonical ordering for atoms. These rules are expressed so that at the point where the *n*th atom is introduced into a partially assembled structure, and some particular allocation of bonds has been made, it is possible to determine whether it would be assigned canonical number *n* in a final canonical form.

A simplified example requirement on canonical form might specify that for two atoms of the same type the lower canonical number must be accorded to the one that has the greater number of bonds to other still lower-numbered atoms. Application of this rule to the data given in Table X-4 shows that this cannot yield a final canonical matrix. In the partial structure (**X-14**) atom 5 has two bonds back to lower-numbered atoms whereas atom 4 has only one such bond. This contravenes the canonicity rule, and the partial structure can be immediately discarded. This check can be made directly without any need to consider permutations of the numberings already associated with atoms in each class.

The most efficient procedures for generating candidate chemical structures are those based on the systematic development of canonical adjacency matrices. As already noted, these procedures are for manipulating atomic fragments only. Consequently, any known substructures must be treated as superatoms. In the generation procedures, the internal structure of superatoms is to be ignored. Subsequent embedding steps are necessary to complete structure generation. Since duplicates are possible during embedding, the embedder must incorporate canonicalization procedures and duplicate-elimination tests even though these are not needed in the generator.

X.B.4. Generation Procedures Based on Vertex Graphs

Lederberg[18–20] has shown how it is possible to define the structure of any chemical compound in terms of "trees" and "vertex graphs". The tree part of these structure definitions specifies all the branchings of acyclic parts, including interconnections among the vertex graph parts. Vertex graphs serve to characterize the form of each separate cyclic component in a molecule. Vertex graphs consist of the atoms, or vertices, which are the points of ring fusions of purely cyclic components together with the edges that interconnect these vertices. An example structure is shown in Figure X-15 together with the vertex graphs that it contains. In this structure two cyclic components are joined by a chain of acyclic bonds; there are two vertex graphs included, one for each cyclic component. In Figure X-15 the atoms of the structure that correspond to vertices in the vertex graphs are tagged by asterisks.

As well as serving as a mechanism for classifying the form of molecules, Lederberg's scheme provides the basis for a method for exhaustive structure generation.[12] This method breaks the overall structure generation task into many separate subtasks.

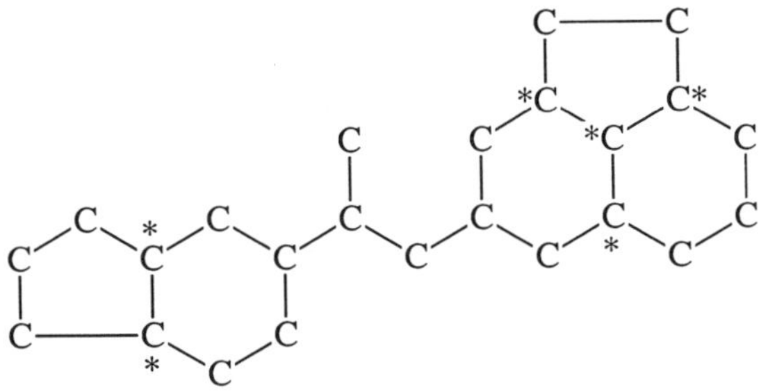

or

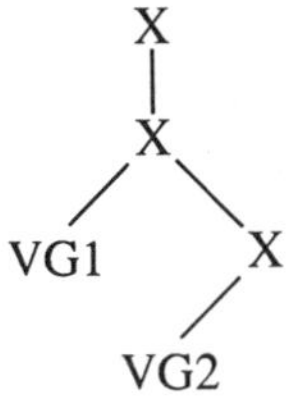

with and

VG1 =  VG2 =

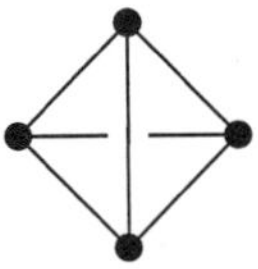

FIGURE X-15. A chemical structure and the vertex graphs that it incorporates.

One subtask involves partitioning of the chemical atoms of a molecule into those that will be used to build acyclic chains and those that will form cyclic components. Another subtask is the generation of acyclic tree structures using both those chemical atoms selected for the acyclic parts and some "ring–superatoms" representing cyclic components.

The various atoms bonded to any given ring–superatom in these acyclic tree structures define the substituents that must be attached to the corresponding cyclic structure(s) developed for that ring–superatom. Consequently, another subtask in this overall generation scheme must entail taking a given cyclic component and set of substituents and determining all the different ways in which those substituents may be placed about the cyclic component. This subtask is obviously a labeling process similar to those described previously.

The most difficult subtask is the generation of all cyclic structures that can be built with the atoms assigned to a given cyclic component. This subtask is accomplished by a process that elaborates vertex graphs.

Generation of structures of the cyclic components or superatoms starts with identification of the vertex graphs appropriate for the set of atoms making up the cyclic component. These can be determined given the valence of the atoms assigned to the component and the number of degrees of unsaturation. From these data, it is possible to determine the number of atoms that must be effectively bivalent within the cyclic component. [The remaining valences on these atoms are used for bonds to (implicit) hydrogens or for connection to acyclic substituents or other separate cyclic components]. The number of nodes (atoms) with higher degree in the cyclic component are thus determined.

The computed degrees of these remaining higher-valent nodes can be used to index into some "catalog" of vertex graphs. This catalog will contain all vertex graphs from one of the standard tabulations.[21–23] The graphs in the catalog can be identified in terms of the number of vertices of specified degree that they incorporate. The example vertex graphs shown in Figure X-15 would be among those indexed under two trivalent nodes (VG1) and four trivalent nodes (VG2). Once identified, the required vertex graphs can be retrieved.

The vertex graphs themselves can be represented in the catalog as simple connection tables showing the bonding between the vertices. Some example graphs with trivalent and quadrivalent nodes are shown, together with their connection table representations, in Figure X-16.

Vertex graphs exhibit symmetries; use of these symmetries is one of the main features of a structure generation scheme based on vertex graphs. It is more useful to define the symmetries in terms of edge permutation groups, rather than vertex permutation groups. The edge permutation groups can either be recorded in the catalog or recomputed as necessary by use of algorithms similar to those described in Chapter IX.

Although the complete vertex graph expansion procedure is complex, a simple illustration of its principles can be obtained by treating, as a special case, one of the structural problems already described. The example $(CH_2)_2(CH)_4$,

Vertex graphs:

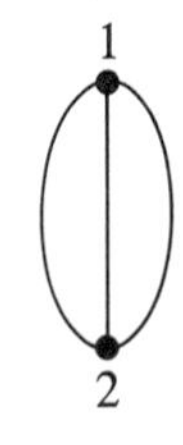

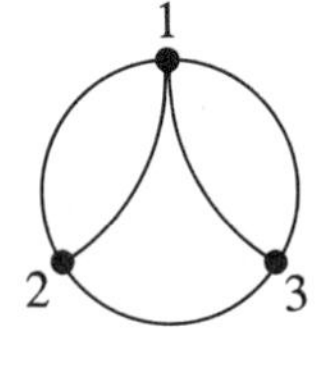

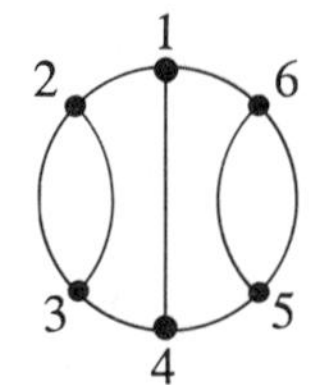

Connection tables:

Atom	Neighbors
1	2 2 2
2	1 1 1

Atom	Neighbors
1	2 2 3 3
2	1 1 3
3	1 1 2

Atom	Neighbours
1	2 4 6
2	1 3 3
3	2 2 4
4	1 3 5
5	4 6 6
6	1 5 5

FIGURE X-16. Connection tables for some vertex graphs.

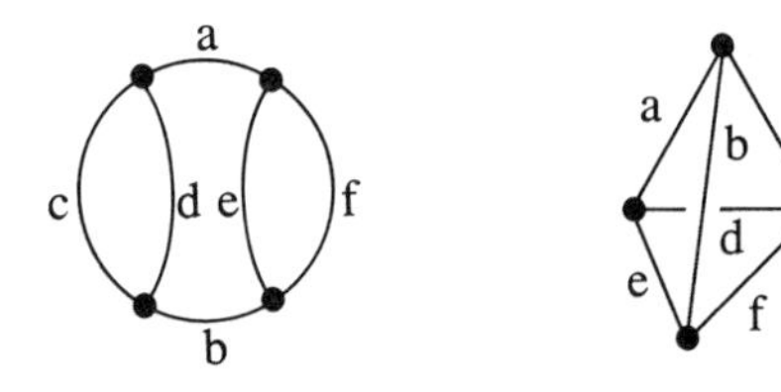

FIGURE X-17. The two vertex graphs incorporating four trivalent nodes. (Edges bear arbitrary labels for subsequent reference).

which causes at least some slight difficulties for the adjacency matrix-oriented structure generators, is trivially simple with the alternative vertex-graph-based approach. Initially, excluding structures with an acyclic bond interconnecting two different ring systems, this problem reduces to the need to find a method for building structures that incorporate two bivalent and four trivalent nodes into a single fused or bridged ring system.

In the first step, the bivalents can be ignored and all graphs comprising four trivalent nodes must be found. The appropriate graphs can be retrieved from a catalog. There are just two such graphs; these are shown in Figure X-17. (Their edges have been arbitrarily labeled for subsequent reference.)

Given such a complete set of vertex graphs, it is possible to build up representations of all structures. Here, the elaboration of the graphs is trivial for the positions of the >CH– trivalent nodes are fixed, and only the two remaining bivalent $-CH_2-$s need be considered.

Vertex graphs can be viewed as being derived from normal structures by disconnecting all acyclic bonds and then, for each cyclic component, eliminating all the bivalent nodes from the component's edges to leave only the nodes of higher degree. Exactly the reverse process is required for recreation of structures. The different ways of distributing the available bivalent nodes to the edges of the vertex graphs must be determined.

This next phase of the structure building process involves simply labeling of a graph using algorithms similar to those discussed Chapter IX. Here, edges rather than nodes are being labeled, and so it is necessary to determine the permutation symmetry groups of edges and then derive all distinct ways of labeling these edges with two identical labels. The labelings of one of the two vertex graphs and the final structures that these labeled graphs represent are shown in Figure X-18.

The three structures that can be obtained from the tetrahedron like vertex graph can be similarly developed. These are obtained by placing two bivalents on edge a, or one bivalent on edge a together with the second either on edge b, or on edge f.

For this limited part of the structure generation process, the vertex graph analysis is much more direct than any adjacency matrix manipulator. The completeness of the catalog of vertex graphs guarantees that all basic structural

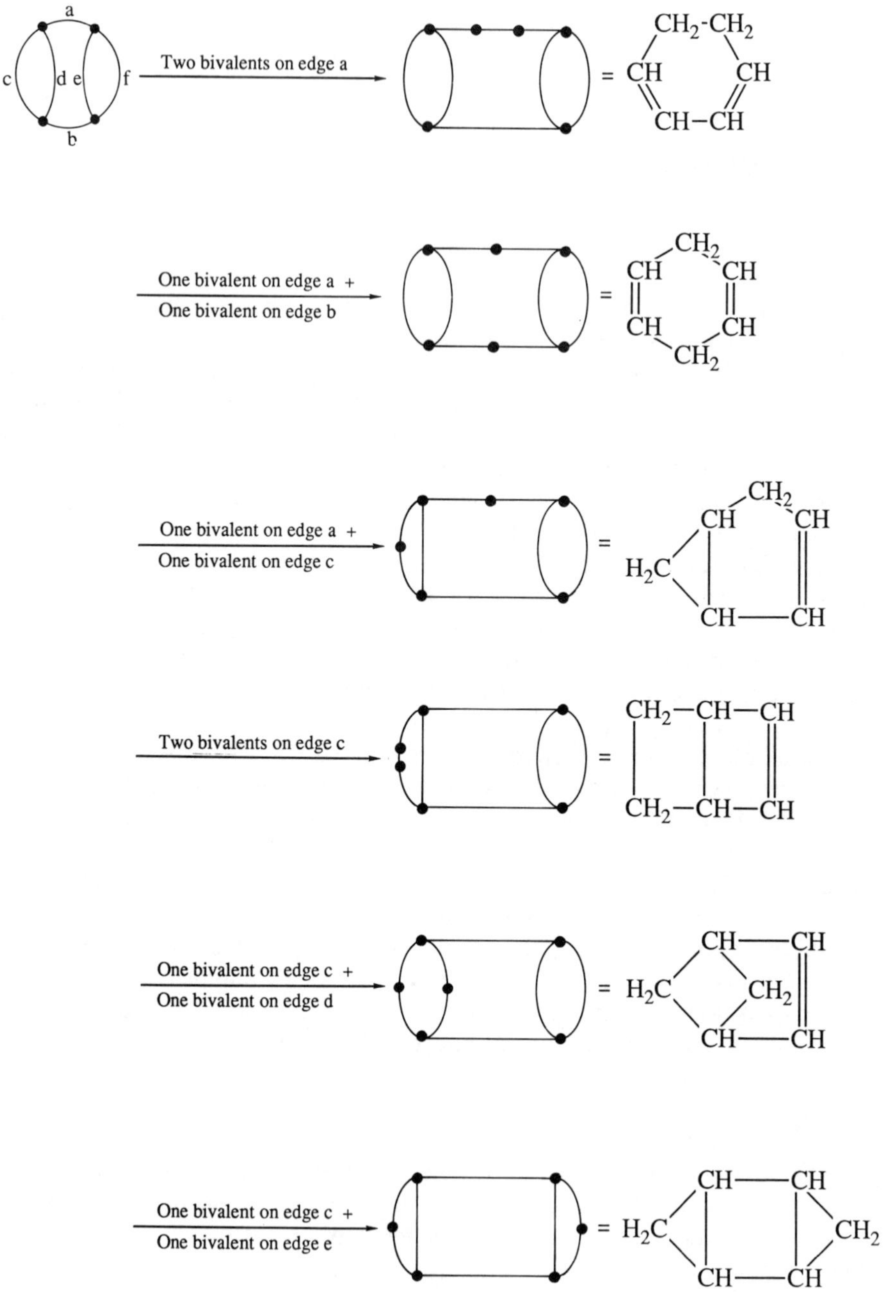

Two bivalents on edge a

One bivalent on edge a +
One bivalent on edge b

One bivalent on edge a +
One bivalent on edge c

Two bivalents on edge c

One bivalent on edge c +
One bivalent on edge d

One bivalent on edge c +
One bivalent on edge e

FIGURE X-18. The various labelings of the first of the two vertex graphs and corresponding structures.

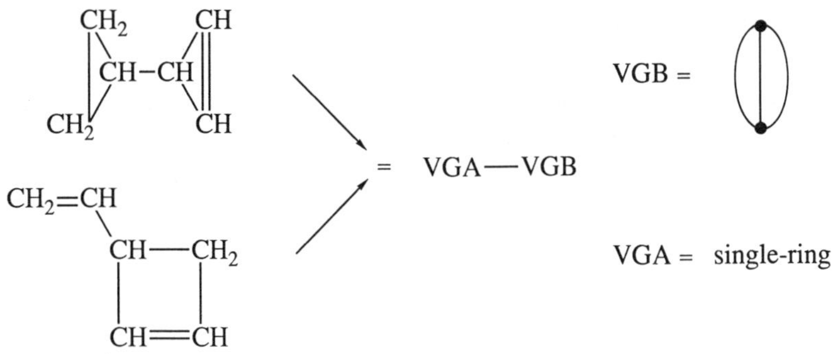

CH_2 CH
CH–CH
CH_2 CH

CH_2=CH
CH—CH_2
CH=CH

= VGA—VGB

VGB =

VGA = single-ring

FIGURE X-19. Example $(CH_2)_2(CH)_4$ structures based on two ring-superatoms and their common vertex graph systems.

forms will be derived. The relatively simple combinatorial processes involved in placement of a set of identical bivalent nodes on edges guarantee that all elaborations to complete structures will be produced with no possibility of redundancy. There is no requirement to recompute symmetries nor any need for canonicalization checks at intermediate stages of structure assembly.

Of course, the complete structural problem involves considerably more work. As well as considering structures that combine all atoms into a single fused-bridged ring system, it would be necessary to consider alternative arrangements with more than one ring-superatom. Given the atom set $(CH_2)_2(CH)_4$, another possible arrangement involves one ring system with a single degree of unsaturation and a second with two degrees of unsaturation. Example structures, together with the vertex graphs that they incorporate, are shown in Figure X-19.

These structures both incorporate the same two vertex graphs (but with different allocations of numbers of atoms to graphs). These are the standard vertex graph with two trivalent nodes and the special case of "single-ring." "Single-ring" is a sort of degenerate case of a vertex graph, one with a cycle but no vertices. (It should be noted that, in vertex graph terms, a double bond is just a small ring system.)

There are other cases where special analysis procedures are needed. These include various forms of spiro fused systems that, when elaborated to a complete structure, would involve a ring with an exocyclic double bond. An example final structure, incorporating some of these features that require special processing, is shown as **X-17**.

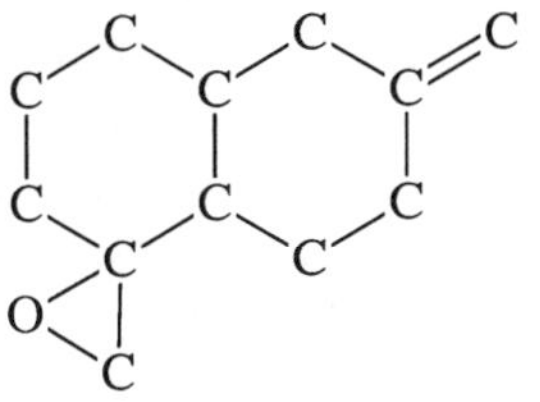

X-17

When reduced to a vertex graph form (**X-18**), this structure involves two "loops" (as shown in structure **X-18**). Graphs with self-loops on nodes need not be held in the catalog of standard vertex graphs. They can be developed, as necessary, by labeling edges of standard vertex graphs with special loop nodes.

X-18

Looped graphs require tetravalent nodes and at least two degrees of unsaturation in a ring–superatom. Consequently, none of these graphs are generated for the $(CH_2)_2(CH)_4$ problem.

The example of $(CH_2)_2(CH)_4$ is greatly simplified in that the valences of the atoms are almost all specified, their only free valences are those needed for building up the cyclic structures. More typically, the allocation of free valences will be less completely defined. A ring-superatom will be given as a set of atoms and its degree of unsaturation. Thus one might have the compositions C_6U_3, where all atoms and unsaturations have been allocated to a single ring-superatom, or C_5U_3 where the sixth carbon has been reserved for some acyclic part of the complete structure. The composition of a ring–superatom part will determine the total number of free valences on the constituent atoms. Most of these free valences will be needed for bonds to (implicit) hydrogens or acyclic substituents. However, there will usually be several different ways in which the remaining free valences could be distributed.

For example, the composition C_6U_3 allows a distribution to give one tetravalent, two trivalent, and three bivalent nodes [i.e., $(CH_2)_3(CH)_2C$] as well as the distribution with four trivalent and two bivalent nodes [i.e., $(CH_2)_2(CH)_4$] already discussed. Other vertex graph systems must be considered for the case with three degrees of unsaturation, two trivalents, and one tetravalent node. Here, as illustrated in Figure X-20, both looped and non-looped vertex graph-based structures are possible.

The complete vertex graph based analysis is thus elaborate. First, atoms must be partitioned in all possible ways between acyclic parts and one or more ring-superatoms. The total free valence of each ring-superatom must be calculated and all distinct ways of distributing the free valences among its constituent atoms determined. Each distribution defines a different problem to be separately analyzed by use of vertex graphs. Appropriate vertex graphs must be selected from the catalog; if unsaturations and vertex degrees permit, looped variants of standard vertex graphs must be considered. All the possible looped and nonlooped vertex graphs must then be elaborated by placing bivalent nodes and labeling the various nodes with atom names.

Application of the vertex graph-based procedures to chemical problems is

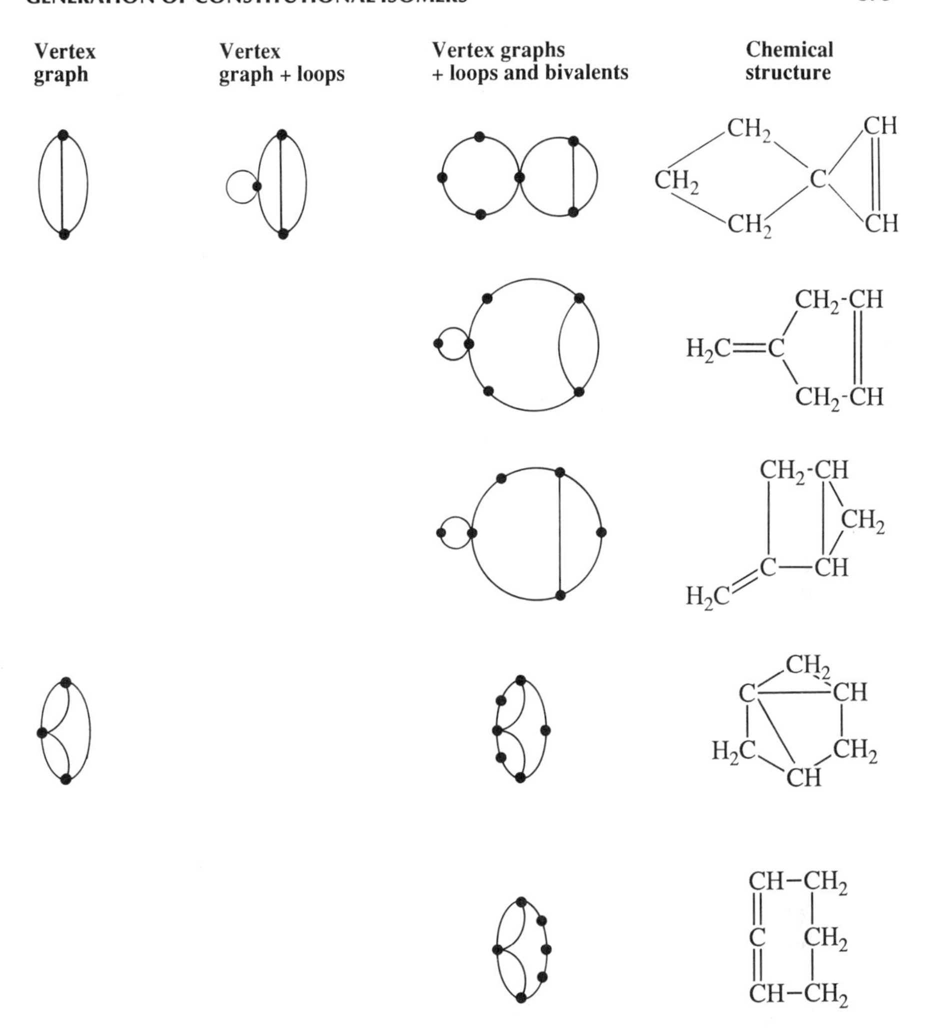

FIGURE X-20. Some alternative vertex graph systems, and example structures, possible for a more general C_6H_8 problem.

further complicated by the need to accommodate known structural constraints. These generation procedures are defined solely in terms of graph theoretic concepts and are remote from chemical concepts of groups of bonded atoms with restrictions on their further interconnections. The only way in which known substructures can be incorporated into this scheme is to treat them as atoms (i.e., superatoms) forming a part of the initial composition list.

This immediately raises a new difficulty, the valence of such a superatom will often exceed four (e.g,. the superatom END, in Figure X-5, has a valence of 7), and, in principle, the valence of a superatom could be arbitrarily large. The catalogs of vertex graphs available to such a generator must thus be extended to include nodes of higher valence.[23] Additional analysis steps are also required to deal with the possibility of some of the valences of a superatom being used for "internal bonds."

Of course, the structural forms involving polyatomic superatoms are incompletely specified. As with the more recent version of **CONGEN**, final representations of chemical structures must be developed by embedding these superatoms.[1]

The original vertex graph programs served a useful purpose in providing the first provably correct method for exhaustive structure generation. The results obtained from these programs have been of value in verifying the correct implementation of other subsequently developed exhaustive algorithms.[11] Vertex graphs themselves do constitute a useful method for classifying chemical structures. However, the complexities of the procedures involved render a vertex graph based approach to structure generation inappropriate for routine chemical structure elucidation problems.

X.C. GENERATION OF CONFIGURATIONAL STEREOISOMERS

Once a set of candidate constitutional isomers has been generated and then perhaps further pruned in some review process for the identification and elimination of undesired candidates, it becomes appropriate to consider configurational stereochemistry. The set of remaining constitutional candidate structures will normally be passed to some separate program where the possible configurational stereochemistry of each constitutional isomer can be analyzed. The analysis of configurational stereoisomerism in any given structure is a two step process: (1) the stereocenters must be identified; and (2) all *distinct* configuration designations must be assigned to these centers (subject to any additional stereochemical constraints).

The first stage of the analysis might begin by assuming that every tri- and tetrasubstituted carbon represents a potential stereocenter, as do nitrogens at bridgeheads or in double bonds. Not all such atoms will be true stereocenters. The true stereocenters must be determined by some filtering process. This filtering process will entail an examination of the various substituent groups on the atoms and double bonds being considered as potential stereocenters.

When looking at a single tetravalent carbon in a simple acyclic structure, one normally determines whether it is a true stereocenter by checking whether it bears four distinct substituents; if two of its substituents are equivalent, that central quaternary carbon atom is discarded from any list of potential stereocenters. The analysis of a structure with many potential stereocenters is inevitably more complex. Although more elaborate, the checks necessary for a complex structure do involve simply consideration of the equivalences of substituent groups at the various potential stereocenters.

Any equivalences among the substituents of an atom are, of course, readily identified from the topological symmetry group of the structure (which may be either recorded with the structure or recomputed as needed using algorithms discussed in Chapter IX.) Equivalences among substituents, which do not themselves incorporate any potential stereocenters, may suffice to eliminate a potential stereocenter. The entire process is iterative. The elimination of some potential stereocenters requires a reanalysis of the symmetries at other potential stereocenters and, possibly, further elimination and reanalysis steps.

This first stage of the stereochemical analysis might identify, say, n atoms as being true stereocenters. With n stereocenters, there are 2^n different stereochemical designations possible for the structure (because each of the n individual stereocenters can be accorded an appropriate *R/S*, *cis/trans*, or 0/1 label). Consequently, there are potentially 2^n stereoisomers for the molecular structure. However, if that structure has symmetry, there will generally be fewer than 2^n true stereoisomers.

The second stage of the stereochemical analysis entails an examination of how the symmetries in the structure interrelate the stereocenters. Symmetries in a structure will act to partition the 2^n different stereochemical designations, or "potential stereoisomers", into a number of equivalence classes, for example, N classes ($N <= 2^n$). *These equivalence classes of stereochemical designations correspond to distinct* stereoisomers; thus, if the symmetry analysis identifies N equivalence classes of stereochemical designations, the structure has N stereoisomers. This symmetry analysis is complex and difficult to perform manually. It is in this stage that computer-aided analysis is most useful.

The algorithms for the generation of stereoisomers are thus based largely on the examination of various representations of molecular symmetry.[24-26] The ordinary topological symmetry will suffice for identification of potential stereocenters. However, the procedures for finding the equivalence classes of potential stereoisomers require a different symmetry group. This new representation of molecular symmetry, the "Configuration Symmetry Group" (CSG),[24] defines the effects of the symmetries on configurations of atoms. Some symmetries represented in the CSG will preserve configurations, and others will invert configurations. The derivation of the CSG is the major task of the second stage of a configurational stereoisomer generator. The final development of the distinct stereoisomers is a relatively simple process involving merely manipulations of bit-patterns (representing sets of atom configurations) in accord with the permutation and inversion operations defined in the CSG.

Stereochemical constraints can be applied during these generation processes. A few simple constraints apply even for identification of potential stereocenters. For instance, sometimes one may wish to include nonbridgehead/nonmultiply bonded nitrogen atoms as potential stereocenters. More elaborately, there might be restrictions on the relative configurations of neighboring stereocenters (as, possibly, inferred from $^1H-^1H$ NMR couplings) or a requirement such that the structure be chiral. Such constraints can be used to limit the generation procedures or applied in subsequent filtering processes that discard unsatisfactory stereoisomers.[26]

The various algorithms required for configurational stereoisomer generation have been presented in detail elsewhere.[24–26] The major stages are outlined briefly in the subsequent sections. The examples presented are mostly simplified and do not include any involving double bonds; the extension of the analysis to the stereochemistry of double bonds is presented in the original papers.

X.C.1. Identification of Stereocenters

The first step in the procedures for identifying "true stereocenters" will involve the application of "chemical constraints" that exclude some atoms from the category of potential stereocenters. Thus mono- and bivalent atoms would be tagged as nonstereocenters, as would atoms in triple bonds or those flagged as being in some aromatic system. Atom-type constraints, such as inclusion or exclusion of nitrogens, can also be applied at this stage. (Such "atom type" constraints may necessitate an analysis of all the rings in a structure because one might wish to allow bridgehead nitrogens to be stereocenters, and to assume facile inversion of any other nitrogens in a structure.) The atoms that would thus be eliminated from the category of "potential stereocenters" are marked by asterisks in the example structures **X-19–X-23**.

Next, the main iterative filtering cycle is applied. At each stage of this iterative cycle an attempt is made to exclude more atoms from the set of potential stereocenters. The environment of each remaining potential stereocenter is examined in turn to identify possible equivalences among its bonded

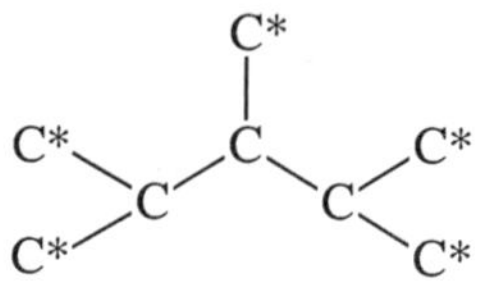

X-19

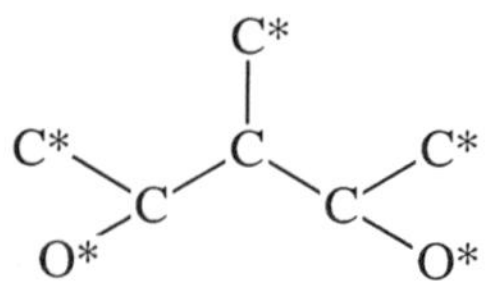

X-20

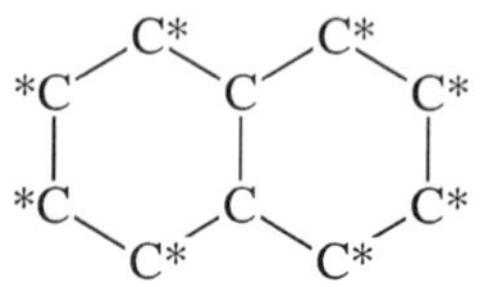

X-21

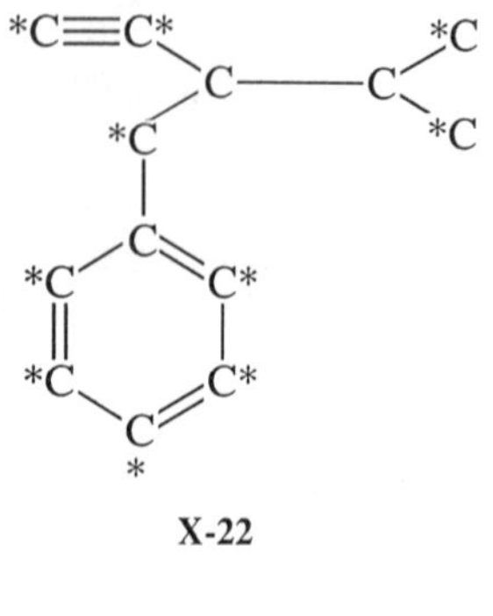

X-22

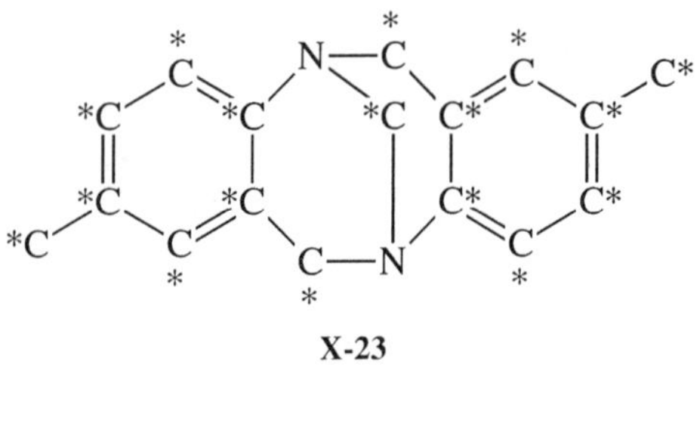

X-23

TABLE X-7. Permutation Symmetry Group for Atoms in Structure X-19 with Numbering as Shown in Structure X-24

1	2	3	4	5	6	7	8
1	2	3	4	5	7	6	8
1	2	3	5	4	6	7	8
1	2	3	5	4	7	6	8
1	3	2	6	7	4	5	8
1	3	2	7	6	4	5	8
1	3	2	6	7	5	4	8
1	3	2	7	6	5	4	8

neighbors. Thus for structure **X-19**, shown numbered as **X-24**, the environments of atoms 1, 2, and 3 would be further analyzed. As shown by the symmetry data, summarized in Table X-7, each of these atoms has two equivalent neighbors.

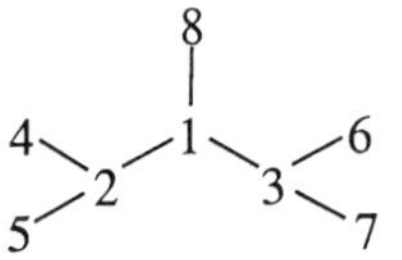

X-24

Atom 1 would not be eliminated from the set of potential stereocenters in the first iteration of the filtering process. Although its neighbors 2 and 3 are topologically equivalent, at this stage in the analysis they are still themselves rated as potential stereocenters. However, when atom 2 is analyzed, the equivalence of its two methyl substituents, atoms 4 and 5, is sufficient to eliminate it from the class of potential stereocenters. Similarly, atom 3 is rejected as a potential stereocenter because of the equivalence of its neighbors 6 and 7. Because some potential stereocenters are eliminated through this first iteration of the filtering process, the cycle is repeated. In the second iteration, atom 1 would also eliminated because of the equivalence of its neighbors 2 and 3, which would by then be tagged as nonstereocenters.

A similar analysis applied to structure **X-20**, with numbering as shown in **X-25**, would terminate after a single cycle with atoms 1, 2, and 3 all remaining as true stereocenters. Although atom 1 in **X-20** does bear topologically equivalent substituent atoms, these are themselves true stereocenters.

As illustrated in structure **X-26**, the checks for potential stereocenters in

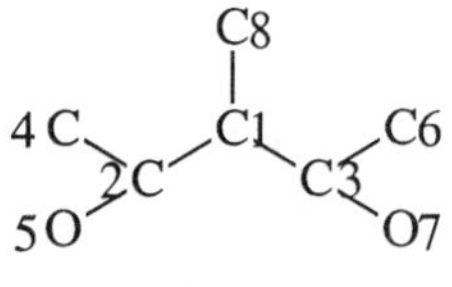

X-25

equivalent substituents on an atom may necessitate some searching. In **X-26**, the central methine would not be eliminated from the set of potential stereocenters because its two equivalent substituent chains eventually lead to other stereocenters. The discovery of any potential stereocenters can be performed by using a simple recursive graph traverser to search out from one of an atom's topologically equivalent neighbors. The search terminates when all paths have been explored or another potential stereocenter has been found.

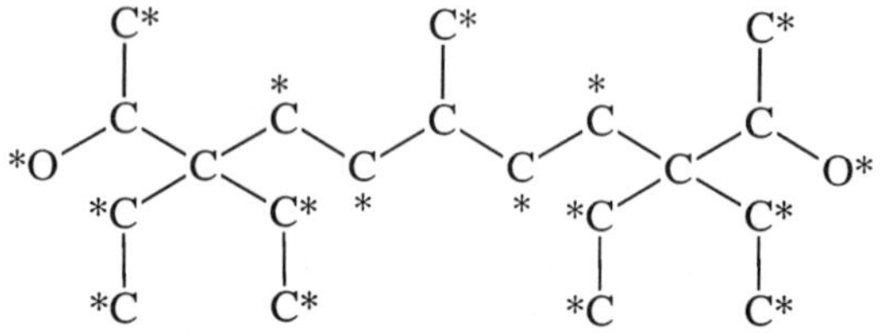

X-26

In a structure such as decalin (**X-21**) this search may lead around the ring. Carbon C(5) of the decalin skeleton has as equivalent substituents C(4) and C(6). Search out from C(4) reveals C(10) as a connected potential stereocenter in a topologically equivalent substituent group. The presence of potential stereocenter C(10) prevents C(5) from being eliminated from the set of potential stereocenters. Similarly, when C(10) of the decalin is analyzed, will be found to be a true stereocenter.

As discussed in the original paper[25], there are a number of different possible types of topological equivalence among substituents at a potential stereocenter. These different types are distinguished according to the number of equivalent substituents that are permuted. It is possible for an atom to possess two pairs of identical substituents or three identical substituents and yet still be a true stereocenter.

The only cases where a potential stereocenter can be eliminated are those where there exist some permutations in the symmetry group that either exchange one pair of substituents or permute four identical substituents. The algorithm for finding which potential stereocenters can be eliminated involves first checking the symmetry group of the structure for a permutation that exchanges a pair of substituents (or all four substituents) at a center and then verifying that the permuted substituents do not themselves contain other potential stereocenters.

The true stereocenters in a structure are simply those atoms that survive both the "chemical constraint" filters and the iterative elimination cycles.

X.C.2. Symmetry Analysis to Find Equivalence Classes of Possible Stereoisomers

As discussed in Chapter VII on the computerized representation of structures and substructures, the configuration at a stereocenter is most conveniently defined in terms of a parity tag with an implicit (or explicit[27]) ordering of its bonded neighbors. A standard representation of the two possible configurations at a tetravalent center are shown as **X-27** and **X-28**. (The neighbors of the unlabeled stereocenter are presumed to be ordered, from 1 to 4, according to their numberings in the constitutional structure.)

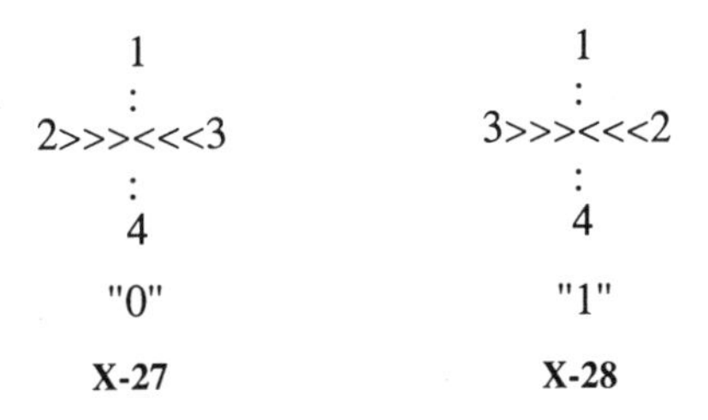

X-27 **X-28**

The permutations in a structure's symmetry group define those renumberings of the atoms that preserve connectivity. *Any renumbering of the neighbors of a stereocenter may change the represented configuration*. For example, if an explicit ordering of neighbors were employed, the configuration in structure **X-28** would be defined by the sequence (1 2 4 3) (reading clockwise round the center). If the symmetry group of the structure included a permutation that exchanged number labels 2 and 3, the sequence of neighbors in a renumbered structure would be changed to (1 3 4 2), that is, to the sequence for the configuration in **X-27**. The particular symmetry associated with this permutation of atom labels 2 and 3 thus implies inversion at this stereocenter.

More complex symmetries can permute several of the atom number labels on the neighbors of a stereocenter. It is possible to determine whether any particular symmetry operation belonging to a structure changes the implied configuration at a stereocenter. The number of pairwise exchanges necessary to restore the original ordering of neighbors must be found. If this number is odd, the symmetry element corresponding to a particular permutation inverts the stereocenter; however, if the number is even then it retains configuration. For example, a permutation that renumbered the neighbors of the stereocenter as (1 4 3 2) would retain configuration because it entails two pairwise exchanges.

The first stage in the process of finding how symmetries in a structure might interrelate potential stereoisomers requires the identification of how the various symmetries affect configurations at each stereocenter. A complete analysis of atom and double-bond configurations requires data on both the atom ("node") symmetry group and the bond ("edge") symmetry group of a structure. The required "graph" symmetry group is a "product" of the atom symmetry group and a subgroup of the edge symmetry group. (This subgroup comprises those "edge" permutations that interchange the two "edges" corresponding to a double bond.) None of the following examples concern double-bond configurations; consequently, the atom permutation group will suffice as a representation of constitutional, or "graph" symmetry. The devel-

TABLE X-8. Atom Permutation Symmetry Group for Structure X-20 with Numbering of X-25

1	2	3	4	5	6	7	8
1	3	2	6	7	4	5	8

opment of graph symmetry groups for structures with both double bonds and atomic stereocenters is detailed in the original literature.[24,25]

The effects of symmetries on atom configurations can be determined algorithmically by a process that examines the renumberings of atoms implied by each symmetry operation. A few simple examples are provided by the diol structure (**X-20**), tartaric acid (**X-29**), and decalin (**X-21**). The atom symmetry group for **X-20** (with the numbering as in **X-25**) is shown in Table X-8. [The first entry in each of the atom permutation symmetry groups shown in subsequent tables is the "identity" element defining the given (canonical) numbering of the structure.]

The data in Figure X-21 summarize the effect of the permutation in Table X-8 on atoms labeled 1 and 2 in the original structure. The neighbors of these atoms are listed, in increasing order, for the original structure (H ranks less than any numbered neighbor in these orderings).

The corresponding numberings of atoms are also given for the transformed-renumbered structures. The renumberings can be simply read from the permutation; thus the neighbors of atom 1 match 3–2, 2–3, and 8–8. One pairwise exchange of labels would be necessary to restore the original ascending order of

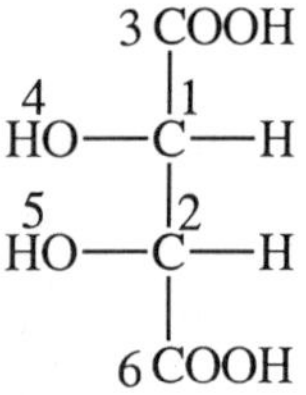

```
          3 COOH
             |1
      4 HO—C—H
             |2
      5 HO—C—H
             |
          6 COOH
```

X-29

C(1) Original	C(1) Transformed	C(2) Original	C(2) Transformed
H	H	H	H
2	3	1	1
3	2	4	6
8	8	5	7
Inverted		Retained	

FIGURE X-21. Effects of the permutation given in Table X-8 on the configurations at C(1) and C(2) of structure X-20.

TABLE X-9. Atom Permutation Symmetry Group for Structure X-29

1	2	3	4	5	6
2	1	5	6	3	4

neighbors at atom labeled 1; consequently, the symmetry operation corresponding to this permutation inverts this stereocenter. For atoms originally labeled 2 and 3, the transformed neighbor lists remain in ascending order; since no exchanges are necessary, the configurations at these atoms are retained.

As shown by the data in Figure X-22, the effect of the (2 1 5 6 3 4) permutation for structure **X-29** (see also Table X-9) is to retain configuration at both stereocenters. Looking at stereocenter originally numbered 1, one sees that the relabeling of neighbors 2–1, 3–5, and 4–6 preserves the ascending number sequence and hence configuration. For the decalin **X-21**, with the atom symmetry group given by the permutations in Table X-10, two of the permutations (b and c) act to invert configurations at both stereocenters C(5) and C(10); permutation d, which requires two pairwise exchanges to restore the neighbors to ascending order, retains configurations. The effects of the permutations on C(5) are summarized by the data in Figure X-23.

C(1)		C(2)	
Original	Transformed	Original	Transformed
H	H	H	H
2	1	1	2
3	5	5	3
4	6	6	4
Retained		Retained	

FIGURE X-22. Effects of permutation given in Table X-9 on configurations at C(1) and C(2) of tartaric acid

TABLE X-10. Atom Permutation Symmetry Group for Decalin Structure X-21

a	1	2	3	4	5	6	7	8	9	10
b	9	8	7	6	5	4	3	2	1	10
c	4	3	2	1	10	9	8	7	6	5
d	6	7	8	9	10	1	2	3	4	5

Original	Transformed	Transformed	Transformed
a	b	c	d
H	H	H	H
4	6	1	9
6	4	9	1
10	10	5	5
	Inverted	Inverted	Retained

FIGURE X-23. Effects of permutations given in Table X-10 on configurations at C(5) of a decalin.

Simple cases such as those illustrated in these examples can often be analyzed intuitively or by appeal to models of specific stereoisomers. Such analyses are less successful in structures with more complex symmetries; tetramethylcyclobutane (**X-30**) provides examples of some more complex symmetries. Some of the symmetries of this structure preserve all atom configurations, others invert all stereocenters, and some of the symmetries invert some stereocenters while retaining configurations at others. The affects of some of the permutations, listed in Table X-11, on C(1) and C(2) in **X-30** are summarized in Figure X-24.

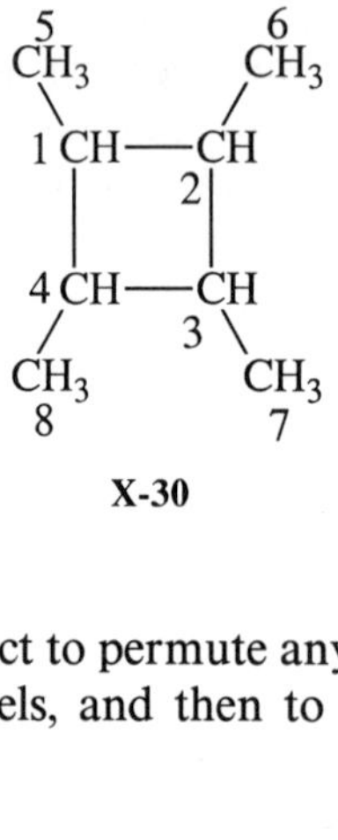

```
   5          6
  CH3        CH3
    \        /
  1 CH——CH
      |   2|
      |    |
  4 CH——CH
    /   3 \
  CH3      CH3
   8        7
```

X-30

The symmetry operations act to permute any labels on atoms, such as any preassigned configuration labels, and then to invert the configurations on

TABLE X-11. Atom Permutation Symmetry Group for Structure X-30

a	1	2	3	4	5	6	7	8
b	1	4	3	2	5	8	7	6
c	2	1	4	3	6	5	8	7
d	4	1	2	3	8	5	6	7
e	3	2	1	4	7	6	5	8
f	3	4	1	2	7	8	5	6
g	2	3	4	1	6	7	8	5
h	4	3	2	1	8	7	6	5

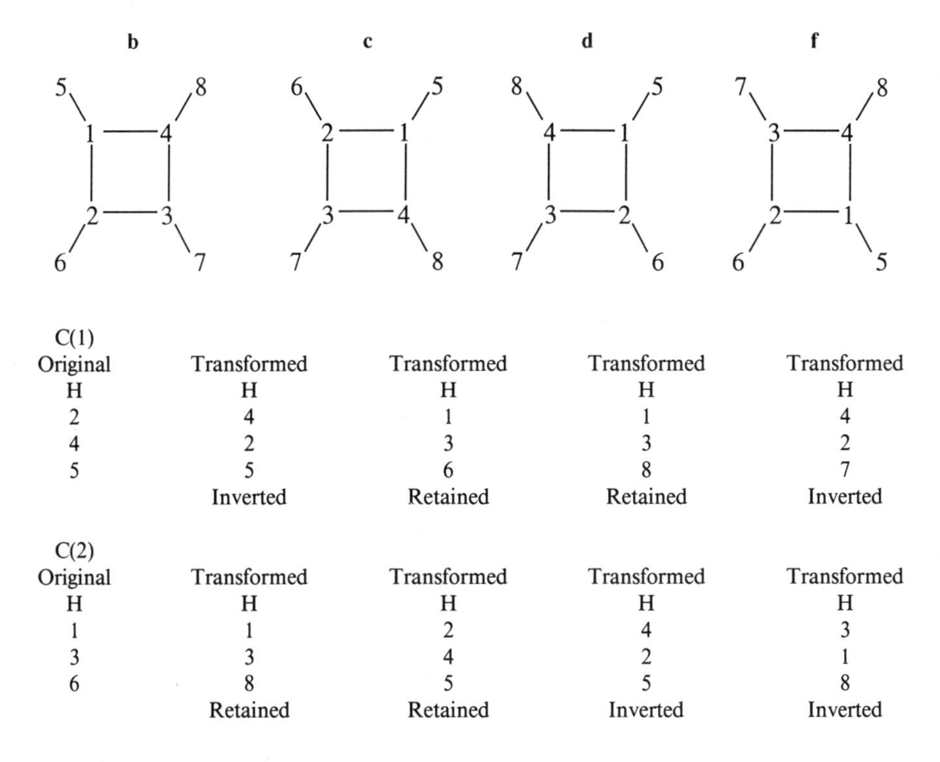

C(1)

Original	Transformed	Transformed	Transformed	Transformed
H	H	H	H	H
2	4	1	1	4
4	2	3	3	2
5	5	6	8	7
	Inverted	Retained	Retained	Inverted

C(2)

Original	Transformed	Transformed	Transformed	Transformed
H	H	H	H	H
1	1	2	4	3
3	3	4	2	1
6	8	5	5	8
	Retained	Retained	Inverted	Inverted

FIGURE X-24. Effects of permutations given in Table X-11 on configurations at C(1) and C(2) of tetramethylcyclobutane.

some of the stereocenters. Once the effects of the permutations on configurations at various stereocenters have been determined, it is no longer necessary to hold the complete permutation symmetry group of all atoms. The configuration symmetry group[24] summarizes the essential data identifying how stereocenters are permuted and how these permutations affect configurations. The entries in the CSG consist of a permutation of stereocenters supplemented by tags indicating inversions (a very compact notation for CSGs is given in the original paper.[24])

The entries in the CSG can be used as operators that take a stereoisomer, with arbitrary configurations specified at all stereocenters, and can then generate from this all symmetrically equivalent stereoisomers. An example, using tartaric acid, is shown by the data in Figure X-25. There are four potential stereoisomers of tartaric acid; the two stereocenters can be assigned configurations 0–0, 0–1, 1–0 and 1–1. Apart from the identity element, the symmetry group of tartaric acid (Table X-9) contains only one permutation that, as discussed above, exchanges labels but does not invert configurations. The CSG for this structure thus consists of the identity and a simple permutation of the stereocenters.

Configurations in potential stereo-isomer C(2) C(1)	Permutation-inversions of CSG		
	Identity	Permute 1 & 2	
0 0	0 0	0 0	Class 1
0 1	0 1	1 0	) Class 2
1 0	1 0	0 1	) Class 2
1 1	1 1	1 1	Class 3

FIGURE X-25. Permutation-inversion operations of the CSG for tartaric acid act to partition its four potential stereoisomers into three equivalence classes.

The permutation in the CSG for tartaric acid interconverts the 0–1/1–0 potential stereoisomers; the 0–0 and 1–1 potential stereoisomers are unaffected by the operations of the CSG. In consequence, the four potential stereoisomers of tartaric acid are grouped into three equivalence classes. The 0–0 and 1–1 stereoisomers are enantiomeric (enantiomeric forms can be obtained by inverting all configuration designations); in fact, they represent the *d* and *l* forms of tartaric acid. The 0–1 stereoisomer is its own enantiomer and represents the *meso* form of tartaric acid.

As analyzed above, the atom permutation symmetry group of decalin had, in addition to the identity element, three different permutations. Two of these acted to invert both of the C(5) and C(10) configurations while one permuted atom labels but retained configuration. The operations of the CSG for decalin factor its four potential stereoisomers into two classes as shown in Figure X-26. One of the two equivalence classes represents *cis*-fused decalins, and the other corresponds to *trans*-fused forms.

The original paper[24] illustrates how the eight elements of the CSG for tetramethylcyclobutane factor its 16 potential stereoisomers into four equivalence classes. Consequently, this structure can exist in only four distinct stereoisomeric forms.

Configurations in potential stereo-isomer C(10) C(5)	Permutation-inversions of CSG				
	Identity	10 5 (Invert both)	5 10 (Permute & invert)	5 10 (Permute)	
0 0	0 0	1 1	1 1	0 0	Class 1
0 1	0 1	1 0	0 1	1 0	Class 2
1 0	1 0	0 1	1 0	0 1	Class 2
1 1	1 1	0 0	0 0	1 1	Class 1

FIGURE X-26. Permutation-inversion operations of CSG for decalin act to partition its four potential stereoisomers into two equivalence classes.

In addition to establishing the number of true stereoisomers a structure possesses, the process of identifying equivalence classes of potential stereoisomers also determines a canonical identification for a stereoisomer. The true stereocenters in a structure can be assigned unique sequence numbers [on the basis of their numbering in the original (canonical) constitutional structure.] Different potential stereoisomers can then be represented by specific sequences of 1s and 0s. The individual 1/0 bits in these sequences define the configurations at the corresponding stereocenters in any arbitrarily specified potential stereoisomer. Such bit sequences can be interpreted as representing binary numbers and thus admit simple comparison and ordering. The operations of the CSG permute and invert bits in these sequences and may convert a given arbitrarily specified configurational stereoisomer into some symmetrically equivalent form characterized by a lower binary number. The canonical form of any potential stereoisomer can be taken as that member of the same equivalence class that is characterized by the least binary number.

Some examples, based on tetramethylcyclobutane, are shown in Figure X-27. Stereo-isomers with configurations [C(4)–0, C(3)–0, C(2)–1, and C(1)–0] or [C(4)–1, C(3)–0, C(2)–1, and C(1)–1] are, under the operations of the CSG, equivalent to the stereoisomer with configuration [C(4)–0, C(3)–0, C(2)–0, and C(1)–1]. The stereoisomer with the configurations 0001 represents the canonical form for 8 of the 16 potential stereoisomers of tetramethylcyclobutane.

This process for identifying canonical stereoisomers is applicable to substructures as well as complete structures.[28] There are many applications where it is useful to correlate some activity (e.g., ^{13}C chemical shift) with a local stereochemical environment. Data-bases of substructure-activity pairs require that substructures be in canonical form.

Exhaustive stereoisomer generators also depend on these mechanisms for identifying the canonical representations of equivalence classes of stereoisomers. The final generation process is quite simple. The potential stereoisomers of a structure with *n* true stereocenters can, as already noted, be put in correspondence with the integers in the range $0 \cdots (2^n-1)$. Each of these integer

Arbitrary Configuration Designation					
C(4)	C(3)	C(2)	C(1)		
0	0	1	0	c, permute 1&2, 3&4, retain configs →	0 0 0 1
	(Value = 2)				(Value = 1)
1	0	1	1	f, permute 1&3, 2&4, invert all →	0 0 0 1
	(Value = 11)				(Value = 1)

FIGURE X-27. Equivalence of different potential stereoisomers, under operations of CSG, provides the basis for a canonical naming scheme.

values can be considered in turn with the operations of the CSG applied to its corresponding bit pattern. If any CSG operation should yield a bit pattern having a lower integer value, it is established that that particular potential stereoisomer is noncanonical and thus may be discarded.

An example of the generation procedure is shown in Figure X-28 for the diol structure **X-20**. This has three true stereocenters; its potential stereoisomers can be represented by the integers 0 · · · 7. Apart from the identity element, the only operator of this structure's CSG is the permutation that exchanges the labels on atoms C(2) and C(3) and inverts the configuration at C(1).

Potential stereoisomers 0, 2, 3, and 6 are accepted as a true stereoisomers because the operations of the CSG on their representations yield larger valued forms. The remaining potential stereoisomers can be discarded. Since stereoisomers 2 and 3 are converted by the CSG operations into their enantiomers, they must be achiral.

An exhaustive generator of configurational isomers can thus be constructed from the various routines already outlined. First, the n true stereocenters in a given (canonical) constitutional isomer are identified and assigned some unique sequence numbers. (The original numbering of the atoms in the canonical constitutional structure can serve as the basis for this sequencing.) Second, the CSG must be derived. Finally, the sequence of integers $0 \ldots (2^n - 1)$ must be tested for canonicity under the permutation inversion operations of the CSG.

Stereochemical constraints can act as additional filters during the stage where the program generates and tests the sequence of integers that represent potential stereoisomers. For example, if a compound is known to be chiral, any stereoisomer that is converted by the CSG operations into its enantiomeric form can be discarded (e.g., stereoisomer 2 in Figure X-28). Other constraints might restrict the relative configurations of two or more of the stereocenters; the bit patterns representing the various potential stereoisomers can be checked against these restrictions.

Configurations			Bit pattern	Value	Permute-invert operation of CSG	Value	Action
C(3)	C(2)	C(1)					
0	0	0	000	0	001	1	Keep
0	0	1	001	1	000	0	Discard
0	1	0	010	2	101	5	Keep
0	1	1	011	3	100	4	Keep
1	0	0	100	4	011	3	Discard
1	0	1	101	5	010	2	Discard
1	1	0	110	6	111	7	Keep
1	1	1	111	7	110	6	Discard

FIGURE X-28. Generation of distinct stereoisomers for structure X-20.

Substructural constraints are of considerable importance, particularly in cases where the molecular composition has a substantial degree of unsaturation. Corresponding structures will then often have double bonds in rings or fused rings or bridged ring systems. An unconstrained stereoisomer generator will create many stereoisomers containing energetically disfavored substructural features. Thus stereoisomers with *trans*-double bonds in small rings will be created, as will stereoisomers with highly strained fused ring systems. A stereoisomer generator would, for example, find seven stereoisomers for twistane (**X-31**); only two of these, the *dl* pair with all *cis*-bridges, are reasonable.[26]

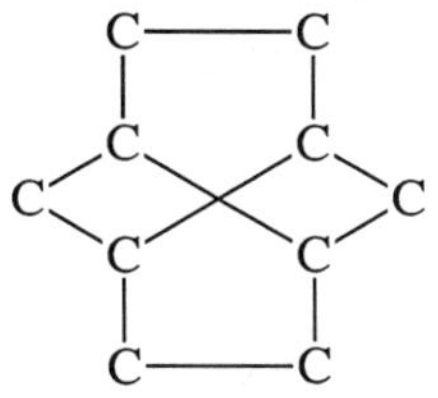

X-31

Constraints on the stereochemistry of various substructures reflect considerations of energetics. In some applications one might be interested in the higher energy species such as in–out bicyclic structures. It is better to define constraints as needed rather than attempt to build presumptions about energetics into a stereoisomer generator. Constraints that eliminate high-energy forms can typically be defined in terms of substructures with configurational designations on some atoms. Thus, if *trans*-double bonds are not to be allowed in small rings, one can define and then employ a substructure such as **X-32**. The **STEREO** program, that incorporates a constrained configurational stereoisomer generator, can employ such constraints during the generation process (potential stereoisomers that violate substructural constraints can be prospectively avoided) or in some subsequent stereoisomer pruning procedure.[26] The original paper describing the **STEREO** program contains several illustrative examples of the definition and use of such substructural constraints.

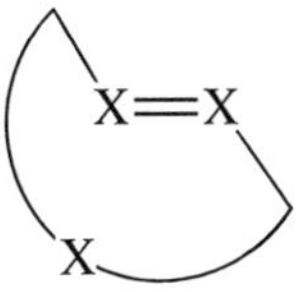

(Link node)

X-32

X.D. GENERATION OF CONFORMATIONAL FORMS

X.D.1. Exhaustive Generation–Canonicalization of Conformations

The final step in the development of model structural candidates for an unknown would involve the specification of their possible conformational forms. A specification of conformation is necessary for the counting and generation of possible conformers and their canonicalization and registration-in and retrieval-from data bases. The definition of conformation necessary for such applications is not an explicit geometric model; rather, it is a specification of the connectivity, supplemented by configuration tags at stereocenters and some description of the torsional state of rotatable bonds. (Of course, a geometric model of a defined conformer can then always be created by supplementing a canonical conformation description with data on standard bond lengths and angles.)

Unlike either constitution (which is specified in terms of bonds that are either present or absent) or configurations (which can be specified as a two-valued property of stereocenters), the description of conformation involves a continuous variable, namely, the torsional angle about a single bond. It is necessary to convert this into a discrete property of the bond; values for this property should be assignable independently to the various rotatable bonds of a structure.

As illustrated in Figure X-29, it is possible to define the rotameric state of any particular single bond between sp^3 atoms in terms of positions on a "grid." The six-position grid shown in Figure X-29 allows for the definition of eclipsed and staggered rotamers.

The rotameric state of bond in a given configurational isomer can be defined by conceptually "viewing" along the axis of the bond and specifying how substituent atoms are positioned. The positions can be described uniquely

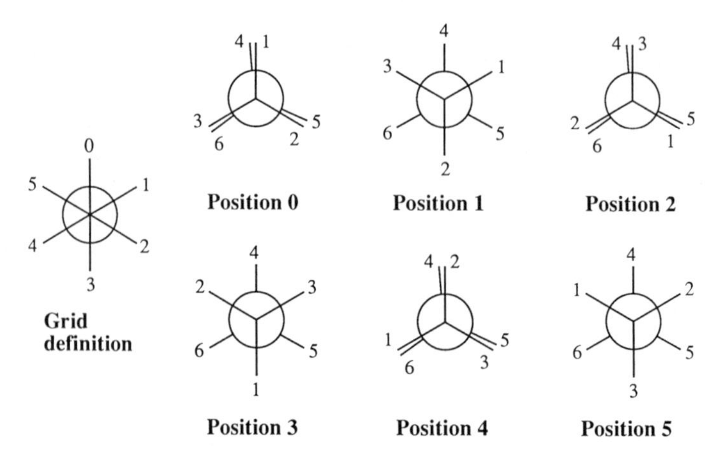

FIGURE X-29. A six-position grid for definition of eclipsed and staggered rotamers of an sp^3-sp^3 bond.

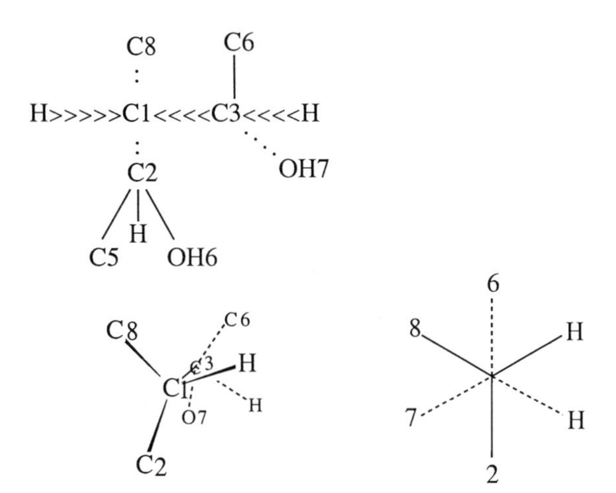

FIGURE X-30. Definition of rotameric state of one bond in an example structure.

in terms of the atom numbers by basing the conformation definition on the relationship between the two smallest numbered atoms connected to the atoms of the central bond. An example definition of the rotameric state of a bond is shown in Figure X-30.

The rotameric state of the C(1)–C(3) bond is defined by the relative positions of C(2) and C(6). (These are the lowest numbered substituents on C(1) and C(3), respectively). In the example illustrated, when projected onto the standard six-position grid, these substituents are in position 3 and consequently the rotameric state of the bond is 3.

Different types of grid are necessary for different types of bond (bonds between two sp^3 atoms, bonds connecting sp^2 and sp^3 atoms, etc.). It is useful if a grid can be rendered finer, so as to capture more accurately the exact shape of a molecule, and yet still allow comparison with conformations defined on a coarse grid. These various requirements can be met and rotameric states of bonds defined in terms of "labels" that are simply integer sequences defining positions on grids of specified precision.[29]

An arbitrary conformation of a structure can be specified by supplementing its (canonical) configurational stereochemical definition with labels defining the rotameric state of each rotatable bond. (If the process is intended to yield canonical conformers, it is necessary to start with a definition of the structure that is canonical with respect to both constitution and configuration.) If the configurational stereoisomer being processed has no symmetry, each possible labeling of the bonds represents a distinct conformer (although not necessarily one that is physically realizable). If, however, there are symmetries present in the configurational stereoisomer, some of the alternative arbitrary conformation designations may be equivalent.

The main part of any conformer generator–canonicalizer program consists of procedures for creating an appropriate representation of symmetries and using these data to determine equivalences among arbitrary conformation descriptions.[30] The analysis is similar to that used in developing configurational stereoisomers. The effects of any symmetries on the rotameric states of bonds can be determined from the atom permutations and can be represented in terms of a "Conformation Symmetry Group" (CFSG). The various elements of the CFSG define permutations of bonds and rotameric states.

Like the elements of the CSG, the elements of the CFSG can be used as operators that act on specified conformations and generate equivalent forms. It is again possible to identify equivalence classes of related conformations. The number of equivalence classes defines the number of distinct conformers of a structure. As in the case of configurations, it is possible to define an ordering for the various conformation designations that together comprise an equivalence class. The "least-valued" conformation designation in any equivalence class can be taken as the canonical representation for all its symmetrically related conformers.

Thus a *conformation generator* can take a form closely analogous to the configurational stereoisomer generator. First, a canonical ordering for the rotatable bonds must be defined,[29] and the CFSG of the structure determined.[30] Each possible allocation of rotameric labels to bonds must be created and tested by use of the operations of the CFSG; if any operation of the CFSG converts a given rotameric state into a lower-valued form, the given conformer has been established as being noncanonical and thus can be discarded.

Such a conformer generator has only limited application in structure elucidation. Most of the conformers generated would represent high-energy or otherwise unacceptable forms. Some might be physically unfeasible because of implicit overlaps of atoms or intersections of bonds. Most of the conformers generated for structures with ring systems would be physically unrealizable. The presence of rings in a structure implies complex restrictions on the rotameric states of the various bonds comprising each ring. If, as in the generation scheme outlined above, rotameric states are allocated independently to the various bonds, few of the conformations created will allow for ring closure at normal bond lengths. It is difficult to express such physical constraints in terms of restrictions on the relative rotameric states of sequences of bonds.

A more immediate practical application of these procedures is for the canonical naming of structures and substructures. A conformer canonicalization program would take some specification of the shape of the molecule and from this generate canonical names for the entire structure and possibly also for various atom-, or bond-centered substructures. One current system, an extension of the **DENDRAL** programs, takes a standard specification of a molecule (as a connection table with configurational stereochemical tags) together with atomic coordinates (as obtained from X-ray data or molecular modeling programs).[29] The atomic coordinates are converted into specifications of rotameric states of bonds. The CFSG of the structure is derived and used to determine the equivalence class of and canonical name for the given molecular structure.

Canonical names can also be generated for selected atom-, or bond-centered substructures of the given molecule. These canonical substructure names define the environment of some central atom or bond in terms of "shells" (usually corresponding to different bond radii).[29] At each shell level, it is possible to specify, separately, details of constitution configuration and rotameric state. The substructure names, or "codes," are simply strings of alphanumeric symbols with distinguishable sub-strings corresponding to different shell levels and different properties at each shell level. In this way, these conformer canonicalizers make it possible to encode the shape of substructures in a manner that allows these data to be easily stored in and retrieved from data bases. Such encodings should assist in studies on correlations between biochemical activity and molecular shape.

X.D.2. Empirical Approaches to Selective Conformation Generation

An alternative approach to the *generation of conformations* starts with the physical constraints. Most of these will simply define limitations on how closely the atoms of a structure may be placed in space. Thus bonded atoms should be separated by standard bond distances; the separations of 1–3 atoms, linked through two bonds, are determined by standard bond lengths and bond angles; nonbonded atoms should normally be no closer than 2 A. There exists a basic repertoire of such constraints on atom separations; sometimes, these constraints may be supplemented by additional more specific data characterizing a particular compound. The problem of conformation generation then reduces to the need to find X, Y, Z coordinates for the compound's atoms such that all these distance constraints are satisfied.

The *distance geometry* method due to Crippen and Havell[31–33] provides an approach to solving such problems. The first step in this method for generating atomic coordinates involves estimation of minimum and maximum limits on the distance between each pair of atoms. 1–2 and 1–3 interatomic distances can be fixed precisely from standard bond lengths and angles (as can 1–4 distances of pairs of substituents on double bonds). Maximum and minimum values (dependent on dihedral angles) can be specified for other 1–4 distances. (Atoms in three- and four-membered rings must be treated as special cases.) The minimum separation of nonbonded atoms can be set to some default value (or estimated from Van der Waals radii); maximum bounds on interatomic distances can be related to the number of atoms in the molecule. These initial upper and lower limits can then be somewhat refined.[33]

A distance matrix can then be constructed with all interatomic separations lying within the computed bounds. This distance matrix is subjected to several further transforms. First, the interatomic distances are converted to distances from the center of mass of the molecule. Then, a "metric matrix" is constructed and resolved into its eigenvalues and eigenvectors.[33] The eigenvectors associated with the three largest eigenvalues constitute trial X, Y, Z coordinates for the atoms.

These eigenvector coordinates may satisfy neither the original interatomic distance constraints nor the constraints on atom placements that are implicit in configuration designations at the various stereocenters. The problem now involves the need for iterative modification of the coordinates, as obtained from the eigenvector analysis, until all constraints are satisfied to the best possible degree.

It is possible to define a function that expresses the degree of mismatch between the requirements of the distance and configuration constraints and the form of a particular structure as defined by any given set of atomic coordinates. The structure's atomic coordinates must be varied so as to minimize this mismatch–error function. This is a conventional computational problem; there are several algorithms, available in standard mathematical packages, for minimizing a function of many variables. If the iterative minimization process converges with a (near-) zero function value, the final computed coordinates should have all atoms at satisfactory separations and in appropriate configurations.

This approach has been implemented in a program, **BUILD3D**, that can create three-dimensional representations of any structures specified by a chemist user or created through **STEREO** and **GENOA** (or **CONGEN**).[34] The **BUILD3D** program is based on the **EMBED** distance geometry program due to Crippen and Havell with various modifications to ensure the correct processing of atoms in defined configurations.

The **BUILD3D** program also incorporates standard programs, from the Quantum Chemical Program Exchange Library (Bloomington, IN), that allow three dimensional representations of the molecule to be presented on appropriate graphics devices. The chemist user can view generated conformations and save the coordinates of any chosen conformations. It is also possible to further refine a chosen set of coordinates. There exist standard procedures for empirically calculating strain energies of molecules in given conformations and then varying the atomic coordinates to minimize this energy. Such standard force-field molecular-energy minimization programs can be invoked from within **BUILD3D**.

The major difficulty with this approach to conformer generation is that it is serendipitous. There is currently no mechanism for the systematic exploration of all conformations that satisfy the distance constraints. The different conformations that are generated simply result from different initial arbitrary placements of atoms at random distances (subject to the minimum and maximum limits on separations).

By reviewing sequences of arbitrarily generated conformations and their nearest energy minimized forms, a chemist can identify plausible conformations for any given structure. It is possible to then use the derived X, Y, Z coordinates in programs that search for the largest three-dimensional substructure common to a group of related compounds.[35] As briefly discussed in

Chapter II, this approach has been demonstrated in relation to the identification of the common pharmacophore in warburganal and related drimane-based sesquiterpene heliocides.[36]

X.E. SUMMARY

There are a number of algorithms for exhaustive generation of constitutional structures. The increased use of very specific spectral data, such as ^{13}C-^{13}C couplings, may well permit the wider use of the simplest structure generators, those that simply fill in a connection matrix using specific bonding constraints directly derived from the data. More typically, structure elucidation programs will have to combine diverse structural fragments inferred from many different types of spectral, chemical, or biochemical data. The various inferred substructures and presumed skeletal fragments will normally overlap to varying and unknown degrees. As well as requirements for the presence of particular substructures, there will usually be several other detailed constraints on prohibited structural forms. For such structure elucidation problems, the process of structure generation involves investigation of all acceptable ways of overlapping and fitting together the known substructures. Such problems would appear to be most readily solved by the constructive graph matching and simple generation algorithms of **GENOA**.

If available structural data are less comprehensive, however, there will be many different valid combinations of the fragments, that is, many cases for **GENOA**. Each of these individual cases will be less well specified, comprising several disjoint fragments and with frequent occurrences of identical fragments. The final generation algorithms of **GENOA** then prove computationally expensive as, inevitably, many duplicate structures are assembled only to be eliminated by the final canonicalization procedures.

When data are limited, the generation of large numbers of structural candidates may be better performed by some alternative approach. These alternative generators achieve greater computational efficiency through detailed analyses of each bond formation step; these analyses can minimize, sometimes even eliminate, structure duplication.

Methods exploiting the analysis of symmetries in partially assembled structures have tended to be based on heuristic rather than mathematically proven techniques. If the analysis is heuristic, there is always some small possibility that exhaustive structure generation will not be achieved (because of either nonexistent "symmetries" being perceived or "prohibited-ring avoidance" or other heuristics prematurely curtailing explorations).

The most effective and reliable methods for structural problems with large numbers of (possibly identical) fragments appear to be those based on the development of a canonical connection matrix or canonical "connectivity stack." (The types of structural problem most efficiently solved by a vertex graph analysis should not arise in routine chemical structure elucidation.)

However, canonical matrix generators (and vertex graph analyzers) do necessitate a more complex multistep approach to structure generation. In these methods, the generation step is limited to the combination of atoms and superatoms; known constituent substructures must be manipulated as if they were atomic entities. Final structures can be obtained only through further analysis in which these superatom substructures are embedded.

Algorithms for the definition of configurational stereochemistry, as presented in the chemical literature, provide a simple and direct approach to constrained generation of stereoisomers. Generation of stereoisomers really requires only the computation of the configurational symmetry group of the structure, as expressed in terms of the node permutation symmetry group of the possible stereocenters and the edge permutation group for the double bonds. These permutation groups can be found by the same procedures as previously discussed in relation to the problem of canonicalizing structures. The algorithms for stereochemical analysis could be adapted for use in any structure elucidation system, thus widening the availability of stereochemical analysis.

A possible long-term development might be the integration of some stereochemical analysis with the initial candidate structure generation procedures. Hydrogen-1 NMR couplings, and to some extent ^{1}H and ^{13}C NMR shifts, can be interpreted to yield stereochemical as well as topological information. In current structure elucidation systems, these two aspects of the structural inferences are used separately, the topological information constraining the structure generator and the stereochemical data constraining the stereoisomer analysis.

Methods for the constrained generation of conformational forms can be expected to form the focus of considerable further research efforts. The major problem here is the specification of constraints for and effective implementation of a constraint mechanism in a conformer generator.

REFERENCES

1. R. E. Carhart, D. H. Smith, H. Brown, and C. Djerassi, "Applications of Artificial Intelligence for Chemical Inference. XVII. An Approach to Computer-Assisted Elucidation of Molecular Structure," *J. Am. Chem. Soc.*, **97** (1975), 5755.
2. D. H. Smith and R. E. Carhart, "Structural Isomerism of Mono- and Sesquiterpenoid Skeletons," *Tetrahedron*, **32** (1976), 2513.
3. Y. Kudo and S. Sasaki, "The Connectivity Stack, a New Format for Representation of Chemical Structures," *J. Chem. Doc.*, **14** (1974), 200.
4. Y. Kudo and S. Sasaki, "Principle for Exhaustive Enumeration of Unique Structures Consistent with Structural Information," *J. Chem. Inf. Comput. Sci.*, **16** (1976), 43.
5. S. Sasaki, I. Fujiwara, H. Abe, and T. Yamaski, "A Computer Program System --- New Chemics --- for Structure Elucidation of Organic Compounds by Spectral and other Structural Information," *Anal. Chim. Acta*, **122** (1980), 87.

6. T. Oshima, Y. Ishida, K. Saito, and S. Sasaki, "Chemics-UBE, a Modified System of Chemics," *Anal. Chim. Acta*, **122** (1980), 95.
7. R. E. Carhart, D. H. Smith, N. A. B. Gray, J. G. Nourse, and C. Djerassi, "GENOA: A Computer Program for Structure Elucidation Utilizing Overlapping and Alternative Substructures," *J. Org. Chem.*, **46** (1981), 1708.
8. C. A. Shelley, T. R. Hays, M. E. Munk, and R. V. Roman, "An Approach to Automated Partial Structure Expansion," *Anal. Chim. Acta*, **103** (1978), 121.
9. C. A. Shelley, M. E. Munk, and R. V. Roman, "A Unique Computer Representation for Molecular Structures," *Anal. Chim. Acta*, **103** (1978), 245.
10. C. A. Shelley and M. E. Munk, "CASE, a Computer Model of the Structure Elucidation Process," *Anal. Chim. Actal,* **133** (1981), 507.
11. R. E. Carhart, *A Simple Approach to the Computer Generation of Chemical Structures*, Memorandum MIP-R-118, Machine Intelligence Research Unit, University of Edinburgh, 1977.
12. L. M. Masinter, N. S. Sridharan, J. Lederberg, and D. H. Smith, "Applications of Artificial Intelligence for Chemical Inference. XII. Exhaustive Generation of Cyclic and Acyclic Isomers," *J. Am. Chem. Soc.*, **96** (1974), 7702.
13. L. M. Masinter, N. S. Sridharan, R. E. Carhart, and D. H. Smith, "Applications of Artificial Intelligence for Chemical Inference. XIII. Labeling of Objects Having Symmetry," *J. Am. Chem. Soc.*, **96** (1974), 7714.
14. D. H. Smith, "The Scope of Structural Isomerism," *J. Chem. Inf. Comput. Sci.*, **15** (1975), 203.
15. C. A. Shelley and M. E. Munk, "Computer Perception of Topological Symmetry," *J. Chem. Inf. Comput. Sci.*, **17** (1977), 110.
16. R. E. Carhart, "Erroneous Claims Concerning the Perception of Topological Symmetry," *J. Chem. Inf. Comput. Sci.*, **18** (1978), 108.
17. H. L. Morgan, "Generation of Unique Machine Description for Chemical Structures, a Technique Developed at Chemical Abstracts Service," *J. Chem. Doc.*, **5** (1965), 107.
18. J. Lederberg, *DENDRAL-64: A System for Computer Construction, Enumeration and Notation of Organic Molecules as Tree Structures and Cyclic Graphs,* Technical Report CR57029, NASA, 1964.
19. J. Lederberg, *DENDRAL. A System for Computer Construction, Enumeration and Notation of Organic Molecules as Tree Structures and Cyclic Graphs, Part II, Topology of Cyclic Graphs*, Technical Report CR68898, NASA, 1965.
20. J. Lederberg, "Topological Mapping of Organic Molecules," *Proc. Natl. Acad. Sci.* (USA), **53** (1965), 134.
21. N. S. Sridharan, *Computer Generation of Vertex Graphs*, Technical Report STAN-CS-73-381, Computer Science Department, Stanford University, 1973.
22. A. T. Balaban, "Chemical Graphs XVIII. Graphs of Degrees Four or Less, Isomers of Anulenes, and Nomenclature of Bridged Polycyclic Structures," *Rev. Roumaine Chim.*, **18** (1973), 635.
23. R. E. Carhart, D. H. Smith, H. Brown, and N. S. Sridharan, "Applications of Artificial Intelligence for Chemical Inference. XVI. Computer Generation of Vertex-Graphs and Ring Systems," *J. Chem. Inf. Comput. Sci.*, **15** (1975), 124.
24. J. G. Nourse, "The Configuration Symmetry Group and Its Application to Stereoisomer Generation, Specification and Enumeration," *J. Am. Chem. Soc.*, **101** (1979), 1210.
25. J. G. Nourse, R. E. Carhart, D. H. Smith, and C. Djerassi, "Exhaustive Generation of Stereoisomers for Structure Elucidation," *J. Am. Chem. Soc*, **101** (1979), 1216.
26. J. G. Nourse, D. H. Smith, R. E. Carhart, and C. Djerassi, "Computer-Assisted Elucidation of Molecular Structure with Stereochemistry," *J. Am. Chem. Soc.*, **102** (1980), 6289.

27. W. T. Wipke and T. M. Dyott, "Stereochemically Unique Naming Algorithm," *J. Am. Chem. Soc.*, **96** (1974), 4834.

28. N. A.B. Gray, J. G. Nourse, C. W. Crandell, D. H. Smith, and C. Djerassi, "Stereochemical Substructure Codes for 13C Spectral Analysis," *Org. Magn. Reson.*, **15** (1981), 375.

29. A. L. Fella, J. G. Nourse, and D. H. Smith, "Conformation Specification of Chemical Structures in Computer Programs," *J. Chem. Inf. Comput. Sci.*, **23** (1983), 43.

30. J. G. Nourse, "Specification and Unconstrained Enumeration of Conformations of Chemical Structures for Computer-assisted Structure Elucidation," *J. Chem. Inf. Comput. Sci.*, **21** (1981), 168.

31. G. M. Crippen and T. F. Havel, "Stable Calculation of Coordinates from Distance Information," *Acta. Crystallogr., Sect. A*, **A34** (1978), 282.

32. G. M. Crippen, *Distance Geometry and Conformation Calculations*, Research Studies Press, 1981.

33. G. M. Crippen, "A Novel Approach to Calculation of Conformation: Distance Geometry," *J. Comput. Phys.*, **24** (1977), 96.

34. J. C. Wenger and D. H. Smith, "Deriving Three-dimensional Representations of Molecular Structure from Connection Tables Augmented with Configuration Designations Using Distance Geometry," *J. Chem. Inf. Comput. Sci.*, **22** (1982), 29.

35. D. H. Smith, J. G. Nourse, and C. W. Crandell, "Computer Techniques for Representation of Three-Dimensional Substructures and Exploration of Potential Pharmacophores," in *Structure Activity Correlation as a Predictive Tool in Toxicology*, L. Goldberg, ed., Hemisphere Publishing Corp., New York, 1982, 171, Chapter 11.

36. C. Djerassi, D. H. Smith, C. W. Crandell, N. A.B. Gray, J. G. Nourse, and M. R. Lindley, "The DENDRAL project: Computational Aids to Natural Products Structure Elucidation," *Pure Appl. Chem.*, **54** (1982), 2425.

XI

EVALUATION OF STRUCTURAL CANDIDATES

XI.A. INTRODUCTION

In most applications of structure generation programs, the initial constraints are not sufficient to uniquely define the form of the unknown molecule. Instead, several candidate structures will be produced. Usually, there will be 20–40 generated candidates but occasional applications do arise where it is useful to proceed to generate structures even though the available constraints are relatively weak. Consequently, it is sometimes necessary to contend with several hundred candidates. The chemist must then perform such additional experiments as may be necessary to eliminate inappropriate candidate structures and thus reveal the correct molecular structure.

Chemical transformations can provide additional data on the structure of an unknown molecule. New compounds are frequently degraded and the simpler products obtained, through degradation, are identified. The nature of such degradation products provides structural information that can be transformed back into the context of the original structural problem. As discussed in Chapter XII, the results of actual experiments can be used to eliminate some candidate structures.

Obviously, it is advantageous to minimize the number of experiments that must be performed. Programs for manipulating sets of candidate structures provide a mechanism whereby a chemist may explore the range of structural types apparently consistent with existing data and may, thereby, be able to identify the most effective additional experiments. The expected results from such experiments may then also be simulated. The computer thus serves as a kind of "dry laboratory" for gedankenexperiments.

The most helpful information for planning additional experiments is some categorization of the various candidates according to their common and distinct structural features and the identification of those candidates most consistent with all of the available data.

The categorization of candidate structures is an extension of the various substructure matching procedures considered earlier. The first step in the evaluation of candidates is, typically, the performance of some form of *survey* for standard molecular skeletons or particular combinations of substructural features. Such an analysis requires only the matching of predefined substructures (representing standard functional groups and appropriate molecular skeletons) against a set of candidate structures as either hypothesized by the chemist or derived through the use of a structure generation program.

Rather more sophisticated categorizations, independent of any prespecified substructures or skeletons, can be obtained through an analysis of the candidate structures themselves. For example, the various candidates can be reduced to their cyclic skeletons (in effect, reversing the assembly procedure of a vertex graph structure generator); the various fundamental cyclic skeletons thus identified can then be used to group the structures. These cyclic skeletons can be presented to the chemist along with statistics detailing their frequency of occurrence.

Another useful technique for grouping candidates is to determine the different ways in which a particular constituent substructure has been embedded (i.e., incorporated) in the various generated candidates. For example, an initial interpretation of the available spectral data might have revealed the presence of a >CH-OH in the structure; this fragment would then have been defined as a constraint and, consequently, would be present in all generated structures. The larger environment of this methine carbon would differ among the candidates. Such a methine should be fairly readily recognized in either the ^{1}H NMR or ^{13}C NMR spectra and could be used as the starting point for a series of decoupling experiments. Before such experiments are attempted, however, it would useful to identify the ways in which the >CH-OH had been embedded. The first structural differences might prove to be some three or four bonds remote from the methine carbon. In such circumstances this methine would prove a poor choice of starting point for a decoupling study.

Usually, structure generators employ only topological (constitutional) structural descriptions. Stereochemistry should certainly be considered in the evaluation stage if it had not been analyzed earlier. One reason for proceeding to a description of stereochemistry is, of course, that a structure is not completely determined until its configurational stereochemistry has been defined. A further reason is that stereochemical constraints can sometimes eliminate particular topological candidates. Every stereoisomer of some constitutional structure may incorporate some unnaturally strained arrangement of atoms such as a *trans* double bond in a small ring or *trans* fused small rings.

Such stereochemical features can be analyzed algorithmically and used to eliminate unsatisfactory constitutional isomers. A related, but much more empirical, analysis is based on computation of "strain" energies for various candidates; those rated as highly strained may be discardable.

The final aspect of evaluation, and the main topic for this chapter, is the assessment of the consistency of the various candidates with the spectral data characterizing the unknown. Some use will have been made of spectral information in deriving substructural constraints prior to candidate generation; much more detailed spectral analysis is possible in the evaluation process. The interpretation of spectral data relies on subspectral-substructural correlations; at best, such correlations capture only relatively local effects and, even then, are crude. In the evaluation stage, where candidate structures (possibly including stereochemical information) are available, it is practical to both consider those spectral properties dependent on the complete molecular structure and reevaluate other spectral characteristics now taking into account longer range and stereochemical influences.

Evaluation of candidates through spectral analysis involves (1) the prediction of the appropriate spectral property for each candidate, (2) the calculation of some score expressing the consistency of predicted and observed properties, and (3) the ranking of the candidates according to the scores obtained. The objective is to try to partition the candidates into two groups. One group consists of those structures for which spectral predictions are not incompatible with the observed data; the second group comprises those candidates that can not adequately rationalize the experimental observations.

Most spectral prediction methods are fairly empirical. It is not normally appropriate to exclude candidate structures from further consideration merely on the basis of such spectral predictions. It is frequently useful to combine the results of various different spectral predictions and, by intersecting the various derived groups of "reasonable" structures, find which structures appear to be most plausible overall. Such identifications of most plausible candidates are particularly helpful in problems yielding large numbers of possible solutions; the identification of the most likely structure from among hundreds can frequently suggest analogous known compounds and can help to focus the search for additional discriminating experiments.

XI.B. EXAMPLES OF EVALUATION–TEST COMPONENTS IN STRUCTURE ELUCIDATION SYSTEMS

Most computer-based systems for assisting in the structure elucidation process include some *evaluation*, or *test* component. Candidate structures created through the use of structure generation programs, or possibly explicitly defined by the chemist, are stored in a file on the computer system. This file of candidate structures constitutes the primary input to the evaluation component of the overall system.

Evaluation programs should preferably be interactive with the user able to invoke such different processing as may be appropriate in a specific application, able to view partial results and to partition the overall set of candidate structures into various distinct subgroups. Some systems are rather more batch oriented, with standard processing steps performed for all candidates.

The following discussion is based mainly on the evaluation components in Munk's **CASE** system,[1,2] and in the **DENDRAL** system.[3] In both the **CASE** and **DENDRAL** systems, details of candidate structures have to be kept in files. Many of the operations involve selection of subsets of these structures according to specified properties; a few operations can require comparisons of the various structures. Such manipulations of a set of candidates would be simpler if all information were stored in the main memory of the computer. Usually, however, such storage usage is impractical and data must be read from file as they are required. Both the **CASE** and **DENDRAL** candidate evaluation programs are interactive with the various processing options invoked at user request.

In **CASE**, evaluation is performed through the use of the ***EDITOR*** module.[2] As shown schematically in Figure XI-1, ***EDITOR*** allows for the partitioning of the set of candidate structures among a number of subfiles on the basis of various properties selected by the user. Partitionings can be based on the result

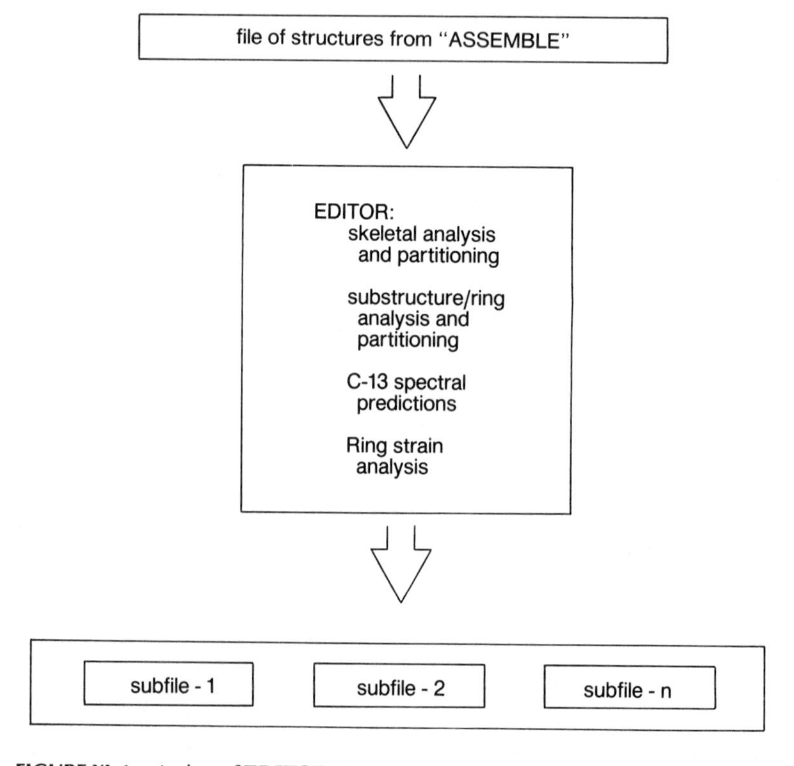

FIGURE XI-1. A view of **EDITOR**, the structure evaluation component from **CASE**.

of simple tests for the presence of specific user-defined or library substructures, multiple-bonds, and ring systems. The ***EDITOR*** module allows expression of some of these substructural constraints in terms of spectral properties; thus constraints can be defined in terms of the number of vinylic, cyclopropyl, and heterobonded hydrogen atoms. A more elaborate spectral based partitioning uses a prediction of the number of resonances in the ^{13}C spectrum (as derived through an analysis of topological symmetry). Further partitionings of the set of candidates can be made according to a categorization by carbocyclic skeletons or through assessment of ring strain. The ***EDITOR*** module allows logical operations to be performed on the various subfiles thus created. Thus by appropriate intersection operations it is possible to identify structures with different combinations of features.

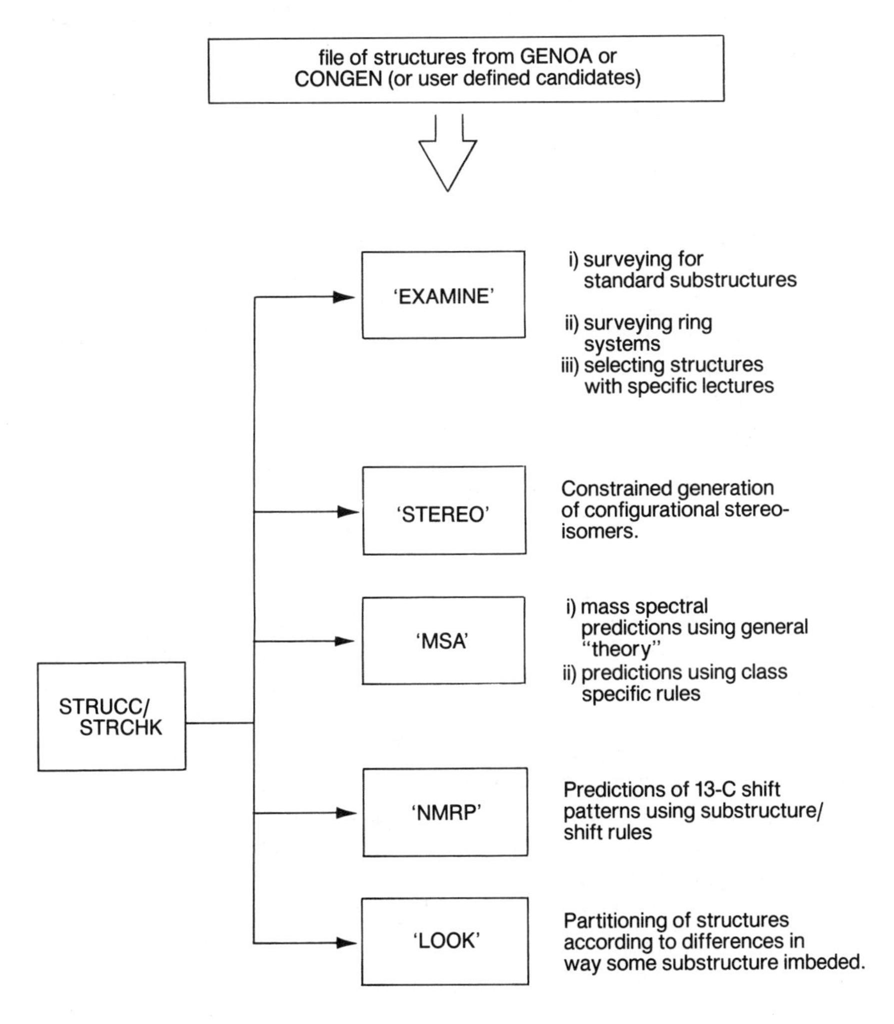

FIGURE XI-2. Structure evaluation systems for **DENDRAL**.

The evaluation component within **DENDRAL** has evolved as different structure generators have been developed. The earliest systems, coupled with the vertex graph generators, allowed for simple surveying of structures and a restricted form of mass spectral prediction and also served as an interface to the REACT programs discussed Chapter XII.[4–6] Two more developed versions of these evaluation components were the programs **STRUCC**[7] and its later version **STRCHK** (both mnemonics expressing the STRUCture CHecKing role of the programs).[3] These are *executive* programs through which specific processing options can be invoked. The data shown in Figure XI-2 illustrate some of the options available through these executive programs. All the various programs for structure evaluation work on the same single structure file; files of selected subsets of structures can be created as required.

Programs available through the **STRUCC-STRCHK** executives include ***EXAMINE*** and ***LOOK*** functions for surveying structures,[7] the **STEREO** system for stereochemical analyses,[8] and spectral prediction subsystems such as ***MSA***[9,10] and ***NMRA***[11]. The surveying options, illustrated in the example analysis of warburganal, are implemented through a restricted version of the ***EXAMINE*** program; more elaborate versions of ***EXAMINE*** allow for surveys according to ring systems in addition to checks for defined substructures. The ***EXAMINE*** program involves three distinct processing phases. The structural features (ring systems and substructures) of interest must first be specified. Then all candidates in the structure file are processed to determine which features they incorporate. A report summarizing the frequency of occurrence of the various features is produced. Finally, the user is enabled, through an interactive control program, to inspect structures possessing specific combinations of features. Typically, this interactive part of the program is used to identify the frequencies of occurrence of particular combinations of structural features and obtain identifications and, optionally, displays of example structures with specified properties. It is also possible to eliminate candidates with unsatisfactory combinations of features from the main structure file.

Using methods described previously in Chapter X, the **STEREO** program allows for the counting and generation of configurational stereoisomers of each candidate constitutional isomer. This analysis of stereochemistry can exploit both general and specific stereochemical constraints. General constraints can be used to prohibit the creation of stereoisomers containing highly strained systems, such as *trans* fused small rings; specific constraints can express known stereochemical relationships between various groups within the molecule (as may be inferred from proton couplings and similar data). Once created, stereochemical information can be saved as a supplement to the main structure file, where these data are available for subsequent use in, for example, programs predicting spectral properties subject to significant steric influences. Changes to the main structure file can also be made from **STEREO** if, for example, stereochemical constraints can exclude particular constitutional isomers.

The ***LOOK*** program provides a mechanism for partitioning structures according to differences in the way in which a substructure was embedded. Spectral analysis is concerned mainly with the prediction of ^{13}C shifts using ***NMRA*** and mass spectral fragmentations using ***MSA***. Results from these various methods of analyzing and ranking structures are, primarily, tabulations of the relative plausibilities of the various candidates and, if required, details of predicted spectral properties. Results from spectral rankings can be used to delete unsatisfactory candidates from the structure file. The necessary file manipulations can be carried out directly from the analysis programs or in combination with processing options in ***EXAMINE***.

Both **STRUCC** and **STRCHK** also serve as interfaces to many of the other more specialized, or more experimental, modules in the **DENDRAL** system. These other modules have included **BUILD3D** (the system for manipulating and displaying conformational representations),[12] ***HNMRP*** (a function for "predicting" proton resonance spectra by methods similar to those used in the ^{13}C analysis),[13] ***PRUNE*** (a more restricted version of ***EXAMINE*** allowing additional structural constraints to be specified and used to eliminate inappropriate candidates),[5,14] **REACT** (the chemical reaction simulator),[4,15] and ***RESONANCECHECK*** [a function identifying simple differences among candidates in respect to gross features of ^{1}H NMR and ^{13}C NMR spectra (e.g., different numbers of exchangeable protons, differences in the expected multiplicities of signals in ^{13}C spectra)].[7] The **STRCHK** system is also the executive through which the functions for assigning ^{13}C spectra and manipulating data bases may be invoked.[16]

XI.C. STRUCTURE SURVEYING AND PARITITIONING FUNCTIONS

XI.C.1. Surveying for Standard Structural Fragments

Commonly, the initial constraints available in a structure elucidation problem may leave the bonding of some heteroatoms incompletely specified and may have some residual degrees of unsaturation (corresponding to undefined ring systems). In such circumstances it is quite possible for extra functional groups to be created, including relatively unusual groups such as peracids, cumulenes, and enols. Such groups are generally contraindicated by available spectral and chemical data. However, it is difficult, and possibly unnecessary, for the chemist to have to determine all prohibited groups and specify prohibiting constraints. The first evaluation step can simply be the checking of generated candidates against a library of standard functional groups. Structures containing unacceptable functionality, or unwanted combinations of functional groups, may thus be identified and discarded ("pruned").

The other common application of surveying techniques is the determination of which, if any, candidates incorporate a known skeleton. The most common structure elucidation problem is the identification of "natural products" from some plant, or other organism. Related organisms give rise to similar structures often either incorporating a standard skeleton or derived through varying transformations, such as ring openings, of a standard skeleton. Libraries of, for example, standard sesquiterpenes can be constructed and then used to aid in the analysis of subsequently isolated sesquiterpenes.

In the example of the elucidation of the structure of warburganal, in Chapter II, the constraints were such as to fully define the molecular functionality. Heteroatoms were all incorporated in specific groups, and constraints from ^{13}C data limited the number of Csp^2 carbons so that the number of multiple bonds was also defined. Consequently, there was no need to survey for standard functional groups. The example illustrates the use of a library of standard skeleta. Since the unknown was suspected to be either a standard or modified sesquiterpene a predefined library of relevant skeletons was available. Such a library can be supplemented or superseded by a set of user-defined skeletons; this might be necessary for elucidation of the structure of some more novel molecule. The example also illustrates simple use of the interactive part of the program for selecting and displaying structures of particular interest.

It is also useful to survey generated candidates for various characteristic ring systems. The complete ***EXAMINE*** program could be used to detect spiro atoms, user-specified ring fusions, and bridged and "isthmus" structures.[7] The constraints available for warburganal required that each tetravalent carbon be bonded to at least one terminal group (from the set $-CH_3$, $-CH{=}O$ and $-OH$); consequently, no ring systems incorporating spiro atoms could be generated. A fair number of the 42 candidates are based on isthmus graphs; two examples are **XI-1** and **XI-2**.

Isthmus graphs involve disjoint ring systems connected by a bond or a chain of atoms. There are naturally occurring compounds that incorporate isthmuses; an example is fructilone (**XI-3**), which also possesses a spiro junction.

However, most isthmus structures, as generated for an unknown, represent biochemically implausible forms. Although isthmus structures cannot be rejected, it is usually reasonable to concentrate the search for the correct structure among the other generated candidates.

Information on ring fusions and bridged structures provides a useful general perspective of the range of structural types that must be considered. Among the

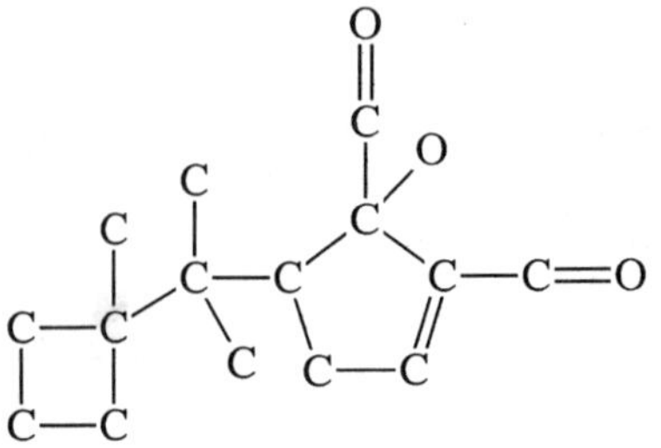

XI-1 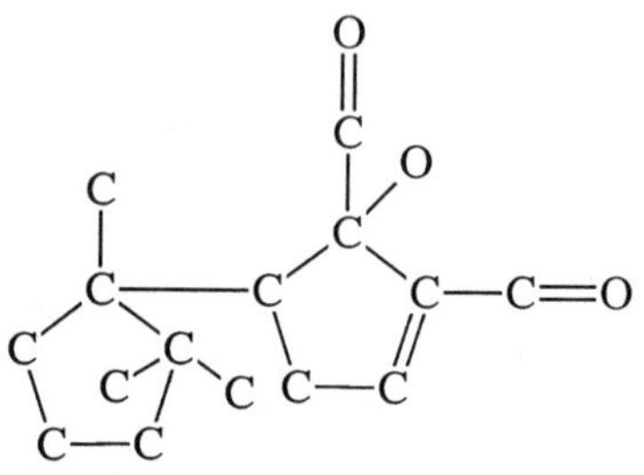 **XI-2**

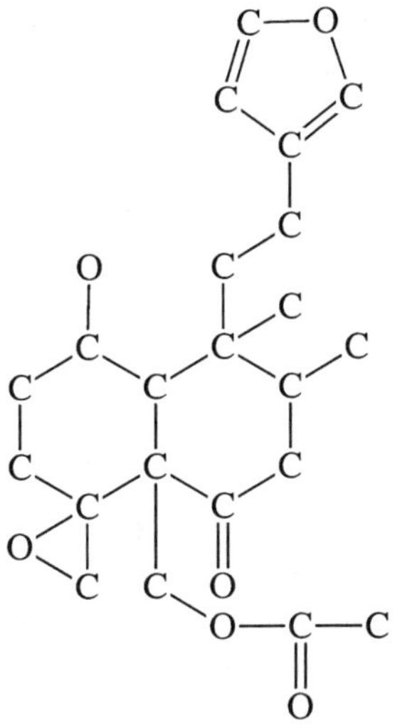

XI-3

candidates for warburganal, there are instances of 4–8, 5–6, 5–7 and 6–6 ring fusions. There are also a couple of bridged systems, an example is **XI-4**.

Apart from specifications of substructures and ring systems, there are other useful criteria for selecting subsets of structures. Scores measuring compatibility with mass spectra or ^{13}C data could conceivably be used in combination with requirements for specific substructures. The *EXAMINE* program allowed also for scores associated with individual substructures. Often, there are structural features whose presence or absence cannot be stated with certainty but to which some plausibility score may be assigned. Positive scores can be accorded to features for which there is reasonable, although not definitive, evidence; negative scores can be correlated with less likely substructures. In *EXAMINE* the overall score accrued by a candidate structure was simply the sum of scores associated with all its constituent substructures. Structures containing combinations of the more plausible substructural features obtain

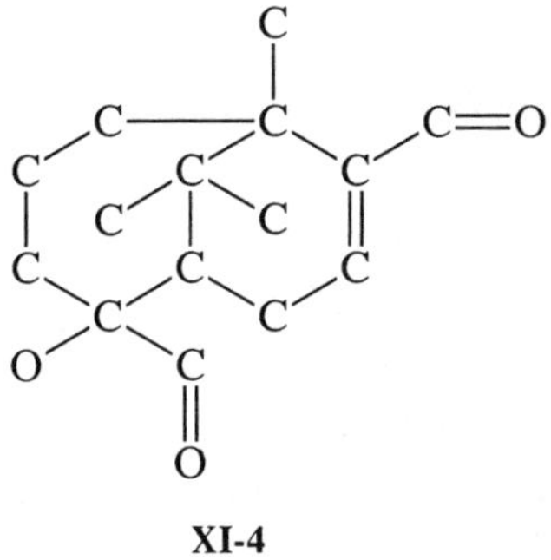

XI-4

the best scores. A score range could constitute one of the terms in the selection request specifying required structures. Through the use of selections based on scores it is possible to isolate and display subsets of structures containing either predominately favored, or disfavored, substructural features. Such scores should be used only as an aid to scanning through large sets of structures to find examples, seemingly of particular interest, for display to the user. It is not appropriate to use such scores as the basis for a scheme for eliminating candidates.

XI.C.1.a. Implementation of Survey Functions

The three phases of a system for surveying candidate structures can certainly be considered separately and, possibly, may be encoded as separate programs communicating through files. The first phase comprises merely an interface through which the user identifies those features that are to be searched for in each candidate. These desired features may represent substructures taken from a standard library or defined by use of the standard structure editing facilities of the structure-elucidation system.

The main processing function reads in each candidate from the structure file and attempts to match each defined substructure onto that candidate. Usually, a simplified graph-matching routine for detecting the presence of a substructure will suffice. Then the survey program can dispense with the more elaborate code of a graph matcher such as that used for maintaining details of each distinct match and for counting the number of matches. Occasionally, it may be useful to distinguish candidate structures according to the number of instances of some substructure that they incorporate. Generally, such needs can be met by defining a series of substructures: one containing a single instance of the feature desired, another comprising two disjoint and unconnected instances of that feature, and so forth. If merit values or scores are associated with substructures, the successful matching causes an overall candidate score to be updated.

If selection criteria are also to be expressed in terms of ring systems, the structure testing routine must perform a full ring analysis. All rings in a candidate must be found and their constituent atoms identified. The lists of ring systems can then be analyzed to determine the presence or absence of requested features. Isthmus graphs can be detected by checking whether the ring atoms form two or more distinct groups with no ring containing atoms from more than one such group. Spiro junctions are evidenced by a pair of rings that possess one common atom and are otherwise completely disjoint (with no third ring system containing atoms taken from both rings). Specified ring fusions can be identified by pairwise comparison of rings of appropriate size; if only two atoms are common to the rings, and if these atoms are directly bonded, an instance of the specified fused ring system exists. Although simple, this analysis sometimes gives results that may not quite correspond to the chemist's intent. Thus the structure:

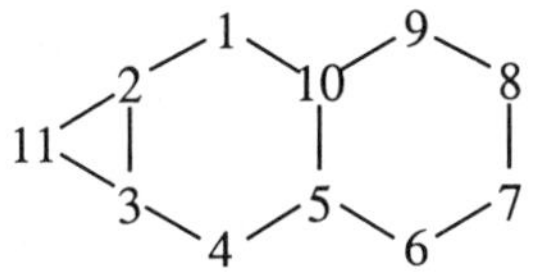

XI-5

would be, quite correctly, characterized as an example of a 3–10 ring fusion. There are a variety of simple definitions for "bridged" structures; as with the ring fusions, most such definitions will occasionally identify bridges, in polycyclic fused structures, that a chemist might not regard as quite fitting to the intended request.

If the survey system does not require the counting of the number of instances of each substructure, the results for each candidate take the form of a set of binary "yes"/"no" flags, each indicating the presence, or absence, of a specific substructure or ring system of interest. Such data are conveniently encoded in a *bit-map* record. A data structure with a number of bits equal to the number of features being surveyed (or, probably, equal to the next-higher power of 2) can be allocated and used to encode the results for each candidate. This record may be stored internally in the program or on some temporary file. Subsequent selection requests for identifications of structures with particular features can then be processed solely in terms of *boolean* test operations on these bit-map records.

Most selection requests are simple requiring two or three boolean tests (e.g., ***DESIRED FEATURES>ALLENE OR KETENE***). It is, however, advantageous to allow for arbitrarily complex boolean expressions constructed from the basic operators AND, OR and NOT combining parenthesized subexpressions of terms and operators. There are standard algorithms for "parsing" such expressions and deriving interpreted "code" for their evaluation. The **DENDRAL** survey program uses algorithms taken from the text by Gries.[17] The request entered by the user must be "parsed" and interpretable code generated. The bit-map records, associated with each candidate, may then be processed by this code, and those satisfying the request may thus be identified. (Selection requests based on scores, either those accrued by matching substructures or those assigned through spectral analysis, require special processing.) In the **DENDRAL** system, it is possible to review the structures selected by the current criterion, obtain reports on the frequency of other features by which these structures might be discriminated, then specify additional selection criteria. The refinement of search requests allowed by this interactive approach permits structures of particular interest to be easily identified and displayed.

XI.C.2. Partitioning Candidates by Structural Similarities

Surveying for standard structural forms can be very useful, but it is really limited to an expression of those differences that the chemist expects to find. Of course, the chemist can add to the standard libraries of functional groups and skeletons used for surveying the candidates and thus partially tailor the system to specific applications. However, the identification of such additional discriminating features requires that the chemist first inspect large numbers of the candidates. Automatic methods for finding discriminating substructures and using these to partition the set of candidates are necessary when large numbers of candidate structures must be analyzed.

The basis for such methods is an algorithm for abstracting, from a given structure, one or more substructures. The selected substructures are converted to some canonical form. These canonical substructures then serve to characterize the given structure. The substructure selection algorithm is applied to each candidate in turn; candidate structures are grouped according to the various (canonicalized) substructural features that they exhibit. Details of the discriminating substructural features and the structures that these characterize can be then be presented to the chemist.

Methods of this type have been most developed in the ***EDITOR*** of Munk's **CASE** system.[2] The algorithms, used within ***EDITOR*** to identify characteristic substructural features, involve manipulations of the skeletal structures of candidates. The ***EDITOR*** module defines three different algorithms, ***HETERO***, ***BRANCHES***, and ***ACYCLIC***. The ***HETERO*** option eliminates any heteroatoms, thus possibly fragmenting the molecule, and uses the largest remaining group of connected carbon atoms to characterize the molecule. The ***BRANCHES*** and ***ACYCLIC*** options offer different approaches to pruning away terminal and acyclic atoms to leave the cyclic skeleton. These options, in effect, recreate something resembling the "cyclic skeleton" of the vertex graph analysis. The vertex graphs underlying a structure could themselves be easily recreated and could constitute a simple, although possibly rather crude, basis for categorizing candidates.

The published descriptions of the various algorithms used in ***EDITOR*** lack detail, but some impressions of their nature can be realized by the following partitioning of the candidates generated for warburganal. This partitioning, shown in Figure XI-3, was derived by identifying the unique skeleta remaining after all terminal atoms had been deleted (incidentally eliminating also all heteroatoms) and bond orders had all been reduced to one (i.e., something close to ***EDITOR***'s "***BRANCHES***" option).

Each one of these 10 different cyclic skeleta was present in at least 2 of the 42 candidates for warburganal; the most common is the 7–5 fused ring system. Such a partitioning could be derived through conventional surveys for isthmus graphs, bridged rings, and specified ring fusions. The automated analysis of ***EDITOR*** is not only much less laborious than a manually directed survey, but is also more likely to capture and detail all of the structural variants.

Another approach to identifying characteristic substructures works by determining the various ways in which a specific substructure has been embedded in the candidates. The essence of the method is first to find the substructure in each candidate structure; expand the substructure by incorpo-

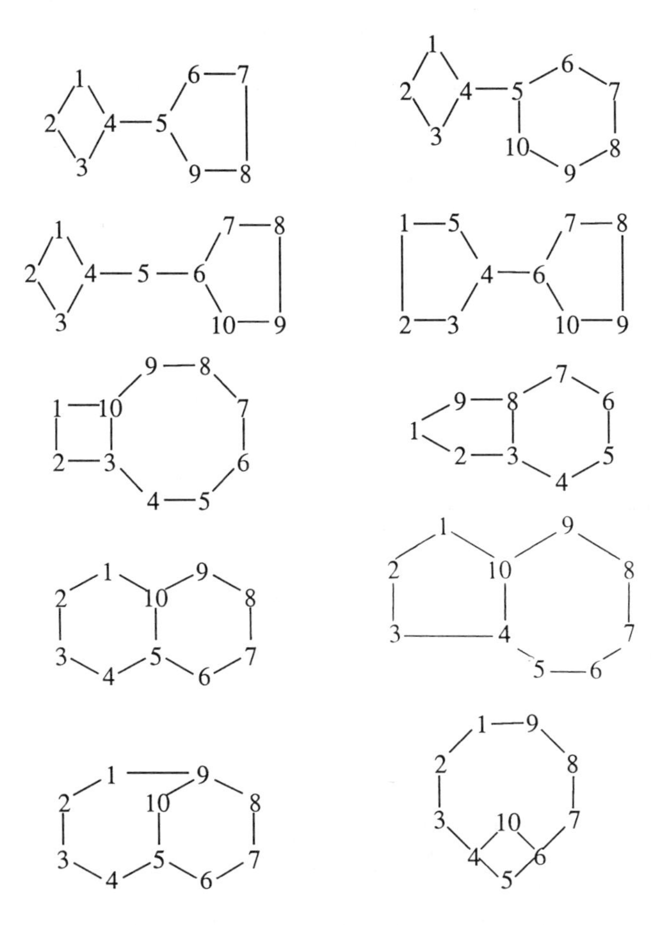

FIGURE XI-3. A partitioning of the candidates for warburganal based on **EDITOR**'s "BRANCHES" analysis.

rating neighboring atoms; canonicalize the enlarged substructure; and, finally, use the canonical features thus obtained as a basis for partitioning the set of candidates. The chemist must identify the starting substructural fragment of interest (normally, but not necessarily, one of those used originally to constrain the generation process). Consequently, this approach is not as completely automatic as the skeletal analyses of the **CASE** system; however, this form of substructural analysis can be much more directly coupled with possible discriminating experiments.

One implementation of this basic approach was the ***LOOK*** function of **STRUCC**.[7] The ***LOOK*** function provided for a number of alternative substructure expansion options designed to accommodate different needs. The simplest expansion procedure merely incorporates all the bonded neighbors of the constituent atoms of a given substructure. Frequently, this is not optimal; for example, if the objective is to help in planning of proton decoupling experiments, expansions through atoms other than carbons bearing protons only add irrelevant parts of the structure. The ***LOOK*** expansion options included constraints such as one limiting expansion to paths that might possibly be identified by proton decoupling. The options also provided for extensions causing incorporation of methyl groups, heteroatoms, or multiply bonded atoms at a two-bond radius, in addition to all neighbors at the one-bond radius.

Within ***LOOK***, the entire matching and expansion cycle was repeated automatically until different features were found among the candidate structures. (It is possible for the immediate one-bond, two-bond, or larger environment of a substructure to be identical in all candidates.) These discriminating features were then displayed along with details of their frequency of occurrence. Additional information could be obtained concerning the particular structures incorporating a discriminating feature or further analysis steps invoked. Useful forms of analysis include further expansion of a substructure and categorization according to any differences in the way in which the discriminating feature is incorporated into ring systems.

The ***LOOK*** function is illustrated here through an application to data from CONGEN's analysis of (+)-Palustrol.[18] Details of the example are summarized in Figure XI-4. This problem involved the determination of the structure of a $C_{15}H_{26}O$ sesquiterpenol isolated from a marine organism. Available spectral data established the presence of a cyclopropyl system, a tertiary alcohol–⟩C–OH, and other groups. As described in the original paper, constrained generation using **CONGEN** yielded 272 structures. Results from a ^{1}H NMR decoupling experiment provided more information on the structural relationship between one of the methyl groups and a methine in the cyclopropyl system. Use of this additional structural information eliminated many of the structures but still left 88 possible candidates. Of these 88 final structures, 12 could be discarded as they contained ethyl groups contrary to ^{1}H NMR data.

The major remaining structural question concerned the environment of the tertiary alcohol, substructural constraint B of the original problem. The ***LOOK*** function was used to partition the 76 candidates according to how the–⟩C–OH group had been embedded; ***LOOK*** was required to include methyl groups at a two-bond radius in the embeddings it reported. The results are illustrated in Figure XI-4.

As described in the original paper, the dehydration of this compound was known to yield three products: one with a vinyl methyl group, one with a vinyl hydrogen, and one with neither.[18] These data are consistent only with the fourth of the five ways reported by ***LOOK*** for incorporating the tertiary alcohol. The set of 76 structures could be pruned to give only the 12 appropriate structures.

Another example, summarized through the data in Figure XI-5, is provided by the candidates produced for warburganal. One possible approach to obtaining more detailed structural information would be a limited ^{13}C-^{13}C coupling study using double quantum coherence or some similar experimental techniques. The –⟩C–CH=O group should constitute a readily identifiable signal

```
# LOOK
Simple uniform expansion method? N
Specify expansion control options:
>METHYLS
>DONE
Feature to be investigated: B

5 structural classes were generated.
```

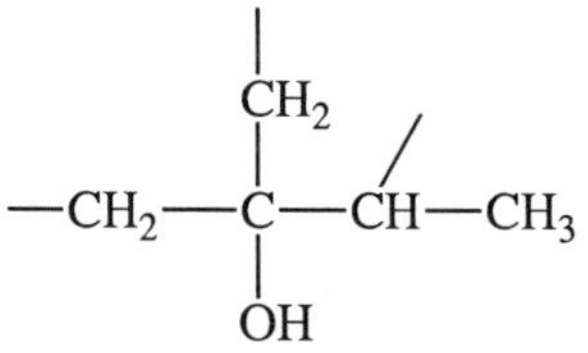

```
            |
           CH2
            |   /
   —CH2—C—CH—CH3
            |
           OH
```

(1) 26 instances,

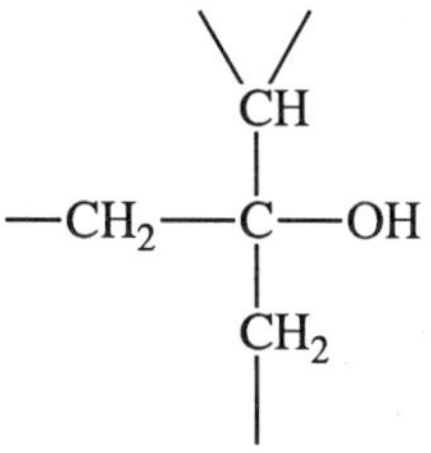

```
          \ /
          CH
           |
   —CH2—C—OH
           |
          CH2
           |
```

(2) 17 instances,

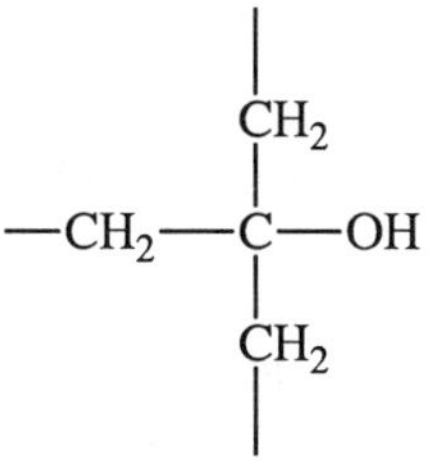

```
           |
          CH2
           |
   —CH2—C—OH
           |
          CH2
           |
```

(3) 13 instances,

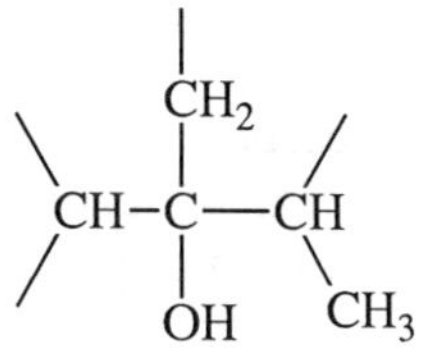

```
            |
           CH2
   \        |   /
    CH-C—CH
   /        |   \
           OH   CH3
```

(4) 12 instances,

```
            |
           CH2
   \        |   /
    CH-C—CH
   /        |   \
 CH3      OH   CH3
```

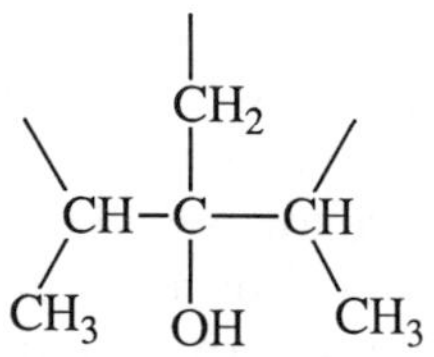

(5) 8 instances,

FIGURE XI-4. Partitioning the candidates for the sesquiterpenol palustrol according to embedding of a tertiary alcohol group (known to the system as substructure B).

```
 \ /                      |
  CH                      CH2
  |                       |
HO—C—CH=O               HO—C—CH=O
  |                       |
  C=                     —C—
 /                        |

  |                       |
  CH2                    —C—
  |                       |
HO—C—CH=O               HO—C—CH=O
  |                       |
  CH                     —C—
 / \                      |

 \ /                     \
  CH                      C=
  |                       |
HO—C—CH=O               HO—C—CH=O
  |                       |
 —C—                     —C—
  |                       |
```

FIGURE XI-5. Some of the embeddings of the >C-CHO group as found among the candidate structures for warburganal.

pattern from which to explore further bonding. Application of the analysis of the ***LOOK*** function to the 42 candidates, reveals some 14 distinct environments for this starting group. In six of these, the quaternary alkyl carbon is bonded to a hydroxy group; in the other embeddings the quaternary carbon is bonded to either one or two methyl groups. The coupling of the -CHO carbon would identify that it was bonded to the quaternary alkyl carbon whose shift implied a substituent hydroxyl group. Further couplings of that quaternary carbon could then identify which of the six distinct embeddings was correct.

XI.C.3. Empirical Ring Strain Analysis

It is common for an underconstrained generation process to result in structures containing highly congested and strained cyclic systems. Cyclopropenes, unsaturated epoxy ring systems, three-membered rings with one or more exocyclic double bonds, and fused pairs of three and four membered rings are all commonly encountered. Usually, such structural forms are incompatible with the biochemical origin of an unknown compound and such candidates are immediately discounted by the chemist.

Various approaches have been designed either to prevent the generation of candidates containing undesired ring systems, or to filter the candidates produced. The simplest approach is to use some form of library of forbidden structural fragments. Substructures representing disallowed systems (e.g., with a carbon-carbon triple bond in a three-, four- or five-membered ring) can be predefined in such a library. These forbidden fragments are used, automatically, as constraints in the structure generation process; any partially assembled or final structure containing a forbidden fragment is eliminated.

It is also possible to compute an estimated "strain" energy for a molecule with the use of empirical potential functions. Computation of strain energy avoids dependence on any predefined (and limited) set of disfavored substructures and thus provides a more universal approach to analysis. Once strain energies have been computed, candidates with excessively high strain can be identified and discarded.

The disadvantage of these analyses is their totally empirical nature. Occasionally, structures are encountered that contain "forbidden fragments" or that are, indeed, quite highly strained. If such structures are automatically eliminated on the basis of some empirical analysis, the structure elucidation system will, on occasion, discard the correct candidate and may thereby fall into disrepute. These rather *ad hoc* analyses should not be applied to eliminate candidate structures entirely; rather, they should be limited to facilitate partitioning of a large set of candidates into groups of plausible and implausible structures. The objective of such a partitioning is again simply to focus the chemist's attention on the most worthwhile candidates.

XI.D. EXPLOITING STEREOCHEMICAL CONSTRAINTS

Quite apart from its intrinsic importance in fully defining a structure and providing a model for detailed spectral predictions, an analysis of configurational stereochemistry can provide additional constraints that may exclude particular candidate structures. Ideally, stereochemical constraints would be exploited during the candidate generation process. However, until stereochemistry is utilized within structure generation procedures, the exploitation of stereochemical constraints must be deferred to the evaluation stage.

Some stereochemical constraints express molecular symmetries. The analysis of molecular symmetry presents problems. Experimental data define the properties of either a specific conformation of a molecule or of some equiliberating set of conformers. The symmetry of a conformer may be lower than that of a representation of the corresponding configurational isomer, whose symmetry may again be less than that apparent in the constitutional structure. The **STEREO** program allows symmetry constraints to be expressed in terms of equivalent atoms, equivalent substructures, or chirality. Such experimental data provide a lower bound on the symmetry of the configurational isomers manipulated by the **STEREO** program. Configurational isomers possessing more than the observed number of instances of equivalent substructures or atoms cannot be excluded.[3,8]

The **STEREO** program provides also for constraints excluding configurational structures incorporating systems such as

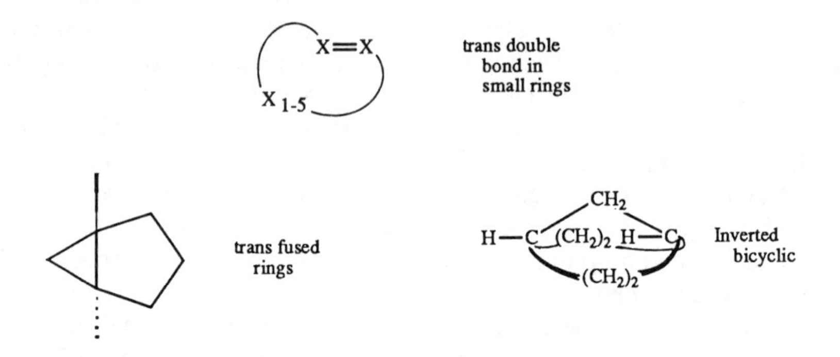

It is possible for such undesired stereochemical features to be found in every configurational stereoisomer in some complex, polycyclic, fused ring structures. Examples have been presented of isomers and stereoisomers of various $(CH)_n$ compounds.[8] Isomers and stereoisomers were generated exhaustively and then the stereoisomerism was reinvestigated assuming various stereochemical constraints. In the $(CH)_8$ system, for example, three out of twenty constitutional isomers can be rejected if inverted bicyclics and *trans* double bonds in and *trans* fusions of small rings are all prohibited. An example constitutional structure rejected by such criteria is shown as **XI-6**.

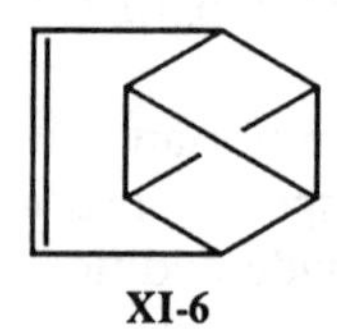

XI-6

XI.E. SPECTRUM PREDICTION AND STRUCTURE RANKING

XI.E.1. Introduction

The interpretive processes, employed prior to structure generation, should have thoroughly exploited simple subspectral-substructural relationships in establishing the presence of specific constituent molecular fragments. Subsequently, the surveying processes, as applied to generated candidates, will have identified any inappropriate functionality for which negative constraints have been omitted. Candidate structures containing combinations of functional groups obviously contraindicated by spectral evidence will have been discarded in the course of this postgeneration analysis. The remaining candidates should, therefore, all be consistent with the gross spectral data. Detailed analyses of spectral properties are necessary to achieve any further discrimination amongst the candidates.

A number of issues must be considered in prediction of spectral properties for candidates and using these predictions for evaluative purposes. First, an appropriate structure representation must be determined for the candidates. This may be the topological structure representation, but often this will be insufficient because important factors determining spectral properties will not be represented. Second, a prediction method must be chosen. Here, there are several possibilities ranging from empirical correlation charts through to quantum-mechanical calculations. Finally, some method must exist for assessing the compatibility of the spectral data, as recorded for the unknown, with the predictions made for each candidate. This assessment process must be conservative. Candidate structures that apparently have spectral properties inconsistent with observed data should not be eliminated if there is any doubt concerning the quality of predictions.

Such general issues are reviewed briefly below, and then some examples of predictive methods are presented.

XI.E.1.a. Structure Analysis for Spectrum Prediction

The various candidate structures must be analyzed and, on the basis of results of this analysis, expected spectral properties derived. Such an analysis may be limited to considering the constitutional (topological) structure. Mass spectra are only slightly influenced by steric effects; thus a constitutional analysis could well suffice for these data. A stereochemical representation for each candidate structure, incorporating configurational details, might be employed. If configurational stereochemistry is represented, the analysis can detect and utilize differences such as the *cis-trans* relationships of substituents on a ring or double bond. Such a more detailed structural analysis permits a more realistic assessment of the various factors that influence ^{13}C chemical shifts and similar spectral properties. An analysis extending to the consideration of the conformational forms of candidate structures might well be necessary in attempts to derive probable ^{1}H NMR coupling patterns.

XI.E.1.b. Prediction Methods

There are constraints on the methods that can be employed for estimation of spectral properties for candidate structures. Typically, there will be many dozen different constitutional isomers to be considered; if configurational stereochemistry is also represented, hundreds of candidate stereoisomers may have to be processed. Consequently, the methods employed for spectral prediction must be relatively inexpensive computationally to be of practical value. And yet, the prediction methods must be sufficiently detailed so as to capture subtle differences in the spectral patterns as would be obtained for different, but generally similar, candidate structures.

Prediction methods take many forms. The simplest are really applications of very detailed correlation charts or tables. These methods employ tables containing definitions of substructures, possibly incorporating appropriate stereochemical information, and their associated spectral features. Each entry in a such a correlation table can be conceived as being a simple substructure–spectral-feature rule. The spectral properties of a candidate structure are determined through identification of its constituent substructures and combination of their individual subspectral features. The basic algorithm involves working through a table of substructure definitions, attempting to match each substructure onto the candidate structure, and then—if a match is found—"predicting" the corresponding spectral feature. In many ways, such an analysis is similar to that employed, in reverse, in inferring substructures from spectral patterns, but the predictive analysis can allow for more detail in the form of the substructures and for steric and other long-range (three-, four-bond) influences.

Additivity rules are commonly used by chemists for prediction of molecular spectra. Such rules define a relationship between a spectral feature and a set of enumerable substructural properties; thus a chemical shift might be defined in terms of the number of carbon atoms, hydrogen atoms, and heteroatoms at one-, two-, and three-bonds radii. The relationship is typically expressed as a linear formula combining appropriately weighted contributions from each of the substructural properties. The weights are derived from a statistical analysis of the spectra and substructural properties of reference compounds. Additivity formulas provide a very compact way of encoding the information from several hundred pairs of substructures and subspectra. Rather than employ a very detailed correlation chart, the chemist can simply count the number of occurrences of each of a set of substructural properties and combine these counts using a dozen or so weighting factors.

Such additivity formulas can, of course, be used within computer programs. However, little advantage is realized through their use. Additivity formulas benefit the chemist by eliminating the need to search through lengthy tables of substructures and subspectra. Once properly programmed, such searches are readily performed on a computer. Although all additivity rules involve similar forms of analysis, the substructural features, which are counted and combined, vary with the specific application. Different programs are required for each application, whereas a single program should suffice for all directly expressed substructural-subspectral correlations. Furthermore, as more data are acquired, the weighting factors initially chosen for an additivity formula are liable to prove inadequate in certain cases. The statistical analysis of the available reference data must then be repeated to derive new, more widely applicable weights. In contrast, substructure–spectral-feature rules may be organized so that no costly statistical analysis or "learning" is necessary. Some rule systems allow new substructures and subspectra to be immediately combined into an existing correlation table and used directly in subsequent applications. It appears unlikely that additivity formulas will play a major role in computer-based prediction systems.

Predictions based on substructure–spectral-feature rules require only standard graph-matching procedures for finding a (stereochemical) substructure in a (stereo)isomer. The processing entailed in the use of additivity formulas is more varied. Depending on the particular structural features considered and combined by the formula, the processing might require investigations of ring systems, enumeration of atoms at various bond radii, or possibly even some limited forms of substructure matching. Still more elaborate analyses of the forms of the candidate structures are required in prediction methods that use simplified theories, or models of the physical processes giving rise to the spectral patterns.

Such simplified theories attempt to describe the primary factors that determine some aspect of a compound's spectrum. A "theory" might take the form of a statement that, apart from diastereotopic methyls, *topologically equivalent carbon atoms should show identical resonance shifts in the* ^{13}C *spectrum*. Given this theory, it becomes appropriate to analyze the topological symmetry group of each candidate structure to determine the number of topologically distinct atoms and, hence, the number of expected resonances. Another simplified theory might presume that *if a particular observed ion can be rationalized in terms of the fragmentations of each of two candidate structures, the more plausible structure is that for which the presumed fragmentation process is simpler*. If mass spectra are to be analyzed on the basis of this theory, it is necessary to determine all possible ways of fragmenting a structure (subject to chosen limits on the number of bonds broken, the number of hydrogen atoms transferred in or out of an ion, and other factors determining the complexity of a fragmentation process).

Finally, it is possible to apply conventional semiempirical quantum-mechanical calculations, or classical methods, for spectrum prediction. Such methods have not been widely utilized. One problem with quantum-mechanical approaches is that they require a coordinate representation of the molecule; consequently, some specific conformation, of some particular configurational stereoisomer, would have to be presumed for each candidate constitutional structure. Both the quantum-mechanical approaches and the classical methods are computationally costly and thus are not well suited to the analysis of dozens of candidate structures for some complex organic molecule.

XI.E.1.c. Evaluation of Predicted Spectral Properties

There are significant problems relating to the assessment of compatibility of predicted spectral properties for candidates and data obtained for an unknown. Spectrum matching methods, as employed in file search compound identification procedures, are rarely appropriate. When identifying a structure by a file-based spectrum matching procedure, one is comparing the recorded spectrum of the unknown with spectral data, acquired under similar experimental conditions, characterizing known reference compounds. In this context it is appropriate to measure some simple "distance" between the two sets of spectral data and then use these distances to measure compatibility. The data characterizing each reference compound are comparable, and a simple distance measurement, between the unknown spectrum and reference data, should not be biased. However, the predicted data characterizing each of a set of candidate structures may not be strictly comparable, either among themselves or with observed data.

The various methods used to predict spectral properties of different candidates typically introduce certain biases. Consider first the rule-based approaches using sets of substructure–spectral-feature rules. Such rules are derived from data characterizing known reference structures and, typically, define different substructures and spectral features at varying degrees of precision. Some rules may incorporate very detailed substructure definitions, with correspondingly precisely specified spectral patterns. When applicable, such detailed substructures permit accurate spectral predictions. In other cases, the only applicable rules may be those in which simple substructures are associated with generalized and approximate spectral descriptions. Predictions based on such rules can be biased in favor of candidates similar in structure to those reference compounds used in creating the rule set.

Biases and limitations are also characteristic of the simplified spectrum prediction theories. For example, it is neither correct to assume that topologically equivalent carbon atoms must give identical ^{13}C resonance shifts nor to presume that equivalence of shift implies topological equivalence. This basic ^{13}C prediction theory has to be modified in order to encompass special cases such as diastereotopic methyls or methylenes in long alkyl chains. As is illustrated below, a bias favoring structures with disjoint ring systems is consequent on the presumption that the structure with the simpler fragmentation process always provides a better rationalization for an observed ion.

The formula used to assess a compatibility score for observed and predicted data should make allowance for any intrinsic biases of the prediction method. In the rule-based systems, for example, a candidate structure whose predicted properties do not well match the observed data should not be accorded a poor score if these predictions used only very general rules. In such a case, the fault lies not in the candidate, but in the inadequacy of the rule set employed.

The final ranking process, based on the assessed scores, is simple. Given the various biases and limitations of the theories used for spectral prediction, the absolute values of scores and specific rankings of candidates are of limited significance. However, the scoring and ranking process should at least divide the set of candidates into two groups. The first group should contain those candidates whose predicted properties are compatible with the observed data, together with those candidates for which the prediction theory has been assessed as being unreliable. The second group should include structures, for which the theory appears adequate, whose predicted properties show some marked discrepancies with observed data. The correct structure should then be included in the first group, either because it does serve to rationalize well the observed data or because a poor score has not been assigned since the prediction theory was manifestly inadequate.

XI.E.2. Rule-Based Spectrum Prediction

Rule-based spectrum prediction represents the simplest, most adaptable approach and hence has the widest applicability. Rule-based systems require a set of substructure–spectral-feature rules and an interpreter that can apply these rules in the context of a particular structure. In the simplest examples, the right-hand side of each rule is merely a description of an expected spectral pattern. This pattern should be observed in the spectrum of any compound containing the substructure identified on the left-hand side of the rule. In such cases, the interpreter consists simply of a graph-matching routine. The substructures on the left-hand sides (usually called the "premise parts") of each rule are matched in turn onto a given candidate structure. If the match is successful, the corresponding spectral feature can be predicted.

Some systems incorporate more elaborate right-hand sides in their rules. The rules' premises still define substructures that must match onto a candidate structure for the rule to be applicable. Successful matching of the rule premise not only establishes the rule as applicable, but also provides a specific context for interpreting the right-hand side of the rule. This right-hand side will, in effect, take the form of some algorithmic function that defines a spectral property dependent on the larger structural environment surrounding the matched substructure. In such systems the interpreter must incorporate the graph-matching functions for matching the substructures of the rules' premises. Additionally, the interpreter program must incorporate some procedures for applying the functional definitions on the right-hand sides of rules that yield the expected spectral properties of a given structure.

As the size of the rule set is increased, the graph-matching processes needed to match the substructures of the rule premises can become inordinately costly. Yet, some method must be found for searching through the rule set to find those rules appropriate to a given structure. If the substructures in the rules' premise parts can be arbitrarily complex and varied, as is the case for some mass spectral prediction rules, the only approach is the costly procedure of graph matching rules' substructures to candidates. In other applications there may be intrinsic constraints on the form of substructures that can be meaningfully employed in prediction rules. Given such constraints on the form of rules' premise substructures, it may be possible to direct the search through the rule set for relevant prediction rules. An analysis of the form of each candidate structure may be used to directly identify such appropriate rules.

For example, ^{13}C resonance shifts are most sensitive to the immediate bonded neighbors of the resonating carbon atom; neighbors two bonds remote still have considerable influence on shifts; more distant atoms generally have less importance. This hierarchy of influence can be reflected in hierarchically defined sets of substructures. The rule set can incorporate (1) general rules relating one-bond environments to broad shift ranges, (2) more detailed rules based on substructures that define a two-bond environment and a more limited shift range, and (3) highly specific rules relating the three- and four-bond stereochemical environment of a carbon atom to a narrow shift range. If these substructures are canonical in form and are appropriately represented, they may be used as a basis for ordering the rules in the rule set. A candidate structure can be analyzed to derive canonical representations of the one-, two-, three-, and four-bond substructural environments of each of its constituent carbon atoms. These various canonical representations can then be used to "key into" the rule set to find and abstract rules incorporating identical substructures. These are the rules appropriate for predicting the resonance of an atom in the defined substructural environment. Rules based on larger substructures should be used in preference to rules matching smaller portions of the atom's molecular environment because of their greater specificity and precision. Provided the method for keying into the rule set can be made efficient, the cost of applying the rule-based prediction scheme should be largely unaffected by the increasing size of the rule set.

The following examples, all drawn from work in the **DENDRAL** project, illustrate various aspects of rule-based spectrum prediction. The first example, ^{13}C spectrum prediction for constitutional structures by the method of Mitchell and Schwenzer,[22,23] illustrates a straightforward rule-based approach. The rules define substructures, describing the molecular environment of a resonating carbon atom, and the corresponding shift range wherein the resonance should be observed. The substructures must be graph matched to a given candidate; successful matching results in the prediction of the ^{13}C shift range for the appropriately identified carbon atom. This study did not include any extensive investigation of possible problems resulting from inadequacies of the rule set as applied to particular structures. However, the formula used to compute the quality of match between predicted and observed spectra was designed to reduce the importance accorded to predictions based on very general rules. For this system, Mitchell devised an elaborate search procedure for creating substructure–spectral-feature rules that rationalized observations in some training set of data from reference compounds. An appropriately modified and extended version of Mitchell's rule-formation procedure could be used in other similar applications.

The work by Lavanchy et al.[10] on rule-based prediction of mass spectra provides an example of a system wherein the right-hand sides of rules are really algorithmic functions. These functions yield mass spectral patterns that are determined by the specific overall structural environment found in a candidate molecule for a rule's premise substructure. Rules for this program could be created through automatic programs or could be based on a chemist's own analysis of data characterizing known reference compounds incorporating some common skeleton.

The final example concerns the more elaborate **DENDRAL** system for predicting ^{13}C spectra for both constitutional isomers and stereoisomers. This is an example of a system with a very large rule set. Appropriate prediction rules are determined through analysis of candidate structures. Rules, which define hierarchically related substructures, are derived automatically from data characterizing reference compounds. The evaluation scheme takes explicit account of the quality of the rule set in relation to the form of a specific candidate structure.

XI.E.2.a. The Production Rule Model for Spectrum Prediction

Although several rule-based spectrum prediction schemes had been reported earlier, the work by Mitchell and Schwenzer provides the clearest instance of such a system. Their system was devised to predict ^{13}C spectra for (topological) candidate structures. Knowledge concerning ^{13}C shifts was encoded in rules of the form

$$\text{(Substructure description)} \xrightarrow{\text{implies}} \text{(range of characteristic } ^{13}\text{C shift)}$$

(This work originated in a study of possible applications of artificial intelligence (AI) techniques; in keeping with the standard AI terminology, Mitchell and Schwenzer describe their system as being a "production rule" approach.) An example production rule, derived through automated analysis of the spectra of approximately 70 alkanes and acyclic amines, is

$$CH_3^*\text{–CH}_2\text{–CH}_2\text{–CH}_2\text{–X–} \longrightarrow 14.0 \text{ ppm} \leqslant \delta(\text{C}) \leqslant 14.7 \text{ ppm}.$$

This rule specifies that the shift of the tagged methyl carbon in the substructure shown (where X represents any non-hydrogen atom) should lie in the range 14.0 to 14.7 ppm.

The ^{13}C spectrum of a candidate structure was predicted by an interpreter program that utilized a set of these production rules. The interpreter first tested the premise part of each rule by attempting to graph match the specified substructure onto a given candidate structure. When a match was found, the shift range in the right-hand side of the rule was ascribed to the associated carbon.

Often, several rules could be found that all applied to the same carbon; these different rules would incorporate substructures defining differing aspects of the structural environment of the resonating atom. The shift range predicted for an atom was taken as the narrowest range consistent with the predictions from the various rules matched to that carbon.

The predicted spectrum of a candidate structure was assigned a score based on its similarity with the spectrum recorded for the unknown compound. This entailed putting predicted and observed resonances into correspondence; then a score measuring the consistency of predicted and observed shifts was computed.

This consistency measure took account of the precision of the prediction and thus, indirectly, the appropriateness of the prediction rule in the context of the given structure. Rules incorporating general substructural descriptions yield broad prediction ranges, whereas rules based on highly specific substructures are characterized by narrow ranges for resonance shifts. The mismatch between a predicted shift and the presumed corresponding observed resonance shift was inversely weighted by the shift range of the prediction rule. Thus less importance was accorded to those predictions based on generalized substructures.

The system, as implemented, has a number of limitations, most of these result from the method used to derive rules for spectrum prediction from data characterizing a training set of known reference compounds. The main objective of Mitchell's work was not really the analysis of chemical data but rather the development of a general model of concept formation wherein a concept, such as one of these spectrum prediction rules, was "learned" through the analysis of successive examples. This model of concept formation is based on the idea of a search for rules that explain particular example observations in the training set data.

In the chemical application to alkane and amine ^{13}C spectral analysis, rule generation involved a search for substructures correlated with specific shift ranges. Starting from a substructure representing only a single carbon atom, a generation program explored all possible ways of building up larger substructures by adding neighboring atoms, restricting atom types, or specifying numbers of substituent hydrogens and so forth. In the context of alkanes and amines, for example, one more specific substructure would be the methyl group. Then the program might explore methyls bonded to carbons C^*H_3-C, to nitrogens C^*H_3-N, or to any effectively bivalent atom (i.e.,–CH_2–or–NH–) C^*H_3–X–, or any effectively trivalent atom (>CH–or >N–) C^*H_3–X<.

The training set compounds were examined to identify all occurrences of each new substructure produced by this generation procedure. If there were insufficient examples of a substructure in the training set, that substructure would be discarded. Otherwise, the rule formation program would analyze the distribution of shifts associated with the resonating atom, corresponding to the focal atom of the substructure, in each reference structure containing the substructure.

Associated with the initial one-atom substructure was the "seed" rule: $C \longrightarrow -\infty < \delta_C < \infty$. New rules were created through the combination of each new, more detailed substructure from the generator and the distribution of shifts found from examples in the training set. If a new rule could be used to rationalize a sufficient number of observations and if it had a shift range narrower than the parent (less detailed) rule from which it was derived, it would be adopted by the program. Attempts would then be made to further refine the new prediction rule by generating subsequent rules incorporating still more detailed elaborations of the same substructure. If these refined rules permitted still more precise predictions for sufficient examples in the reference data, the substructure elaboration and testing processes would again be repeated. The rule-formation program assembled a final set of rules, for use in subsequent spectrum prediction, by taking those generated rules that explained sufficient data with adequate precision and could not be further refined by the search process.

In the example application to alkanes and amines, this procedure generated, among others, the level 1 rules

$$C^*H_3\text{–} \longrightarrow 8.1 \leqslant \delta(C) \leqslant 45.6 \quad \text{and} \quad \text{–}C^*H_2\text{–} \longrightarrow 17.9 \leqslant \delta(C) \leqslant 60.1.$$

Further elaboration of the first of these gave five level 2 rules, including

$$C^*H_3\text{–}N \longrightarrow 28.8 \leqslant \delta(C) \leqslant 45.6 \quad \text{and} \quad C^*H_3\text{–}X\text{–} \longrightarrow 8.1 \longleftarrow \delta(C) \longleftarrow 36.7.$$

These all represented improvements over their parent rules and were subject to further analysis.

The problem with this approach to rule formation is that it does not adequately reflect the physical basis of the factors inducing ^{13}C chemical shifts. The program can easily be misled into exploring inappropriate elaborations of some substructure that, fortuitously, happen to give rules applicable to some specific training set of compounds. One of the rules generated from a training set of alkanes and amines has the form

$$C^*H_3\text{–}X\text{–}X\text{–}CH < \longrightarrow 14.3 < \delta(C) < 14.7$$

The search procedure found that limitation of the alpha neighbor of the resonating methyl to being bivalent (i.e.,–CH_2–or–NH–in the context of alkanes and amines) resulted in a rule that explained sufficient data points with adequate precision. This substructure was further elaborated to require first a chain of two bivalent atoms and then a chain of two bivalents and a $>CH$– methine. Each elaboration gave a rule that, in the context of the training set data, showed improved performance.

The final rule developed is, of course, incorrect. It implies that the methyl in $CH_3\text{–}NH\text{–}CH_2\text{–}CH<$ should resonate in this region from 14.3 to 14.7 ppm; this prediction being made despite the fact that, among the training set compounds, no CH_3–N resonance was observed below 28.8 ppm. This error results from the program's over-generalization from reference examples where a substructure was found to match. In the training set of alkanes and amines, all instances of $CH_3\text{–}X\text{–}X\text{–}CH<$ substructures actually involved the substructure $CH_3\text{–}CH_2\text{–}X\text{–}CH<$. Because the rule-formation program explored substructures based on CH_3–X–, it ignored the specific detail of the examples found and, in effect, overgeneralized the atom types allowed for the alpha neighbor of the methyl group.

Apart from erroneous examples such as this, the rule-formation procedure results in large numbers of distinct rules based on similar, partially overlapping substructures. Thus the final set of rules abstracted from the reference alkane and amine data included the following:

$$\mathrm{C{-}X{-}C^{*}H_2{-}X(C)(C){-}C} \longrightarrow 44.7 \leqslant \delta(\mathrm{C}) \leqslant 44.9$$

$$\mathrm{{-}X{-}C^{*}H_2{-}X({-})({-}){-}} \longrightarrow 41.1 \leqslant \delta(\mathrm{C}) \leqslant 45.1$$

$$\mathrm{C{-}C^{*}H_2{-}C} \longrightarrow 17.9 \leqslant \delta(\mathrm{C}) \leqslant 56.9$$

(There are no restrictions on the numbers of substituent hydrogen atoms for most of the atoms in the substructures of these rules.) Each of these rules would match the structural environment of C(3) in 4,4-dimethylheptane and would provide an alternative shift range prediction. These different predictions must be resolved; depending on the particular rules and training set compounds from which they were derived, it is possible for the predictions to correspond to partially disjoint or completely nonoverlapping spectral ranges. There is no hierarchical organization to the rules; each describes a slightly different aspect of the environment of the resonating methylene carbon atom. The failure to achieve a complete match between the substructure in one rule and the environment of an atom in some candidate cannot be used to avoid the unsuccessful testing of all other similar rules for possible matching onto that atom in the structure.

In the specific case of ^{13}C data, these problems could be alleviated by restricting the rule-formation process so that the substructure definitions produced were strictly hierarchical in form. All neighbors at a one-bond radius would have to be specified before those at the two-bond radius could be considered. A rule generation procedure must be constrained to take account of known physical influences on spectra. The nature of a carbon's alpha neighbor is known to be more important than branching three bonds remote. The search procedure can "learn" this fact only if the training data are sufficiently comprehensive. If the objective is the development of useful rules rather than a demonstration of learning, it is inappropriate not to provide known structural relationships to the rule-formation program. Mitchell's general approach to rule formation, but with a constrained rule generator, should be applicable to other spectral-structural relationships. Details would necessarily differ, for distinct types of physically derived constraint would dictate distinct methods for substructure expansion. A particular advantage, however, is that this overall approach is **not** limited to applications involving atom-, or bond-centered, hierarchically defined substructures.

Apart from the need for a more physically based rule-formation procedure, the main problem is the cost of applying such a prediction system. Each rule's substructure must be matched onto each candidate structure. In their system, Mitchell and Schwenzer employed some 138 rules when predicting spectra for alkanes and amines; on structures such as the C_9 alkanes and C_6 amines, processing times were of the order of 1 min per structure. These times were for a preliminary implementation, in interpreted LISP, and thus represent an overpessimistic estimate of the cost of a practical system. However, graph matching of substructures onto candidate molecules is intrinsically costly; the approach is practical only when at most a few hundred substructures must be considered. Different methods for identifying appropriate rules must be developed for applications where there may be several hundred or thousands of detailed substructure–spectral-features rules.

XI.E.2.b. Rule based Mass Spectrum Prediction

In the Mitchell-Schwenzer ^{13}C system the right-hand part of each rule comprises simply an assertion regarding a specific spectral feature that should be directly observed. Such relationships, between substructures and specific spectral features, cannot always be defined. For example, when considering mass spectra, one can make use of a rule such as *"if you see a carbonyl group then you can expect cleavage of the bonds alpha to that group,"* but such a rule does not directly specify what ions should then be observed.

In such a case, to determine the composition and mass of the expected fragment ions, one must examine the larger structural environment of the carbonyl group found in a candidate molecular structure. First, it must be determined whether the suggested bond break would indeed cleave the structure; breaking only one bond alpha to a carbonyl group would not cleave a molecule in which the carbonyl was a part of a ring system. If the specified bond break does cleave the molecule, the atoms constituting the two resulting parts must be identified. Only then can one determine the expected fragment compositions and masses.

The earliest structure evaluation schemes were based on predictions of mass spectral fragmentations. Mass spectra were favored partly because there is reasonable physical justification for analyzing mass spectra solely in terms of topological structures as were obtained from the earliest structure generators. Mass spectral prediction methods were never so simple as a substructure–spectral-feature production rule interpreter. Inherently, the prediction of a compound's mass spectrum does involve the application of some functional description of a fragmentation process.

General fragmentation rules, expressing concepts such as alpha cleavage and McLafferty rearrangements, are of limited efficacy. Practical applications have demonstrated that, in order to use mass spectral data to discriminate among the candidate structures for some molecule, one requires quite specific rules characterizing the skeletal fragmentations of compounds of that particular class.

Applications of such detailed, class-specific fragmentation rules have been published by Lavanchy et al.[10] These published results concern structure elucidation of marine sterols using mass spectral data. This mass spectral analysis was twofold. First, class-specific rules concerning the fragmentations of sterol nuclei were employed. On the basis of this first-stage analysis, the appropriate molecular skeleton could normally be identified. Second, rules concerning fragmentations of side chains were exploited in an attempt to restrict the number and variety of side chains being considered. Most of the ion current is due to fragments formed by cleavages of the steroid nuclei; the side-chain fragmentations produce only subtle, although distinctive, spectral patterns of low-intensity high-mass ions. If nuclear and side-chain fragmentations were processed simultaneously, the variations due to different side chains would have little influence on candidates' final scores, because they would be swamped by the much larger changes in other spectral regions. This rule-based approach to mass spectral analysis does permit consideration of only selected types of fragmentation process, thus making it possible to examine minor fragmentation modes in the presence of other much more intense ion patterns. Through this two-stage analysis of mass spectra, the number of candidates possible for an unknown marine sterol could typically be reduced from hundreds down to a dozen or so.

Each rule, as used by this mass spectral prediction system, defines a substructure, identifies the bonds that break, and indicates concomitant neutral losses and hydrogen transfers involved in the fragmentation process. A typical fragmentation rule, taken from the set used in the analysis of marine sterols, is shown in Figure XI-6.

This particular rule describes the standard D-ring cleavage of a steroid. The six-atom substructure represents the premise part of the rule. This must match onto a particular candidate structure. The test for matching the rule premise

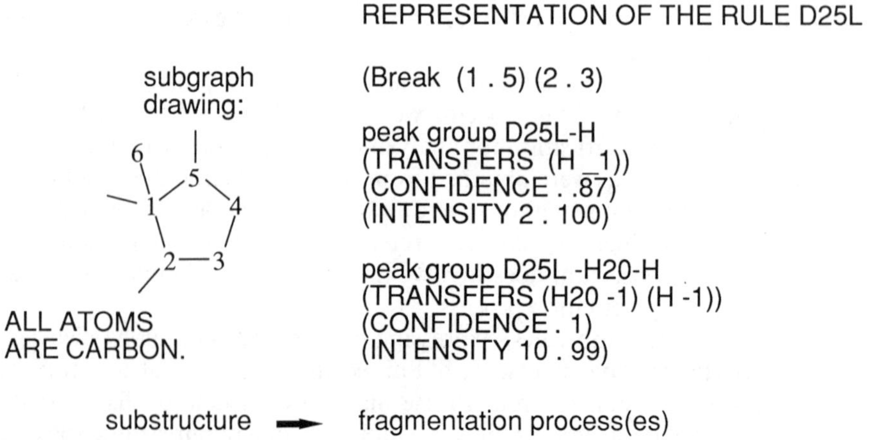

REPRESENTATION OF THE RULE D25L

subgraph drawing:

6, 5, 1, 4, 2, 3

ALL ATOMS ARE CARBON.

(Break (1 . 5) (2 . 3)

peak group D25L-H
(TRANSFERS (H _1))
(CONFIDENCE . .87)
(INTENSITY 2 . 100)

peak group D25L -H20-H
(TRANSFERS (H20 -1) (H -1))
(CONFIDENCE . 1)
(INTENSITY 10 . 99)

substructure → fragmentation process(es)

FIGURE XI-6. Rule for predicting mass spectral fragmentations of steroidal and similar compounds.

requires only a standard graph-matching routine working on a topological representation of the candidate. The substructure would, for example, match as shown onto the structure of 4α,24S-dimethylcholestanol (**XI-7**). Atom C(1) of the substructure is matched to C(13) of the steroid nucleus, C(2) onto C(14), and so forth.

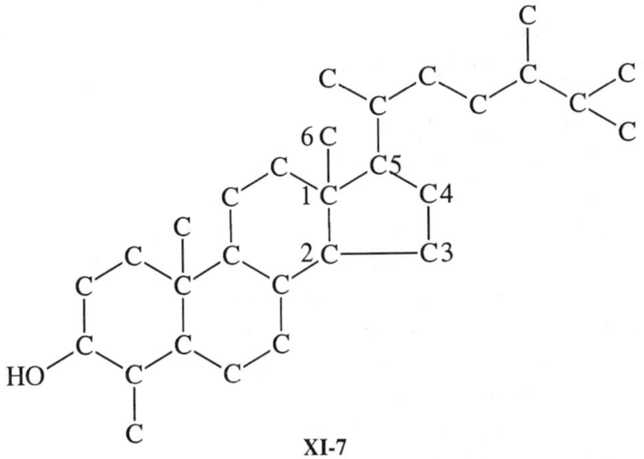

XI-7

If the substructure of the rule premise can be matched onto a candidate structure, the action part of the rule is evaluated. First, it must be determined whether the molecule is cleaved by the given bond breaks as defined in the rule. The interpretation of the ***BREAK*** statement in the example rule is that the bonds are to be broken between those structural atoms matched to C(1) and C(5), and also between those matched to C(2) and C(3), of the substructure. These correspond to the C(13)-C(17) and C(14)-C(15) bonds of the steroid nucleus. The ***BREAK*** statement further implies that charge should reside on that fragment of the structure containing the atoms matched to C(1) and C(2) of the substructure.

This charged portion of the molecule must be identified. A copy of the connection table of the candidate structure is made, omitting the bonds supposedly cleaved in the fragmentation process. A spanning tree, similar to those generated when searching for ring systems, is then grown through this connection table. This tree is rooted on one of the atoms matched to a substructural atom identified as being in the charged fragment. Thus, in this example, the root for the spanning tree would correspond C(13) of the steroid. The atoms of the candidate structure incorporated in the derived spanning tree are those comprising the basic fragment ion. In the example, C(1)-C(14), the OH oxygen, and the three methyl groups would all be identified as being in the charged fragment, giving a basic composition of $C_{17}H_{28}O$.

It is possible that the spanning tree comprises all atoms of the structure. This would occur in a candidate structure containing an additional cyclic system such as a lactone ring system joining C(18) of the steroid to C(21) of the side chain. Even after omission of those bonds in the connection table that correspond to bonds broken in the fragmentation process, a path would still exist between all atoms of the candidate structure. If the spanning tree proves to contain all the atoms of the original molecule, the fragmentation process is recognized as being irrelevant in the context of the given candidate structure, and further processing of that rule and candidate structure may be abandoned.

Most fragmentation processes entail more than simple bond breaks. Usually, at least some hydrogen atoms are transferred either into or out from the resulting charged fragment. Commonly, other eliminations occur such as the loss of water or substituent methyl groups. The rules employed in this mass spectral analysis system typically define several *peak groups*. Each peak group entails the same bond breaks but differs with respect to accompanying neutral losses. The example rule details two such peak groups; the first involves a transfer of one hydrogen from the ion and the second involves both a hydrogen transfer and loss of a water molecule from the ion. These transfers are defined, in the rules, purely in terms of the composition changes that they engender. The spectrum prediction program derives the final compositions and masses of the expected fragment ion(s) by applying the composition changes specified in the peak groups to the basic composition found for the fragment ion. In the case of the example, two fragment ions, $C_{17}H_{27}O$ and $C_{17}H_{25}$, would be predicted for 4α,24S-dimethylcholestanol.

Each peak group specification includes also an expected intensity range, ***INTENSITY***, for resulting ions, and a ***CONFIDENCE FACTOR*** for the process. The ***INTENSITY*** specification summarizes data typifying those reference spectra used for devising the prediction rules. The ***CONFIDENCE FACTOR*** is a measure of how typical a particular fragmentation process is for the class of compounds being analyzed. These various data are used in the procedures for evaluating the match between the observed and predicted spectra. For the example, the prediction is that the $C_{17}H_{25}$ ion must be present and should be at least 10% of the base peak in the observed spectrum; the intensity range for the other fragment ion is such that, in reality, the only requirement is that an ion be observed with that composition.

Mass spectral prediction rules, as used in this system, are chosen to characterize the majority of structures in some training set of reference compounds used in their development. Consequently, even some quite characteristic fragmentation processes may not be included in the rule set; certainly, many minor fragmentation processes will be ignored. Thus in this approach there is a tendency to *underpredict* the mass spectrum of a structure. The failure to predict some particular observed ion should not count heavily against a candidate. However, strong evidence for rejecting a candidate can be adduced from the prediction of a fragment ion that is not actually observed. The function used to assess candidates' spectra is based on these characteristic limitations of this prediction method.

The matching score of each candidate structure is taken as the sum of scores derived for each predicted ion. These individual ion scores are determined by a function of the form

$$W^* (X \text{ or } 0 \text{ or } -Y),$$

where X = contribution when a predicted ion is observed and its observed intensity is within a predicted range

0 = contribution when an ion is observed but its intensity is not compatible with the predicted range

Y = contribution when an ion is predicted but not observed

W = a weighting factor expressing the relative importance of the ion or of the process leading to it (such as the ion mass or the confidence factor associated with the prediction rule).

(The results from these ranking functions are fairly insensitive to the specific values chosen for the parameters X and Y; generally, to reflect the importance accorded to invalid predictions, $Y > X$.)

As illustrated in the published application, rules used in the mass spectral prediction system can either be created on the basis of a chemist's own analysis of available data from reference compounds or be derived through an automated spectral analysis system. In the work of Lavanchy et al. the dozen or so rules defining side-chain fragmentations were formulated by chemists experienced with marine sterol spectra. Those rules describing the fragmentations of sterol nuclei were derived with the help of the **INTSUM** program.

The INTerpretation and SUMmary (**INTSUM**)[19,20] program can assist the chemist in identifying fragmentations of a common skeleton for which substantial evidence exists in the spectra of many example reference compounds. This program identifies those fragmentation processes characterizing some minimum portion (e.g., 75%) of available reference structures and spectra. It first determines all possible fragmentations of a given molecular skeleton (subject to defined limits on the overall complexity of the processes considered). Recorded spectra of reference compounds are then interpreted in terms of these possible fragmentations and accompanying neutral losses and hydrogen transfers. A summary is provided detailing all those processes that could explain ions observed in the reference spectra, together with their frequency of occurrence and the range of recorded ion intensities. The example rule given above would have been defined, in **INTSUM**'s summary, in terms of the cleavage of the C(13)-C(17) and C(14)-C(15) bonds of a complete steroid nucleus. From the data provided by **INTSUM**, the chemist can abstract information needed to define rules as used by the spectrum prediction program.

Results from **INTSUM** can be used in the **RULEGEN-RULEMOD** programs[20,21]. These two programs abstract more general descriptions of fragmentation processes from the very specific definitions of cleavages provided by **INTSUM**. The analysis performed in **RULEGEN-RULEMOD** is similar, in principle, to the approach used by Mitchell in developing rules for ^{13}C prediction. The primary use of the **RULEGEN-RULEMOD** analysis is to identify, in general terms, the principal fragmentations characteristic of some class of compounds. Emphasis is accorded to those fragmentation processes common to all example structures. Fragmentation rules derived through **RULEGEN-RULEMOD** are of limited utility in attempts to discriminate among different candidate structures that all incorporate the same structure and exhibit the same principal fragmentations.

Although this rule-based approach to mass spectral analysis can achieve a high performance level at discriminating among candidates, its scope of application is limited. The pertinent mass spectral prediction rules have to be encoded prior to their use in structure elucidation. Much detailed mass spectral data, on similar known reference compounds, must be analyzed in order to develop these rules. Consequently, use of this approach is limited largely to those research projects involving long-term studies of restricted classes of compounds.

XI.E.2.c. Dealing with Multiple Rules and Predicting C-13 Spectra of Stereoisomers

With only a few rules, accurate mass spectral predictions can be achieved for compounds from some limited class, and subtle structural differences can be detected and exploited. Several extensions to the previously described method of Mitchell and Schwenzer,[22,23] are necessary before one can achieve comparably accurate predictions of ^{13}C shifts of atoms in complex molecules.

Rules using constitutional (topological) substructures can provide some degree of discrimination among different candidate structures. The shift ranges of such ^{13}C prediction rules are generally fairly broad but may still suffice. For example, a particular candidate may incorporate methyl groups only in substructures predicted as yielding resonances in, say, the 17–26 ppm region. Such a candidate obviously will be rejected if the only quartet resonances in the observed spectrum are at 14 and 28 ppm. The effectiveness of rules using constitutional substructures, and consequent broad shift ranges, depends on the variety of structural types remaining at the stage where the ^{13}C analysis is performed. If there are many structures of quite varied types, a constitutional analysis may be appropriate. Such a limited analysis may allow some candidates to be discarded prior to generation of stereoisomers.

Obviously, the final determination of the correct stereochemical structure of a compound requires the representation of its stereoisomers and predictions based on stereosubstructures. However, the importance of steric shifts in ^{13}C analysis is such as to make it appropriate to consider stereochemistry even at the stage where one is simply trying to identify the correct structure among a number of candidates.

A simple example of how constitutional analysis may be insufficient is provided by the spectrum and structure of a derivatized monoterpenoid, **XI-8**: The shifts for the two topologically equivalent methyls, C(1) and C(9), are 25.7 and 17.6 ppm, respectively. A spectrum prediction rule, describing purely the topological environment of these methyls, would have to be something like

$$C^{*}H_3\text{–}C(CH_3)\text{=}CH\text{–}CH_2\text{–} \longrightarrow 17 \text{ ppm} < \delta(C) < 26 \text{ ppm}$$

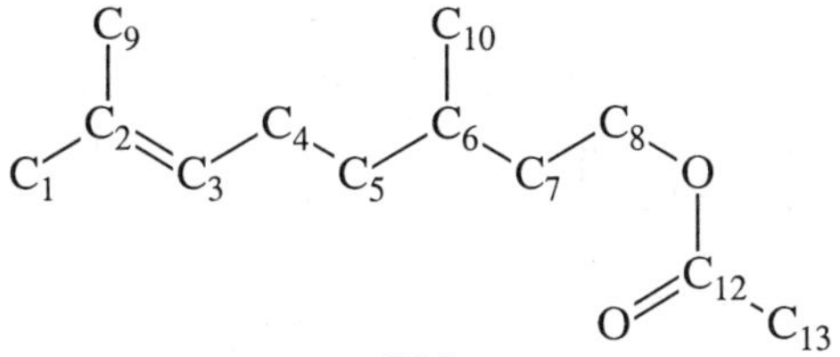

XI-8

Inevitably, a prediction involving so broad a range of chemical shift lacks discriminating power. A candidate structure incorporating this vinylic substructure would be predicted to have two quartet resonances in this general spectral region. These predicted resonances could seemingly match adequately to a couple of observed resonances at, say, 20.5 and 21 ppm. In the true structure, these resonances might be due to an aromatic methyl and an acetoxy methyl group. Although the candidate and true structure might have a number of significant differences in form, their predicted spectra might be effectively identical. Spectra based on rules using only constitutional substructures may be so "fuzzy" that useful discrimination cannot be achieved.

However, if configurational stereochemistry can be represented in both candidate structures and in the rules' substructures, two distinct rules are obtained, each having a quite precise shift range:

$$\mathrm{C^{*}H_3{-}C(CH_3){=}CH{-}CH_2}(\textit{trans}){-}\!\!\longrightarrow 25\ \mathrm{ppm} < \delta(\mathrm{C}) < 26\ \mathrm{ppm}$$

$$\mathrm{C^{*}H_3{-}C(CH_3){=}CH{-}CH_2}(\textit{cis}){-}\!\!\longrightarrow 17\ \mathrm{ppm} < \delta(\mathrm{C}) < 18\ \mathrm{ppm}$$

These predictions would no longer match to observed quartet resonances at around 20–22 ppm; consequently, the candidate would be revealed as being inappropriate. Thus a requirement for an effective system for analysis of ^{13}C data is that it can utilize stereochemical information if available.

As noted in the discussion of Mitchell's rule-formation procedure, there are physical constraints on the form of substructures used in ^{13}C prediction rules. General substructural forms involving arbitrarily complex ring systems or other ramifications, such as may well be necessary in a mass spectral prediction rule, are not required in ^{13}C analysis. Rather, one requires a substructure that defines the "spherical" environment of the resonating atom out to some maximum "bond radius." (In the following discussion, "bond radius" is used to describe not some through-space distance, but rather the length of a path through the bonds of a structures.)

A hierarchy of such substructure definitions can be created; the simplest describes only the one-bond environment, and others extend to two, three, and four bonds and incorporate stereochemical detail at the point where distinction between stereoisomers is possible. (It is advantageous if the stereochemical detail is separate from the definition of the constitutional substructure. If these data can be separated, the same set of substructure–shift rules can apply for both processing of constitutional candidates and configurational stereoisomers).

If the substructures used in the prediction rules define strictly hierarchical, spherical environments of a central atom (or bond, as might be appropriate in some other application), there is no need for any elaborate "learning" process. Instead of "learning" abstractions that describe the reference data, these data are used to describe themselves. Each reference structure can be analyzed, and the substructural environments of carbon atoms, with assigned resonances, can be characterized out to various bond radii. Generated descriptions of these substructural environments must be put into some canonical form so that the same environment can be recognized in any structure. The canonicalized substructure representations, and associated shift values, may then be used directly to update, or expand a prediction rule set. In the rule set, each canonical substructure is associated with a shift range spanning all shift values as observed for that substructure in any reference compound. The data characterizing each reference compound are used to update rules specifying one-, two-, three-, and four-bond substructural environments. Since all substructures are canonical in form, the appropriate rules requiring modification are readily identified.

This set of rules can then be used to predict spectra for candidate isomers or stereoisomers. The substructures of the rules must be matched onto each candidate. As the rules' substructures are canonical in form, only a single rule, at a specific value of the bond radius, will match the environment of each constituent carbon atom in any particular candidate structure. The most reliable prediction will normally be that based on the most detailed match, with the rule and candidate substructures corresponding to the greatest bond radius. (If the shift range characterizing, say, a four-bond matching substructure has been derived from only two or three instances in known reference structures, it may be more appropriate to use a prediction based on a three-bond match with a shift range derived from a larger, more representative set of example reference compounds.)

Such a system allows for accurate predictions of ^{13}C shifts. The inclusion of stereochemistry in the rules' substructures permits precise shift range specifications. The hierarchical, canonical forms for the substructure definitions avoid problems that arise when alternative substructure-shift predictions are found for an atom and also allow for a consistent approach to generalization. If an exact matching substructure is not found for the environment of an atom in some candidate, a more general rule can be identified by reducing the bond radius of the substructure representing the atom's environment.

The problem besetting this prediction method is the inherently vast size of the rule set. In their limited study of acyclic alkanes and amines, Mitchell and Schwenzer employed more than 100 rules to characterize the spectra of seventy reference compounds.[22,23] Obviously, a much larger number of more diverse substructure–shift range rules are required for characterization of atom environments in compounds of more varied functionality and with more complex structural forms incorporating stereochemical detail. Thousands of distinct environments are possible if the substructure description extends to a four-bond radius as is, for example, necessary to relate the stereochemistry of the ring substituents to that of the ring fusions in steroids. The results from a **DENDRAL** study provide an illustration of the numbers of substructures required. Some 22,000 distinct four-bond stereochemical atom environments were generated from less than 2000 reference compounds. (These references were mainly natural products varying in size from C_{10} monoterpenes to C_{30} alkaloids.) A system requiring the graph matching of thousands of substructures could not provide a practical basis for spectral prediction.

Independently, both Bremser[24,25] and Jezl and Dalrymple[26] had devised solutions to this problem. Their main concerns were the provision of some assistance in the "manual" interpretation and analysis of ^{13}C data. Their systems could be used to obtain crude shift range predictions for chosen atoms in some structure hypothesized by the chemist. In their systems, the structural environment of an atom was represented as a "code"—that is, a sequence of alphanumeric symbols.

The codes used by Jezl and Dalrymple were somewhat *ad hoc*, but those of Bremser define in canonical form a hierarchically organized spherical environment for each atom.[27] Bremser's encoding system does not express stereochemistry but, otherwise, provides the basis for a practical ^{13}C spectrum prediction system for either manual or automated use. The substructures, defining the atom environments in a candidate structure and the premise parts of the prediction rules, are both represented as alphanumeric codes. Rather than proceed through the rule set testing for premise substructures that match to an atom's environment, one can simply encode the environment and look up the resulting code to obtain a shift range. The comparison of two alphanumeric strings is much simpler than a full graph-matching test for substructural equivalence.

The **DENDRAL** ^{13}C spectrum prediction functions use the substructure encoding concepts of Bremser[24, 25] and Jezl and Dalrymple.[26] The encoding scheme employed is such that the substructure definitions are hierarchical at each bond-radius level and stereochemical detail is encoded separately from the constitutional description.[28] During processing of the substructural environment of an atom, the full four-bond code is created (including stereochemistry whenever the structure being processed has designated configurations on stereocenters and double bonds). The full code can simply be truncated to obtain the corresponding three-, two-, or one-bond environment if required.

The prediction process is simple. The substructural environment of each carbon atom in a hypothesized structure is analyzed, out to a four-bond radius, and the appropriate substructure code generated. The code is used, by means of a standard hashing procedure, to key into a *random-access* data base. There are standard procedures for linking together all entries in a "hashed" file that possess the same hash key.[29] The linked entries with common key value can be searched to identify the one with a code identical to that of the search request. The same hash key can arise by chance for quite different substructure codes, so some form of final identity matching of alphanumeric codes is always required. It is convenient to omit stereochemical detail during generation of the hash key from a substructure code. This omission allows the same random-access file to be used for prediction of spectra of constitutional isomers as well as stereoisomers; shift data characterizing different stereoisomers can be combined to yield a shift range for a constitutional isomer.

When the appropriate entry in the data base of substructure–shift rules has been identified, the corresponding shift data can be retrieved. With any limited limited data base of prediction rules, it is unusual to find rules with substructures corresponding to the complete four-bond atom environment of every atom in a candidate structure. If a four-bond substructure is not found, the **DENDRAL** ^{13}C spectrum program proceeds by eliminating the outermost shell of surrounding atoms and trying, successively, the three-, two-, or even one-bond models until a matching prediction rule is found.

An example of a predicted spectrum, for androstan-1α-ol (**XI-9**) is shown in Table XI-1. (All androstanol structures had been removed from the the data base before it was used for prediction.)

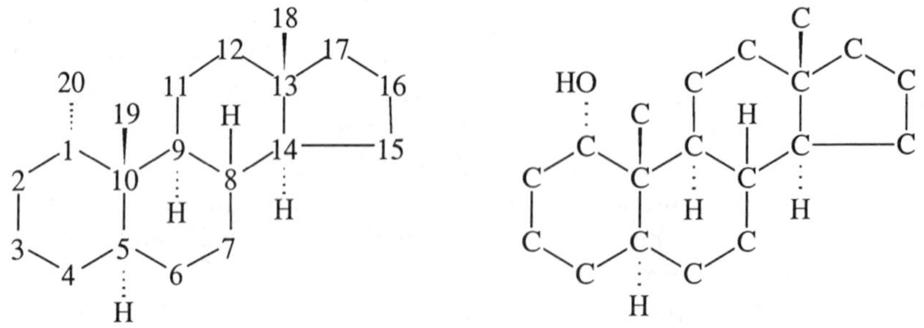

XI-9

The "shell" given for each atom (Table XI-1) defines the bond radius of the matching substructure in the rule used for shift range prediction. As in any typical application, the quality of the rules varies widely. Predictions for atoms such as C(17) and C(18) are based on substructures matching to a four-bond radius. For these atoms, the substructural environment is identical to that in the unsubstituted compound androstane (**XI-10**) (or to any androstane with only a ring-A substituent group). Predictions using detailed shell-3 and shell-4 substructures result in narrow, well-defined ranges for the expected chemical shift and close agreement between predicted and observed resonances. For most of the atoms, such detailed matching substructures result in predicted chemical shift ranges limited to within less than 1 ppm. Predictions for other atoms [e.g., C(1), C(2), C(10), and C(19)] must be based on more general substructures and result in much broader predicted ranges. The crudest prediction, with a range exceeding 23 ppm, is for C(10). The shift prediction for this atom is based on a one-bond environment, that is, prediction for a quaternary alkyl carbon bonded to a methyl and three methines.

TABLE XI-1. Predicted and Observed ^{13}C Spectra for Androstan-1-α-ol

Node	Shell	Predicted Chemical Shifts			Observed Shift
		Minimum	Maximum	Average	
C(1)	2	71.5	79.7	76.5	71.5
C(2)	2	28.9	33.4	31.1	29.0
C(3)	2	15.9	26.8	23.7	20.3
C(4)	2	27.2	29.7	28.8	28.6
C(5)	2	39.0	46.3	42.6	39.0
C(6)	3	28.9	29.0	28.9	29.0
C(7)	3	30.5	32.6	31.8	32.2
C(8)	2	34.1	36.4	35.4	35.9
C(9)	2	46.1	55.4	49.3	47.5
C(10)	1	35.0	58.6	42.6	40.2
C(11)	2	18.1	27.3	21.2	20.1
C(12)	3	38.5	39.3	38.9	38.8
C(13)	4	40.2	41.0	40.7	40.8
C(14)	3	54.3	54.8	54.6	54.5
C(15)	4	25.5	25.8	25.5	25.6
C(16)	4	20.4	20.6	20.5	20.5
C(17)	4	40.2	40.6	40.4	40.5
C(18)	4	17.3	17.8	17.5	17.5
C(19)	2	5.9	24.5	12.7	12.9

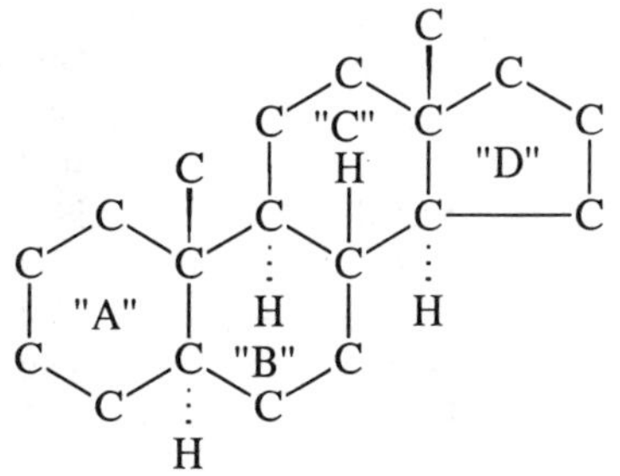

XI-10
Androstane

The quality (as measured by the matching shell level) of the rules used for spectrum prediction determines the subtlety of discrimination among different structures that can be achieved. In this example the fact that only one- and two-bond models are available for ring A and B atoms means that the predicted spectra cannot be used to resolve the stereoisomerism at C(1). An identical spectrum is predicted for both androstan-1α-ol (**XI-9**) and androstan-1β-ol (**XI-11**). In fact, as shown in Table XI-2, the spectra of these epimers exhibit a number of quite significant differences. Substantial differences (e.g., 6 ppm) in

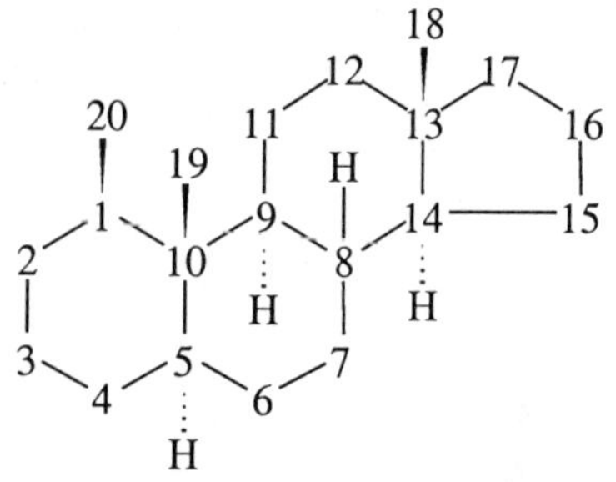

XI-11

TABLE XI-2. Observed Spectra for Androstan-1α-ol and Androstan-1-β-ol

	Androstan-1α-ol	Androstan-1β-ol
C(1)	71.5 d	78.8 d
C(2)	29.0 t	33.4 t
C(3)	20.3 t	24.7 t
C(4)	28.6 t	28.7 t
C(5)	39.0 d	46.3 d
C(6)	29.0 t	28.8 t
C(7)	32.2 t	32.5 t
C(8)	35.9 d	35.1 d
C(9)	47.5 d	55.4 d
C(10)	40.2 s	42.6 s
C(11)	20.1 t	24.7 t
C(12)	38.8 t	39.3 t
C(13)	40.8 s	40.2 s
C(14)	54.5 d	54.6 d
C(15)	25.6 t	25.8 t
C(16)	20.5 t	20.4 t
C(17)	40.5 t	40.6 t
C(18)	17.5 q	17.4 q
C(19)	12.9 q	6.7 q

shifts occur with the atoms [C(1), C(2), C(3), C(5), C(9), C(10), and C(19)] that, in one or another epimer, are sterically close to the hydroxyl group. Atoms that interact less with the hydroxyl group [e.g., C(4), C(6), C(8)] do not exhibit significant shift differences between the epimers. The predicted shift ranges, shown in Table XI-1 for the α-epimer, can be seen to span the range of shifts associated with the atoms in the two epimers.

Although imperfect, the quality of the spectral predictions is generally sufficient to help to identify the correct isomer. The objective of the entire spectrum-prediction/structure-ranking process is, as always, simply to separate a set of candidate structures into those that may be eliminated and a smaller subset, including the correct structure, that must be considered further.

As an example, the 22 possible stereoisomers of 5-α-androstane incorporating one secondary alcohol substituent (>CH-OH) were generated, using **CONGEN** and **STEREO**, and their spectra were predicted and compared with the observed data for the 1α-ol stereoisomer. Results are abstracted in Table XI-3 showing the predicted spectra for the two 1-hydroxy stereoisomers and some of the more similar stereoisomers such as the 12- (**XI-12** - **XI-13**) and 17-hydroxy androstanes (**XI-14** and **XI-15**).

Sometimes, the differences among hypothesized candidate structures correspond to substructures that result in very similar resonance shifts. Thus, although the atom environments of C(9) in androstan-17-β-ol and C(14) in androstan-1α-ol differ at the two-bond radius, the two substructures are fortuitously associated with almost identical shifts (54.9 and 54.6 ppm) and match about equally well with the observed resonance at 54.5 ppm. In general, other features of the hypothesized candidates result in distinguishable resonances. For example, the predictions for the C(18) and C(19) methyl groups differ markedly between these two stereoisomers. Thus the fact that certain quite different substructures can result in similar individual resonances is not necessarily an impediment to the prediction and ranking process.

Structures are ranked by a measure of dissimilarity between the predicted spectra and observed data.[30] The mean of the predicted spectral shifts for each atom in a candidate structure is used to define its resonance shift for comparison purposes. Predicted and observed resonances are placed into correspond-

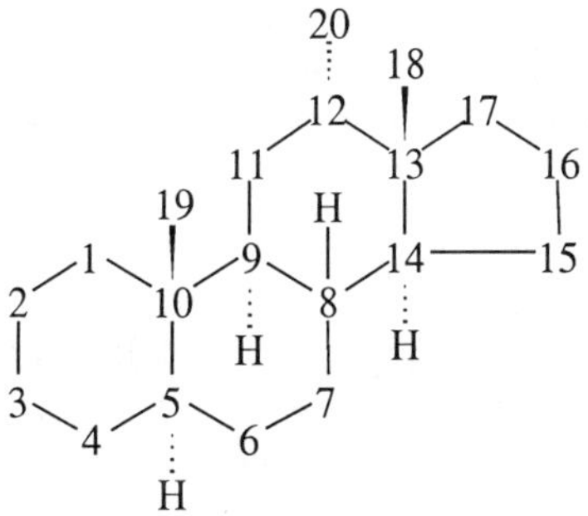

XI-12

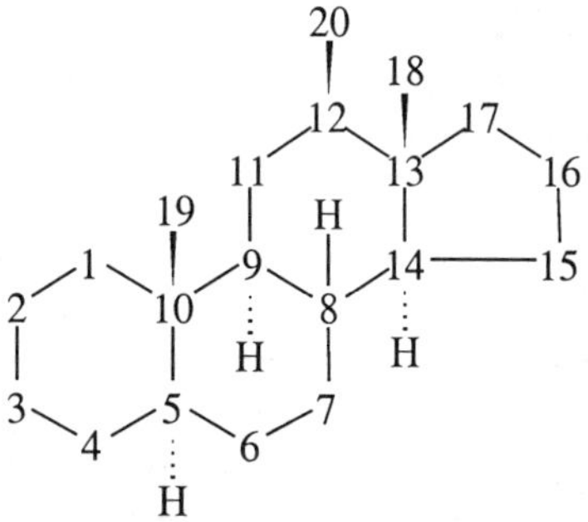

XI-13

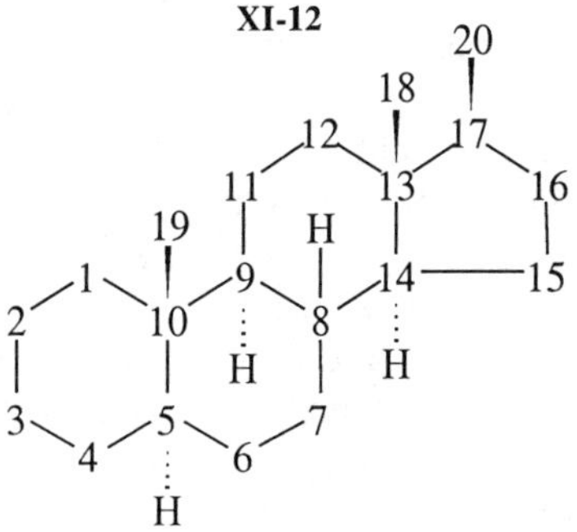

XI-14

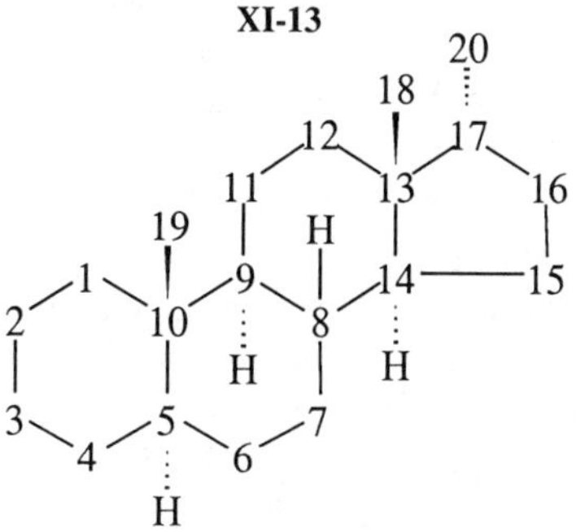

XI-15

TABLE XI.3. Comparison of Spectra Predicted for a Variety of Androstanols with Observed Data for Androstan-1α-ol

	Comparison of Observed and Predicted Spectra[a]				
Observed	Androstan-12α-ol	Androstan-12β-ol	Androstan-17α-ol	Androstan-17β-ol	Androstan-1α-ol (also 1β-ol)
71.5(d)	75.8(d,12,1)	75.8(d,12,1)	75.8(d,17,1)	81.2(d,17,4)	76.5(d, 1,2)
54.5(d)	50.1(d,14,1)	50.1(d,14,1)	54.6(d, 9,3)	54.9(d, 9,4)	54.6(d,14,3)
47.5(d)	47.1(d, 5,3)	47.1(d, 5,3)	47.1(d, 5,4)	51.2(d,14,4)	49.3(d, 9,2)
40.8(s)	39.3(s,13,1)	39.3(s,13,1)	39.3(s,13,1)	43.2(s,13,4)	42.6(s,10,1)
40.5(t)	36.8(t,17,1)	36.8(t,17,1)	38.7(t, 1,4)	38.7(t, 1,4)	40.4(t,17,4)
40.2(s)	35.7(s,10,2)	35.7(s,10,2)	36.3(s,10,4)	36.3(s,10,4)	40.7(s,13,4)
39.0(d)	40.0(d, 9,2)	40.0(d, 9,2)	49.2(d,14,2)	47.1(d, 5,4)	42.6(d, 5,2)
38.8(t)	38.6(t, 1,3)	38.6(t, 1,3)	37.6(t,12,2)	37.2(t,12,4)	38.9(t,12,3)
35.9(d)	35.4(d, 8,2)	35.4(d, 8,2)	35.4(d, 8,2)	35.7(d, 8,4)	35.4(d, 8,2)
32.2(t)	31.8(t, 7,3)	31.8(t, 7,3)	31.8(t, 7,3)	32.0(t, 7,4)	31.8(t, 7,3)
29.0(t)	29.0(t, 4,4)	32.2(t,11,1)	29.0(t, 4,4)	30.6(t,16,4)	31.1(t, 2,2)
29.0(t)	29.0(t, 6,4)	29.0(t, 4,4)	29.0(t, 6,4)	29.0(t, 4,4)	28.9(t, 6,3)
28.6(t)	28.7(t,11,2)	29.0(t, 6,4)	26.8(t,16,1)	29.0(t, 6,4)	28.8(t, 4,2)
25.6(t)	26.7(t, 3,4)	26.7(t, 3,4)	26.7(t, 3,4)	26.7(t, 3,4)	25.5(t,15,4)
20.5(t)	25.5(t,15,2)	25.5(t,15,2)	24.7(t,15,2)	23.5(t,15,4)	23.7(t, 3,2)
20.3(t)	22.1(t, 2,4)	22.1(t, 2,4)	22.1(t, 2,4)	22.1(t, 2,4)	21.2(t,11,2)
20.1(t)	20.5(t,16,2)	20.5(t,16,2)	20.8(t,11,3)	20.9(t,11,4)	20.5(t,16,4)
17.5(q)	16.7(q,19,3)	16.7(q,19,3)	14.9(q,18,2)	12.2(q,19,4)	17.5(q,18,4)
12.9(q)	14.9(q,18,2)	14.9(q,18,2)	12.2(q,19,4)	11.3(q,18,4)	12.7(q,19,2)

[a]The predicted spectra are given as a shift value followed, in parenthesis, by the signal multiplicity, the atom number, and the shell level of the matching substructure used in the prediction process.

ence on the basis of shift and multiplicity, and a dissimilarity value for each resonance is calculated. This dissimilarity value is taken as the square of the difference between predicted and observed shift values. The overall dissimilarity score between the observed spectrum and that predicted for a particular candidate structure is basically a sum of the dissimilarity values contributed by each atom. (This dissimilarity score is really a simple distance measure between two points in n-dimensional space with the numeric shift values defining the coordinates of the points.)

A prediction based simply on the one-bond environment of an atom would be the weighted mean of a set of resonances that might span a range of as much as 70 ppm; obviously, such a mean value would contribute a very substantial dissimilarity score if matched with an observed resonance from either extreme end of the allowed range. Candidate structures incorporating novel features, and consequently having spectral predictions derived from generalized models, would tend to be discriminated against by any simple scoring function. Yet, in such cases the problem is not in the form of the hypothesized candidate structure, but rather in limitations of the data base of prediction rules. Since the main objective of the structure evaluation process is to identify and eliminate those structures that are incompatible with experimental observations, candidates for which no reliable predictions can be made should be accorded *low* dissimilarity scores. The scoring function must, therefore, take into account the quality of the model used for predicting the individual resonances in each hypothesized candidate structure.

A direct measure of the suitability of a prediction rule is given by the shell level at which that rule's substructure matches the environment of an atom in a candidate structure. This shell level can be used as a weighting factor for combining the individual resonance dissimilarity values to derive an overall dissimilarity score for a predicted spectrum. Thus, with the use of the shell level, a crude shell-1 prediction of 81 ppm matches an observed resonance at 75 ppm "better" than does a shell-4 prediction of 71 ppm and contributes less to the overall dissimilarity score for a compound. Although completely empirical, the shell weighted dissimilarity score has the desired characteristics for a scoring function for these ^{13}C spectral predictions. This function has been shown to provide reasonably good discrimination among stereoisomers for which reliable predictions can be made while avoiding bias against structures containing novel substructural features.[30]

Of the 22 possible stereoisomers, the two 1-hydroxy isomers are best matched with the observed data for the androstan-1α-ol; as noted earlier, the predictions cannot differentiate between the two epimers. The two androstan-12-ols are third and fourth ranked. Similar analyses of the observed spectra for other isomers give equivalent results, with the appropriate stereoisomer ranked within the top four candidates in all cases. These results are summarized in Table XI-4.

The effectiveness of this approach to structure evaluation is determined by the comprehensiveness of the rule set. If the rule set contains substructures representing atom environments, as found in the various candidates, quite accurate shift range predictions are achieved. These predictions usually suffice

TABLE XI-4. Summary of Results Obtained When Ranking Androstanols According to the Match Between their Predicted Spectra and Data Observed for Androstan-1α-ol

Results of Scoring Functions			
Compound	Shell	SDIS2[a] Score	Rank
Androstan-12α-ol	2.4	21.9	3
Androstan-12β-ol	2.3	23.3	4
Androstan-17α-ol	2.8	47.5	6
Androstan-17β-ol	4.0	124.9	21
Androstan-1α-ol	2.6	14.9	1
Androstan-1β-ol	2.6	14.9	1

[a]SDIS2—shell-weighted dissimilarity score.

to greatly reduce the number of candidates that the chemist should really need to consider. However, it is quite rare for the rule set to contain all necessary rules. Often, candidates incorporate unusual features not observed in available reference compounds. Accurate predictions cannot be made for such candidates; consequently, they cannot be eliminated.

XI.E.3. Simplified Spectrum Prediction Theories

Rule-based procedures provide a generally simple, convenient approach to the analysis of structures within their domain. As discussed above in relation to the rule-based ^{13}C prediction systems, discrimination among structures does depend on the rule base containing appropriate models. More general procedures, capable of handling arbitrary structures, are also required.

Such procedures must be based on theoretical models of the physical processes giving rise to spectral absorptions. A theory can provide the basis for some algorithmic structural analysis that computes properties of defined structures. There are two main examples of theoretical approaches to spectrum prediction for structural analysis: (1) there are programs that undertake an analysis of ^{13}C data through a simplified theory deriving from considerations of topological symmetry; and (2) there is **DENDRAL**'s half-order theory of mass spectrometry, which serves as a basis for rationalizing observed data in terms of fragmentations of given structural candidates.

XI.E.3.a. Theories for ^{13}C Spectra Prediction Based on Topological Analyses

Programs for analyzing the ^{13}C NMR spectra of candidate structures have been developed for both Munk's **CASE** system[31] and Sasaki's **CHEMICS** system.[32] These two ^{13}C analysis programs are very similar. Both are based on a simple theory for predicting the number of distinct resonances that should be observed in the ^{13}C spectrum of a compound.

The theoretical basis of the analysis is that, whereas structurally distinct atoms should have distinct resonances, atoms that are symmetrically equivalent should show identical resonances. Therefore, the number of distinct signals in a compound's spectrum should correspond to the number of distinct structural environments for carbon atoms in that molecule. If a structure is completely asymmetric, the number of observed resonances should equal the number of constituent carbons. In a molecule with some symmetry, such as a 1-,4-disubstituted phenyl system, the symmetrically equivalent carbons should have identical shifts; consequently, there will be fewer distinct signals than carbon atoms.

A discrepancy between the number of distinct resonances as in the recorded spectrum of an unknown and the number predicted for a candidate structure establishes that that candidate has inappropriate symmetry. Consequently, such a candidate structure can be eliminated. In both the **CHEMICS** and **CASE** systems, candidate structures resulting from the generation phase are screened by predicting and checking resonance numbers.

Obviously, there will be occasional problems due to chance degeneracies of different ^{13}C resonance shifts. A particularly common case is that of methylene carbons in long alkyl chains; although each such methylene may be in a unique structural environment, the resonances of these methylenes seldom can be resolved. In Sasaki's **EXAMINE C13** program, such problems are partially catered for by treating these methylenes as a special case; distinct resonances are predicted only for those methylenes within a three-bond radius of a terminal or substituent group. Additionally, neither the **CASE** nor the **CHEMICS** analysis system requires exact correspondence of resonance counts. Instead, a limit on the difference in the predicted and observed number of signals is specified by the user.

A more significant problem with these theoretical approaches relates to the form of the symmetry analysis. These theories are based on an analysis of topological symmetry similar to that described in earlier chapters. The symmetry group of the molecule is determined and, from the symmetry permutations, equivalent carbon atoms identified. The number of distinct groups of equivalent carbon atoms in a candidate defines the expected number of resonances for that structure.

Such a topological analysis is, however, inappropriate. An example is provided by the two C(3) methyls in 3,3,5-trimethylcyclohexanone (**XI-16**). These carbons are topologically equivalent and would, on the basis of the simple theory, be predicted to have identical resonances.

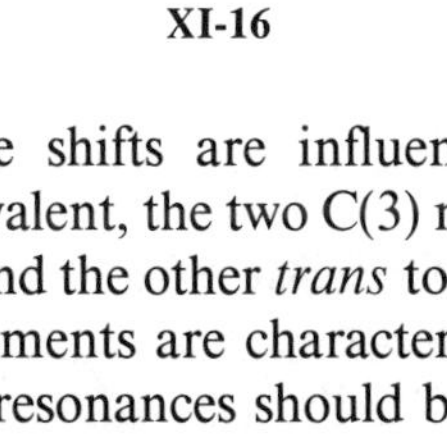

XI-16

However, carbon resonance shifts are influenced also by steric factors. Although topologically equivalent, the two C(3) methyls are in differing steric environments with one *cis* and the other *trans* to the C(5) methyl. Carbons in such differing steric environments are characterized by quite distinct shifts. Consequently, nine distinct resonances should be expected for this structure rather than the eight predicted on the basis of the simple topological theory.

Ad hoc extensions to the basic theory allow obvious cases, such as these diastereotopic methyls, to be detected and accorded special treatment. Both versions of this ^{13}C analysis method include subroutines to identify topologically equivalent geminal methyls and then find any stereocenters within some limited bond radius. If a possible stereocenter is found, distinct resonances are predicted for the methyls.

A more satisfactory approach would be to take stereochemistry into explicit account. This would require extension to the **CASE** and **CHEMICS** programs, which do not currently represent molecular stereochemistry. If configurational stereoisomers were represented, the symmetry analysis could be based on the configurational symmetry group, and thus *ad hoc* processing could be avoided. Even a configurational symmetry group analysis would not resolve all problems. The problem of chance degeneracies would remain. In a few cases conformational effects might break the effective symmetry of configurationally equivalent atoms and hence result in an observed number of resonances apparently inconsistent with a configurational representation of the correct structure.

XI.E.3.b. A Half-Order Theory of Mass Spectra

Chemists can often infer the specific positions of substituents on ring systems and other similar structural details from a simple analysis of a compound's mass spectrum. Such an analysis involves the examination of various hypothesized candidate structures to determine whether their fragmentation could be expected to result in ions with the compositions, or perhaps only masses, actually observed in the recorded mass spectrum of the unknown. Usually, the mass spectral processes, as considered by the chemist, would be relatively simple. Thus the most elaborate process considered might comprise a ring cleavage, with accompanying hydrogen transfers, and possibly further loss of substituent methyls or neutral moieties such as H_2O.

It is easy to determine whether an ion of given composition would result from application of a particular postulated fragmentation process to a specific candidate structure. The plausibility of a candidate structure may be reduced if its fragmentation cannot result in an ion observed as having significant intensity in the spectrum of the unknown. However, it is not sufficient to consider only one postulated fragmentation process before the plausibility of a candidate can actually be downgraded. It must be determined that there are no alternative mechanisms, involving different skeletal fragmentations, that might yield an ion of the observed composition. All processes of comparable complexity must be examined and their implications regarding ions resulting for a given structure developed. A candidate is inconsistent with the spectral data only if there is no reasonable mechanism whereby it could fragment to give the observed ion.

This form of mass spectral analysis must involve a combinatorial search for all the possible fragmentations of each candidate structure. The combinatorial nature of the processing makes the task more suited to a computer than to human analysis. If the analysis task is delegated to a computer, two further refinements of the approach are practical. First, rather than select a single ion from the observed spectrum, the processing can apply to the complete data. Candidate structures are thus assessed according to how good a rationalization they can provide of the full spectrum. A second refinement is to distinguish among the different fragmentation processes that can be found to account for an observed ion. A particular ion may be interpretable in terms of a simple cleavage of a single acyclic bond in one candidate. Another candidate structure might provide a rationalization of the same observed ion involving cleavage of a fused ring system, loss of an acyclic chain as a secondary process, further loss of some neutral moiety, and various hydrogen transfers. Although both structures may thus serve to rationalize the ion, the structure with the simpler postulated fragmentation mechanism would, in general, be held more plausible. The **DENDRAL** Mass Spectral Analysis (*MSA*) system incorporates functions implementing this generally applicable approach to spectral analysis.

In the *MSA* functions, all possible fragmentation processes subject to overall complexity constraints are first identified for each of the candidate structures. This analysis determines those ion compositions and masses that can reasonably be rationalized in terms of a given candidate's structure. Of course, not all the ions thus predicted even for the correct structure would be observed. For example, in the correct structure there might be some plausible process involving a simple ring cleavage, with bond breaks adjacent to quaternary and tertiary carbons. The recorded spectrum might not show any appropriate fragment ions. Charge may have been localized, and all fragmentation may have been directed by some constituent group in another remote part of the molecule.

Another factor leading to "overprediction" of the spectrum is the uniform application of hydrogen transfers and neutral losses. The analysis might, for example, allow transfer a maximum of two hydrogens either out of or into an ion. Although there might be instances of each such hydrogen transfer in the complete spectrum, the majority of ions will be produced by fragmentations processes in which only one or two of these different hydrogen transfers occurs. The combinatorial analysis might predict five ions from each fragmentation process, one corresponding to each possible hydrogen transfer, whereas only one or two ions may be observed.

Inevitably, analysis of *all possible fragmentations* of a structure results in many more ions being "predicted" than are actually observed. Moreover, ion intensities can not reasonably be predicted. The basic combinatorial analysis of possible fragmentation processes does not allow for the effects of any initial charge localization. Various extensions might allow some empirical treatment of charge localization, but the approach would still fail to address an even more important problem. Fragmentations are, of course, competitive. The various rates for different processes are of considerable importance in determination of ion intensities.

The spectrum predicted for a structure, through a combinatorial analysis of fragmentations, cannot correspond closely to the observed spectral patterns. Consequently, the direct comparison of predicted and observed data cannot serve as the basis for evaluating different candidates. Instead, candidates can be evaluated according to the plausibilities of their postulated fragmentation processes. For each candidate, the most plausible process leading to each observed ion is identified. A combination of the plausibility estimate for this best process, together with an importance rating for the particular ion, contributes to a cumulative score for the candidate.

The combinatorial analysis of possible molecular fragmentations constitutes **DENDRAL**'s so-called "half-order" theory of mass spectrometry. (The name for the theory is intended to acknowledge the fact that it ignores many of the important factors determining the actual mass spectral patterns obtained for a compound.) The spectral analysis procedures allow molecular fragmentations to be defined in terms of one or more cleavage steps (each step involving one or more bond cleavages), hydrogen transfers, and neutral losses. There are constraints imposed on the overall complexity of any fragmentation process considered. These constraints include limits on the total number of bonds that may be cleaved in a process, limits on the maximum allowed number of consecutive cleavage steps, and restrictions on the complexity of any individual cleavage step. The user also selects those hydrogen transfers that are to be considered as possibly accompanying any given molecular fragmentation, and what—if any—further neutral losses may also be invoked (e.g. loss of CO or H_2O).

It would be possible for a program to operate by considering all one-bond cleavages, all combinations of pairs of bond cleavages, all sets of three bond cleavages, and so on. However, in structures of chemically interesting size and complexity, such an approach would fail because of the "combinatorial explosion" that would occur. Only a small subset of all two- and three-bond combinations will constitute genuine, distinct cleavages of a structure. For an efficient approach, it is necessary to determine which combinations of bond-cleavages will result in actual fragmentation of a given molecular structure; these combinations of bonds are the "cutsets" of that structure.

Identification of the cutsets for some structure requires an analysis of its ring systems. Each acyclic bond in a structure constitutes a one-bond cutset; two-bond cutsets correspond to simple ring cleavages, whereas three-bond cutsets are equivalent to cleavages in fused or bridged ring systems. The half-order model allows the user to specify which cutsets are required; for most applications, one- and two-bond cutsets (acyclic and simple ring cleavages) are sufficient. This user control is expressed in terms of constraints as to whether the most complex single step be an acyclic cleavage (only one-bond cutsets needed), a simple ring-cleavage (one- and two-bond cutsets needed), or a cleavage of a fused ring system (one-, two- and three-bond cutsets needed).

Each step in a fragmentation process constitutes a cleavage of a molecule into two parts *independent* of the number of bonds cleaved in order to form the fragments. The user selects the maximum number of steps that may be allowed for any fragmentation process. In most examples one- or two-step processes will suffice to explain all the major fragmentations observed. If only one-step processes are to be considered, each cutset will correspond to a fragmentation process. Each cutset gives rise to two ions, depending on which side of the cut the charge is formally placed. If multistep processes are to be permitted, then cutsets must be combined, in ways that yield only allowed fragmentations.

The half-order theory allows plausibility values in the range 0–1 to be associated with each aspect of a fragmentation process. For example, cleavage of adjacent bonds is energetically unfavorable (for it formally results in the creation of a carbene species), and thus a lower plausibility can be associated with such processes. Similarly, processes involving more than one step, or complex steps such as cleavage of fused ring systems, can also be assigned lower plausibility values.

The half-order theory also allows for variation in the relative plausibility of cleavage of bonds dependent on their character. The basic model distinguishes only between the cases of single bonds, aromatic bonds, and multiple bonds, permitting a different plausibility value to be assigned to each such class of bonds. Finer discrimination, such as distinction between vinylic and allylic single bonds, is achieved through the use of substructural templates. These substructural templates define the structural environment of a bond for which some specific cleavage plausibility should be used. The templates are matched to a structure, by a conventional node-by-node graph-matching algorithm, and the appropriate bonds are identified; these bonds are assigned the cleavage plausibility given with the substructure. Unmatched bonds in a structure are assigned the default cleavage plausibility for their bond class.

The overall plausibility of a particular fragmentation process is taken as the product of each of its component factors—plausibilities of cleavages of the bonds involved, hydrogen transfers, neutral losses invoked, and any appropriate modifying factors such as reduced plausibility associated with multistep processes or adjacent cleavages.

The processing involved in application of the half-order theory to a molecular structure involves the following steps:

1. Identify a set of fundamental cycles and store representations of these cycles, a spanning tree of the structure, and related data.
2. Determine the relative cleavage plausibility of each bond. If substructures are being used to modify bond cleavage plausibilities, the graph-matching would be performed at this stage. [Bond cleavage plausibilities should be determined before the cutsets are found (next step), as any bond with zero cleavage plausibility can then be ignored.]
3. Identify cutsets.
4. Predict all ions that can be created, using the computed cutsets and the given hydrogen transfers and neutral losses:
   a. Select combinations of fragmentations, considering in turn processes involving a total of 1, 2,...maximum number of bond cleavages.
   b. A standard integer partitioning algorithm assigns the current number of bonds in all legal and distinct ways to different steps in the multi-step process. (Obviously no one step can involve more than three bond cleavages, and must have less than three unless cleavage of fused or bridged rings is allowed.) This partitioning determines the number of one-, two-, and three-bond cutsets required in the process.

c. A recursive function is then used to create all possible distinct selections of cutsets of each size from the sets of available cutsets. Some checks are made here to ensure that the same bond is not cleaved twice at different steps of the process and to eliminate many of the selections of cuts that will, in fact, lead to disconnected fragments.

d. Apparently valid selections of cutsets are checked to verify that the ion is actually connected. Then, the composition and basic plausibility of the process are computed. All variants on the ion due to the allowed hydrogen transfers are created and checked; the checks involve composition and valence constraints on both the ion and the neutral fragment(s) lost.

e. As each hydrogen transfer variant of the ion is generated, all possible single neutral losses are used to construct new variants. Composition and valence checks are made on the ion part and legal variants are passed to the final processing function (compositions are converted to nominal masses if only low-resolution data are required).

f. The final processing varies according to the operation mode of the program. When simply predicting spectra, any ions predicted are stored. (If the same ion is predicted by many different processes, a record is kept of its largest plausibility value.) When ranking structures, any ions predicted but not present in the observed spectrum used for comparison may be ignored; otherwise, the maximum plausibility value for all those processes leading to an observed ion is recorded.

Details of algorithms for identifying cutsets and finding valid combinations of cutsets have been presented.[9] Examples of predicted spectra and results from practical applications have also been illustrated.[7,9]

When interpreting results based on the half-order approach to spectrum prediction, it must always be realized that cases can arise where its simplifying assumptions may be inappropriate. The kind of problems that arise can be illustrated by considering the fragmentation behavior of some monoterpenes. Example structures could be δ-1-menthene (**XI-17**) and (*endo*)-isocamphane (**XI-18**). These compounds have quite similar spectra: m/z 138 (28%, 16%), m/z 123 (14%, 14%), m/z 110 (0%, 6.5%), m/z 109 (3.5%, 67%), m/z 96 (17%, 18%), m/z 95 (100%, 100%), m/z 94 (8%, 9%) (the two values given in parenthesis represent the ion intensities in the two spectra). With these structures and other $C_{10}H_{18}$ monoterpenes, the spectra all show the same fragment ions but subtle intensity variations can permit interpretation and identification. However, here it is not necessarily valid to assume that the correct structure is the one that explains an ion by a "simpler" and, therefore, higher plausibility process. For δ-1-menthene the base peak at m/z 95 is due to a simple acyclic cleavage

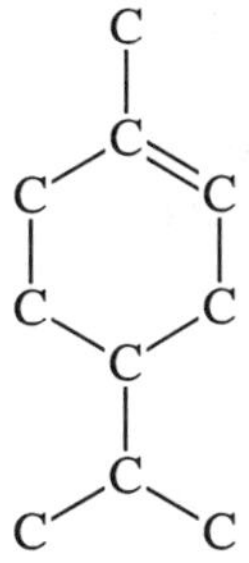

XI-17 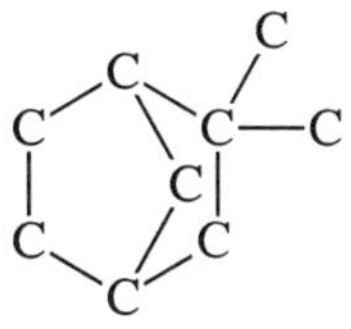 **XI-18**

leading to loss of the isopropyl substituent group; no such simple fragmentation mode is possible for isocamphane. For isocamphane, the base peak is explained as originating through expulsion of the *gem*-dimethyl group (i.e., ring scission with cleavage of adjacent bonds) and a transfer of a hydrogen. Given any reasonable parameterization, δ-1-menthene will constitute the more plausible structure for not only its own spectrum, but for the spectrum of isocamphane as well. Thus uncritical application of the half-order model can lead to a structurally simpler candidate, such as δ-1-menthene, always being favored over more complex candidates, such as isocamphane. When analyzing structures, where such subtle spectral differences are important, it is necessary to use class-specific fragmentation rules.

This example provides a specific instance of an inherent tendency to favor structures where cleavages of a single bond can result in an ion of given composition, as opposed to structures where ring bonds must be cleaved. This bias results in isthmus graph structures often accruing higher scores than the true molecular structure when an observed spectrum is analyzed. For example, consider the structures and cleavages shown as **XI-19** and **XI-20**:

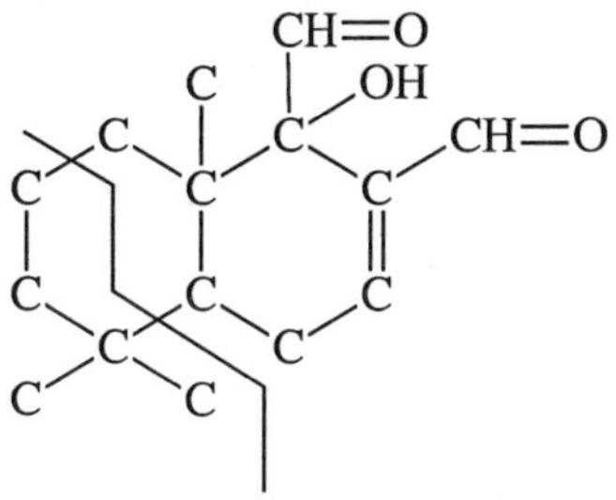

XI-19 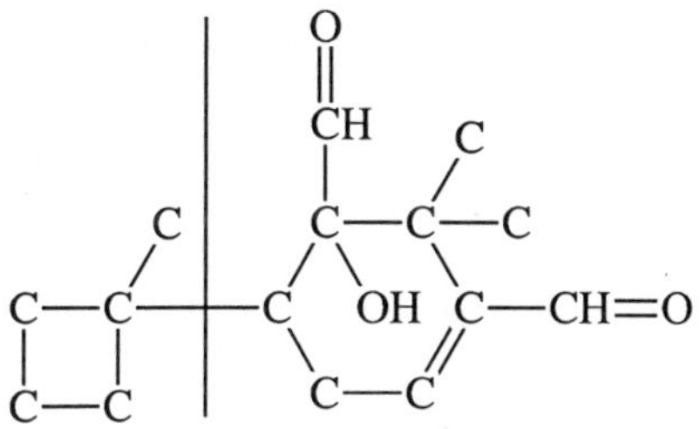 **XI-20**

These cleavages of the example structures would, with appropriate hydrogen transfers, give rise to fragments of identical composition. In the case of **XI-20** only a single acyclic bond need be broken, and thus the fragmentation will be rated as highly plausible. In **XI-19** the fragmentation is more complex, requiring two bonds in a ring to be cleaved, and will typically be accorded a lower plausibility. Given a fortuitously appropriate distribution of functionality between the "islands" of an isthmus graph, it is possible for cleavage of the isthmus to provide the basis for plausible processes leading to all major observed ions. Quite commonly, a disproportionate fraction of isthmus graphs are found among those candidate structures most highly ranked on the basis of this half-order spectral analysis.

XI.E.4. Classical and Quantum-Mechanical Methods for Spectrum Prediction

As yet, there are no reports of practical applications in structure elucidation of either classical or quantum-mechanical methods for predicting spectra. However, Gribov's group has reported on preliminary investigations of the use of both approaches. Structures produced by the **STREC** structure elucidation system can be passed to a model building program that utilizes libraries of standard bond lengths and atom environments.[33] An initial geometry is obtained for a structure through the use of these standard models. This initial geometry is optimized by minimizing a total molecular energy as estimated from atom and bond potentials. Final conformations of structures are developed by use of semiempirical quantum-mechanical energy calculations.

In the **STREC** system, IR absorption spectra for candidate structures can be calculated by semiempirical force-field calculations.[34,35] The calculation program uses a library containing information concerning the geometry, force constants, and electrooptical parameters of molecules and standard fragments. This library is used for formulating the system of equations necessary for calculation of the overall spectra. Spectral curves are constructed as envelopes of the absorption bands corresponding to separate normal vibrations. Illustrative examples of computed and observed absorption curves have been presented for many small molecules.

This work on classical and quantum-mechanical spectrum prediction has not yet addressed in detail such issues as (1) the scoring of a match between predicted and observed data; (2) the assessment of the effects, with particular structural candidates, of any limitations of empirical spectrum-prediction parameters, and (3) an appropriate ranking method.

X.F. SUMMARY

The surveying of candidate structures to identify constituent standard functional groups and skeletal fragments is a straightforward application of the graph-matching techniques discussed earlier. Once the graph-matching has been performed, the identification of structures with particular combinations of features requires only simple manipulations of bit-map records identifying the particular substructural constituents of each candidate. Interactive programs have the advantage of allowing the chemist to select desired structures, display examples, and then elaborate and refine the selection criteria. As well as standard substructural features, it is useful if the surveying system allows inquiries concerning ring systems. Although details of constituent rings can be helpful, information about ring fusions, spiroatoms, isthmuses, and bridged ring systems is generally more valuable.

For the most part, structures of interest can be specified by defining some appropriate boolean expression requiring the presence of certain combinations of substructures and the absence of other combinations structural features. Scores, either associated with specific matched substructures or derived by spectrum prediction and matching functions, can be employed as selection criteria. The use of specific score values is, however, somewhat dubious. In particular, it would not appear appropriate to attempt to combine scores, as derived by different structural analysis techniques, into a single overall score. The most appropriate use for scores might be the identification of those candidate structures with better than median values for each scored property.

Analyses of sets of structures, in order to find partitioning features, can again be based almost entirely on algorithms considered earlier. Most of the analyses performed by ***EDITOR*** in Munk's **CASE** system involve some form of ring-finding and canonicalization procedures. These are used together to identify, for example, a characteristic carbocyclic ring system of a molecule. Methods based on partitioning the set of candidate structures according to the manner of embedding of a particular substructure require simply standard graph-matching and canonicalization routines. The value of these various analyses and partitionings of the set of candidate structures can be determined only by much additional practical application.

Rule-based methods for spectrum prediction could easily be developed for additional applications. Some limited work on ^{1}H NMR has been completed.[13] The ^{1}H NMR analysis program was closely similar to that described for ^{13}C analysis, as it was based on the encoding of the configurational stereochemical environment of the resonating hydrogen atom and the use rules defining *coded configurational-stereosubstructure/shift* relationships. More extensive analysis of ^{1}H NMR spectra is probably dependent on some method for encoding conformational forms for substructures.[36]

Both the ^{13}C and preliminary ^{1}H NMR systems have been concerned with atom properties and have employed atom-centered substructures. It is, of course, possible to define canonical bond-centered substructures. Rules relating bond-centered substructures and associated spectral patterns could be derived from reference compounds in a manner analogous to that used for the atom-centered systems. Possible applications of such bond-oriented rules might include IR and raman spectroscopy, or analyses of ^{1}H NMR coupling patterns.

There are other applications that could exploit substructure-prediction function rules, similar to those used in the detailed mass spectral analyses described above. A possible example would be the prediction of the UV-visible absorption spectra of a compound. The substructural match would define the context of some particular type of conjugated system and the functional part could evaluate the specific influence of auxochromes.

Although additional applications for rule-based systems can thus be readily identified, some cautions must be expressed. For the results to be of real significance, appropriate structural models must be employed. It may be sufficient to consider only the constitutional forms of candidate structures, either because the spectral property of interest is only marginally influenced by steric factors or because some initial degree of discrimination can be reliably achieved through crude spectral predictions. However, most spectral properties are quite markedly influenced by steric factors, and one must generally consider the configurational, if not also conformational, form of a structure when attempting precise spectrum prediction. Only precise spectral predictions will effectively discriminate among candidates closely similar in structural form. The stereochemistry of the candidate structures may well have to be represented.

Most computer-assisted structure elucidation systems do not incorporate any stereochemical representations and thus cannot provide appropriate structural models for detailed spectral analysis. Even if stereochemical representations can be generated, there are further problems. A typical structure elucidation problem will result in 20 to 50 different candidate constitutional isomers. For each constitutional isomer, there may be several configurational stereoisomers. For every stereoisomer, there will be many plausible conformers. Thus very large numbers of stereochemical candidates may have to be generated and analyzed.

The scope of application for any rule-based method is limited by the set of prediction rules available. Some systems, such as Lavanchy's rule-based mass spectrum predictions for sterols, may be designed to handle limited cases using separate libraries of class-specific rules. Obviously, such rules are applied only to structural problems falling within their domain. The disadvantage of such an approach is that separate rule sets have to be created in advance for each planned area of application. The rule creation process is costly and practical only if considerable class-specific data are already available.

Other rule-based systems, such as the **DENDRAL** ^{13}C spectral prediction procedures, attempt to utilize a general-purpose library of rules. This does permit somewhat wider application but also necessitates the identification of those structures for which there are no appropriate prediction rules (so that these structures are not accorded poor scores). The ^{13}C predictions are of little value in examples where the unknown structure and the various generated candidates are different from any of the reference molecules used to create the rules. The results from the ^{13}C prediction and ranking processes would not in any way discriminate among all those candidates incorporating novel structural features.

A final caution relating to the development of rule-based systems is that it should first be determined whether the frequency of practical application of these systems is likely to be commensurate with the cost of their development. A quite effective UV-visible spectrum prediction procedure could probably be developed along the lines indicated earlier. However, few structure elucidation problems are likely to arise in which an analysis of the UV-visible spectrum is of importance.

Simplified theories of the physical processes underlying spectral absorptions, avoid some of the problems besetting the empirical rule-based approaches. Such theories allow spectra to be predicted algorithmically. All structures should, in principle, be equally readily analyzed without resort to class-specific and *ad hoc* analyses. Of course, programs using simplified theories to predict spectra will still have to deal with the problems of the choice of an appropriate structure representation and of large numbers of candidates when configurational and conformational structure is considered.

There are, as noted earlier, significant limitations to existing theories for ^{13}C signal number prediction. Furthermore, the majority of steroidal, alkaloidal, and other structures of biochemical interest possess little symmetry, even in their topology. When stereochemistry is considered, all their carbon atoms should be distinct and the number of signals should, with the exception of chance degeneracy, correspond to the number of constituent carbons. Few of the candidates generated for such structures will show any symmetry. It would be unlikely that significant numbers of candidate structures would ever be eliminated on the basis of discrepancies in counts of distinct resonances. Consequently, the scope for practical application of this "symmetry" approach to ^{13}C analysis is limited.

The half-order theory of mass spectrometry is inadequate in that it does not represent the effects of charge localization or the competitive nature of different fragmentation modes. Strictly, it cannot *predict* mass spectra; ion plausibilities, derived through the half-order analysis, bear little relationship to observed intensities. The half-order analysis does identify whether a particular observed ion can be rationalized in terms of fragmentations of a given structure, but this evaluation is intrinsically biased toward certain structural types.

More elaborate and exacting spectral theories could allow for better evaluation and ranking of candidates. Mass spectrometry is probably the area where further developments would be of optimal value. Mass spectral data are currently underutilized in automated spectral interpretation and structure elucidation systems; this under-utilization is likely to continue until effective interpretive schemes for handling MS-MS data are developed. Accurate mass spectral predictions for candidates would provide a means for exploiting the structural information present in a compound's conventional mass spectrum. Mass spectral data have the further advantage of allowing analyses to be based primarily on the constitutional structure as available from all structure generators.

REFERENCES

1. C. A. Shelley, H. B. Woodruff, C. R. Snelling, and M. E. Munk, "Interactive Structure Elucidation," in *Computer-Assisted Structure Elucidation*, D. H. Smith, ed., American Chemical Society, Washington, DC, 1977, 92, Chapter 7.
2. C. A. Shelley and M. E. Munk, "CASE, Computer Model of the Structure Elucidation Process," *Anal. Chim. Acta.*, **133** (1981), 507.

3. D. H. Smith, N. A. B. Gray, J. G. Nourse, and C. W. Crandell, "The DENDRAL Project: Recent Advances in Computer Assisted Structure Elucidation," *Anal. Chim. Acta*, **133** (1981), 471.
4. T. H. Varkony, R. E. Carhart, and D. H. Smith, "Computer-Assisted Structure Elucidation. Modeling Chemical Reaction Sequences Used in Molecular Structure Problems," in *Computer-Assisted Organic Synthesis*, W. T. Wipke and J. Howe, eds., American Chemical Society, Washington, DC, 1977, 188, Chapter 9.
5. R. E. Carhart, T. H. Varkony, and D. H. Smith, "Computer Assistance for the Structural Chemist" in *Computer-Assisted Structure Elucidation*, D. H. Smith, ed., American Chemical Society, Washington, DC, 1977, 126, Chapter 9.
6. D. H. Smith and R. E. Carhart, "Structure Elucidation Based on Computer Analysis of High and Low Resolution Mass Spectral Data," in *High Performance Mass Spectrometry: Chemical Applications*, M. L. Gross, American Chemical Society, Washington, DC, 1978, 325, Chapter 18.
7. N. A. B. Gray, *STRUCC—Structure Checking Program Manual*, Department of Chemistry, Stanford University, Stanford CA 94305, 1979.
8. J. G. Nourse, D. H. Smith, and C. Djerassi, "Computer-Assisted Elucidation of Molecular Structure with Stereochemistry," *J. Am. Chem. Soc.*, **102** (1980), 6289.
9. N. A. B. Gray, R. E. Carhart, A. Lavanchy, D. H. Smith, T. Varkony, B. G. Buchanan, W. C. White, and L. Creary, "Computerized Mass Spectrum Prediction and Ranking," *Anal. Chem.*, **52** (1980), 1095.
10. A. Lavanchy, T. Varkony, D. H. Smith, N. A. B. Gray, W. C. White, R. E. Carhart, B. G. Buchanan, and C. Djerassi, "Rule-Based Mass Spectrum Prediction and Ranking: Applications to Structure Elucidation of Novel Marine Sterols," *Org. Mass Spectrom.*, **15** (1980), 355.
11. N. A. B. Gray, C. W. Crandell, J. G. Nourse, D. H. Smith, M. L. Dageforde, and C. Djerassi, "Computer Assisted Structural Interpretation of Carbon-13 Spectral Data," *J. Org. Chem.*, **46** (1981), 703.
12. J. C. Wenger and D. H. Smith, "Deriving Three-Dimensional Representations of Molecular Structure from Connection Tables Augmented with Configuration Designations Using Distance Geometry," *J. Chem. Inf. Comput. Sci.*, **22** (1982), 29.
13. H. Egli, D. H. Smith, and C. Djerassi, "Computer Assisted Structural Interpretation of ^{1}H-NMR Spectral Data," *Helv. Chim. Acta*, **65** (1982), 1898.
14. R. E. Carhart, D. H. Smith, H. Brown, and C. Djerassi, "Applications of Artificial Intelligence for Chemical Inference. XVII. An Approach to Computer-Assisted Elucidation of Molecular Structure," *J. Am. Chem. Soc.*, **97** (1975), 5755.
15. T. H. Varkony, R. E. Carhart, D. H. Smith, and C. Djerassi, "Computer-Assisted Simulation of Chemical Reaction Sequences. Applications to Problems of Structure Elucidation," *J. Chem. Inf. Comput. Sci.*, **18** (1978), 168.
16. M. R. Lindley, N. A.B. Gray, D. H. Smith, and C. Djerassi, "A Computerized Approach to the Verification of C-13 NMR Spectral Assignments," *J. Org. Chem.*, **47** (1982), 1027.
17. D. Gries, *Compiler Construction for Digital Computers* Wiley, New York, 1971.
18. C. Cheer, D. H. Smith, C. Djerassi, B. Tursch, J. C. Braekman, and D. Daloze, "Applications of Artificial Intelligence for Chemical Inference. XXI. Chemical Studies of Marine Invertebrates. XVII. The Computer-Assisted Identification of (+)-Palustrol in the Marine Organism Cespitularia *sp.*, aff. Subvirdis,," *Tetrahedron*, **32** (1976), 1807.
19. D. H. Smith, B. G. Buchanan, W. C. White, E. A. Feigenbaum, J. Lederberg, and C. Djerassi, "Applications of Artificial Intelligence for Chemical Inference. X. INTSUM—Data Interpretation and Summary Program Applied to the Collected Mass Spectra of Estrogenic Steroids," *Tetrahedron*, **29** (1973), 3117.

20. R. K. Lindsay, B. G. Buchanan, E. A. Feigenbaum, and J. Lederberg, *Applications of Artificial Intelligence for Organic Chemistry: The DENDRAL Project*, McGraw-Hill, New York, 1980.

21. B. G. Buchanan, D. H. Smith, W. C. White, R. Gritter, E. A. Feigenbaum, J. Lederberg, and C. Djerassi, "Applications of Artificial Intelligence for Chemical Inference. XXII. Automatic Rule Formation in Mass Spectrometry by Means of the Meta-DENDRAL Program," *J. Am. Chem. Soc.*, **98** (1976), 6168.

22. T. M. Mitchell and G. M. Schwenzer, "Applications of Artificial Intelligence for Chemical Inference. XXV. A Computer Program for Automated Empirical 13-C Rule Formation," *Org. Magn. Reson.*, **11** (1978), 378.

23. G. M. Schwenzer and T. M. Mitchell, "Computer-Assisted Structure Elucidation Using Automatically Acquired 13-C NMR Rules," in *Computer-Assisted Structure Elucidation*, D. H. Smith, ed., American Chemical Society, Washington, DC, 1977, 58, Chapter 5.

24. W. Bremser, M. Klier, and E. Meyer, "Mutual Assignment of Subspectra and Substructures—A Way to Structure Elucidation by C-13 NMR Spectroscopy," *Org. Magn. Reson.*, **7** (1975), 97.

25. W. Bremser, "The Importance of Multiplicities and Substructures for the Evaluation of Relevant Spectral Similarities for Computer Aided Interpretation of C-13 NMR Spectra," *Z. Anal. Chem.*, **286** (1977), 1.

26. B. A. Jezl and D. L. Dalrymple, "Computer Program for the Retrieval and Assignment of Chemical Environments and Shifts to Facilitate Interpretation of Carbon-13 Nuclear Magnetic Resonance Spectra," *Anal. Chem.*, **47** (1975), 203.

27. W. Bremser, "HOSE—a Novel Substructure Code," *Anal. Chim. Acta*, **103** (1978), 355.

28. N. A. B. Gray, J. G. Nourse, C. W. Crandell, D. H. Smith, and C. Djerassi, "Stereochemical Substructure Codes for 13C Spectral Analysis," *Org. Magn. Reson.*, **15** (1981), 375.

29. D. E. Knuth, *The Art of Computer Programming.3. Sorting and Searching*, Addison-Wesley, Reading, MA, 1975, pp.506–549.

30. C. W. Crandell, N. A.B. Gray, and D. H. Smith, "Structure Evaluation Using Predicted C-13 Spectra," *J. Chem. Inf. Comput. Sci.*, **22** (1982), 48.

31. C. A. Shelley and M. E. Munk, "Signal Number Prediction in Carbon-13 Nuclear Magnetic Resonance Spectrometry," *Anal. Chem.*, **50** (1978), 1522.

32. I. Fujiwara, T. Okuyama, T. Yamasaki, H. Abe, and S. Sasaki, "Computer-Aided Structure Elucidation of Organic Compounds with the CHEMICS System: Removal of Redundant Candidates by 13-C NMR Prediction," *Anal. Chim. Acta*, **133** (1981), 527.

33. L. A. Gribov, M. E. Elyashberg, and M. M. Raikhshtat, "A New Approach to the Determination of Molecular Spatial Structure Based on the Use of Spectra and Computers," *J. Molec. Struct.*, **53** (1979), 81.

34. L. A. Gribov, M. E. Elyashberg, and V. V. Serov, "On the Solution of One Classical Problem in Vibrational Spectroscopy," *J. Molec. Struct.*, **50** (1978), 371.

35. L. A. Gribov, V. A. Dementiev, and A. T. Todorovsky, "Calculation of Spectral Absorption Curves for Polyatomic Molecules" *J. Molec. Struct.*, **50** (1978), 389.

36. J. G. Nourse, "Specification and Enumeration of Conformations of Chemical Structures for Computer-Assisted Structure Elucidation," *J. Chem. Inf. Comput. Sci.*, **21** (1981), 168.

XII

STRUCTURE TRANSFORMATIONS

XII.A. INTRODUCTION

There are two main ways in which an analysis of structural transformations can facilitate structure elucidation.

First, one can, in the laboratory, use real chemical transformations to degrade an unknown structure into one or more simpler molecules. Structural data characterizing these simpler molecules can then be reworked to provide constraints on the form of an unknown. A list of candidates structural forms for an unknown can thus be "pruned" to yield only those compatible with the results of chemical degradation reactions. Computer programs can aid in a few of aspects of the analysis of such chemical degradations. Programs can model the application of various proposed alternative degradation reactions to each of a given set of structures. These programs thus can provide data indicating how effective a particular reaction might be at yielding information that would discriminate among the candidates. It is also possible for a chemical reaction modeling program to take experimental results, partially characterizing the products of some degradation reaction actually performed in the laboratory, and use these to prune away any initial candidates that are incompatible with the types or distributions of products observed. Such programs can avoid the requirement for the chemist to derive the constraints on the form of the starting material.

The second use of structure transformation data for structure elucidation arises in those applications where an "unknown" results from some sequence of chemical modifications of a known starting material. In such an application, a program for modeling chemical interconversions can serve as a *generator* of structural candidates. Although this type of complete structure elucidation problem will occasionally occur, more typically a structure transformation system will serve best as a generator of some component part of the unknown structure. For example, one might be able to presume that an unknown's skeleton must have been derived from a known structure through some sequence of well-characterized biochemical reactions. One could then generate all skeletons that would be produced through some finite number of biochemical conversion steps. One of these generated skeletons would then, presumably, be the correct skeleton for the unknown. The rest of the structure elucidation problem would then involve determination of the nature of substituents and derivation of constraints that would limit the ways in which these substituents might be bonded to the molecule's skeletal structure. The results of the modeling program might be exploited in various ways. Essentially, one requires a kind of intersection of (1) the file containing the results defining the skeletons identified by the modeling of chemical interconversions, with (2) the file containing the structures generated as being compatible with other chemical-spectral data characterizing the unknown.†

Both these analyses of chemical transformations will, if implemented in computer programs, require similar processing routines:

- Both require some subsystem that permits the chemist to define the chemical reaction(s) that are to be applied (either to a single supposed starting material or to each of a set of candidate structures generated for an unknown). These reaction definitions will need to specify a reaction site together with details of the bonds formed and/or cleaved by the reaction.
- A reaction site specification will generally take the form of a substructure definition (similar to any other substructure that might be defined and used as a constraint in structure generation) and some additional constraints on its molecular environment.
- Another component in such programs is a routine for applying a reaction definition to a single structure.
- Application of a reaction involves first a *graph-matching* step in which the reaction site(s) is (are) found in the structure, and checks are made to verify that any additional constraints are satisfied in each particular matching found for the reaction site substructure.

†The **GENOA** structure generator could make "prospective" use of the results from a process modeling the chemical interconversions. The structures derived through an analysis of the chemical transformations could be used as a presupposed set of alternative skeleta.

- The next step entails simulation of the bond formation and deletion steps (with possibly the introduction of new atoms and/or elimination of some of the original atoms of the structure). The algorithms and data structure manipulations needed in this step are simple involving just small changes to connection tables representing the structure (the atoms whose bonding will be affected will have been identified by the graph matching).
- Such programs will also involve extensive "bookkeeping" routines, canonicalization routines, and utilities such as routines for manipulating files of structures.

Normally, one will require that the products of reactions be represented in their canonical forms (for this makes it simpler to recognize when the same structure would occur on several different biochemical routes or would result from the degradation of several alternative candidates). Apart from standard canonicalization routines, most of the code and data structures for these parts of reaction modeling programs will be special purpose, designed simply to keep track of how compounds are interrelated. Data defining these relationships are particularly important if a system is to be capable of accepting laboratory acquired structural information characterizing the products of some degradation sequence and using these data to eliminate some of an initial set of candidate reactants.

The interrelationships of structures may be quite complex because any given candidate compound may be capable of undergoing reaction in several different ways even in the first step of some sequence of chemical transformations. Each of these initial products may, in subsequent steps, yield a plurality of further products. Additionally, any given product structure may result from a number of different initial candidate reactants. In such circumstances, complex analyses can be necessary to fully explore the ramifications of any data requiring or prohibiting products with particular structural features.

Although possibly complex in their overall structure, programs for modeling chemical transformations do not require any additional graph manipulation algorithms. Most of their necessary components already exist in structure elucidation systems. Thus a reaction definition subsystem can be constructed by simply extending a structure editor such as would normally be used for defining substructural constraints. The graph matchers, symmetry analysis routines, and canonicalization procedures can all be adopted unchanged.

Once developed, programs for modeling chemical interconversions can be adapted to other tasks unrelated to simple structure elucidation. There are many mechanistic problems, for example problems involving rearrangements of a given structure, where it is useful to be able to keep track of a molecule's individual constituent atoms during the course of some hypothesized reaction. Only simple changes are necessary to convert a structure transformation program to one that can follow the fate of individual atoms. (Essentially, one omits the canonicalization routines!) The ability to follow the fate of individual atoms also makes it possible to predict the position of isotopic label atoms that would be derived through some sequence of reactions applied to an initial substrate with an isotopic label at some known position. Such predictions can make it possible to plan costly isotopic labeling experiments so that the results will mostly clearly differentiate among several different hypothesized reaction mechanisms.

Finally, the modeling and analysis of chemical transforms can be extended to that most difficult of problems—the planning of an effective synthesis scheme for a structure postulated as representing an "unknown."

However, the extension to synthesis planning does involve a radical change in the nature of the problem being analyzed on the computer. In all the other stages described in this text, one is interested in *exhaustiveness*. The best algorithms (the only algorithms!) are those based on some combinatorial process capable of generating **all solutions** of some aspect of a given structural problem. Thus, one requires all existing matchings of substructure into a structure, all ways of extending a partial match to a complete match through further bond formation steps, all numberings of a structures that are equivalent and that identify symmetry elements, and so on. For synthesis planning, however, one wants those syntheses that use reactions in which the synthetic chemist has some confidence and experience, those that involve readily available starting materials and reagents, those that can be expected to proceed with reasonable yield. It is inherent in the nature of the synthesis planning problem that one must define and apply various heuristic functions for selecting and then evaluating the effectiveness of postulated chemical transformation steps.

A good synthesis planning program should not explore all alternatives; rather, it should focus its attention on the practical routes. Obviously, such a program must be able to choose possible chemistries and be able to evaluate how well individual steps in a proposed synthesis might proceed. Furthermore, a synthesis planning program must be capable of encompassing some sense of synthetic *strategy*, some rules or heuristics that will guide its selection of chemistries at each successive step.

In systems for exploring mechanisms of chemical interconversions or for modeling chemical degradations, each step involves the application of some single, user-defined chemical transformation to a set of reactants in order to obtain some set of products. It is inherent in the nature of these problems that the chemist should retain full control of the development of the reaction sequence or tree. In synthesis planning, however, it is the program that must select the reactions or "transforms" to apply at each step. It is these data, defining appropriate chemistries, that form the output required by the chemist. Sometimes, it may be appropriate for the chemist-user to suggest that only a restricted subset of the program's repertoire of chemistries be considered at a particular step; however, even when the program's search is so restricted, there will usually remain large numbers of reactions and/or several ways of applying each reaction that the program must evaluate.

New chemical structural analysis routines not needed for structure elucidation are required for synthesis planning. The first phase of structural analysis, as performed at each step in developing a synthesis plan, entails the "perception" of synthetically important features. This perception process comprises an analysis of chemical functionality and structural form. The analysis of functionality identifies individual functional groups and combinations of groups having particular relationships. The analysis of form identifies ring systems and points at which acyclic appendages are joined to the molecular skeleton. The perception processes are special case extensions of general graph matching and ring-finding procedures. The data derived through the perception step provides keys for accessing a data base of transforms and some additional constraints for evaluating possible transforms.

The knowledge of synthetic chemistry available to the planning program will be organized in some reaction data base. For synthesis planning, it is more convenient for the entries in this data base to describe antithetic transforms rather than the corresponding synthetic reactions. When accessing the data base, one has information characterizing a desired product and requires data defining how its particular structural features may be created and how to identify the forms of reactant(s) from which it might be synthesized. Except for this difference, the data for individual entries in the data base will be similar to, though generally somewhat more elaborate, than the simple reaction descriptions used in a system for modeling chemical degradations.

Each transform in such a data base will incorporate some detailed description of a substructure that must be found in the target molecule, together with a description of the bonding changes necessary for deriving the forms of the precursor structures. A typical transform will also involve a large number of tests that provide some semiempirical measure as to how effective the corresponding synthetic reaction is likely to be.

The data base of transforms must be indexed. It is really not practical to attempt to apply each transform to every target; only those transforms likely to be applicable should be retrieved and evaluated. The various data derived in the perception step can serve as keys if the transforms are similarly indexed according to the type of functionality that they involve and their general character, namely ring breaking, bond or group migrating, ring expansion or contraction, and so on.

Once a transform has been retrieved from the data base, its evaluation can proceed using simplified and specialized forms of graph matching. Most of the evaluation tests require checks for possible interfering functionality elsewhere in a molecule. The "perception" processes already applied will have recorded the positions and nature of all functional groups in the structure; these recorded data can be used to simplify the "graph matching" required at the evaluation stage. Structural changes specified in a transform are again easily implemented, requiring simple insertions and deletions of entries in some form of connection table. Further evaluation steps must generally be applied subsequent to the generation of the precursor structures; these further tests will involve application of the perception routines and checks for undesired combinations of functional groups or ring systems in the precursor structure(s).

Once created, precursor structures will normally be canonicalized. Some synthesis planning programs check whether generated precursors are listed in a library of available starting materials. These checks use the canonical structure name to key into a file indexed and ordered by canonical names. Even if checks are not being made against libraries of available starting materials, precursor structures are required in canonical form so that duplicates arising on different paths can be identified.

Like other chemical transformation modeling programs, synthesis planners require bookkeeping routines to record the interrelationships among generated structures. The initial target structure is the molecule whose synthesis is to be developed. Several different transforms may be relevant, some of which may apply in more than one way; consequently, after the first step in the analysis many possible precursor structures will have been generated. Some subset of these, chosen automatically by the program or selected by the chemist, are further analyzed in a similar manner. With syntheses requiring some 10 or so successive steps, the relationships among generated structures can grow quite complex. (The interrelationships of structures are normally described in terms of a "reaction tree"; however, because identical structures can occur on different paths a "reaction directed-graph" might be a more apposite term).

Many of the synthesis steps will involve combination of two precursor structures to give a desired target. Sometimes one of these precursors will be a simple molecule and can be presumed available; more generally, both precursors will be complex and will have to be analyzed individually so as to determine how they too might be synthesized. Such precursor nodes are ANDed together in the synthesis tree. The planning program must recognize that effective syntheses must be found for both before this route can be used to synthesize the desired target; estimates of the effectiveness of the synthesis route have to take account of the difficulties in synthesizing all ANDed nodes.

The difficulties in synthesis planning do not lie in the various structure manipulation steps; these can now all be handled by standard algorithms. Even the evaluation of the efficacy of any one individual step can be reasonably well approximated by various scoring rules based on the nature of the reaction and taking into account appropriate modifying factors. Heuristic functions can be used to combine the estimates of efficacies for individual steps into some overall estimate of the effectiveness of some proposed synthetic route. The problem in synthesis planning is to give a program an adequate sense of chemical strategy.

Frequently, some particular feature in a target molecule could be formed in many ways, all perhaps involving precursor structures at least as complex as the desired target. A program should be able to avoid even the generation of those precursor structures that could not be on any feasible synthesis route. An effective step in the development of a synthesis plan will be one where the precursors are substantially simpler; the corresponding synthetic reactions involved at such a step are those closing rings to give structures with specified stereochemistry or attaching acyclic appendages to ring systems. Such reactions form the "strategic" bonds that can be identified through an analysis of the desired target.

An analysis of strategic bonds can thus give a program at least some minimal sense of strategy. However, more elaborate analyses are necessary. Frequently, none of the transforms in a library will be applicable to a particular target. It is then necessary to consider several successive minor structural transformations involving the migration, exchange, or introduction of functionality. These successive small changes will convert a target molecule into a form that satisfies the requirements of one of the transforms for effecting major structural simplifications. A transform library will contain many transforms capable of achieving simple migrations and functional group exchanges and introductions. However, it is hopeless to apply these transforms haphazardly in the hope that some serendipitous sequence will set the stage for a major simplifying transform. Instead, the search for small sequential structural changes must be strongly guided by the requirements of the major transform whose application is desired.

The chemical strategy modules represent the most difficult portion of a synthesis planner. Currently, the planning programs are still weak in this area; it is, for example, difficult to get a synthesis planning program to exploit information on structures, already synthesized for some related study, that could reasonably serve as starting points in some new synthesis. Although progress has been made toward systems whereby subgoal transformations can be automatically invoked to set the stage for major simplifying transforms, these approaches are complex and not completely generalized.

XII.B. DENDRAL'S "REACT" PROGRAM: EXPLOITING STRUCTURE TRANSFORMATIONS AS AN AID TO STRUCTURE ELUCIDATION

DENDRAL's original CONGEN structure generator[1] was almost immediately extended by systems for modeling chemical interconversions. The first such extensions were special purpose, tailored to particular applications such as exploring structural isomerism of mono- and sesquiterpenoid skeletons.[2] Developments and refinements lead eventually to a system known as **REACT**[3,4,5,6†]

The **DENDRAL-REACT** program was developed primarily as a system for modeling degradation reactions in structure elucidation and analyzing mechanistic problems, particularly those relating to the biosynthesis of natural

†The name "REACT" has been used for other programs that explore, model, or analyze chemical transformations; the most widely known of the various REACTs is probably the system of Govind and Powers[7] for synthesis applications in chemical engineering.

products. When used to model degradations, the program would take as its initial input a set of candidate structures generated through **CONGEN**. In mechanistic studies or applications where **DENDRAL–REACT** was used as a generator, the initial starting material would be defined by using a teletype-oriented structure editor. The **DENDRAL–REACT** program was comprised of a number of subprograms; the principal ones were[5] EDITREACT, REACT, SEPARATE, and PRUNE.

EDITREACT

The **EDITREACT** command allowed the user to define reactions by specifying a reaction site, the bonding changes, and additional constraints. Once defined, a reaction could be immediately applied to a selected set of substrates. Reaction definitions could be saved for use in any similar structural problems that might subsequently arise; more typically, the reactions were defined interactively and immediately applied to the structures involved in some particular problem.

REACT

The **REACT** command was used to detail how a particular defined reaction (or, possibly, a chosen set of defined reactions) should apply. In applying reactions, it was possible to distinguish between single-step reactions, multiple application of the same reaction at each of several different reaction sites in a molecule, exhaustive reaction, reaction to equilibrium conditions [continued application of the defined reaction(s) until no new products were derived], or incomplete reactions (where any assertions as to structural features inferred for the products had to be tempered by the possibility that one of the "products" was in fact the unchanged substrate compound). A special variant on the **REACT** command **MREACT** (Mechanistic REACT), inhibited the canonicalization of reaction products, thus allowing the fate of individual atoms to be followed.

SEPARATE

Typically, a substrate molecule being reacted will give several products. Many degradation reactions involve cleavage of a molecule resulting in at least two products. In addition, there may well be more than one possible reaction site in a structure and more than one way in which the reaction may proceed, and with multistep reactions products may result from different extents of reaction. In the laboratory, chromatographic and other techniques are used to separate the products of reaction; the laboratory separation techniques should yield a number of pure compounds and possibly some "tar" residue containing products resulting from any number of competing side reactions.

The **DENDRAL–REACT** program allowed such separation procedures to be modeled in the computer. The user could detail the results of some laboratory separation process by defining the number of "flasks" containing distinct isolatable compounds. In itself, this specification of the number of isolatable products serves as a constraint on the candidate structures for the initial substrate. If, for example, two products are isolated in the laboratory, any candidate structure for the initial substrate that yielded only onc product could be eliminated from further consideration [assuming, of course, that the reaction definition applied in the computer program was an adequate representation of the reaction(s) actually occurring in the laboratory].† The possibility of side reactions, incomplete reactions, and kinetically unfavorable minor reactions could be catered for by allowing a "tar flask" to be specified. Thus, for example, any candidate substrate structure that could conceivably yield three or more products would be acceptable if **DENDRAL-REACT** were told to allocate two product flasks and a tar flask.

As well as serving to provide some constraints on candidate structures for the real substrate molecule, the SEPARATE command established the context for subsequent pruning steps or for the modeling of further reaction steps.

PRUNE

The PRUNE command in **DENDRAL-REACT** was a sophisticated version of conventional pruning functions available in the evaluation components of conventional structure elucidation systems. The PRUNE command permitted constraints to be applied to the structures in a given "flask." Usually, these constraints would take the form of requirements for particular substructures that had been inferred from spectral or chemical data characterizing products assigned to that flask. The required substructures would first be defined, using the standard structure editor, and then requirements for their presence or absence would be stated.

One effect of the application of constraints to product flasks is to limit the possible distribution among the flasks of the products computed for each candidate reactant. The SEPARATE command places a (minimal) requirement on the number of products that should be derived for candidate reactants to be consistent with observed results. The possible products from any one candidate reactant could, however, be equally readily allocated to any of the flasks. For example, if the degradation reaction were an ozonolysis, cleaving a double bond and yielding carbonyl compounds, some candidate reactants (those with the double bond in a ring or symmetrically substituted) would yield only one product, and others would yield two products (two ketones, two aldehydes, or an aldehyde and a ketone). A SEPARATE command specifying two product flasks would eliminate candidates that would have yielded a single product; the two products from any candidate that would yield, say, an aldehyde and a ketone would simply be marked as being possibly in either one of the two

†The DENDRAL–REACT program did not include stereochemical detail in its structure representation. A reaction that introduced new stereocenters and that could yield stereochemically distinct but topologically identical products would not be properly modeled.

product flasks. A subsequently applied pruning constraint requiring that the first flask contained an aldehyde would clarify their allocation.

A constraint, such as the requirement for at least one aldehydic product in the ozonolysis example, generally has further implications. As in the example, candidate reactants that would yield only ketonic products could be eliminated. Candidates that yielded two aldehydic products would not be eliminated; a constraint requiring that one of the product flasks contained an aldehyde does not prevent the other flask from also containing an aldehyde. The **DENDRAL–REACT** system was capable of exploring such "ramifications" of all specified constraints.

The **DENDRAL–REACT** program had the normal file manipulation and utility routines for saving and displaying partial results.

The use of the **DENDRAL-REACT** program is best illustrated by example. The data used here concern the structure elucidation of [+]-palustrol; this structural problem had originally been solved by use of **CONGEN**.[8] The structure elucidation process, as performed in the laboratory, did involve one chemical degradation step; in the original **CONGEN**-based solution, data derived from this chemical degradation step were exploited in a somewhat *ad hoc* manner. The problem was reanalyzed subsequent to implementation of the **DENDRAL–REACT** system.[5]

This structure elucidation problem concerned a $C_{15}H_{26}O$ sesquiterpene alcohol isolated from extracts of a marine Xeniid. There were considerable structural data available from IR, ^1H NMR, and ^{13}C NMR studies. The complete degree sequence of carbons was known from the ^{13}C NMR (4 × $-CH_3$, 4 × $-CH_2$, 5 × >CH–, >C<, and >C–OH). A series of proton decoupling studies had established a fairly large substructural component, shown as **XII-1**. Nevertheless, some 80-odd structural candidates were compatible with these data; a few examples are shown as **XII-2–XII-9**.

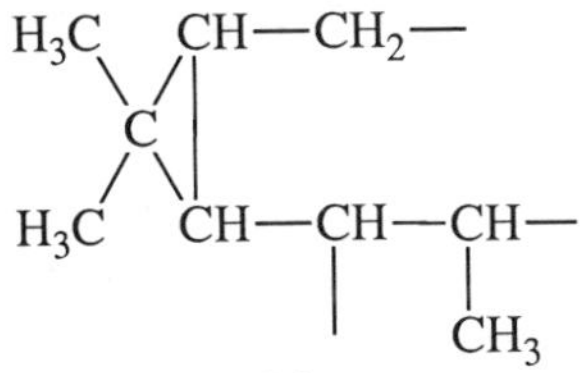

XII-1

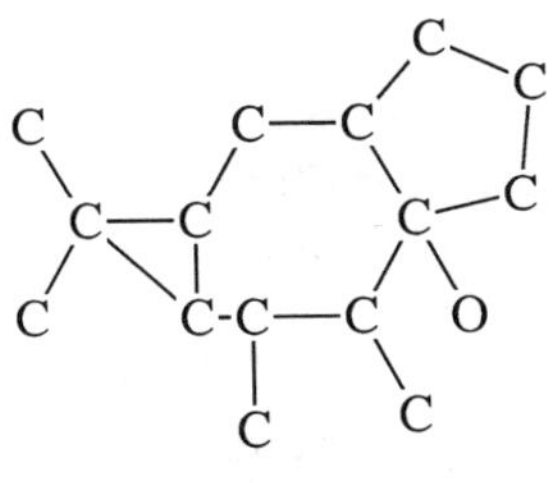

XII-2

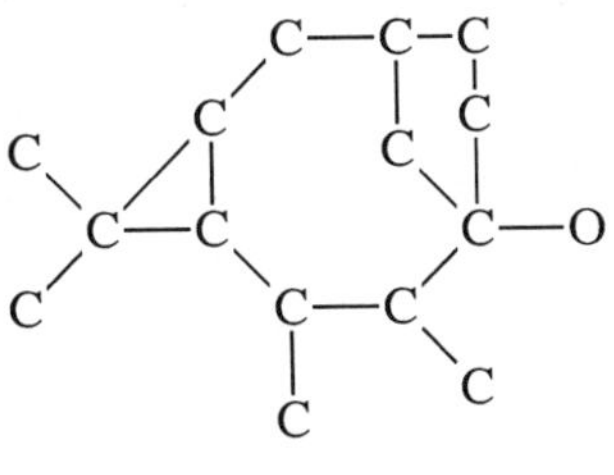

XII-3

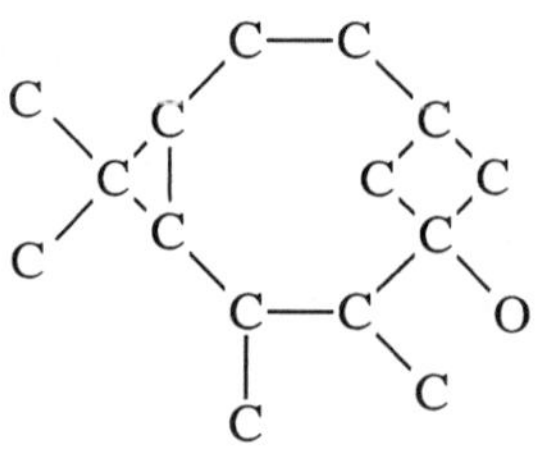

XII-4 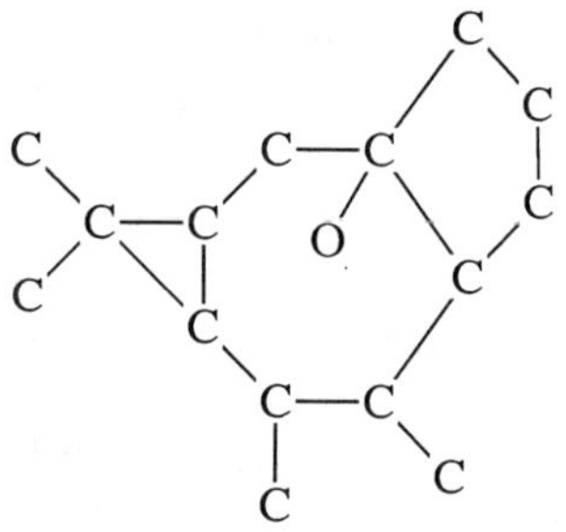 **XII-5**

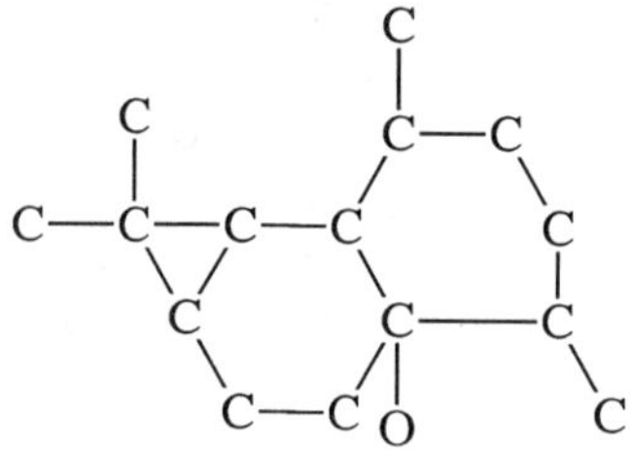

XII-6 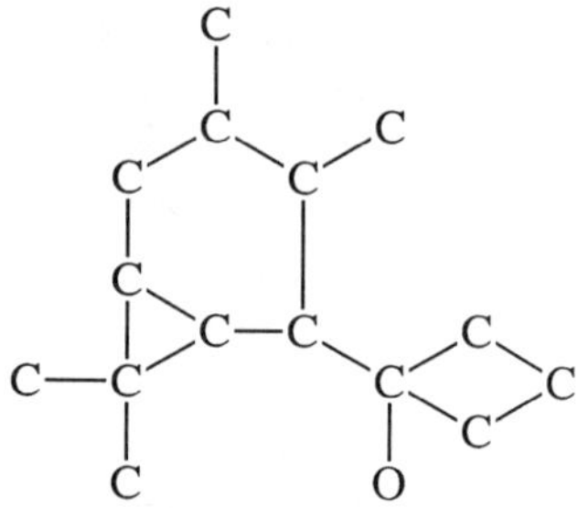 **XII-7**

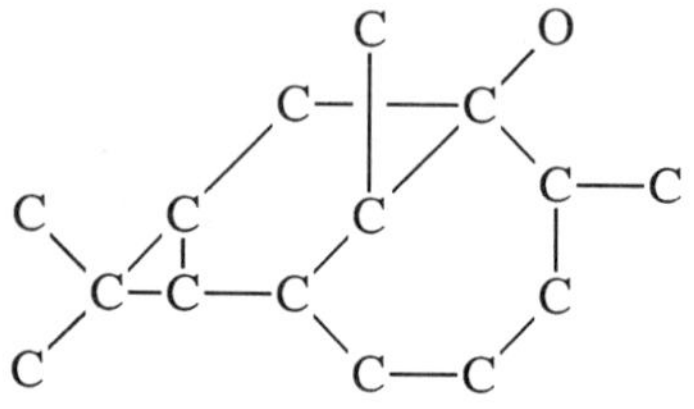

XII-8 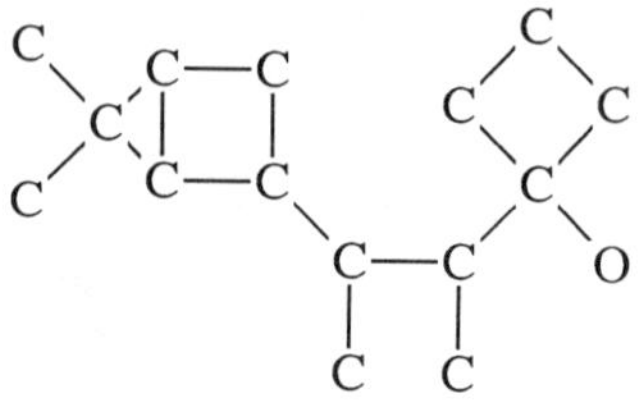 **XII-9**

Further structural information was obtained by performing a dehydration reaction on the unknown alcohol (phosphoryl chloride–pyridine; steambath, 30 min). This reaction produced a mixture of three products isolated by preparative thin-layer chromatography. These reaction and separation steps can be modeled by using the **DENDRAL–REACT** program as shown in Figures XII-1 and XII-3.

Figure XII-1 shows how a reaction can first be defined and then applied to the set of 88 candidates structures, as derived by **CONGEN** for the unknown alcohol. In this example the reaction specification is simple. The dehydration reaction must occur at the site of the $\rangle$C–OH group; the complete reaction site specification defines a three–atom group, C(1)–C(2)–O(3), with hydrogen range restrictions requiring a hydroxyl group and at least one (and as many as three) hydrogens substituted on C(1). The bonding changes effected by the reaction are defined in the TRANSFORM section of the reaction definition. The reaction cleaves the bond between C(2) and O(3) and increases the bond order between C(2) and C(1). The oxygen atom, atom 3 of the substructure, is deleted in the course of the reaction. Usually, some further constraints would be imposed on the reaction; in this example there was no possibility of competing reactions, and thus no constraints were defined. It would have been possible and might have been appropriate to define a constraint that would prohibit the formation of products containing a double bond at a bridgehead atom. Such a constraint would take the form of a substructure definition and the requirement that this substructure not be matched in any of the generated products.

Also detailed in Figure XII-1 are the commands specifying how this reaction was to be applied and the set of structures that were to be processed. Sets of structures are identified by their flask name; the candidates created by **CONGEN** are initially all allocated to a flask called STRUCS. Application of the reaction is again particularly simple in this example; only one possible type of reaction was to be considered and, since the candidate substrate molecules

```
#EDITREACT
REACTION NAME:DEHYDRATION
*SITE
>CHAIN 3
>ATNAME 3 O
>HRANGE 1
>MINIMUM NUMBER OF HYDROGENS:1
>MAXIMUM NUMBER OF HYDROGENS:3
>HRANGE 3 1 1
>DONE


**TRANSFORM
>UNJOIN 2 3
>JOIN 1 2
>DELATS 3
>DONE

#REACT
FLASK NAME OF STARTING MATERIALS:STRUCS

REACTION NAME(S):DEHYDRATION
NUMBER OF STEPS:1

FLASK NAME FOR PRODUCTS:DEHYD

241 PRODUCTS WERE GENERATED
```

FIGURE XII-1. Definition and application of a dehydration reaction in **DENDRAL–REACT**.

possessed only one possible reaction site, only a single reaction step had to be followed.

Actual application of the reaction definition requires the reading of each candidate structure from a file (or possibly from an internal list) and the finding of all possible matchings of the reaction site in it. Each candidate structure, as generated by **CONGEN**, exhibits three possible ways of matching the specified reaction site. The site incorporates the –>C–OH group and one of the carbons neighboring the quaternary carbon atom. Since the only other quaternary >C< in the structure is embedded in the cyclopropyl system; all candidates will have three hydrogen-bearing carbons as neighbors to the –>C–OH.

However, some of the topological candidate structures exhibit symmetries, and for these candidates the three possible matchings are not all distinct. The procedures for matching reaction sites to structures must take into account symmetries (of both structure and reaction site). A structure transformation program should process only the unique matchings in its development of possible product structures. Whereas most of the candidates would be expected to yield three products, structures possessing elements of symmetry may yield less. It is also possible for different candidates to yield identical products. Structures with symmetries and structures resulting in identical products reduce the number of possible products from 264 to 241 as reported by the program. Some of the possible matchings of the reaction site to candidates and the resulting products are illustrated in Figure XII-2.

The modeling of the chromatographic separation step is shown in Figure XII-3. The flask **DEHYD**, that was specified in the application of the reaction (Figure XII-1) as the receptacle for unseparated products, was selected and three new flasks, **D1–D3**, were specified to contain separated products. No additional products (for "tar flask") were permitted. This requirement for three products eliminated 16 of the original eighty eight candidates that resulted from **CONGEN**. The structures eliminated were those with topological symmetries, such as the third structure shown in Figure XII-2. The remaining 72 candidate structures for palustrol yielded 210 distinct products; again some candidates yield identical products. Without additional data, any of the products could be assigned to any of the flasks **D1–D3**. The **DENDRAL–REACT** program could produce simple displays illustrating relationships among the sets of structures assigned to different flasks. Diagrams representing the initial reaction step and the subsequent separation step are also illustrated in Figure XII-3.

The actual structure elucidation process, as performed in the laboratory, involved study of the ^{1}H NMR spectra of the products obtained by dehydration. The major product exhibited a new methyl resonance at 1.59 ppm, indicative of a vinyl methyl group; the second product showed a vinyl hydrogen at 5.1 ppm but no vinyl methyl; the minor product exhibited neither vinyl methyl nor vinyl hydrogen resonances. These data can be expressed in terms of substructures, a CH_3–C=C substructure and a –CH=C< substructure, whose

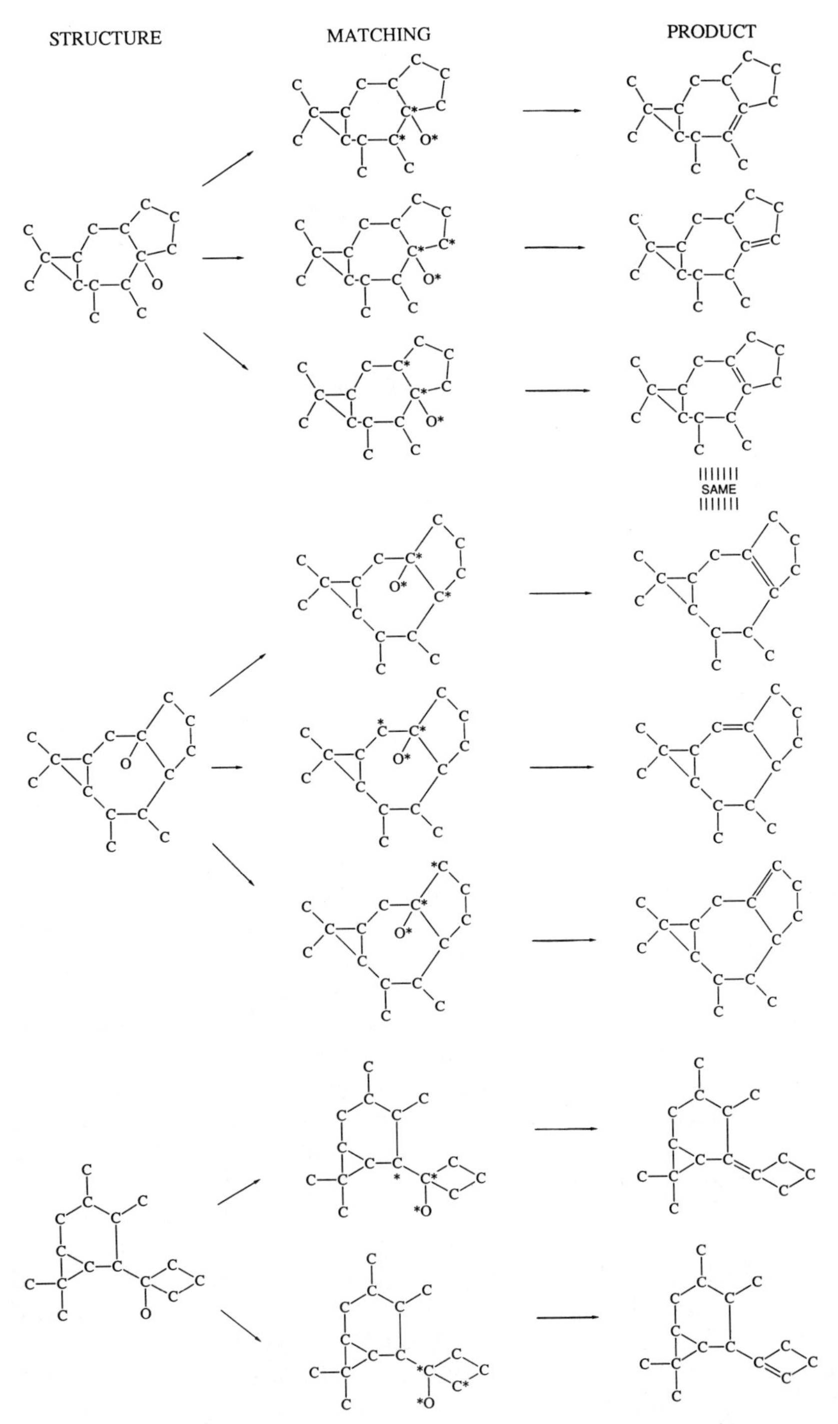

FIGURE XII-2. Alternative matchings of reaction site onto some of the candidate structures.

```
#SEPARATE
NAME OF FLASK TO BE SEPARATED:DEHYD
NEW FLASK NAME:D1
NEW FLASK NAME:D2
NEW FLASK NAME:D3
NEW FLASK NAME:
MAXIMUM NUMBER OF ADDITIONAL PRODUCTS:0

210 STRUCTURES SURVIVED SEPARATION

        STRUCS=88
        !
        *DEHYDRATION→DEHYD=241

        STRUCS=72
        !
        *DEHYDRATION→DEHYD=210-s- !D3=210
                                  !
                                  !D2=210
                                  !
                                  !D1=210
```

FIGURE XII-3. DENDRAL–REACT's SEPARATE command allowed modeling of the application of separation procedures to products obtained in a reaction. These data illustrate how the observation of three products in the dehydration reaction helped to prune away some of the initially possible reactant structures for palustrol.

presence can be required or prohibited in those generated product structures assigned to the various flasks. Actual application of the constraints was simple, a flask was selected, and the required number of instances of the substructure was specified. The DENDRAL–REACT program completely automated the processes involved in exploring all the ramifications of each such constraint.[5] The data summarized in Figure XII-4 illustrate the various numbers of structures surviving after the use of each constraint on the products.

The application of the first constraint, in Figure XII-4, the requirement that the product assigned to flask D1, must incorporate a vinyl proton does not eliminate any structure. All the remaining 72 candidates for palustrol included a $-CH_2-C(OH)<$ substructure and consequently yielded a dehydration product with a $-CH=C<$ substructure. The constraint does clarify somewhat the allocation of products. Of the total of 210 product structures generated by the program, only 129 contained $-CH=C<$ groups; the true structure for the dehydration product assigned to flask D1 must correspond to one of these 129 structures. (There are more than 72 possible $-CH=C<$ containing structures because some of the candidates produced by **CONGEN** for palustrol incorporate a $-CH_2-C(OH)-CH_2-$ substructure and can result in two different $-CH=C<$ containing dehydration products).

```
STRUCS=72
!
*DEHYDRATION→DEHYD=210-s- !D3=187
                          !
                          !D2=187
                          !
                          !D1=129
```

Application of the first constraint; the product in flask D1 was restricted to contain one vinyl proton and no vinyl methyl substructure

```
STRUCS=45
!
*DEHYDRATION→DEHYD=135-s- !D3=76
                          !
                          !D2=52
                          !
                          !D1=69
```

Application of the second constraint; the product structures assigned to flask D2 was limited to those containing a vinyl methyl substructure and no vinyl proton—this constraint eliminates 27 of the 72 structures

```
STRUCS=14
!
*DEHYDRATION→DEHYD=42-s- !D3=14
                         !
                         !D2=14
                         !
                         !D1=14
```

Application of the final product constraint; the third product contains neither a vinyl proton nor a vinyl methyl group—only 14 of the candidate structures for palustrol are consistent with all three constraints.

FIGURE XII-4. DENDRAL–REACT's reaction tree after application of constraints that one product contain a vinyl proton and no vinyl methyls, the next product contain a vinyl methyl but no vinyl protons, and the final minor product contain neither of these substructures.

The second pruning constraint, the requirement that flask D2 contain a structure incorporating a vinyl methyl group but no vinyl proton, now helps to eliminate some of the initial candidates generated for palustrol. Thus some of the candidates cannot yield the correct combinations of products; some of the structures illustrated in Figure XII-2 will not result in a dehydration product with a vinyl methyl group. The **DENDRAL–REACT** program could identify initial structures that would not yield an appropriate distribution of products and eliminate these from the STRUCS flask; consequently, some of the structures previously assigned to flasks D1, D2, and D3 would also be eliminated. In the palustrol example, some 27 of the candidates were eliminated through the first two product constraints. Data characterizing the third product eliminated another 30 or so of these candidates.

The **DENDRAL–REACT** program has been demonstrated on a number of other structure elucidation and mechanistic problems. In practice, the system was used mainly as a generator of biochemically plausible structural components. One application involved the generation of all C_7–C_{11} side chains that could be derived from an initial C_8 (24,25-unsaturated) side chain.[6] The set of reactions applied included *C*-methylation of a double bond, this first step was followed by proton elimination to form a double bond, or cyclization to a cyclopropyl ring, or quenching; generated olefins were permitted to undergo further reaction. The set of side chains thus generated included many novel forms as well as side chains already identified in plant or marine sterols. The generated library of side chains was subsequently exploited in other studies on the automated interpretation of the mass spectra of sterols.[9]

The use of a simple reaction modeling system, such as **DENDRAL–REACT**, is limited in structure elucidation. Applications involving generation of structures should have wider currency. Schemes similar to those used to generate the biochemically plausible sterol side chains, could be defined so as to simulate the likely biodegradation reactions of drugs. Plausible metabolic paths for given drug molecules could thus be explored and the generated metabolites screened, automatically, for substructures correlated with undesired biochemical activity. Such preliminary analyses could well provide data that might help direct experimental studies on drug activities.

XII.C. ASPECTS OF COMPUTER-ASSISTED PLANNING OF CHEMICAL SYNTHESES

XII.C.1. Perception

In the **DENDRAL–REACT** program, and as in all the structure elucidation programs, the basic processing step involves a straightforward graph matching of a substructure onto a current structure (or partially assembled structure). Very little preliminary analysis of the current structure is performed; a symmetry group may be derived (for this simplifies matching) and the substructure will also be analyzed so as to detect symmetry and determine an optimal order for attempting the graph-matching steps. In contrast to this limited analysis, synthesis planning programs devote considerable effort to a *perception* step wherein extensive preliminary analyses are performed on a current target structure. Through these analyses, structural data are abstracted from the connection table representation of the molecule and re-encoded in forms that allow for more rapid subsequent access. [The original connection table may itself be redundant, comprising both a table indexed by bond number (with entries identifying the atoms joined) as well as a connection table indexed by atom number (whose entries give the index numbers of neighboring atoms); this redundancy is again arranged so as to allow for more rapid access to structural information].

The first stage of *perception* is to derive various "atom sets" and "bond sets" (or, equivalently, to fill in entries in some "structure attribute table"). These sets comprise numerous 0/1 binary data fields, with one such field for each atom (bond) in the structure. Typical example sets include the set of all carbon atoms, the set of all oxygen atoms, the set of atoms with double bonds, the set of spiro atoms, the set of bridgehead atoms, the set of all stereocenters, the set of all single bonds, and so forth. Many of these sets can be derived without any detailed analysis, being filled in as the connection table for a structure is assembled or read in by the program. Some sets, such as the sets of spiro atoms or bridgehead atoms, can be completed only subsequent to an analysis of the ring systems in the structure. Arrays of sets can also be defined; one example is the array, indexed by atom number, that contains sets representing those other atoms that are alpha to each index atom. The total number of different atom and bond sets can reach into the hundreds. (At one stage in its development, **LHASA** used some 200 sets.[10])

The investment of computational effort in deriving all these sets for each structure is repaid through the much greater simplicity of subsequent tests that may be necessary in evaluation of a chemical transformation. For example, if the program were considering whether to generate a structure with a >C=C< group as a precursor for a target with a >C(OH)–CH< substructure, it could check that neither carbon of the target's >C(OH)–CH< was in the set of bridgehead atoms (which would imply an undesirable double bond at a bridgehead in the precursor). This check would require only a test of the settings of 2 bits in the already composed "set of bridgehead atoms." The results for many other structural queries can be found by taking logical combinations of sets. An example might be the identification of carbons doubly bonded to oxygens; the set of such atoms could be derived through a variety of set manipulation steps. One derivation involves the ANDing of the set of doubly bonded atoms with the set of oxygen atoms (to give the doubly bonded oxygens), the derivation of the set of all atoms alpha to these doubly bonded oxygens, and the intersection of this set with the set of carbon atoms. Using sets, such a structural query can be answered by one set-ANDing operation, a simple loop ORing the alpha sets of the doubly bonded oxygen atoms, and final set-ANDing; without set data, the query would require much more costly processing of the connection table representation of the structure (effectively, graph matching of O=C onto the structure). Synthesis planning programs make extensive use of set data in the procedures for perceiving functional groups and for applying and evaluating possible transforms.

The next step in the perception process is typically the analysis of rings. Methods for identification of all rings in a structure have been described in

Chapter IX. A synthesis planning program will typically need only some subset, the "synthetically significant" or "real" rings. Although it is possible to incorporate the heuristic ring selection rules into the ring-finding routine, a more satisfactory approach is to use something such as the efficient Wipke–Dyott[11] ring finder and apply the heuristic ring selection rules to the set of all rings found in the structure. Sets will normally be used to record the identity of all atoms and bonds that are included in rings. The synthetically significant rings will be recorded as list structures, ordered by size and containing the index numbers of constituent atoms. The sets of spiro atoms and bridgehead atoms can be constructed by performing appropriate intersections of sets of atoms in rings or bonds in rings.

The ring-finding process is also instrumental in identification of the *strategic bonds* in a structure. Again, the identity of the strategic bonds can be recorded in sets. Synthesis planning systems distinguish several different categories of strategic bond: fusion network strategic bonds, bridged network strategic bonds, and appendage strategic bonds.

The functionality of the target must be defined in the next step of perception. Although the various atom and bond sets derived in the first phase provide much implicit data concerning functionality, more explicit representations are required. The data structures representing functional groups will have to contain the index numbers of atoms and bonds comprising each group. Details of the environment of each functional group must also be recorded as they are identified. A structure may contain several instances of a functional group differing substantially with respect to steric hindrance or electronic environment. These differences need to be recorded so that it will subsequently be possible to test for selective reactivity.

A functional group recognition procedure could be based on a standard graph matcher (which could be arranged to exploit some of the atom and bond set data already derived). Such an approach would allow for arbitrarily complex functionalities, such as a vinylogous amide –CO–C=C–N<, to be readily defined. More typically, synthesis planning programs use data table-driven functional group recognition schemes. In these schemes the graph-matching steps are described in detail. Pensak[10] has provided a (simplified) portion of the scheme for identifying the nature of a cabonyl based functional group:

```
        IF BOND IS NOT CARBONYL THEN GO TO...
        IF HYDROGEN COUNT IS ONE THEN IDENTIFIED AS ALDEHYDE
        IF CARBON COUNT IS TWO THEN IDENTIFIED AS KETONE
        IF SINGLY BONDED OXYGEN IS NOT ATTACHED THEN GO TO A1
        IF CARBON ON NEW ORIGIN THEN IDENTIFIED AS ESTER
        IF HYDROGEN ON NEW ORIGIN THEN IDENTIFIED AS ACID

A1      IF SINGLY BONDED NITROGEN NOT ATTACHED THEN GO TO...
        IF TWO HYDROGEN ON NEW ORIGIN THEN IDENTIFIED AS AMIDE*1
        .
        .
```

The "program" detailing the steps must be written for the system; it must then be converted from this source form into an encoded scheme for an interpreter. The interpreter will have routines for testing the atom type of neighbors, testing bond types, counting atom types, and determining how to perform conditional branches. There are usually quite complex restrictions on the sequence in which tests for groups should be performed. Less conventional types of functionality—acetals, hemiketals, enamines, carbamates, and some combinations of adjacent groups—can cause considerable difficulty and must be dealt with by *ad hoc* mechanisms.

Once such a group recognition "program" has been written and thoroughly tested, it will usually suffice for many years before it needs to be extended to accommodate other types of functionality. There are obvious disadvantages in having to write these pieces of code to find substructures, in having to to have a special "compiler" to convert the code into the tables used by an interpreter, and in having a special interpreter—when all could be handled by a standard graph-matching subroutine. The extra investment in the special-purpose routines is claimed necessary to provide adequate performance in a planning program. The hand optimization of the steps for finding functional groups are felt necessary to deal with the tens of thousands of structure analyses performed during development of a synthesis.

The final step in perception will normally be the identification of stereocenters. These will be recorded in the sets used to represent stereochemical information. Various quite complex analyses may be necessary to determine relative stereochemistry of different stereocenters from the structural diagrams (with their dotted and wedged bonds) drawn in by the chemist.[12,13]

XII.C.2. Transforms

The objective of current practical synthesis planning systems is to determine sequences of known synthetic reactions that will convert supposedly available compounds into a desired target structure. The utility of such programs depends very much on the comprehensiveness of the chemical transformation "data bases" or "libraries" that they access. The construction of libraries requires major efforts by synthetic chemists (who will not normally have much computer expertise) together with some assistance by the developers of the synthesis planning system.

Data for the libraries are usually held in two different forms. There will be an original "source" form describing the transformations and some "compiled" form where these data have been converted into a representation more suited to rapid computer access and manipulation. The original source form should be readable by the synthetic chemists collaborating in the development of the system. Preferably, synthetic chemists should be able to compose their own transforms and add these to existing libraries; however, some technical assistance is normally necessary for transforms to be optimally encoded. Details have been published for the "source languages" of **SYNCHEM**, **LHASA**, and **SECS**.

XII.C.2.a. "Source" Representation of Transforms

Of the three systems—**SYNCHEM**, **SECS**, and **LHASA**—the **SYNCHEM** transforms[14,15] are the simplest. The transforms, as defined and read by chemists, take the form of simple tabular data structures.† The example of Figure XII-5 is derived from examples in Yen[15] and Sridharan[14] (with some simplifications).

The first section of these schema contained identification data; the example schema in Figure XII-5 is based on schema number 4 from Chapter 21 of the library. In **SYNCHEM**, Chapter 21 of the library contained all transforms that could be used to form an olefinic bond in the target structure. Schema 4 in this chapter concerned the Wittig reaction.

The TRANSFORMATION section of these schema contained the definition of the structural changes. The descriptions of the substructure sought in the target molecule and the substructure(s) in precursor molecule(s) were given in a slightly modified form of Wiswesser linear notation (the modification allowed for unspecified substituent groups represented by the various *$i*s.) The

Field of record	Example
IDENTIFICATION	Olefin Bond Library Library# = 21; Schema# =4; Name='WITTIG REACTION'
TRANSFORMATION DEFINITION	Target Group ='$2Y$4UY$6$8' Precursor Group ='$2V$4' Precursor Group ='$6Y$8E'
PARAMETERS	Ease = 1.0; Confidence = 1.0; Steps = 3;
Variations on basic reaction	Number of Methods = 1; Name =' 1. R3P. 2. Base'
VALIDATION	ATEST: 'HALIDE', 'KETONE', 'OXIDE'.
ADJUSTMENT	NTEST: 'ALDEHYDE', CONSEQUENCE = 'PROTECTION', DSTEPS = 2; ATEST: (several given in original schema) . IF '$2&$4&$8=ALKYL!ARYL!VINYL' DEASE=-0.05 . .

FIGURE XII-5. A reaction schema as used in SYNCHEM.

†**SYNCHEM**2[16] uses a reaction input language more complex than the simple fixed format **SYNCHEM** schema.

```
R1      R3                R1              R3
  \    /                    \               \
   C==C        <----         C==O    +       CH==Br
  /    \                    /               /
R2      R4                R2              R4

  Target                       Procursor (s)

$2      $6                $2              $6
  \ 'U' /                   \               \
   C==C        <----         V       +       Y—E
  /    \                    /               /
$4      $8                $4              $8
```

Target substructure: $2$4U$6$8
Prescursor substructure (s): $2V$4, $6Y$8E.

data defining the target's substructure are essentially equivalent to a reaction site substructure definition in **DENDRAL–REACT**; specification of precursor substructures provides an implicit definition of the bonding changes.

The **PARAMETERS** section of a schema detailed various semiempirical measures of the utility of the synthetic reaction; these measures were used to guide **SYNCHEM**'s automated search system. In the example schema in Figure XII-5, there is only one variation defined for the Wittig reaction—the use of triphenyl phosphine and a base. More typically, there would be several alternative methods for accomplishing the same structural change; these alternatives would be listed as distinct methods in the reaction schema. For example, the schema detailing methods for producing halides from alcohols included reactions with thionyl chloride, tosylation and reaction with a sodium salt, as well as direct reaction with HBr–HCl, and so on. These different methods involve different reaction conditions and are applicable to different types of precursor molecule; subsequent evaluation tests in a schema were separately defined for each of the different methods that it specified.

These schema incorporated a variety of evaluation clauses. The simplest of these, the ATESTS of the **VALIDATION** section of a schema, allowed for "go/no-go" tests. The ATESTS in the example transform specify that the desired target structure was to be tested for the presence of a halide, a ketone, or an epoxide substructure. (Such tests simply entailed looking in an attribute vector already derived for the target structure to determine whether appropriate fields were set.) If any of these groups were present, the corresponding synthetic reaction would not work and this transform would be discarded by the program without further analysis. The system allowed tests for classes of functional groups, including all ACID SENSITIVE groups, to be specified in the ATESTS section as well as tests on individual groups.

The **ADJUSTMENTS** portion of a schema described how the intrinsic merit of the reaction might be modified by secondary factors. Two distinct types of test could be specified in this section. The NTESTs again tested for global attributes of the target molecule. These allowed expression of how particular functionality elsewhere in the molecule might modify the ease reaction, eliminate certain of the alternative methods, or require additional steps such as protection. In the example transform, the single NTEST specifies that an aldehyde group, elsewhere in the target (and, consequently, in one of the precursors), would necessitate protection prior to the Wittig reaction step (and hence would add two steps, protection and removal of the protecting group, to the overall synthesis).

The **SYNCHEM** transforms could also involve tests on the immediate environment of the reaction site. Before such tests could be applied, the target substructure had to be matched to the structure undergoing analysis. Although WLN was used for communication with the chemist, internally **SYNCHEM** used "Topological Structure Descriptors," which were a form of connection table representation. The WLN descriptions of substructures were converted into and held in connection table form in the actual data base files. Their matching onto the current target structure involved standard graph-matching mechanisms. All alternative matchings had to be found; each distinct matching would yield a different set of atoms for the *$i* groups of the transform's target substructure.

In the example reaction, the nature of the substituent R$_i$ groups on the double bond can significantly affect the ease of reaction. The actual transform used in **SYNCHEM** included some six different qualifying tests that variously modified the EASE parameter of the reaction. The example in Figure XII-5 illustrates the test for a trisubstituted double bond (three of the four *$i* groups being alkyl, aryl, or vinyl); such an arrangement was held to decrease the ease of reaction by 0.5.

For those who already know WLN, the transform representations of **SYNCHEM** are fairly simple to read and write. (Definitions of elaborate substructures are, however, somewhat unintelligible; the substructure L6 BUTJ A$2 B$4 C$6 D$8 E$10 F$14. F$16. represents the unsaturated ring system that must be sought in a target structure when seeking to apply a Diels–Alder reaction.) However, these reaction schema are limited and cannot readily allow expression of all factors relevant to a synthesis plan. For example, in **SYNCHEM**, the Diels–Alder reaction is recorded as only one way to produce a double bond; the reaction is considered whenever a target structure possesses a

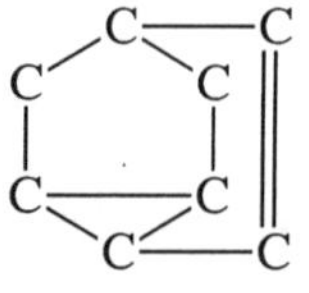

XII-10 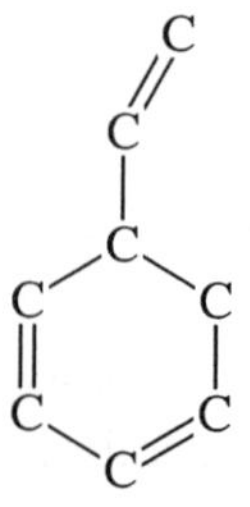 **XII-11**

double bond and used whenever that double bond is in an otherwise saturated six-membered ring. The only conditional tests defined in the transform serve simply to reduce the estimated ease of reaction if there are certain substituent groups at locants D and E of the ring (not the doubly bonded atoms). Sometimes such a simple analysis will suffice. The **SYNCHEM** program will discover the utility of a Diels–Alder reaction for the conversion of **XII-11** into **XII-10**.

Unfortunately, there will also be attempts to apply the Diels–Alder transform to quite inappropriate target compounds such as **XII-12** and **XII-13**.

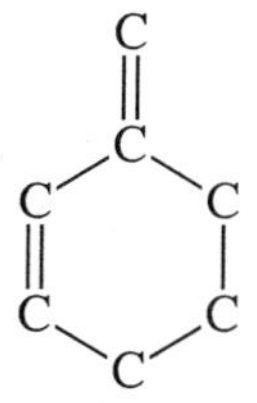

XII-12 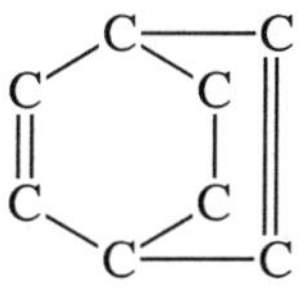 **XII-13**

[For structure **XII-12**, the program would construct some form of allene for its "diene"; with **XII-13**, the "diene" precursor would actually be a readily available (aromatic) compound, a result that might well make the synthetic step appear favorable.]

One serious deficiency is that all the transforms are keyed primarily by single functional groups, without any further data defining the kind of structural change that they produce. Such a limited keying inevitably leads to consideration of many transformations of little or no value in the development of a synthesis for any particular target structure. A synthesis planning program using these transforms works by generation of precursors that differ from the target by some change of functional groups or by the (synthetic) formation of a functional group. However, the majority of such interchanges of functionality will not represent significant steps toward the development of a practical synthesis.

For example, synthetic reactions forming a double bond by Wittig or Diels–Alder reactions are of major value. In the synthetic direction, such reactions assemble a more complex structure from two simple molecules (or join together two separate parts of a single molecule thus producing rings). But other reactions yielding a double bond in the target molecule, such as dehydration of an alcohol or dehydrohalogenation of a halide, represent simply an interchange of functionality. Selection of the corresponding schema would not significantly aid in the development of a synthesis plan.

The transforms of **SECS** illustrate some of the more sophisticated descriptions necessary in a system that is to attempt to restrict its development to the more plausible of synthesis routes. Two **SECS**'s transforms are illustrated in Figure XII-6.[17,18] Many of the components in **SECS**'s transforms are similar in concept to those of the simpler **SYNCHEM** descriptions. After all, all transforms must provide names and references, details of target and precursor structures, an intrinsic merit rating, validation tests for features that disqualify the transform, and further tests that adjust the intrinsic merit rating.

In **SECS**, transform descriptions are given in a highly regular and stylized pseudocode form. Almost all the statements in this pseudocode are situation→action rules. A test is made for some feature (usually this will correspond to a simple check on one of the sets describing the current target structure) and, based on the result of this test, some appropriate action is invoked. Actions can involve the application of more specific tests, changes in the intrinsic priority of the transform or specifications of structural manipulations (as in the placement of charge on a nitrogen in the second example in Figure XII-6). Some statements describing structure manipulations (the bond breaks, etc.) are unconditional. (As illustrated in the first of the examples in Figure XII-6, "comments" can be included in these pseudocode descriptions.)

The features required in the target structure can be described simply in terms of a functional group (and possibly its relationship to a bond broken in the antithetic transform), a pair of functional groups bearing some defined relationships, or as a "pattern" (i.e., a general substructure definition as in the

```
IDENTIFICATION/             PROXIMITY GUIDED EPOXIDATION
REFERENCES                  ALCOHOL GROUP CIS TO EPOXIDE ON RING
                            E.COLVIN, J CHEM SOC PERKIN I 1989 (1973)

SUBSTRUCTURE                O-C-C-@1<1,3,2>/

INTRINSIC PRIORITY          PRIORITY 0

CHARACTER                   CHARACTER ALTERS GROUP

VALIDATION TESTS            IF STEREOCENTER IS CARBON OFFPATH THEN
  AND                         BEGIN IF ALCOHOL IS WITHIN GAMMA TO ATOM 2 (1) THEN
  ADJUSTMENT TO                BEGIN IF BOND 1 AND (1) ARE CIS THEN ADD 50
  INTRINSIC                          ELSE KILL; epoxidation would have wrong stereochem
  PRIORITY                     IF (1) IS ONRING OF SIZE 5-6 THEN ADD 50
                               DONE
                               IF NITRILE IS EPSILON TO ATOM 2 (2) THEN
                                 BEGIN IF BOND 1 AND (2) ARE TRANS THEN ADD 30
                                       ELSE SUBT 50; epoxide trans to nitrile is favored
                                       DONE
                            DONE

REACTION CONDITIONS         CONDITIONS SLIGHTLY OXIDIZING

EXPLICIT                    DELETE ATOM 1
DEFINITION OF               MAKE BOND FROM ATOM2 TO ATOM 3
STRUCTURAL
CHANGES
```

FIGURE XII-6. Examples of SECS's transform definitions. (Reprinted, in part, with permission from W. T. Wipke, University of California at Santa Cruz).

```
IDENTIFICATION/            ALDOL-COND
 REFERENCES                H.O. HOUSE, 'MOD. SYN. RXNS.', (1972) p 629.

SUBSTRUCTURE               DGROUP WGROUP PATH 3

INTRINSIC PRIORITY         PRIORITY 100

CHARACTER                  CHARACTER BREAKS CHAIN BREAKS RING

VALIDATION TESTS           IF GROUP 1 IS ESTERX THEN KILL
  AND                      IF GROUP 2 IS A NITRILE THEN SUBT 20
  ADJUSTMENT TO            IF AN RGROUP IS ALPHA TO ATOM 2 THEN SUBT 10 FOR EACH
  INTRINSIC
  PRIORITY

REACTION CONDITIONS        CONDITIONS BASIC AND NUCLEOPHILIC OR
                           CONDITIONS ACIDIC AND NUCLEOPHILIC THEN SUBT 20

EXPLICIT                   BREAK BOND 1
DEFINITION OF              IF ATOM 1 IN GROUP 1 IS NITROGEN (1) THEN
STRUCTURAL                    BEGIN IF (1) IS QUATERNARY THEN KILL
CHANGES                             ELSE IF (1) IS TERTIARY THEN ADD + TO (1)
AND                           DONE
FURTHER CONDITIONAL        MAKE BOND FROM ATOM 1 IN GROUP 1 TO ATOM 2 IN GROUP 1
EVALUATION TESTS           IF ATOMS ALPHA TO ATOM 1 OFFPATH ARE ALL ATTACHED
                              TO 0 HYDROGENS THEN ADD 50
                           IF ATOM 1 IS A STEREOCENTER (1) THEN
                              BEGIN IF ATOM IS ALPHA TO ATOM 1 OFFPATH (2)
                                      PUT (1) OR (2) INTO (1)
                                    IF ATOM 2 IS ON MOST HINDERED SIDE OF (1)
                                      THEN SUBT 50
                              DONE
```

FIGURE XII-6. (*Continued*)

epoxide transform). (Generalized classes of functionality, such as electron-withdrawing groups **WGROUP**, can be used or particular functional groups can be specified when appropriate.)

Structural manipulations are explicit, defined in terms of atoms added or deleted, bonds made or broken. The **CONDITIONS** descriptions in the transform definitions are used in validation tests. In **SECS** there are tables of functional groups that define their sensitivity to various reaction conditions. Transforms can be "killed" if there is some group in the structure that could not survive the reaction conditions; alternative conditions can be specified as in the ALDOL transform, where it is possible, although less effective, to perform the reaction in acidic rather than basic conditions. If there are sensitive but protectable groups, **SECS** will tag the use of the transform with the required protection steps.

Any particular test on the transformation requirements can establish context data for subsequent more detailed tests. In the example in Figure XII-6, the test "IF ATOM 1 IN GROUP 1 IS NITROGEN (1)" provides a tag "1" by which the matched nitrogen atom can subsequently be referenced. It is then possible to refer back to this atom in the more detailed test "IF (1) IS QUATERNARY...". The atom or group tested can be quite remote from the actual transform site, as in the example of the nitrile group of the epoxide transform.

The **SECS** system applies many checks on stereochemical requirements of reactions. Some such checks are illustrated in the examples in Figure XII-6, where a validation test can eliminate the epoxidation transform if the corresponding synthetic step would yield the wrong stereoisomer.

Tests on the character of functional groups at a reaction site or in its near neighborhood will often fail to sufficiently constrain a synthesis planner. Often a hypothesized precursor structure will have more than one potential reaction site and thus be capable of yielding a plurality of compounds. A synthesis route involving a structure that could yield a mixture of products would normally be of little practical value; the yield of the desired product could well be small and extensive separation steps might be necessary. By default, such synthetic routes would normally be pruned away from a generated synthesis tree. However, sometimes one particular product (hopefully the desired product) may be kinetically favored. For example, the kinetics of the reaction may be controlled by steric factors, as when approach to a functional group is simpler from one side than the other (with the consequence that most of the product of reaction will be in one particular steric form). In other cases the way in which a reaction proceeds may be determined from the relative electronegatives of various adjacent atoms. It is possible to accommodate such qualifying factors in additional test procedures that are applied to the precursor structures derived in the application of a transform. If these tests suggest that the predominant product of the synthetic reaction would be other than the desired target structure, the transform's intrinsic merit can be downgraded or the transform can be discarded.

Such additional analysis procedures have been incorporated into **SECS** and **LHASA**. If a synthesis planning program can generate a valid three-dimensional model of a precursor structure, it is possible to estimate the effects of steric congestion and thus maybe determine whether the actual synthetic reaction would yield the desired stereochemically defined product. The **SECS** system allows for tests on an empirical measure of congestion at various sites, such as "IF STERIC AT GROUP 1 BETTER THAN KETONE ANYWHERE THEN...". The example ALDOL transform in Figure XII-6 checks to determine whether the target structure has a stereocenter where the C-C bond has been built and, if so, uses a test that measures the relative degrees of congestion on different sides of the double bond for a hypothesized precursor. Similarly, electronic effects can be estimated by means of simple molecular orbital calculations. With the use of Huckel molecular orbital methods, localization energies can be calculated for atoms in conjugated systems. These empirically estimated energies can then be used in evaluation tests comparing the susceptibility of various atoms to radicals, nucleophiles, and electrophiles.

The **CHARACTER** field of a transform definition is an important aspect of transforms in **SECS**. This field identifies the nature of the changes produced by a transform. Selection of transforms is restricted to those producing changes as required by the current *strategy*.

Both **SECS** and **LHASA** derive from the original **OCSS** (Organic Chemistry Simulation of Synthesis) program and share many of their basic strategies. In both programs, transforms are grouped into classes. The simplest are those that interchange a functional group in the target for some different group in the precursor, or which yield a precursor with additional functionality (where the corresponding synthetic step removes a group, e.g., reduction of a ketone >C=O to a methylene $-CH_2$). The basic requirements of transforms that introduce functionality will almost always be matched (the antithetic transform yielding a precursor with an extra ketone group will match any structure with a methylene); the presence of any particular functional group in the target can trigger a functional group interchange reaction (in much the same way that the detection of a group in **SYNCHEM** could invoke schema from a particular chapter of its reaction library). Such simple transforms are considered applicable by **LHASA** and **SECS** only if invoked as *subgoals* in attempts to apply transforms that effect major structural changes. Functional group interchange transforms are characterized as **ALTERS GROUP**, possibly with additional characterization such as **INVERTS STEREO** as may be appropriate.

There are some transforms that achieve major structural change whose applicability can be suggested by a single functional group in the target. These "single-group" transforms correspond to synthetic reactions that build carbon-carbon bonds and leave a characteristic functional group at or near the point where the bond was created. Examples of such synthetic reactions include the Wittig reaction (which leaves the new carbon-carbon bond as an olefinic functional group) and Grignard reactions (that leave an alcohol or some other group at the point where the carbon-carbon bond was formed). The antithetic transforms can be characterized as **BREAKS CHAIN**, **BREAKS RING**, and so on; some may be more specific and can be characterized in **SECS** as being capable of breaking rings of a particular size (e.g., BREAKS 4RING). If the current strategy of the program has established the goal of breaking some four-membered ring in a target structure, the ring-breaking transforms, particularly those specifically for breaking four-membered rings, are selected for possible application. The target structure must still, of course, possess appropriate functionality, with the required relationship to the bond broken, for any particular transform to be selected.

Other transforms in **SECS** and **LHASA**, that require pairs of functional groups or some arbitrary substructure, will again be characterized by the structural changes they induce as well as by the features that must be matched by the target structure. These transforms will again be capable of cleaving rings and chains; some will be able to expand or contract a ring, and others may serve to reconnect appendage chains in the target to produce a ring in the precursor (such reconnective transforms may be particularly useful in problems where the target has chiral centers on appendage chains).[19]

Often, none of the transforms appropriate to the current strategy will match the target. The strategy may require a transform, keyed by a single functional group, that will produce ring cleavage at some selected bond. The target may not possess the right kind of functionality with required relationship to the bond chosen by the strategy module. In other cases, the target might possess a functional group somewhere remote from the desired reaction site that would interfere with the proposed synthetic reaction and that cannot be readily protected. In **SECS**, if a transform could in principle produce desired structural changes and has a high intrinsic priority, details of any failures to completely match a target can be saved. Subsequently, if **SECS** transform application procedures cannot directly achieve the goals dictated by the strategy module, these saved data on partially matched transforms can be reanalyzed. Subgoals can be created that would, by exchange of functional groups, clear away those features that proved an impediment to otherwise attractive transforms.[17]

The **SECS** approach to the use of transforms is closely related to a "production rule" scheme. The transforms comprise a *premise part* and an *action part*. The premise part includes the specification of their **CHARACTER**, the substructural features required (as an arbitrary pattern, a pair of groups, etc.), and all the qualifier tests that together determine whether the appropriately adjusted transform priority exceeds some minimum cutoff value. The action part specifies the structural changes and results in the creation of the precursor structure(s). The **SECS** program works as a normal production rule interpreter searching through its rules for those that match the current context. This context is defined by the attributes of the current target structure and the goals established by the strategy module. The retrieval of rules from the rule base is filtered by the context data so that only potentially relevant transforms need actually be read from disk. The transforms that are fetched and evaluated may either be discarded (if of too low priority), succeed and yield precursors, or partially match and thus modify the current context by adding new subgoals.

If **SECS** is a "production rule interpreter," **LHASA** is a "programming language interpreter" for the specialized language **CHMTRN.** The **LHASA** transforms are typically "programs." This is apparent even in some of the simpler transforms such as the single group and group pair transforms. Two **LHASA** transforms are shown in Figure XII-7.[10,20] The first transform, an example of a single group transform that could involve antithetic cleavage of a strategic bond, is rather similar to the **SECS** transforms. The transform is defined in a simple pseudocode form. First, the essential requirements in the target are specified (almost any carbonyl group) and an intrinsic priority is given. A series of validation tests and rating adjustment tests follow. The structural changes are explicitly defined; reaction conditions are specified, again the program will check for sensitive groups and automatically deal with

```
IDENTIFICATION/          ALKYLATION OF CARBANION
 REFERENCES              ... O=C-C-C => O=C-C + X-C

SUBSTRUCTURE             GROUP *1 MUST BE KETONE OR ALDEHYDE OR ESTER OR NITRILE
                                 OR AMIDE*3 OR LACTAM OR LACTONE OR ACID
                         SUBGOAL*1 MUST BE KETONE OR ALDEHYDE OR NITRILE
                                 OR LACTONE OR AMIDE*3 OR LACTAM OR ESTER OR ACID

                         EXTENDED SUBGOAL

INTRINSIC PRIORITY       RATING 30

VALIDATION TESTS         KILL IF ATOM*2 IS MULTIPLY BONDED
  AND                    KILL IF ATOM*3 IS MULTIPLY BONDED ...vinyl,aryl
  ADJUSTMENT TO          KILL IF NO HYDROGENS ON ATOM*3
  INTRINSIC              ADD 5 FOR EACH HYDROGEN ON ATOM*3
  PRIORITY               ADD 5 IF ATOM*3 IS BENZYLIC
                         SUBTRACT 5 IF BOND*2 IS IN A RING
                         ADD 15 IF BOND*2 IS IN A RING OF SIZE 5 THROUGH 6
                         SUBTRACT 24 IF BOND*2 IS IN A RING OF SIZE 8 THROUGH 12
                         .
                         .
                         .
                         SUBTRACT 20 IF THERE IS A QUATERNARY*CENTER AT ATOM*2
                                 AND IF FEWER THAN TWO HYDROGENS ON
                                 ALPHA TO ATOM*2 OFFPATH
                                 AND IF ATOM*3 IS A TERTIARY*CENTER
                         .
                         .

REACTION                 CONDITIONS Enolate AND IN FRAGMENT*1 NaNH2
CONDITIONS               IF WITHDRAWING BOND ON ATOM*2 THEN CONDITIONS PH9:10

EXPLICIT                     ATTACH BROMIDE TO ATOM*3
DEFINITION OF                BREAK BOND*2
STRUCTURAL                   INVERT AT ATOM*3
CHANGES

FURTHER CONDITIONAL      KILL IF ATOM*3 IS IN A RING OF SIZE 3
EVALUATION TESTS         KILL IF ATOM*2 IS NOT ENOLIZABLE
ON PRECURSORS            IF ATOM*1 IS NOT A KETONE THEN FINISHED
                         SAVE 1 AS ALPHA TO ATOM*1 OFFPATH
                         IF SAVE*ATOM 1 IS NOT ENOLIZABLE THEN FINISHED
                         .
                         .
                         .
```

FIGURE XII-7. Examples of **LHASA** transforms.

```
IDENTIFICATION/        MANNICH REACTION
 REFERENCES            ... N-C-C-C=O => N + C=O + C-C=O

SUBSTRUCTURE           GROUP*1 MUST BE AMINE*1 OR AMINE*2
                       GROUP*2 CAN BE KETONE OR ALDEHYDE OR ESTER
                               OR AMIDE*2 OR AMIDE*3 OR NITRILE OR VINYLW
                       SUBGOAL*1 CAN BE AMINE*1 OR AMINE*2
                       SUBGOAL*2 CAN BE KETONE OR ALDEHYDE OR ESTER
                               OR AMIDE*2 OR AMIDE*3 OR NITRILE OR VINYLW

INTRINSIC PRIORITY     RATING 15

VALIDATION TESTS       GLOBAL*10 KILL IF CARBON*2 IS A QUATERNARY*BRIDGEHEAD
  AND                          ADD 5 IF CARBON*1 IS IN AN AMINE*2
  ADJUSTMENT TO                ADD 5 IF BOND*1 IS A RINGBOND
  INTRINSIC                    SUBTRACT 5 FOR EACH ALKYL GROUP ON CARBON*1
  PRIORITY                     SUBTRACT 20 FOR EACH WITHDRAWING GROUP ANYWHERE
                               ADD 5 IF BOND*1 IS A RINGBOND THEN GO TO BLOCK1
                               SUBTRACT 15 IF MINUS CHARGE BETTER ALPHA TO
                                         CARBON*3 OFFPATH THAN ON CARBON*2
                               BLKEND BLOCK1
                               ENDSHARE

EXPLICIT                 BREAK BOND**1
DEFINITION OF            ATTACH CARBONYL TO CARBON*1
STRUCTURAL               BREAK BOND*1
CHANGES
```

FIGURE XII-7. (*Continued*)

protection as required. In **LHASA**, the structural changes are frequently given in terms of macro operations such as "CONVERT THE GROUP TO...," or "ATTACH <group> TO <locant>"; these avoid the need to spell out all changes in full detail and save space in the tables. One difference from a **SECS** transform is the explicit consideration given to *subgoals*. In **LHASA**, the responsibility for invoking subgoals is not placed in the strategy-module or transform-evaluation system; instead, transforms will specify whether they can generate subgoal requests and identify the particular subgoals that should be considered.

The second transform shows more marked differences from those in **SECS**. Here, parts of the code the transform are assigned global labels. In the example, the code from the **GLOBAL*10** statement to the **ENDSHARE** statement can be shared. These transform statements can be invoked from some other unrelated transform that happens to require an identical check on the environment of a three-carbon atom substructure. Essentially, these **CHMTRN** statements form a subroutine that can be invoked by other transforms in the library.

The **LHASA** system's predilection for a programming approach is particularly evident in its *ring-manipulation supertransforms*. **LHASA** ring transforms are keyed by the occurrence of rings of appropriate size. Thus any nonaromatic carbocyclic ring is a candidate for **LHASA**'s Diels-Alder transform (the first of the major ring transforms that was implemented).[12] The **LHASA** model of the Diels Alder transform entails conversion of a "cyclohex-2-ene" into a diene (comprising atoms 1–4 of the cyclohexene) and a dieneophile. In any six-membered ring, there are six ways of selecting a bond as a candidate for being the ring double bond. Each must be tried in turn. In the **LHASA** ring transform schemes, the instructions to try the next possible choice of bond must be included as a part of the transform.

The code of the ring transforms consists mainly of statements that involve a test for a feature followed by a subroutine that performs a subgoal transform or by some operation, such as **REORIENT** (selecting the next possible bond and starting the analysis over again) or **REJECT** (rejecting the entire ring as a candidate for the transform), or by jumps to other portions of the code. Essentially, the code defines the form of binary search tree. A part of the tree and code for the Diels–Alder transform is shown in Figure XII-8.

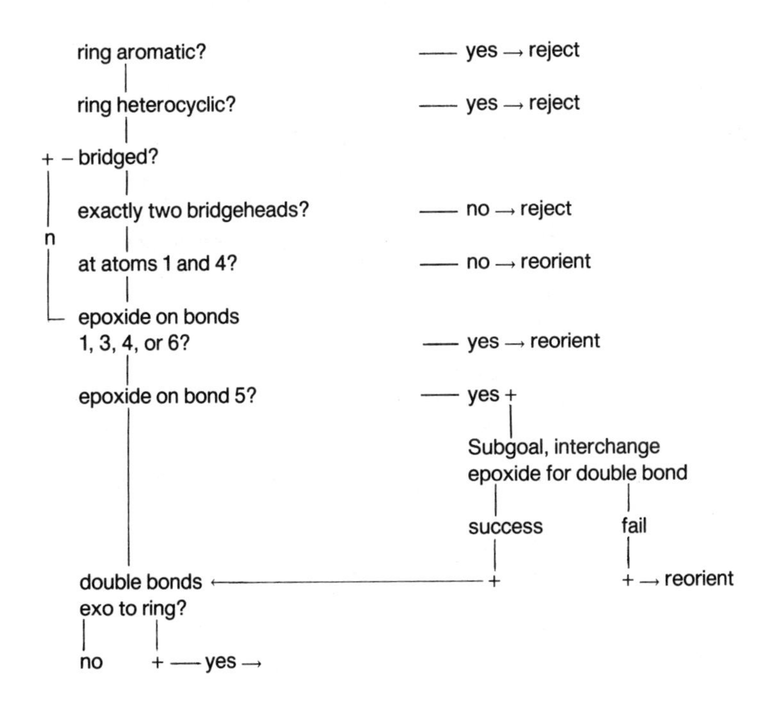

FIGURE XII-8. Parts of LHASA's Diels-Alder "flowchart" and transform code. (Reprinted, in part, with permission from E. J. Corey, et. al., *J. Am. Chem. Soc.* **96** (1974), 7728, 7733; American Chemical Society).

```
    IF THE RING IS AROMATIC THEN REJECT
    IF THE RING IS HETEROCYCLIC THEN REJECT
    IF THE RING IS NOT BRIDGED THEN GO TO A9
    IF THERE ARE NOT TWO BRIDGEHEADS ON*THE*RING THEN REJECT
    IF CARBON*1 IS NOT A BRIDGEHEAD OR CARBON 4*IS NOT A BRIDGEHEAD THEN REORIENT
A9  IF THERE IS AN EPOXIDE ON BOND*1 OR AN EPOXIDE ON BOND*3 THEN REORIENT
    IF THERE IS AN EPOXIDE ON BOND*4 OR AN EPOXIDE ON BOND*6 THEN REORIENT
    IF THERE IS NOT AN EPOXIDE ON BOND*5 THEN GO TO 02
    EXCHANGE IT FOR AN OLEFIN
    IF UNSUCCESSFUL THEN REORIENT
02  IF THERE ARE MORE THAN TWO OLEFINS ALPHA*TO*RING THE REJECT

    .
    .
    .
```

Placing a double bond:

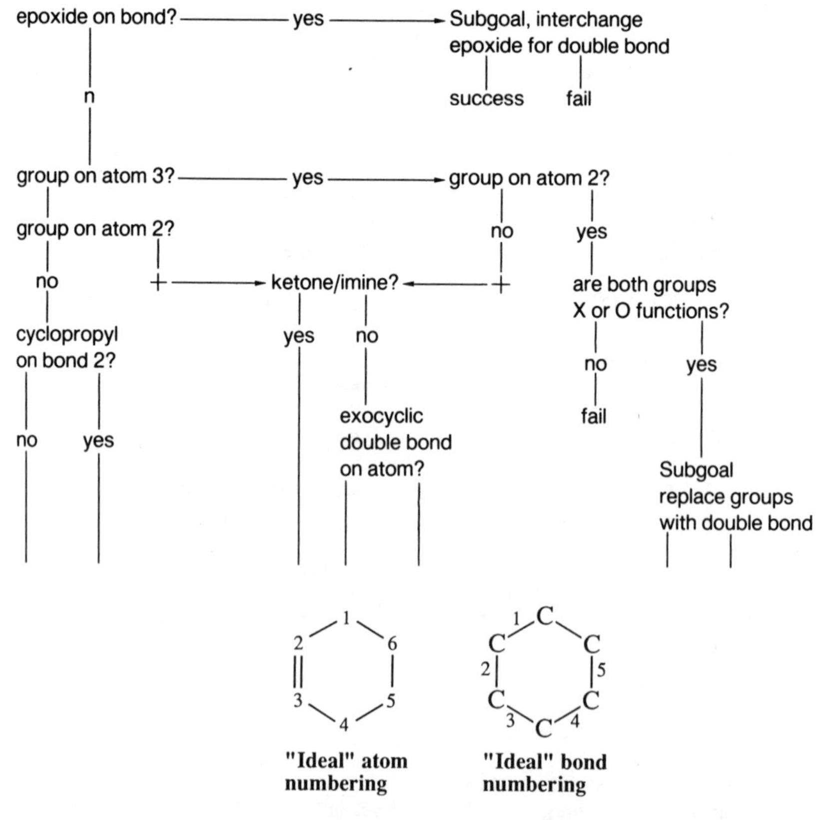

FIGURE XII-8. (*Continued*)

The code of a **LHASA** ring transform details various structural features that might block its application, together with a prescription of subgoals that might clear these blockages. The analysis here is limited only by the perspicacity of the author of the transform who can describe multiple subgoal sequences applicable to particular structural forms.

Another characteristic of **LHASA** supertransforms is the inclusion of some preliminary "prescreening" or "prior procedure evaluation" scheme. These preliminary steps tend to put greater "strategic" sense into the transforms. One relatively simple example is provided by the transforms for stereoselective olefin synthesis.[21] The **LHASA** olefin synthesis package was designed to allow deeper focused searches for sequences of transformation steps than would be possible if olefins were processed in isolation. The **LHASA** analysis is not limited to considering how a particular olefin bond can be produced in the target; it can extend to producing the bond together with several other structural moieties in its immediate environment. With polyenes and similar compounds, the analysis can include strategic decisions regarding how the various olefin bonds might all be created. For some targets, it might be best to proceed synthetically by building up a chain through several olefin formation steps each involving some different kind of reaction. Other targets might best be synthesized by connecting a central C=C as the last step in the synthesis; still other methods might exploit the use of the same reaction repeatedly applied. The olefin "package" incorporates some of these strategies and special-purpose analyses.

The first step of the **LHASA** olefin analysis entails identification of all the olefinic bonds and determination of which transforms are likely to be useful for their creation. The transforms in the library used by the special olefin synthesis package include extensive prescreens (similar to the "ATESTS" of **SYNCHEM** schema). These prescreens are special set variables; additional FORTRAN code computes attributes for each olefin bond found in a target. The olefin bonds–applicable transform correspondence can be determined by appropriate set manipulation operations. The example shown in Figure XII-9 includes the prescreen tests and some of the other code of one of **LHASA**'s olefin transforms. The prescreens, just like ATESTS, identify structural features that would render the transform inappropriate. The data derived through this prescreening step allow for more efficient access to viable transforms and can be used by the special-purpose olefin strategy module.

For simple structures with only one olefin, **LHASA** will proceed to try all the transforms that survived prescreening. These may invoke subgoals. In the example in Figure XII-9, the transform requires that certain groups bearing particular relationships to the double bond be exchanged for an ester or a tertiary amide; if these interchanges can not be accomplished, the transform is killed. In polyenes the **LHASA** olefin package will try each of its three different strategies. Again, the requirements of a strategy, such as the sequential reapplication of the same transform, may not always be immediately satisfied and subgoals may have to be established. These subgoals will be quite specific but

```
IDENTIFICATION/         Claisen Rearrangement
 REFERENCES

INTRINSIC PRIORITY      RATING 30

PRESCREEN FILTERS       RINGBOND TRANS2, FUNCTIONAL*GROUP TRANS1
                        NO*FG*WITHIN*ALPHA TRANS*3
                        OLEFIN*TYPE TETRASUBSTITUTED Z*DISUBSTITUTED
                        SPACING*1*5 WELL*DEFINED*T*FIRST WELL*DEFINED*L*LAST

PART OF CODE            IF THERE IS A FUNCTIONAL GROUP ON ALPHA TO ATOM*3
                           OFFPATH THEN GO TO BLOCK4
                        IF THERE IS A FUNCTIONAL GROUP ON BETA TO ATOM*3
                           OFFPATH THEN GO TO TO BLOCK3
                        IF THERE IS A FUNCTIONAL GROUP ON GAMMA TO ATOM*3
                           OFFPATH THEN GO TO BLOCK4
                      BLOCK3 DESIGNATE THE GROUP AS GROUP ONE
                        IF HETERO*1 IS NITROGEN THEN GO TO BLOCK6
                        IF THE FIRST GROUP IS CARBONYL OR ESTER THEN GO TO BLOCK2
                        GO TO BLOCK4
                      BLOCK6 IF THE FIRST GROUP IS AMIDE*3 THEN GO TO BLOCK2
                        EXCHANGE THE GROUP FOR AN AMIDE*3 AND*THEN GO TO BLOCK5
                      BLOCK4 EXCHANGE THE GROUP FOR AN ESTER ON BETA TO ATOM*3 OFFPATH
                      BLOCK5 IF UNSUCCESSFUL THEN REJECT
                      BLOCK6 DESIGNATE THE GROUP ON BETA TO ATOM*3 OFFPATH AS GROUP 1
                        .
                        .
                        .
```

FIGURE XII-9. An example of part of one of **LHASA**'s olefin transforms, illustrating the prescreen tests and transform code. (Reprinted, in part, with permission from E. J. Corey, et. al., *J. Am. Chem. Soc.*, **43** (1978), 2210; American Chemical Society).

could themselves entail major transforms; for instance, a transform for forming a specific olefin bond could be invoked as a subgoal that would convert the target to a precursor structure where a strategy for forming all the remaining double bonds might apply.

The **LHASA** HALOLACTONIZATION ring transforms entail a much more elaborate form of prescreening.[22] Like other **LHASA** ring transforms, the halolactonization transforms are applied very widely. Any carbocyclic ring of five to seven members is initially a candidate for a halolactonization analysis. The essence of the substructure required in the target for a halolactonization transform is the ring, a starting atom on this ring and one of its ring bonds, and a candidate for a COOH substituent on some face of the (stereochemically) defined ring. For a six-membered ring, there can be as many as 144 different mappings of this essential substructure. The actual halolactonization transform scheme involves attempts to convert, by some sequence of group interchange and introduction steps, the target into one of a set of 10 intermediate structures for which actual halolactonization transforms are immediately applicable. Although fewer matchings usually are possible, there can be more than 1000 different halolactonization sequences to evaluate for each suitably sized ring in a target structure.[22] Exhaustive search through all these possibilities would rarely be worthwhile; consequently **LHASA** uses a "prior procedure evaluation" scheme to assess which are likely to be easiest and limits subsequent searches to these paths.

Conversion of a given target to one of the "generic intermediates," to which a halolactonization transform can directly apply, typically involves steps such as placement of a carbonyl, hydroxyl, or epoxide group on some selected atom(s), dealkylation, allylic transpositions, and so on. These steps are defined in the transforms as *subroutines* that can be called with arguments identifying the atom(s) that are to be manipulated. The atoms of a target are analyzed, during the procedure evaluation step, in order to derive rough ratings as to how readily they might be converted to carbonyls, epoxides, or alternative usable functionalities. Since the structural requirements of the various generic intermediates are known, the changes necessary for conversion of a given target to a particular intermediate can be estimated and the computed atom ratings used to score that possible change. The actual code for the transform becomes complex. A small sample is shown in Figure XII-10. Some parts of this code

Parts of atom categorization

```
PUT THE ATOMS IN THE CURRENT*RING INTO RINGATS
PUT THE BONDS IN THE CURRENT*RING INTO RIGNBDS
TEMP 1 GETS LEAVING GROUPS MINUS RINGATS MINUS
          MULTIPLY BONDED ATOMS
EXODBS GETS THE BONDS*TO*BONDS OF RINGBDS WHICH
          ARE ALSO MULTIPLE BONDS
CARBON*NUCLEOPHILE GETS CARBON ATOMS WHICH ARE
          ALSO ATOMS*TO*ATOMS OF RINGATS MINUS
          ATOMS*TOATOMS OF TEMP 1 MINUS
          BONDS*TO*ATOMS OF EXODBS
.
.
.
```

Estimating a subroutine rating

```
CALC      VAR1 = #1
BLOCK1    IF SAVED*ATOM [ VAR1 ] IS NOT A KETONE
                    THEN GO TO CO1
          SR*GET*CO 0 (VAR1) = #0
          GO TO DE
CO1       IF SAVE*ATOM [ VAR1 ] IS NOT IN GET*CO*EASY
                    THEN GO TO CO3
          SR*GET*CO 0 (VAR1) = #1
          GO TO DE
CO3       .
          .
          .
```

FIGURE XII-10. Fragments from LHASA's halolactonization transform package. (Reprinted, in part, with permission from E. J. Corey, et al., *J. Chem. Inf. Comp. Sci.*, **20** (1980), 227; American Chemical Society).

Evaluation of transform

```
.
.
.
6.      PROCRAT = PRELRAT + BOXRAT
7.      WHO*WHERE (IX) = DIRECTION + FACE#COUNT * #2
                + ORIENTATION * #8 + PROCNO * #64
                + LABEL * #4096

8.      SORT IN ASCENDING ORDER [ PROC*PTR ] ELEMENTS
                OF THE ARRAYS AT PROC*RATING 1 AND
                WHO*WHERE 1

9.      LOOP ON IX VALUES EQUALING FROM (#1) UP*TO
        (PROC*PTR)
        IF UNSUCCESSFUL THEN DISCONTINUE
        DIRECTION = WHO*WHERE (IX) :AND: #1
        FACE*COUNT = ( WHO*WHERE (IX) / #2) :AND: #1
        .
        .

10.     LOOP ON PATH OF [ RING*SIZE] BONDS STARTING
                FROM SAVED*ATOM [ ORIENTATION]
                TOWARDS SAVED*BOND [ ORIENTATION -
                DIRECTION] USING ONLY ATOMS IN RINGATS

11.     JUMP USING [ PROCNO ] TO LOCATIONS PROC1 PROC2
                .
                .

PROC8   CALL GET*ENONE AT ATOM*3 AND ATOM*2 AND ATOM*1
        IF SUCCESSFUL THEN GO TO PROC8B
        IF THERE IS A HETERO ATOM ALPHA TO ATOM*3 THEN
                GO TO PROC8A
        CALL RING3 AT ATOM*1 AND ATOM*2 AND ATOM*3
        IF UNSUCCESSFUL THEN DISCONTINUE
PROC8B  CALL GET*LACTONE AT ATOM*1 AND SAVED*ATOM 1
        IF UNSUCCESSFUL THEN DISCONTINUE
        IF THERE IS AN OLEFIN ON ATOM*2 THEN EXCHANGE
                THE GROUP FOR AN IODIDE DEFINED*CIS TO
                THE BETA FACE
        .
        .
        .
```

FIGURE XII-10. (*Continued*)

concern the building of the sets that categorize atoms according to whether they are easy sites for placement of a carbonyl group. Other parts of the code are concerned with transform evaluation and application.

The entire **LHASA** mechanism of elaborate transform programs is in some ways suspect. The **LHASA** CHMTRN language has been made extremely sophisticated, and it does allow for the highly intricate detailing of multistep schemes, but the complexities of and potentials for interactions among transforms present problems. There are possible analogies to computer operating systems. Like the operating systems of the late 1960s and early 1970s, **LHASA** provides all embracing facilities with built-in support for many types of special purpose processing. In the hands of their creators, such systems can prove very powerful. For most users, however, these systems can present difficulties, with regard to maintenance and modification. In operating systems, at least, the tendency is now towards the implementation of systems that stress simplicity and uniformity rather than multiple levels of elaboration and refinement.

Some aspects of the elaboration of the CHMTRN language and its interpreter have been attempts to accommodate machine efficiency considerations; the relative costs of computer versus human time are changing, and it may well be reasonable now to accept less computationally efficient but clearer methods for representing and interpreting transforms. For example, **LHASA** largely avoids the *graph matching* of arbitrary substructures onto targets; its group-pair and single-group transforms are both selected from data on perceived functional groups, and its ring transforms are applied exhaustively to every ring of appropriate size found in the molecule. The complexity of some of the ring transforms stems in part from the need to derive useful results when attempts are made at very general application. These transforms must specify things such as the need to consider alternative orientations of rings, elaborate sequences of group exchanges to attempt to induce particular patterns of groups around the ring, and so on.

Ring transforms attempt to find sequences of group exchanges that convert rings into particular generic subgoals, such as the 10 generic intermediates of the halolactonization scheme.[22] It could well be easier to define the various generic intermediates as substructures and use a graph matcher to find all matchings in a current target. Small mismatches between a substructure and the target could be allowed to establish group-exchange subgoals as in **SECS**. Where there are standard sequences of several reaction steps that can indirectly convert one group to some other group, these could be defined as immediately applicable "transforms" (even though they might represent whole sequences of group exchanges); such "pseudotransforms" might capture something of the multiple group exchanges sequences derived by **LHASA**.

Such approaches may well be computationally somewhat less efficient and fail to detect situations where two or three successive small changes are needed to obtain the required substructure match. However, much of the code of the interpreter program and the transforms themselves would be greatly simplified. A graph matcher can find all matchings, and so transform code need not explicitly request reorientations of rings. A suitably extended substructure definition (with restrictions on how it may be embedded into a complete structure) will match only onto those places in a target where a transform has some chance of success.

XII.C.2.b. "Compiled" Representations of Transforms

Transform descriptions are "compiled" to produce the libraries actually employed by the synthesis planning programs. This compilation process both creates the machine-interpretable forms for each transform and builds up indices into the transform library that will permit relevant transforms to be found for a given target structure. These indices, which will normally be sufficiently small to be held entirely in memory, contain both the disk addresses of the complete transform definitions and some data that can help to restrict the selection of transforms. Again, **SYNCHEM** provides a simple, clear example. The indices in **SYNCHEM** included the ATESTS (the go/no-go tests) that characterized each transform. Only those transforms known to satisfy the essential validation tests would need to be read from disk. The indices for **SECS** and **LHASA** are naturally more elaborate. Data given in the indices will include its class (group-pair, single group, etc.), together with details of the groups and the "path length" (separation of groups or distance of group from bond made or cleaved). In **LHASA**, additional data in the indices identify those transforms that specify subgoals and provide details of the subgoals that may be requested.

The main body of any transform, with all the conditional qualifying tests, will be held in compiled form on disk. For **SYNCHEM**, these data were really more detailed substructure connection tables which, if matched to the target, would cause changes to a transform's priority. In both **SECS** and **LHASA**, the tests defined in the source transforms are converted to tabular representations with encoded calls for an interpreter routine. Thus a test such as "SUBTRACT 20 IF ALPHA TO ATOM*3 OFFPATH IS ENOLIZABLE" will, in principle, result in a table of entries of something of the following form:

```
  2     / SUBTRACT OPERATOR
 20     / (INCREMENT TO BE ADDED TO SUBTRACTED FROM REACTION RATING)
  1     / ALPHA (1-BOND)
  3     / FROM ATOM 3
  2     / OFFPATH
253     / ENOLIZABLE
```

In actual synthesis planning programs, such data are "packed" into subfields of individual computer words; the encoding of this transform qualifier might use only two words rather than the six words implied by the tabular representation. The packing is again in consideration of machine efficiency; the packed data require far less space and minimize the transfers from disk. For **SECS**, these encoded data provided a very concisely represented set of parameterized calls to the various premise testing functions of the transform evaluator. The **LHASA** system uses something similar to say interpretive PCODE as produced by PASCAL compilers or interpretive code as produced by a BASIC compiler.

XII.C.3. Strategy

The strategy component of a synthesis planning program has to limit the development of the "tree" of transforms and precursor structures. Ideally, a good strategy module will avoid the development of unproductive branches in this tree by *prospective* tests that restrict consideration to effective transforms (or, at least, exclude most of the irrelevant alternatives). The less useful branches in a synthesis tree can alternatively be eliminated by *retrospective* evaluation procedures; a branch of the tree can be pruned away if it has failed to reach sufficiently simple available precursor structures after a certain number of transformation steps, or if some empirical measure of reaction efficacies along the path is too low.

It is, of course, possible to trade off more route generation work by the computer against greater sophistication in the prospective evaluation of possible synthetic routes. A completely automated synthesis planner may be able to get by with a rather poor strategy component, provided that its final synthesis plan evaluation procedures are effective. If the system's retrospective evaluation procedures work well, then all the less satisfactory of the generated synthetic plans will be eliminated by the program and the results given to the chemist will be meaningful. Such an automated synthesis planning program will typically explore thousands of partial routes, tens of thousands of postulated intermediate structures. Efficiency in the basic structure manipulation steps become of prime importance in such systems.[16,23]

The **SYNCHEM** system relied on extremely simplified strategies. The program first analyzed a target structure to derive an "attribute table" that basically defined the various functional groups. It was these individual groups that then served to index into the reaction library. A structure such as **XII-14** would be identified as having an olefin, an epoxide, a carboxylic acid, an ether

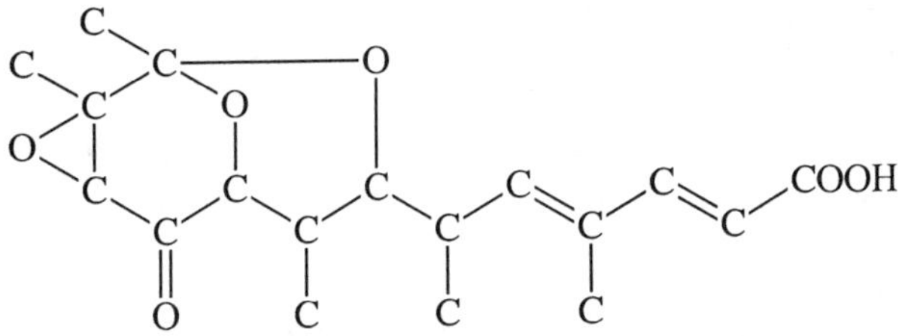

XII-14

and a ketone group; consequently, these five chapters in the **SYNCHEM** library would be accessed for possible transforms.

The **SYNCHEM** system would consider each reaction schema in the selected chapter of its library. It would, for example, find the Wittig reaction (viz. Figure XII-5) in its OLEFIN BOND LIBRARY but, determine that this reaction would not be appropriate for structure **XII-14** when the ATESTs for epoxides and other functionalities were performed. Alternative reactions from the olefin chapter and reactions from chapters for other groups would, however, be found and several precursors generated.

Development of the **SYNCHEM** search tree was guided by empirical estimates of the complexity of the structures at the nodes in the tree and its measures of efficacy of the synthetic reactions. The structure complexity measure was of the form $a \cdot e^{-\alpha \# F} + b \cdot e^{-\beta \# C}$, where $\#F$ was a count of the number of functional groups; $\#c$ was a count of the number of carbons not belonging to stable ring systems; a, b, α, and β were empirical constants. This empirical function provided some crude measure of how difficult it was likely to be to synthesize any particular structure. The nodes in **SYNCHEM**'s search tree were assigned scores that combined measures of the complexity of all structures ANDed together at that node, together with measures of the effectiveness of the synthetic steps already developed that lead from that precursor node to the real target.

The **SYNCHEM** plans were developed by a form of best first search. Each time a new set of precursor structures were created, estimates of the difficulties of completing a synthesis via a reaction sequence involving each new precursor was determined. An initially attractive route might, after three or four transform steps, result in precursors with very unfavorable complexity ratings. These ratings were propagated back up through the tree so as to appropriately revise the scores at earlier steps. **SYNCHEM** would switch attention to some other path if that which it had been most recently exploring fell in merit by some substantial amount.

The earlier **OCSS** program established some basic strategic approaches from which the current methods of **SECS** and **LHASA** have evolved.[24] For **OCSS**, a search for transforms could be restricted to either a "group-oriented" strategy or a "bond-oriented" strategy. Group-based strategies involved pairs of functional groups at some defined separation. The analysis of the target consisted of picking all pairs of functional groups and growing all paths between them. Each group-pair/path combination thus derived would be checked against the indices of the transform library. Through these indices, the program could identify any transform that utilized the same group-pair/path combination in a target. The entries in the transform library for group pair chemistries could all achieve major structural changes; selected transforms could involve simpler preparatory functional group interchange–introduction subgoal transforms.

The bond oriented analyses started with the "strategic" bonds of the target structure. Each such bond was analyzed in turn; paths of up to three additional bonds were grown from the bond's ends, and the presence of functional groups on these paths would be checked. The combination of path length/single–group then served to key into the indices of the single–group transforms. Those transforms with the same path length/single–group would, in principle, be capable of cleaving that strategic bond. These transforms would be retrieved and evaluated in detail. Again, subgoals for functional group interchanges might be invoked if needed to complete the required substructure in the target.

These simple strategies can be substantially refined. For example, **LHASA** exploits an empirical measure of the functional group density of a molecule to facilitate its selection of appropriate transforms.[10] Some of the group-pair transforms used in **LHASA** result in a precursor (or precursors) with fewer functional groups than the target; these transforms are characterized as **SIMPLIFYING** or density-reducing transforms. If a particular target already has a high proportion of its atoms in functional groups, **LHASA** will prefer **SIMPLIFYING** as to derive precursors with less functionality and, consequently, fewer problems of side-reactions and requirements for protection and so on. Conversely, if the target has few functional groups, density reducing transforms should be avoided because they would yield precursors that are likely to be more difficult to synthesize.

The **LHASA** "long range" strategies are implemented partly through special purpose-code and partly in the definitions of the transforms themselves. For example, the stereospecific olefin package has built-in strategies for sequential connection, convergent synthesis, and linear connection. The multilevel subgoals of the ring transforms provide another kind of strategic sense.

The divergent development of **OCSS** to **SECS** has focused more on establishing a clear separation of transforms and strategies. The **SECS** analysis of a target attempts to establish "goals".[17,18] Goals are data structures consisting of requests for (several) desired structural changes. Individual changes may define required group introductions and/or interchanges, bond cleavage, or bond formation steps. These requests are joined together by the use of logical operators (AND, OR, NOT, etc.). The development of these goals is based in part on an analysis similar to that used for identifying strategic bonds but extends into considerations of symmetries and so forth. The goal requests derived for a structure form a part of the context data that **SECS** uses when applying its "production rule"-like transforms.

Unlike **SYNCHEM** or Bersohn's synthesis planning system[23], **LHASA** and **SECS** are not left to automatically generate complete synthesis trees. These programs could be modified to act in an automatic manner. Empirical functions would have to be defined that could estimate the difficulty of synthesizing any given structure. (Such functions could exploit the data on ring systems, functional group density, and other variables already derived; with the use of such data, it should be possible to devise a more chemically significant measure than that used in **SYNCHEM**.) A combination of empirical measures of structure complexity and the already computed measures of reaction "priority" could be used to provide a measure of the value of exploring each generated precursor. A search system similar to **SYNCHEM**'s could then attempt some best–first exploration of possible synthesis trees. The sophisticated strategies for transform selection already used in **SECS** and **LHASA** should allow them to generate practical synthesis routes without additional user control.

Although automated variants of these programs could be useful, **LHASA** and **SECS** function rather as *consultation* or *expert* systems. The chemist is kept involved in the problem-solving process. The programs retrieve potentially relevant transformations and perform many aspects of their evaluation. It remains the prerogative of the chemist to assess the complexity of different generated precursor structures; select the paths worthy of further investigation; and, sometimes, restrict the type of transforms that are to be considered.†

XII.D. CONCLUSIONS

The practical applications of structure transformation systems in structure elucidation are limited. Problems of structural identity are currently solved primarily by the application of spectral techniques, and data from chemical degradations play a very minor role. The computer modeling of the application of chemical degradation reactions to some set of candidates is a complex and costly way of abstracting and exploiting only limited data.

Furthermore, a criterion for useful chemical transforms designed to yield new structural information is that observations on the resulting products be easily translated back to the starting structural possibilities.[4] Degradations satisfying this criterion will be those where only one transformation step is performed and where there is only one reaction site in each structural possibility with only two or three different ways in reaction can occur at that site. Essentially, these reactions group candidates according to the molecular environment of the reaction site. As illustrated in Chapter XI, however, such a grouping can be more readily effected by simply graph matching a definition of the reaction site onto the candidates and then expanding the matched substructure to identify its various alternative embeddings.

The generation of candidates through the application of defined reactions to a given starting material is again a technique of only limited value. Experimental studies that start by reacting some known compound will result in "structure elucidation problems" only when the presumed interconversion reactions don't work out! The structure of the actual product must then be determined before appropriate chemical interconversion steps can be hypothesized and the experimental results rationalized.

The major values of programs for modeling and analyzing structure

†The supertransforms of **LHASA** are actually an intermediate case; when applying these transforms, **LHASA** does work largely in an automatic mode developing multistep parts of total syntheses without user interaction.

transformations are outside structure elucidation, in fields such as mechanistic studies and synthesis planning. Mechanistic analyses provide the focus of work by Salatin and Jorgensen.[25] Jorgensen's **CAMEO** program (Computer Assisted Mechanistic Evaluation of Organic reactions) represents a much more general approach than that embodied in **DENDRAL–REACT**. Whereas **DENDRAL-REACT** allowed for the modeling of mechanistic processes as involved in specific reactions postulated and defined in detail by the user, **CAMEO** is designed to predict products of reactions on the basis of mechanistic principles applicable to broad classes of reactions. The program incorporates code segments for dealing with (1) base-catalyzed and nucleophilic chemistry, (2) acid-catalyzed and electrophilic chemistry, (3) radical chemistry, (4) carbenoid chemistry, (5) thermal pericyclic chemistry, and (6) photochemistry.† **CAMEO**'s interaction with chemists is conventional. The initial structure is entered at a graphics display terminal and then processed by various perception routines. As well as conventional routines for perceiving functional groups (adapted from **LHASA**), **CAMEO** incorporates semiempirical procedures for estimating enthalpies of reactions and relative pK_a values. The code segments defining appropriate chemistries are invoked and alternative reactions explored. Generated products are filtered using constraints that serve to eliminate those with unstable functionality or excessive strain. Reaction trees can be developed by applying similar analysis procedures to the results of the previous steps.

Synthesis planning continues to be an active research area.[26] All the basic structure manipulation algorithms are well defined, and further work can now be concentrated on the problems of synthesis strategy and the expansion of the reaction data bases available to synthesis planning programs. Further developments of synthesis planning systems would appear opportune. Machines with >1 Mbyte of memory, tens of megabytes of disk, reasonable graphics, and local area network connections are now becoming available at approximately $10,000. Such machines have the capacity to run the synthesis planning programs on a single user basis and can be integrated into local networks that would permit many users to access common transform and structure data bases.

Because problems still remain in imparting sufficient chemical strategic sense to a program, synthesis problems are likely to be best solved synergistically with extensive interaction between chemist and computer. The computer program would serve mainly as a consultant accessing data on library reactions and possibly as an evaluator of a chemical transformation selected by the user; the program would also perform all the book-keeping tasks. Although this arrangement reduces the role of the computer, it corresponds more to the typical requirements of chemists than would a fully automated planning system.

†The initial version of the program implemented only the base-catalyzed and nucleophilic chemistry segment.

REFERENCES

1. R. E. Carhart, D. H. Smith, H. Brown, and C. Djerassi, "Applications of Artificial Intelligence for Chemical Inference. XVII. An Approach to Computer-Assisted Elucidation of Molecular Structure," *J. Am. Chem. Soc.,"* **97** (1975), 5755.
2. D. H. Smith and R. E. Carhart, "Structural Isomerism of Mono- and Sesquiterpenoid Skeletons," *Tetrahedron*, **32** (1976), 2513.
3. T. H. Varkony, R. E. Carhart, and D. H. Smith, "Computer Assisted Structure Elucidation: Modelling Chemical Reaction Sequences Used in Molecular Structure Problems," in *Computer-Assisted Organic Synthesis*, W. T. Wipke and J. Howe, eds., American Chemical Society, Washington, DC, 1977, 188, Chapter 9.
4. R. E. Carhart, T. H. Varkony, and D. H. Smith, "Computer Assistance for the Structural Chemist," in *Computer-Assisted Structure Elucidation*, D. H. Smith, ed., American Chemical Society, Washington, DC, 1977, 126, Chapter 9.
5. T. H. Varkony, R. E. Carhart, D. H. Smith, and C. Djerassi, "Computer-Assisted Simulation of Chemical Reaction Sequences. Applications to Problems of Structure Elucidation," *J. Chem. Inf. Comput. Sci.*, **18** (1978), 168.
6. T. H. Varkony, D. H. Smith, and C. Djerassi, "Computer-Assisted Structure Manipulation: Studies in the Biosynthesis of Natural Products," *Tetrahedron*, **34** (1978), 841.
7. R. Govind and G. J. Powers, "A Chemical Engineering View of Reaction Path Synthesis," in *Computer-Assisted Organic Synthesis*, W. T. Wipke and J. Howe, eds., American Chemical Society, Washington, DC, 1977, 81, Chapter 4.
8. C. J. Cheer, D. H. Smith, C. Djerassi, B. Tursch, J. C. Braekman, and D. Daloze, "Applications of Artificial Intelligence for Chemical Inference. XXI. The Computer-Assisted Identification of [+]-Palustrol in the Marine Organism *CESPITULARIA* sp., aff. *SUBVIRIDIS*," *Tetrahedron*, **32** (1976), 1807.
9. N. A. B. Gray, A. Buchs, D. H. Smith, and C. Djerassi, "Computer-Assisted Structural Interpretation of Mass Spectral Data," *Helv. Chim. Acta*, **64** (1981), 458.
10. D. Pensak, *Computer Aided Design of Complex Organic Syntheses*, Ph. D. Thesis, Harvard University, 1973.
11. W. T. Wipke and T. M. Dyott, "Use of Ring Assemblies in Ring Perception Algorithms," *J. Chem. Inf. Comput. Sci.*, **15** (1975), 105.
12. E. J. Corey, W. J. Howe, and D. A. Pensak, "Computer-Assisted Synthetic Analysis. Methods for Machine Generation of Synthetic Intermediates Involving Multistep Look-Ahead," *J. Am. Chem. Soc.*, **96** (1974), 7724.
13. W. T. Wipke and T. M. Dyott, "Simulation and Evaluation of Chemical Synthesis. Computer Representation and Manipulation of Stereochemistry," *J. Am. Chem. Soc.*, **96** (1974), 4834.
14. N. S. Sridharan, *An Application of Artificial Intelligence to Organic Chemical Synthesis*, Ph. D. Thesis, State University of New York at Stony Brook, 1971.
15. S. Yen, *Representation and Implementation of Organic Chemical Reactions in a Heuristic Program for the Discovery of Organic Synthesis Routes*, Ph. D. Thesis, State University of New York at Stony Brook, 1974.
16. H. L. Gelernter, A. F. Sanders, D. L. Larsen, K. K. Agarwal, R. H. Boive, G. A. Spritzer, and J. E. Searleman, "Empirical Explorations of SYNCHEM," *Science*, **197** (1977), 1041.
17. W. T. Wipke, H. Braun, G. Smith, F. Choplin, and W. Sieber, "SECS—Simulation and Evaluation of Chemical Synthesis: Strategy and Planning," in *Computer-Assisted Organic Synthesis*, W. T. Wipke and J. Howe, eds., American Chemical Society, Washington, DC, 1977, 97, Chapter 5.
18. W. T. Wipke, G. I. Ouchi, and S. Krishnan, "Simulation and Evaluation of Chemical Synthesis—SECS: An Application of Artificial Intelligence Techniques," *J. Artif. Intell.*, **11** (1978), 173.

19. E. J. Corey and W. L. Jorgensen, "Computer-Assisted Synthetic Analysis. Synthetic Strategies Based on Appendages and the Use of Reconnective Transforms," *J. Am. Chem. Soc.*, **98** (1976), 188.
20. L. J. Joncas, *Computer Assisted Teaching of Organic Synthesis*, Ph. D. Thesis, Tufts University, Boston, 1981.
21. E. J. Corey and A. K. Long, "Computer-Assisted Synthetic Analysis. Performance of Long-Range Strategies for Stereoselective Olefin Synthesis," *J. Org. Chem.*, **43** (1978), 2208.
22. E. J. Corey, A. K. Long, J. Mulzer, H. W. Orf, A. P. Johnson, and A. P. W. Hewett, "Computer-Assisted Synthetic Analysis. Long-Range Search Procedures for Antithetic Simplification of Complex Targets by Application of the Halolactonization Transform," *J. Chem. Inf. Comput. Sci.*, **20** (1980), 221.
23. M. Bersohn, "Rapid Generation of Reactants in Organic Synthesis Programs," in *Computer-Assisted Organic Synthesis*, W. T. Wipke and J. Howe, eds., American Chemical Society, Washington, DC, 1977, 128, Chapter 6.
24. E. J. Corey and W. T. Wipke, "Computer-Assisted Design of Complex Organic Syntheses," *Science*, **166** (1969), 178.
25. T. D. Salatin and W. L. Jorgensen, "Computer Assisted Mechanistic Evaluation of Organic Reactions. 1. Overview," *J. Org. Chem.*, **45** (1980), 2043.
26. A. K. Long, S. D. Rubenstein, and L. J. Joncas, "A Computer Program for Organic Synthesis," *Chem. & Eng. News*, **1983** (May), 22.

LITERATURE

FILE SEARCH SYSTEMS

Data Base Integrity

Quality Control of Chemical Data Bases. S. R. Heller, G. W. A. Milne, and R. J. Feldmann, *J. Chem. Inf. Comput. Sci.*, **16** (1976), 237.

Error Tabulation by Text Searching of Large Chemical Data Base Compilations—Applications to the ASTM Infrared Spectral Index. J. A. deHaseth, H. B. Woodruff, and T. L. Isenhour, *Appl. Spectrosc.*, **31** (1977), 18.

A Quality Index for Reference Mass Spectra. D. D. Speck, R. Venkataraghavan, and F. W. McLafferty, *Org. Mass Spectrom.*, **13** (1978), 209.

Interactive Simulation of Infrared, Mass and 13-C NMR Spectra. M. Razinger, J. Zupan, M. Penca, and B. Barlic, *J. Chem. Inf. Comput. Sci.*, **20** (1980), 158.

Stereochemical Substructure Codes for 13-C Spectral Analysis. N. A. B. Gray, J. G. Nourse, C. W. Crandell, D. H. Smith, and C. Djerassi, *Org. Magn. Reson.*, **15** (1981), 375.

File Search Techniques

Computerized Searching of Inverted Files. F. E. Lytle, *Anal. Chem.*, **42** (1970), 355.

Effects of Data Compression on Computer Searchable Files. F. E. Lytle and T. L. Brazie, *Anal. Chem.*, **42** (1970), 1532.

Near Optimum Computer Searching of Information Files Using Hash Coding. P. C. Jurs, *Anal. Chem.*, **43** (1971), 364.

Some Comments on Computer Searching of Information Files Using Hash Coding. F. E. Lytle, *Anal. Chem.*, **43** (1971), 1334.

Automatic Identification of Chemical Spectra. A Goodness of Fit Measure Derived from Hypothesis Testing. S. L. Grotch, *Anal. Chem.*, **47** (1975), 1285.

Simple Index for Classifying Mass Spectra with Applications to Fast Library Searching. R. G. Dromey, *Anal. Chem.*, **48** (1976), 1464.

Automatic Computer Construction, Maintenance and Use of Specialized Joint Libraries of Mass Spectra and Retention Indices from Gas Chromatographic Mass Spectrometry System. B. E. Blaisdell, *Anal. Chem.*, **49** (1977), 180.

Inverted File Structure for Molecular Formula and Homologous Series Searching of Large Data Bases. R. G. Dromey, *Anal. Chem.*, **49** (1977), 1982.

Feasibility of a One Dimensional Search System. G. L. Ritter and T. L. Isenhour, *Comput. Chem.*, **1** (1977), 145.

Compilation of Computer Readable Spectra Libraries: General Concepts. R. Buchi, J. T. Clerc, C. H. Jost, H. Koenitzer, and D. Wegman, *Anal. Chim. Acta*, **103** (1978), 21.

Problems in Data Retrieval Systems for Analytical Spectrosccopy. J. Zupan, *Anal. Chim. Acta*, **103** (1978), 273.

Mass Spectrum Dictionary for Library Searching. R. G. Dromey, *Anal. Chem.*, **51** (1979), 229.

Cluster-Analyse Massenspektrometricscher Daten. J. Domokos, E. Pretsch, H. Mandli, H. Konitzer, and J. T. Clerc, *Fresenius Z. Anal. Chem.*, **304** (1980), 241.

Identification and Interpretation of Spectra by Logical Comparison of Information Sets. Part 1. The General Procedure of Spectrum Matching. Part 2. Optimization of Search Methods. Part 3. Combinations of Searches of Different Spectra. J. Kwiatkowski and W. Riepe, *Anal. Chim. Acta*, **135** (1982), 285.

IR

Computer Search System for Retrieval of Infrared Data. D. H. Anderson and G. L. Covert, *Anal. Chem.*, **39** (1967), 1288.

Fast Searching System for the ASTM Infrared Data File. D. S. Erley, *Anal. Chem.*, **40** (1968), 894.

Quantitative Evaluation of Several Infrared Searching Systems. D. S. Erley, *Appl. Spectrosc.*, **25** (1971), 200.

New Computerized Infrared Substance Identification System. R. W. Sebesta and G. G. Johnson, *Anal. Chem.*, **44** (1972), 260.

Computer Storage and Search System for Infrared Spectra Including Peak Width and Intensity. E. C. Penski, A. Padowski, and J. B. Bouck, *Anal. Chem.*, **46** (1974), 955.

Computer Retrieval of Infrared Spectra by a Correlation Coefficient Method. K. Tanabe and S. Saeki, *Anal. Chem.*, **47** (1975), 118.

Similarity Measures for the Classification of Binary Infrared Data. H. B. Woodruff, S. R. Lowry, G. L. Ritter, and T. L. Isenhour, *Anal. Chem.*, **47** (1975), 2027.

A Text Search System Using Boolean Strategies for the Identification of Infrared Spectra. H. B. Woodruff, S. R. Lowry, and T. L. Isenhour, *J. Chem. Inf. Comput. Sci.*, **15** (1975), 207.

Computer Searching of Infrared Spectra Using Peak Location and Intensity Data. R. C. Fox, *Anal. Chem.*, **48** (1976), 717.

A New Retrieval System for IR Spectra. J. Zupan, D. Hadzi, and M. Penca, *Comput. Chem.*, **1** (1976), 71.

Hierarchical Preprocessing of Infrared Data Files. M. Penca, J. Zupan, and D. Hadzi, *Anal. Chim. Acta*, **95** (1977), 3.

Conversion of a Conventional Infrared Library into Computer Readable Form. J. T. Clerc, P. Knulti, H. Koenitzer, and J. Zupan, *Fr. Z. Anal. Chem*, **283** (1977), 177.

Computer-Based Systems for the Retrieval of Infrared Spectral Data. J. T. Clerc and J. Zupan, *Pure Appl. Chem.*, **49** (1977), 1827.

Computer Identification of Infrared Spectra by Correlation Based File Searches. L. A. Powell and G. M. Hieftje, *Anal. Chim. Acta*, **100** (1978), 313.

Numerical Taxonomy and Information Theory Applied to Feature Selection from Filed Infrared Spectra for Automated Interpretation. F. H. Heite, P. F. Dupuis, H. A. van't Klooster, and A. Dijkstra, *Anal. Chim. Acta*, **103** (1978), 313.

Information Theory Applied to Feature Selection of Binary Coded Infrared Spectra for Automated Interpretation by Retrieval of Reference Data. P. F. Dupuis, P, Cleij, H. A. van't Kooster, and A. Dijkstra, *Anal. Chim. Acta*, **112** (1979), 83.

An Algorithm for ASTM Infrared File Searches Based on Intensity Data. K. Tanabe, T. Tamura, J. Hiraishi, and S. Saeki, *Anal. Chim. Acta*, **112** (1979), 211.

Library Retrieval of IR Spectra Based on Detailed Intensity Information. G. T. Rasmussen and T. L. Isenhour, *Appl. Spectrosc.*, **33** (1979), 371.

The Status of Infrared Data Bases. C. L. Fisk, G. W. A. Milne, and S. R. Heller, *J. Chromat. Sci.*, **17** (1979), 441.

Infrared Spectral Search System for Gas Chromatography/Fourie Transform Infrared Spectrometry. S. R. Lowry and D. A. Huppler, *Anal. Chem.*, **53** (1981), 889.

Applications of a Search System and Vapor Phase Library to Spectral Identification Problems. M. D. Erickson, *Appl. Spectrosc.*, **35** (1981), 181.

Fourier Encoded Data Searching of Infrared Spectra. L. V. Azarraga, R. R. Williams, and J. A. deHaseth, *Appl. Spectrosc.*, **35** (1981), 466.

Hierarchical Clustering of Infrared Spectra. J. Zupan, *Anal. Chim. Acta*, **139** (1982), 143.

Boolean Logic System for Infrared Spectral Retrieval. S. R. Lowry and D. A. Huppler, *Anal. Chem.*, **55** (1983), 1288.

Infrared Spectral Compression Procedure for Resolution Independent Search Systems. P. M. Owens and T. L. Isenhour, *Anal. Chem*, **55** (1983), 1548.

Width Enhanced Binary Representation for Library Searching of Vapor Phase Infrared Spectra. F. V. Warren and M. F. Delaney, *Appl. Spectrosc.*, **37** (1983), 172.

Optimization of a Similarity Metric for Library Searching of Highly Compressed Vapor-Phase Infrared Spectra. M. F. Delaney, J. R. Hallowell, and F. V. Warren, *J. Chem. Inf. Comput. Sci.*, **25** (1985), 27.

MS

Mass Spectral Data Processing. 1. Computer Used for Identification of Organic Compounds. B. Pettersson and R. Ryhage, *Arkiv fur Kemi*, **26** (1967), 293.

Computer Recording and Processing of Low Resolution Mass Spectra. R. A. Hites and K. Biemann, *Adv. Mass Spec.*, **4** (1968), 37.

Computer Methods in Analytical Mass Spectrometry. Identification of an Unknown Compound in a Catalog. L. R. Crawford and J. D. Morrison, *Anal. Chem.*, **40** (1968), 1464.

Matching of Mass Spectra when Peak Height is Encoded to One Bit. S. L. Grotch, *Anal. Chem.*, **42** (1970), 1214.

Compound Identification by Computer Matching of Low Resolution Mass Spectra. B. A. Knock, I. C. Smith, D. E. Wright, R. G. Ridley, and W. Kelly, *Anal. Chem.*, **42** (1970), 1516.

Identification of Mass Spectra by Computer Searching of a File of Known Spectra. H. S. Hertz, R. A. Hites, and K. Biemann, *Anal. Chem.*, **43** (1971), 681.

Computer Techniques for Identifying Low Resolution Mass Spectra. S. L. Grotch, *Anal. Chem.*, **43** (1971), 1362.

Small Computer Magnetic Tape Oriented Rapid Search System Applied to Mass Spectrometry. L. E. Wangen, W. S. Woodward, and T. L. Isenhour, *Anal. Chem.*, **43** (1971), 1605.

Conversational Mass Spectral Retrieval System and Its Use as an Aid to Structure Determination. S. R. Heller, *Anal. Chem.*, **44** (1972), 1951.

Computer Identification of Mass Spectra Using highly Compressed Spectral Codes. S. L. Grotch, *Anal. Chem.*, **45** (1973), 2.

Utilization of Automatically Assigned Retention Indices for Computer Identification of Mass Spectra. H. Nau and K. Biemann, *Anal. Lett.*, **6** (1973), 1071.

Computer Aided Interpretation of Mass Spectra. III. A Self Training Interpretive and Retrieval System. K. Kwok, R. Venkataraghavan, and F. W. McLafferty, *J. Am. Chem. Soc.*, **95** (1973), 4185.

Conversational Mass Spectral Search System. IV. The Evolution of a System for the Retrieval of Mass Spectral Information. S. R. Heller, R. J. Feldmann, H. M. Fales, and G. W. A. Milne, *J. Chem. Doc.*, **13** (1973), 130.

Conversational Mass Spectral Search and Retrieval System. II. Combined Search Options. S. R. Heller, H. M. Fales, and G. W. A. Milne, *Org. Mass Spectrom.*, **7** (1973), 107.

Conversational Mass Spectral Search System. Display and Plotting of Spectra and Dissimilarity Comparison. S. R. Heller, D. A. Koniver, H. M. Fales, and G. W. A. Milne, *Anal. Chem.*, **46** (1974), 947.

Comparative Study of Methods of Computer-Matching Mass Spectra. R. J. Mathews and J. D. Morrison, *Aust. J. Chem.*, **27** (1974), 2167.

Probability Based Matching of Mass Spectra. Rapid Identification of Specific Compounds in Mixtures. F. W. McLafferty, R. H. Hertzel, and R. D. Villwok, *Org. Mass Spectrom.*, **9** (1974), 690.

Automated Identification of Mass Spectra by the Reverse Search. F. P. Abramson, *Anal. Chem.*, **47** (1975), 45.

Computer Based Search and Retrieval System for Rapid Mass Spectral Screening of Samples. T. O. Gronneberg, N. A. B. Gray, and G. Eglinton, *Anal. Chem.*, **47** (1975), 415.

Statistical Occurrence of Mass and Abundance Values in Mass Spectra. G. M. Pesyna, F. W. McLafferty, R. Venkataraghavan, and H. E. Dayringer, *Anal. Chem.*, **47** (1975), 1161.

Information Theory Applied to the Selection of Peaks for Retrieval of Mass Spectra. G.van Marlen and A. Dijkstra, *Anal. Chem.*, **48** (1976), 595.

Probability Based Matching System Using a large collection of Reference Mass Spectra. G. M. Pesyna, R. Venkataraghavan, H. E. Dayringer, and F. W. McLafferty, *Anal. Chem.*, **48** (1976), 1362.

Similarity Measures for Binary Coded Mass Spectra. N. A. B. Gray, *Anal. Chem.*, **48** (1976), 1420.

Simple Index for Classifying Mass Spectra with Applications to Fast Library Searching. R. G. Dromey, *Anal. Chem.*, **48** (1976), 1464.

Comparative Study of Methods of Computer-Matching Mass Spectra. II. R. J. Mathews and J. D. Morrison, *Aust. J. Chem.*, **29** (1976), 689.

Computer Aided Interpretation of Mass Spectra. Information on Substructural Probabilities from STIRS. H. E. Dayringer, G. M. Pesyna, R. Venkataraghavan, and F. W. McLafferty, *Org. Mass Spectrom.*, **11** (1976), 529.

Computer Aided Interpretation of Mass Spectra. Increased Information from Characteristic Ions. H. E. Dayringer and F. W. McLafferty, *Org. Mass Spectrom.*, **11** (1976), 543.

Identification de Composes inconnus en Spectrometrie de Masse par Comparison a une Bibliotheque sur Cartouches Digitales. M. Bachir and G. Mouvier, *Org. Mass Spectrom.*, **11** (1976), 634.

Computer Aided Interpretation of Mass Spectra. Increased Information from Neutral Loss Data. H. E. Dayringer, F. W. McLafferty, and R. Venkataraghavan, *Org. Mass Spectrom.*, **11** (1976), 895.

Automatic Computer Construction, Maintenance and Use of Specialized Joint Libraries of Mass Spectra and Retention Indices from Gas Chromatographic Mass Spectrometry System. B. E. Blaisdell, *Anal. Chem.*, **49** (1977), 180.

Comparison of Various k-Nearest Neighbor Voting Schemes with the Self Training Interpretive and Retrieval System for Identifying Molecular Substructures from Mass Spectra. S. R. Lowry, T. L. Isenhour, J. B. Justice, F. W. McLafferty, H. E. Dayringer, and R. Venkataraghavan, *Anal. Chem.*, **49** (1977), 1720.

A Computer Algorithm for Qualitative Identification of Mass Spectral Data Acquired in Trace Level Analysis of Environmental Samples. W. C. Davidson, M. J. Smith, and D. J. Schaefer, *Anal. Lett.*, **10** (1977), 309.

Computer Aided Interpretation of Mass Spectra. STIRS Prediction of Rings Plus Double Bond Values. H. E. Dayringer and F. W. McLafferty, *Org. Mass Spectrom.*, **12** (1977), 53.

Application of a Text Search System Based on Boolean Strategy to Mass Spectra Identification. J. A. deHaseth, H. B. Woodruff, S. L. Lowry, and T. L. Isenhour, *Anal. Chim. Acta*, **103** (1978), 109.

SISCOM—a New Library Search System for Mass Spectra. H. Damen, D. Henneberg, and B. Weimann, *Anal. Chim. Acta*, **103** (1978), 289.

A Quality Index for Reference Mass Spectra. D. D. Speck, R. Venkataraghavan, and F. W. McLafferty, *Org. Mass Spectrom.*, **13** (1978), 209.

Idenfikation der Partialstrukturen von Unbekannten Steroiden mit Eine Schlusselionenkartei und Einem Rechnerunterstutzten Retrieval-System. G. Spiteller, M. Spitteler, M. Ende, and G. H. Hoyen, *Org. Mass Spectrom.*, **13** (1978), 646.

A New Interpretive Mass Spectral Retrieval System, "Doubly Inverted Search," P. Bruck and J. Tamas, *Adv. Mass Spectrom.*, **8B** (1979), 1535.

Optimum Scaling of Mass Spectra for Computer Matching. R. G. Dromey, *Anal. Chim. Acta*, **112** (1979), 133.

Search Strategy and Data Compression for a Retrieval System with Binary Coded Mass Spectral Data. G. van Marlen and J. H. van den Hende, *Anal. Chim. Acta*, **112** (1979), 143.

Identification of Components in Mixtures by a Mathematical Analysis of Mass Spectral Data. G. T. Rasmussen, B. A. Hohne, R. C. Wieboldt, and T. L. Isenhour, *Anal. Chim. Acta*, **112** (1979), 151.

A Combined Forward-Reverse Library Search System for the Identification of Low Resolution Mass Spectra. J. Kwiatkowski and W. Riepe, *Anal. Chim. Acta*, **112** (1979), 219.

Matching of Mixture Mass Spectra by Subtraction of Reference Spectra. B. L. Atwater, R. Venkataraghavan, and F. W. McLafferty, *Anal. Chem.*, **51** (1979), 1945.

Mass Spectral Library Searches Using Ion Series Data Compression. G. T. Rasmussen, T. L. Isenhour, and J. G. Marshall, *J. Chem. Inf. Comput. Sci.*, **19** (1979), 98.

Evaluation of Mass Spectral Search Algorithms. G. T. Rasmussen and T. L. Isenhour, *J. Chem. Inf. Comput. Sci.*, **19** (1979), 179.

Computer Techniques for Mass Spectral Identification. F. W. McLafferty and R. Venkataraghavan, *J. Chromatogr. Sci.*, **17** (1979), 24.

Identification of Lower Terpenoids from Gas-Chromatographic Mass Spectral Data by On-Line Computer Method. R. P. Adams, M. Granat, L. R. Hogge, and E. von Rudloff, *J. Chromatogr. Sci.*, **17** (1979), 75.

Computer Prediction of Molecular Weights from Mass Spectra. I. K. Mun, R. Venkataraghavan, and F. W. McLafferty, *Anal. Chem.*, **53** (1981), 179.

Prediction of Substructures from Unknown Mass Spectra by the Self-Training Interpretive and Retrieval System. K. S. Haraki, R. Venkataraghavan, and F. W. McLafferty, *Anal. Chem.*, **53** (1981), 386.

A Computer Search System for Chemical Structure Elucidation Based on Low Resolution Mass Spectra. K. S. Lebedev, V. M. Tormyshev, B. G. Derendyaev, and V. A. Koptyug, *Anal. Chim. Acta*, **133** (1981), 517.

^{1}H NMR

A Computerized System for Storing, Retrieving and Correlation of NMR Data. H. Skolnik, *Appl. Spectrosc.*, **26** (1972), 173.

Telephone Directory Format for NMR Spectra Retrieval. M. H. Jacobs and L. V. Derslice, *Appl. Spectrosc.*, **26** (1972), 218.

Use of the Computer for the Solution of Structural Problems in Organic Chemistry by the Methods of Molecular Spectrosccopy. V. A. Koptyug, V. S. Bochkarev, B. G. Derendyaev, S. A. Nekoroshev, V. N. Piottukh-Peletskii, M. I. Podgornaya, and G. P. Ul'yanov, *J. Struct. Chem.*, **18** (1977), 355.

Computer Aided NMR Spectral Interpretation. Part 2. Minicomputer Based 13-C/1-H NMR File Search System. V. Mlynarik, M. Vida, and V. Kello, *Anal. Chim. Acta*, **122** (1980), 47.

Development of a New File Search System for Nuclear Magnetic Resonance Spectra. Y. Katagiri, K. Kanohta, K. Nagasawa, T. Okusa, T. Sakai, O. Tsumura, and Y. Yotsui, *Anal. Chim. Acta*, **133** (1981), 535.

^{13}C NMR

Mutual Assignment of Subspectra and Substructures—a way to Structure Elucidation by 13-C NMR Spectrosccopy. W. Bremser, M. Klier, and E. Meyer, *Org. Magn. Reson.*, **7** (1975), 97.

A Computer System for Structural Identification of Organic Compounds from 13-C NMR Data. R. Schwarzenbach, J. Meili, H. Konitzer, and J. T. Clerc, *Org. Magn. Reson.*, **8** (1976), 11.

The Importance of Multiplicities and Substructures for the Evaluation of Relevant Spectral Similarities for Computer Aided Interpretation of 13-C NMR Spectra. W. Bremser, *Z. Anal. Chem.*, **286** (1977), 1.

A Substructure Oriented 13-C NMR Chemical Shift Retrieval System. J. Zupan, S. R. Heller, G. W. A. Milne, and J. A. Miller, *Anal. Chim. Acta*, **103** (1978), 141.

A Carbon-13 Nuclear Magnetic Resonance Spectral Data Base and Search System. D. L. Dalrymple, C. L. Wilkins, G. W. A. Milne, and S. R. Heller, *Org. Magn. Reson.*, **11** (1978), 535.

Spectra-Structure Relationships in Carbon-13 Nuclear Magnetic Resonance Spectrosccopy. Results from a Large Data Base. G. W. A. Milne, J. Zupan, S. R. Heller, and J. A. Miller, *Org. Magn. Reson.*, **12** (1979), 289.

The DARC PLURIDATA System: the 13-C Data Bank. J. E. Dubois and J. C. Bonnett, *Anal. Chim. Acta*, **112** (1979), 245.

Fast Searching for Identical 13-C NMR Spectra via Inverted Files. W. Bremser, H. Wagner, and B. Franke, *Org. Magn. Reson.*, **15** (1981), 178.

An Automated Library Search System for 13-C NMR Spectra Based on the Reproducibility of Chemical Shifts. R. W. Bally, D. van Krimpen, P. Cleij, and H. A. van't Klooster, *Anal. Chim. Acta*, **157** (1984), 227.

CSEARCH: A Computer Program for Identification of Organic Compounds and Fully Automated Assignment of Carbon-13 NMR Spectra. H. Kalchhauser and W. Robien, *J. Chem. Inf. Comput. Sci.*, **25** (1985), 103.

Combined Spectral Data

Strukturaufklarung Organischer Vergindungen durch Computerunterstutzten Vergleich Spektralerdaten. F. Erni and J. T. Clerc, *Helv. Chim. Acta*, **55** (1972), 489.

Combined Retrieval System for Infrared, Mass and Carbon-13 Resonance Spectra. J. Zupan, M. Penca, D. Hadzi, and J. Marsel, *Anal. Chem.*, **49** (1977), 2141.

The Computer Assisted Characterization of Terpenes and Related Compounds by the Use of Combined Spectrometric Data. S. Sasaki, H. Abe, K. Saito, and Y. Ishida, *Bull. Chem. Soc. Jap.*, **51** (1978), 3218.

Minicomputer Oriented Chemical Information System. J. Zupan, M. Penca, M. Razinger, B. Barlic, and D. Hadzi, *Anal. Lett.*, **12A** (1979), 109.

KSIK-a Combined Chemical Information System for a Minicomputer. J. Zupan, M. Penca, M. Razinger, B. Barlic, and D. Hadzi, *Anal. Chim. Acta*, **122** (1980), 103.

The NIH-EPA Chemical Information System in Support of Structure Elucidation. S. R. Heller and G. W. A. Milne, *Anal. Chim. Acta*, **122** (1980), 117.

NIH/EPA Chemical Information System. G. W. A. Milne and S. R. Heller, *J. Chem. Inf. Comput. Sci.*, **20** (1980), 204.

SPEKTREN—a Computer System for the Identification and Structure Elucidation of Organic Compounds. M. Zippel, J. Mowitz, I. Kohler, and H. J. Opferkuch, *Anal. Chim. Acta*, **140** (1982), 123.

Search System for Infrared and Mass Spectra by Factor Analysis and Eigenvector Projection. S. S. Williams, R. B. Lam, and T. L. Isenhour, *Anal. Chem.*, **55** (1983), 1117.

Others

Computer Assisted Structural Interpretation of Fluoresence Spectra. T. C. Miller and L. R. Faulkner, *Anal. Chem.*, **48** (1976), 2083.

Performance of Search Systems

Quantitative Evaluation of Several Infrared Searching Systems. D. S. Erley, *Appl. Spectrosc.*, **25** (1971), 200.

Statistical Method for Prediction of Matching Results in Spectral File Searching. S. L. Grotch, *Anal. Chem.*, **46** (1974), 526.

Comparative Study of Methods of Computer-Matching Mass Spectra. R. J. Mathews and J. D. Morrison, *Aust. J. Chem.*, **27** (1974), 2167.

Comparative Study of Methods of Computer-Matching Mass Spectra. II. R. J. Mathews and J. D. Morrison, *Aust. J. Chem.*, **29** (1976), 689.

Performance Prediction and Evaluation of Systems for Computer Identification of Spectra. F. W. McLafferty, *Anal. Chem.*, **49** (1977), 1441.

Comparison of Various k-Nearest Neighbor Voting Schemes with the Self Training Interpretive and Retrieval System for Identifying Molecular Substructures from Mass Spectra. S. R. Lowry, T. L. Isenhour, J. B. Justice, F. W. McLafferty, H. E. Dayringer, and R. Venkataraghavan, *Anal. Chem.*, **49** (1977), 1720.

Calculation of the Information Content of Retrieval Procedures Applied to Mass Spectral Data Bases. G. van Marlen, A. Dijkstra, and H. A. van't Klooster, *Anal. Chim. Acta*, **112** (1979), 233.

Der Anwendung der Informations Theorie zur Berwertung von Computergestutzen Spektrensuchsystem. K. Schaarschmidt, *Anal. Chim. Acta*, **112** (1979), 385.

Statistical Prediction of File Searching Results for Vapor Phase Infrared Spectrometric Identification of Gas Chromatographic Peaks. M. F. Delaney and P. C. Ulden, *Anal. Chem.*, **51** (1979), 1242.

Evaluation of Mass Spectral Search Algorithms. G. T. Rasmussen and T. L. Isenhour, *J. Chem. Inf. Comput. Sci.*, **19** (1979), 179.

Quantitative Evaluation of Library Searching Performance. M. F. Delaney, F. V. Warner, and J. R. Hallowell, *Anal. Chem.*, **55** (1983), 1925.

PATTERN RECOGNITION

Methods and Evaluation of Performance

Computerized Learning Machines Applied to Chemical Problems. Molecular Formula Determination from Low Resolution Mass Spectra. P. C. Jurs, B. R. Kowalski, and T. L. Isenhour, *Anal. Chem.*, **41** (1969), 21.

Computerized Learning Machines Applied to Chemical Problems. Investigation of Convergence Rate and Predictive Ability of Adaptive Binary Classifiers. P. C. Jurs, B. R. Kowalski, T. L. Isenhour, and C. N. Reilley, *Anal. Chem.*, **41** (1969), 690.

Computerized Learning Machines Applied to Chemical Problems. Multicategory Pattern Classification by Least Squares. P. C. Jurs, B. R. Kowalski, T. L. Isenhour, and C. N. Reilley, *Anal. Chem.*, **41** (1969), 695.

Computerized Learning Machines Applied to Chemical Problems. Interpretation of Infrared Spectrometry. B. R. Kowalski, P. C. Jurs, T. L. Isenhour, and C. N. Reilley, *Anal. Chem.*, **41** (1969), 1945.

An Investigation of Combined Patterns from Diverse Analytical Data Using Computerized Learning Machines. P. C. Jurs, B. R. Kowalski, T. L. Isenhour, and C. N. Reilley, *Anal. Chem.*, **41** (1969), 1949.

Computerized Learning Machines Applied to Chemical Problems. Molecular Structure Parameters from Low Resolution Mass Spectra. P. C. Jurs, B. R. Kowalski, T. L. Isenhour, and C. N. Reilley, *Anal. Chem.*, **42** (1970), 1387.

Mass Spectral Feature Selection and Structural Correlations Using Computerized Learning Machines. P. C. Jurs, *Anal. Chem.*, **42** (1970), 1633.

Machine Intelligence Applied to Chemical Systems. Prediction and Reliability Improvement in Classification of LRMS Data. P. C. Jurs, *Anal. Chem.*, **43** (1971), 22.

Computerized Learning Machines Applied to Chemical Problems. Optimization of a Linear Pattern Classifier by the Addition of a "Width" Parameter. L. E. Wanger, N. M. Frew, and T. L. Isenhour, *Anal. Chem.*, **43** (1971), 845.

Investigation of Fourier Transform Representation of Mass Spectra for Analysis by Computerized Learning Machine. P. C. Jurs, *Anal. Chem.*, **43** (1971), 1812.

The k-Nearest Neighbor Rule (Pattern Recognition) Applied to NMR Spectral Interpretation. B. R. Kowalski and C. F. Bender, *Anal. Chem.*, **44** (1972), 1405.

The Hadamard Transform and Spectral Analysis by Pattern Recognition. B. R. Kowalski and C. F. Bender, *Anal. Chem.*, **45** (1973), 2234.

Improved Discriminant Training and Feature Extraction for the Generation of Simulated Mass Spectra of Small Organic Molecules. J. Schechter and P. C. Jurs, *Appl. Spectrosc.*, **27** (1973), 225.

Information Content of Mass Spectra as Determined by Pattern Recognition Methods. J. B. Justice and T. L. Isenhour, *Anal. Chem.*, **46** (1974), 223.

Multiclass Linear Classifier for Spectral Interpretation. C. F. Bender and B. R. Kowalski, *Anal. Chem.*, **46** (1974), 294.

Pattern Recognition Techniques Applied to the Interpretation of IR Spectra. D. R. Preuss and P. C. Jurs, *Anal. Chem.*, **46** (1974), 520.

Interpretation of Mass Spectrometry Data Using Cluster Analysis. S. R. Heller, C. L. Chas, and K. C. Chu, *Anal. Chem.*, **46** (1974), 951.

Computerized Pattern Recognition Applications to Chemical Analysis, Development of Interactive Feature Selection Methods for k-Nearest Neighbor Technique. M. A. Pichler and S. P. Perone, *Anal. Chem.*, **46** (1974), 1790.

Interpretation of Infrared Spectra Using Pattern Recognition Techniques. R. W. Liddell and P. C. Jurs, *Anal. Chem.*, **46** (1974), 2126.

Bayesian Decision Theory Applied to Multicategory Classification of Binary Infrared Spectra. H. B. Woodruff, S. R. Lowry, and T. L. Isenhour, *Anal. Chem.*, **46** (1974), 2150.

Nonparametric Feature Selection in Pattern Recognition Applied to Chemical Problems. G. S. Zander, A. J. Stuper, and P. C. Jurs, *Anal. Chem.*, **47** (1975), 1085.

Probability Discriminant Functions for Classifying Binary Infrared Spectral Data. S. R. Lowry, H. B. Woodruff, G. L. Ritter, and T. L. Isenhour, *Anal. Chem.*, **47** (1975), 1126.

Classification of Mass Spectra Using Adaptive Digital Learning Networks. T. J. Stonham, I. Alkesander, M. Camp, W. T. Pike, and M. A. Shaw, *Anal. Chem.*, **47** (1975), 1817.

Simplex Pattern Recognition. G. L. Ritter, S. R. Lowry, C. L. Wilkins, and T. L. Isenhour, *Anal. Chem.*, **47** (1975), 1951.

Factor Analysis of Mass Spectra. J. B. Justice and T. L. Isenhour, *Anal. Chem.*, **47** (1975), 2286.

A Comparison of Two Discriminant Functions for Classifying Binary Infrared Spectra. H. B. Woodruff, S. R. Lowry, and T. L. Isenhour, *Appl. Spectrosc.*, **29** (1975), 226.

A Feature Selection Technique for Binary Infrared Spectra. S. R. Lowry and T. L. Isenhour, *J. Chem. Inf. Comput. Sci.*, **15** (1975), 212.

Considerations on the Interpretation of Mass Spectra via Learning Machines. P. Kent and T. Gaumann, *Helv. Chim. Acta*, **58** (1975), 787.

Beurteilungskriterien fur Klassifkatoren zur Automatischen Spektreninterpretation. H. Rotter and K. Varmuza, *Org. Mass Spectrom.*, **10** (1975), 874.

Information Theory Applied to the Selection of Peaks for Retrieval of Mass Spectra. G. van Marlen and A. Dijkstra, *Anal. Chem.*, **48** (1976), 595.

Simplex Pattern Recognition Applied to Carbon-13 NMR Spectrosccopy. T. R. Brunner, C. L. Wilkins, T. F. Lam, L. J. Soltzberg, and S. L. Kaberline, *Anal. Chem.*, **48** (1976), 1146.

Constraints on Learning Machine Classification Methods. N. A. B. Gray, *Anal. Chem.*, **48** (1976), 2265.

Pattern Recognition Methods for the Classification of Binary Infrared Spectral Data. H. B. Woodruff, G. L. Ritter, S. R. Lowry, and T. L. Isenhour, *Appl. Spectrosc.*, **30** (1976), 213.

Reliability of Nonparametric Linear Classifiers. A. J. Stuper and P. C. Jurs, *J. Chem. Inf. Comput. Sci.*, **16** (1976), 238.

Comparison of Various k-Nearest Neighbor Voting Schemes with the Self Training Interpretive and Retrieval System for Identifying Molecular Substructures from Mass Spectra. S. R. Lowry, T. L. Isenhour, J. B. Justice, F. W. McLafferty, H. E. Dayringer, and R. Venkataraghavan. *Anal. Chem.*, **49** (1977), 1720.

Deceptive Correct Separation by the Linear Learning Machine. C. P. Weisel and J. L. Fasching, *Anal. Chem.*, **49** (1977), 2114.

Dimensionality and the Number of Features in Learning Machine Classification Methods. G. L. Ritter, and H. B. Woodruff, *Anal. Chem.*, **49** (1977), 2116.

Classification of Binary C-13 NMR Spectra. C. L. Wilkins and T. R. Brunner, *Anal. Chem.*, **49** (1977), 2136.

Comment on the Application of Feature Selection Method for Binary Coded Patterns. Y. Miyashita, H. Abe, S. Sasaki, and K. Yuta, *Anal. Chem.*, **50** (1978), 1580.

Interactive Pattern Recognition in the Chemical Analysis Laboratory. C. L. Wilkins, *J. Chem. Inf. Comput. Sci.*, **17** (1977), 242.

Evaluation of the Super Modified Simplex for Use in Chemical Pattern Recognition. S. L. Kaberline and C. L. Wilkins, *Anal. Chim. Acta*, **103** (1978), 417.

Four Levels of Pattern Recognition. C. Albano, W. Dunn, U. Edlund, E. Johanssen, B. Norden, M. Sjostrom, and S. Wold. *Anal. Chim. Acta*, **103** (1978), 429.

Classification of Mass Spectra via Pattern Recognition. J. R. McGill and B. R. Kowalski, *J. Chem. Inf. Comput. Sci.*, **18** (1978), 52.

A Comparison of Five Pattern Recognition Methods Based on Classification Results from Six Real Data Bases. M. Sjostrom and B. R. Kowalski, *Anal. Chim. Acta*, **112** (1979), 11.

The Role of Pattern Recognition in Computer Aided Classification of Mass Spectra. W. S. Meisel, M. Jolley, S. R. Heller, and G. W. A. Milne, *Anal. Chim. Acta*, **112** (1979), 407.

On Confidence in the Results of Learning Machines Trained on Mass Spectra. J. A. Richards and A. G. Griffiths, *Anal. Chem.*, **51** (1979), 1358.

Hazards in Factor Analysis. C. G. Swain, H. E. Bryndza, and M. S. Swain, *J. Chem. Inf. Comput. Sci.*, **19** (1979), 19.

Fisher Discriminant Functions for a Multilevel Mass Spectral Filter Network. G. R. Rasmussen, G. L. Ritter, S. R. Lowry, and T. L. Isenhour, *J. Chem. Inf. Comput. Sci.*, **19** (1979), 255.

The Probability of Dichotomization of a Binary Linear Classifier as a Function of Training Set Population Distribution. E. K. Whalen-Pedersen and P. C. Jurs, *J. Chem. Inf. Comput. Sci.*, **19** (1979), 264.

K-Nearest Neighbor Rule in Weighting Measurements for Pattern Recognition. W. A. Byers and S. P. Perone, *Anal. Chem.*, **52** (1980), 2173.

A New Approach to Binary Tree Based Heuristics. J. Zupan, *Anal. Chim. Acta*, **122** (1980), 337.

An Operational Research Model for Pattern Recognition. D. L. Massart, L. Kaufman, and D. Coomans, *Anal. Chim. Acta*, **122** (1980), 347.

Potential Methods in Pattern Recognition. 1. Classification Aspects of the Supervised Method ALLOC. D. Coomans, D. L. Massart, I. Broeckaert, and A. Tassin, *Anal. Chim. Acta*, **133** (1981), 215.

Potential Methods in Pattern Recognition. 2. CLUPOT-- an Unsupervised Pattern Recognition Technique. D. Coomans and D. L. Massart, *Anal. Chim. Acta*, **133** (1981), 225.

Potential Methods in Pattern Recognition. 3. Feature Selection with ALLOC. D. Coomans, M. Derde, D. L. Massart, and I. Broeckaert, *Anal. Chim. Acta*, **133** (1981), 241.

Orthogonal Transforms for Feature Extraction in Chemical Pattern Recognition. L. Domokos and I. Frank, *Anal. Chim. Acta*, **133** (1981), 261.

Applications of Pattern Recognition to Structure-Activity Problems: Use of Minimal Spanning Tree. Y. Miyashita, Y. Takahashi, Y. Yotsui, H. Abe, and S. Sasaki, *Anal. Chim. Acta*, **133** (1981), 615.

Potential Methods in Pattern Recognition. 5. Alloc, Action Oriented Decision Making. D. Coomans, D. L. Massart, and I. Broeckaert, *Anal. Chim. Acta*, **134** (1982), 139.

Alternative K-Nearest Neighbor Rules in Supervised Pattern Recognition. Part 2. Probabilistic Classification on the basis of the KNN Method Modified for Direct Density Estimation. Part 3. Condensed Nearest Neighbor Rules. D. Coomans and D. L. Massart, *Anal. Chim. Acta*, **138** (1982), 145.

Monte Carlo Studies of Classifications Made by Non Parametric Linear Discriminant Functions. T. R. Stouch and P. C. Jurs, *J. Chem. Inf. Comput. Sci.*, **25** (1985), 45.

Applications

Application of Pattern Separation Techniques to Mass Spectrometric Data. Determination of Hydrocarbon Types and the Average Molecular Structure of Gasoline. D. D. Tunnicliff and P. A. Wadsworth, *Anal. Chem.*, **45** (1973), 12.

Generation of Simulated Mass Spectra of Small Organic Molecules by Computerized Pattern Recognition Techniques. J. Schechter and P. C. Jurs, *Appl. Spectrosc.*, **27** (1973), 30.

Interpretation of IR Spectra Using Pattern Recognition Techniques. R. W. Liddell and P. C. Jurs, *Appl. Spectrosc.*, **27** (1973), 371.

Applications of Artificial Intelligence to Chemistry. Use of Pattern Recognition and Cluster Analysis to Determine the Pharmacological Activity of some Organic Compounds K. C. Chu, *Anal. Chem.*, **46** (1974), 1181.

Heuristic Pattern Recognition of Carbon 13 NMR Spectra C. L. Wilkins, R. C. Williams, T. R. Brunner, and P. J. McCombe, *J. Am. Chem. Soc.*, **96** (1974), 4182.

Pattern Recognition. An Application to the Identification of Some Stereoisomeric N-Acetylhexosamines by Mass Spectrometry. J. Vink, W. Heerma, J. P. Kamerling, and J. F. G. Vliegenthart, *Org. Mass Spectrom.*, **9** (1974), 536.

Automated Chemical Structure Analysis of Organic Molecules with a Molecular Structure Generator and Pattern Recognition Techniques. H. Abe and P. C. Jurs, *Anal. Chem.*, **47** (1975), 1829.

Multiple Discriminant Function Analysis of Carbon-13 NMR Spectra: Functional Group Identification by Pattern Recognition. C. L. Wilkins and T. L. Isenhour, *Anal. Chem.*, **47** (1975), 1849.

Large Scale Mass Spectral Analysis by Simplex Pattern Recognition. T. F. Lam, C. L. Wilkins, T. R. Brunner, L. J. Soltzberg, and S. L. Kaberline, *Anal. Chem.*, **48** (1976), 1768.

Classification of Petroleum Pollutants by Linear Discriminant Function Analysis of Infrared Spectral Patterns. J. S. Mattson, C. S. Mattson, M. J. Spencer, and F. W. Spencer, *Anal. Chem.*, **49** (1977), 500.

Computer Assisted Interpretation of Carbon-13 Nuclear Magnetic Resonance Spectra Applied to Structure Elucidation of Natural Products. H. B. Woodruff, C. R. Snelling, C. A. Shelley, and M. E. Munk, *Anal. Chem.*, **49** (1977), 2075.

Computer Aided Interpretation of Steroid Mass Spectra by Pattern Recognition Methods. Part 2. Influence of Mass Spectral Preprocessing on Classification by Distance Measures to the Centers of Gravity. H. Rotter and K. Varmuza, *Anal. Chim. Acta*, **95** (1977), 25.

The Interpretation of Infrared and Raman Spectra Using Pattern Recognition. J. M. Comerford, P. G. Anderson, W. H. Snyder, and H. S. Kimmel, *Spectrochimica Acta*, **33A** (1977), 651.

Computer Aided Interpretation of Steroid Mass Spectra by Pattern Recognition Methods. Part III. Computation of Binary Classifiers by Linear Regression. H. Rotter and K. Varmuza, *Anal. Chim. Acta*, **103** (1978), 61.

Computerized Pattern Recognition for Classification of Organic Compounds from Voltametric Data. D. R. Burgard and S. P. Perone, *Anal. Chem.*, **50** (1978), 1366.

Classification of Crude Oil Gas Chromatograms by Pattern Recognition Techniques. H. A. Clark and P. C. Jurs, *Anal. Chem.*, **51** (1979), 618.

Computerized Pattern Recognition Applied to Gas Chromatographic/Mass Spectrometry Identification of Pentafluropropionyl Dipeptide Methyl Esters. J. N. Ziemer, S. P. Perone, R. M. Caprioli, and W. E. Seifert, *Anal. Chem.*, **51** (1979), 1732.

Application of a Correlation Coefficient Pattern Recognition Technique to Low Resolution Mass Spectra. N. H. Mahle and J. W. Ashley, *Comput. Chem.*, **3** (1979), 19.

Assessment of Oil Contamination in the Marine Environment by Pattern Recognition of Paraffinic Hydrocarbon Contents of Mussels. P. W. Kwan and R. C. Clark, *Anal. Chim. Acta*, **133**, 1981, 151.

Computer-Assisted Structure-Carcinogenicity Studies on Polycyclic Aromatic Hydrocarbons by Pattern Recognition Methods. Y. Miyashita, T. Seki, Y. Takahashi, S. Daiba, Y. Tanaka, Y. Totsui, H. Abe, and S. Sasaki, *Anal. Chim. Acta*, **133** (1981), 603.

Classification of Monosubstituted Phenyl Rings by Parametric Methods Applied to Infrared and Raman Peak Heights. R. Tsao and W. L. Switzer, *Anal. Chim. Acta*, **134** (1982), 111.

Pattern Recognition Applied to Vapour Phase Infrared Spectra. L. Domokos, I. Frank, G. Matolcsy, and G. Jalsovszky, *Anal. Chim. Acta*, **154** (1983), 181.

Structural and Activity Characterization of Organic Compounds by Electroanalysis and Pattern Recognition, W. A. Byers, B. S. Freiser, and S. P. Perone, *Anal. Chem.*, **55** (1983), 620.

EMPIRICAL SPECTRUM ANALYSIS, INTERPRETATION AND CLASSIFICATION

Mass Spectral Data Processing. Computer Used for Identification of Organic Compounds. B. Pettersson and R. Ryhage, *Arkiv fur Kemi*, **26** (1967), 293.

Mass Spectral Data Processing. Identification of Aliphatic Hydrocarbons. B. Pettersson and R. Ryhage, *Anal. Chem.*, **39** (1967), 790.

Computer Methods in Analytical Mass Spectrometry. Empirical Identification of Molecular Class. L. R. Crawford and J. D. Morrison, *Anal. Chem.*, **40** (1968), 1469.

Applications of Artificial Intelligence for Chemical Inference. II. Interpretation of the Low Resolution Mass Spectra of Ketones. A. M. Duffield, A. V. Robertson, C. Djerassi, B. G. Buchanan, G. L. Sutherland E. A. Feigenbaum, and J. Lederberg, *J. Am. Chem. Soc.*, **91** (1969), 2977.

Applications of Artificial Intelligence for Chemical Inference. III. Aliphatic Ethers Diagnosed by the Low Resolution Mass Spectra and Nuclear Magnetic Resonance Data. G. Schroll, A. M. Duffield, C. Djerassi, B. G. Buchanan, G. L. Sutherland E. A. Feigenbaum, and J. Lederberg, *J. Am. Chem. Soc.*, **91** (1969), 7440.

Computer Aided Interpretation of Mass Spectra. R. Venkataraghavan, F. W. McLafferty, and G. E. van Lear, *Org. Mass Spectrom.*, **2** (1969), 1.

Applications of Artificial Intelligence for Chemical Inference. IV. Saturated Amines Diagnosed by Their Low Resolution Mass Spectra and Nuclear Magnetic Resonance Spectra. A. Buchs, A. M. Duffield, G. Schroll, C. Djerassi, A. B. Delfino, B. G. Buchanan, G. L. Sutherland, E. A. Feigenbaum, and J. Lederberg, *J. Am. Chem. Soc.*, **92** (1970), 6831.

A Correlative Notation System for NMR Data. H. Skolnik, *J. Chem. Doc.*, **10** (1970), 216.

Computer Methods in Analytical Mass Spectrometry. Development of Programs for Analysis of Low Resolution Mass Spectra. L. R. Crawford and J. D. Morrison, *Anal. Chem.*, **43** (1971), 1790.

Computer Interpretation of Mass Spectra. D. H. K. Koo and R. D. Sedgwick, *Adv. Mass Spectrom.*, **6** (1972), 1019.

A Compound Classifier Based on Computer Analysis of Low Resolution Mass Spectral Data. Geochemical and Environmental Application. D. H. Smith, *Anal. Chem.*, **44** (1972), 536.

Heuristic Programming as an Ion Generation in Mass Spectrometry. 1. Generation of Primary Ions with Charge Localization. A. B. Delfino and A. Buchs, *Helv. Chim. Acta*, **55** (1972), 2017.

Applications of Artificial Intelligence for Chemical Inference. VIII. An Approach to the Computer Interpretation of the High Resolution Mass Spectra of Complex Molecules. Structure Elucidation of Estrogenic Steroids. D. H. Smith, B. G. Buchanan, R. S. Engelmore, A. M. Duffield, A. Yeo, E. A. Feigenbaum, J. Lederberg, and C. Djerassi, *J. Am. Chem. Soc.*, **94** (1972), 5962.

Raman Spektroskopie und Molekulstruktur. V. Schema vur Bestimmung des Substitutions Musters von Benzolderivaten. B. Schrader and W. Meier, *Z. Anal. Chem.*, **260** (1972), 248,

Computing Methods in Mass Spectrometry. Programming for Aliphatic Amines and Alcohols. J. F. O'Brien and J. D. Morrison, *Aust. J. Chem.*, **26** (1973), 785.

Applications of Artificial Intelligence to Chemical Inference. Part IX. Analysis of Mixtures Without Prior Separation as Illustrated for Estrogens. D. H. Smith, B. G. Buchanan, R. S. Engelmore, H. Adlercreutz, and C. Djerassi, *J. Am. Chem. Soc.*, **95** (1973), 6078.

Applications of Artificial Intelligence to Chemical Inference. Part XI. Analysis of C-13 NMR Data for Structure Elucidation of Acyclic Amines. R. E. Carhart and C. Djerassi, *J. Chem. Soc. Perkin Trans. II*, **1973**, 1753,

Applications of Artificial Intelligence for Chemical Inference. X. A Data Interpretation and Summary Program Applied to the Collected Mass Spectra of Estrogenic Steroids. D. H. Smith, B. G. Buchanan, W. C. White, E. A. Feigenbaum, J. Lederberg, and C. Djerassi, *Tetrahedron*, **29** (1973), 3117.

Structural Interpretation of Proton Magnetic Resonance Spectra by Computer: First Order Spectra. G. Beech, R. T. Jones, and K. Miller, *Anal. Chem.*, **46** (1974), 714.

A Systematic Approach to the Structure Elucidation of Carbon-Hydrogen and Hydroxy Compounds by Means of Raman Spectrosccopy with Laser Sources. J. H. van der Maas and T. Visser, *J. Raman Spectrosc.*, **2** (1974), 563.

Heuristic Programming as an Ion Generator in Mass Spectrometry. II. Generation of Primary Ions from Simple Cyclic Structures. A. B. Delfino and A. Buchs, *Org. Mass Spectrom.*, **9** (1974), 459.

Computer Program for the Retrieval and Assignment of Chemical Environments and Shifts to Facilitate Interpretation of Carbon-13 NMR Spectra. B. A. Jezl and D. L. Dalrymple, *Anal. Chem.*, **47** (1975), 203.

Programs for Spectrum Classification and Screening of GC/MS Data on a Laboratory Computer. N. A. B. Gray and T. O. Gronneberg, *Anal. Chem.*, **47** (1975), 419.

Automatic Classification of Mass Spectra by a Laboratory Computer System N. A. B. Gray, J. A. Zorro, T. O. Gronneberg, S. J. Gaskell, J. N. Cardoso, and G. Eglinton, *Anal. Lett.*, **8** (1975), 461.

A Heuristic Computer Program for Automatic Spectrum Assignment in Microwave Spectrosccopy. A. B. Delfino and K. R. Ramaprasad, *J. Molec. Struct.*, **25** (1975), 293.

Applications of Artificial Intelligence for Chemical Inference. XIV. A General Method for Predicting Molecular Ions in Mass Spectra. R. G. Dromey, B. G. Buchanan, D. H. Smith, J. Lederberg, and C. Djerassi, *J. Org. Chem.*, **40** (1975), 770.

Mutual Assignment of Subspectra and Substructures—a Way to Structure Elucidation by 13-C NMR Spectrosccopy. W. Bremser, M. Klier, and E. Meyer, *Org. Magn. Reson.*, **7** (1975), 97.

A Program for Generating Empirical Spectrum Classification Schemes. N. A. B. Gray, *Org. Mass Spectrom.*, **10** (1975), 507.

Programmierte Gemeinsane Auswertung Charakteristicher Banden in den Raman und Infrarotspektren Organischer Substazen. B. Schrader and W. Meir, *Z. Anal. Chem.*, **275** (1975), 177,

Progressive Filter Network, a General Classification Algorithm. S. R. Lowry, J. C. Marshall, and T. L. Isenhour, *Comput. Chem.*, **1** (1976), 3.

Applications of Artificial Intelligence for Chemical Inference. XXII. Automatic Rule Formation in Mass Spectrometry by Means of the Meta-DENDRAL Program. B. G. Buchanan, D. H. Smith, W. C. White, R. Gritter, E. A. Feigenbaum, J. Lederberg, and C. Djerassi, *J. Am. Chem. Soc.*, **98** (1976), 6168.

Computer Aided Interpretation of Mass Spectra. Information on Substructural Probabilities from STIRS. H. E. Dayringer, G. M. Pesyna, R. Venkataraghavan, and F. W. McLafferty, *Org. Mass Spectrom.*, **11** (1976), 529.

Computer Aided Interpretation of Mass Spectra. Increased Information from Characteristic Ions. H. E. Dayringer and F. W. McLafferty, *Org. Mass Spectrom.*, **11** (1976), 543.

Mass Spectrometry in Structural and Stereochemical Problems. CCXLVI. Electron Impact Induced Fragmentation of Juvenile Hormone Analogs. L. L. Dunham, C. A. Henric, D. H. Smith, and C. Djerassi, *Org. Mass Spectrom.*, **11** (1976), 1120.

Hierarchical Preprocessing of Infrared Data Files. M. Penca, J. Zupan, and D. Hadzi, *Anal. Chim. Acta*, **95** (1977), 3.

Computer Assisted Interpretation of Infrared Spectra. H. B. Woodruff and M. E. Munk, *Anal. Chim. Acta*, **95** (1977), 13.

Uniqueness of Carbon-13 NMR Spectra of Acyclic Saturated Hydrocarbons. H. L. Surprenant and C. N. Reilly, *Anal. Chem.*, **49** (1977), 1134.

Computer Assisted Analysis of Infrared Spectra of Nitrogen Containing Organic Compounds. Y. Miyashita, S. Ochiai, and S. Sasaki, *J. Chem. Inf. Comput. Sci.*, **17** (1977), 228.

A Computerized Infrared Spectral Interpreter as a Tool in Structure Elucidation of Natural Products. H. B. Woodruff and M. E. Munk, *J. Org. Chem.*, **42** (1977), 1761.

Systematic Interpretation of Raman Spectra of Organic Compounds. II. Ethers. T. Visser and J. H. van der Maas, *J. Raman Spectrosc.*, **6** (1977), 114.

Computer Aided Interpretation of Mass Spectra. STIRS Prediction of Rings Plus Double Bond Values. H. E. Dayringer and F. W. McLafferty, *Org. Mass Spectrom.*, **12** (1977), 53.

HOSE—a Novel Substructure Code. W. Bremser, *Anal. Chim. Acta*, **103** (1978), 355.

On the Solution of One Classical Problem in Vibrational Spectrosccopy. L. A. Gribov, M. E. Elyashberg, and V. V. Serov, *J. Molec. Struct.*, **50** (1978), 371.

A Raman/Infrared Interpretation System for CH-Compounds. C. G. A. v Eijk and J. H. van der Maas, *Fresenius Z. Anal. Chem.*, **291** (1978), 308.

Systematic Interpretation of Raman Spectra of Organic Compounds. III. Carbonyl Compounds. T. Visser and J. H. van der Maas, *J. Raman Spectrosc.*, 7 (1978), 125.

Systematic Interpretation of Raman Spectra of Organic Compounds. IV. Nitrogen Compounds. T. Visser and J. H. van der Maas, *J. Raman Spectrosc.*, 7 (1978), 278.

Applications of Artificial Intelligence for Chemical Inference. XXV. A Computer Program for Automated Empirical 13-C Rule Formation. T. M. Mitchell and G. M. Schwenzer, *Org. Magn. Reson.*, **11** (1978), 378.

Idenfikation der Partialstrukturen von Unbekannten Steroiden mit Eine Schlusselionenkartei und Einem Rechnerunterstutzten Retrieval-System. G. Spiteller, M. Spitteler, M. Ende, and G. H. Hoyen, *Org. Mass Spectrom.*, **13** (1978), 646.

Chemical Information from Computer Processed HRMS Data: Determination of Fragmentation Pathways for Single Functional Groups. R. M. Hilmer and J. W. Taylor, *Anal. Chem.*, **51** (1979), 1361.

Chemical Information from Computer Processed HRMS Data: Determination of Fragmentation Patterns of Multifunctional Compounds. L. W. McKeen and J. W. Taylor, *Anal. Chem.*, **51** (1979), 1368.

Computer Program for the Analysis of Infrared Spectra. H. B. Woodruff and G. M. Smith, *Anal. Chem.*, **52** (1980), 2321.

Computer Aided NMR Spectral Interpretation. Part 1. An Artificial Intelligence System. M. Vida, *Anal. Chim. Acta*, **122** (1980), 41.

Systematic Computer Aided Interpretation of Vibrational Spectra. T. Visser and J. H. van der Maas, *Anal. Chim. Acta*, **122** (1980), 357.

Systematic Computer Aided Interpretation of Infrared and Raman Vibrational Spectra Based on the CRISE Program. T. Visser and J. H. van der Maas, *Anal. Chim. Acta*, **122** (1980), 363.

Computerized Mass Spectrum Prediction and Ranking. N. A. B. Gray, R. E. Carhart, A. Lavanchy, D. H. Smith, T. Varkony, B. G. Buchanan, W. C. White, and L. Creary, *Anal. Chem.*, **52** (1980), 1095.

Elucidation Structurale Automatique par RMN du Carbone 13: Methode DARC-EPIOS. Recherche d'Une Relation Discriminante Structure Deplacement Chimique. J. E. Dubois, M. Carabedian, and B. Ancian, *Comptes. Rend. Acad. Sci.* (Paris), **290** (1980), 369,

Elucidation Structurale Automatique par RMN du Carbone-13: Methode DARC-EPIOS. Description de l'Elucidation Progressive pare Intersection Ordonnee de Sous-structures. J. E. Dubois, M. Carabedian, and Bernard Ancian, *Comptes. Rend. Acad. Sci.* (Paris), **290** (1980), 369,

Rule Based Mass Spectrum Prediction and Ranking. Applications to Structure Elucidation of Novel Marine Sterols. A. Lavanchy, T. Varkony, D. H. Smith, N. A. B. Gray, W. C. White, R. E. Carhart, B. G. Buchanan, and C. Djerassi, *Org. Mass Spectrom.*, **15** (1980), 355.

Computer Prediction of Molecular Weights from Mass Spectra. I. K. Mun, R. Venkataraghavan, and F. W. McLafferty, *Anal. Chem.*, **53** (1981), 179.

Prediction of Substructures from Unknown Mass Spectra by the Self-Training Interpretive and Retrieval System. K. S. Haraki, R. Venkataraghavan, and F. W. McLafferty, *Anal. Chem.*, **53** (1981), 386.

A Computer Aided System for Functional Group Determination. M. Farkas, J. Markos, P. Szepesvary, I. Barlta, G. Szalontai, and Z. Simon, *Anal. Chim. Acta*, **133** (1981), 19.

Use of IR and 13-C NMR Data in the Retrieval of Functional Groups for Computer Aided Structure Determination. G. Szalontai, Z. Simon, Z. Csapo, M. Farkas, and Gy. Pfeifer, *Anal. Chim. Acta*, **133** (1981), 31.

Systematic Computer Aided Interpretation of Vibrational Spectra. T. Visser and J. H. van der Maas, *Anal. Chim. Acta*, **133** (1981), 451.

Computer Assisted Structural Interpretation of Mass Spectral Data. N. A. B. Gray, A. Buchs, D. H. Smith, and C. Djerassi, *Helv. Chim. Acta*, **64** (1981), 458.

Computer Assisted Structural Interpretation of Carbon-13 Spectral Data. N. A. B. Gray, C. W. Crandell, J. G. Nourse, D. H. Smith, M. L. Dageforde, and C. Djerassi *J. Org. Chem.*, **45** (1981), 703.

Stereochemical Substructure Codes for 13-C Spectral Analysis. N. A. B. Gray, J. G. Nourse, C. W. Crandell, D. H. Smith, and C. Djerassi, *Org. Magn. Reson.*, **15** (1981), 375.

Generating Rules for PAIRS - a Computerized Infrared Spectral Interpreter. H. B. Woodruff and G. M. Smith, *Anal. Chim. Acta*, **133** (1981), 545.

An Artificial Intelligence System for Computer Aided Mass Spectra Interpretation of Saturated Monohydric Alcohols. C. Damo, C. Dachun, K. Teshu, and C. Shaoyu, *Anal. Chim. Acta*, **133** (1981), 575.

Computer Prediction of Substructures from C-13 NMR Spectra. C. A. Shelley and M. E. Munk, *Anal. Chem.*, **54** (1982), 516.

Table Driven Procedure for Infrared Spectrum Interpretation. M. O. Trulson and M. E. Munk, *Anal. Chem.*, **55** (1983), 2137.

Interactive Fragment Display Program for Interpretation of Mass Spectra. J. Figueras, *Anal. Chim. Acta*, **146** (1983), 29.

C-13 NMR Spectral Interpretation by a Computerized Substitution Chemical Shift Method. H. N. Cheng and S. J. Ellingsen, *J. Chem. Inf. Comput. Sci.*, **23** (1983), 197.

Rules for Computerized Interpretation of Vapor-Phase Infrared Spectra. S. A. Tomellini, J. M. Stevenson, and H. B. Woodruff, *Anal. Chem.*, **56** (1984), 67.

Computer Aided Elucidation of Structure by C-13 NMR. The DARC-EPOIOS Method. Characterization of Ordered Substructures by Correlating the Chemical Shifts of Their Bonded Atoms. J. E. Dubois, M. Carabedian, and I. Dagane, *Anal. Chim. Acta*, **158** (1984), 217.

Combinatorial Problems in Computer Assisted Structural Interpretation of Carbon-13 NMR Spectra. A. H. Lipkus and M. E. Munk, *J. Chem. Inf. Comput. Sci.*, **25** (1985), 38.

Automatic Tracing and Presentation of Interpretation Rules by PAIRS. S. A. Tomellini, R. A. Hartwick, and H. B. Woodruff, *Appl. Spectrosc.*, **39** (1985), 330.

STRUCTURE ELUCIDATION SYSTEMS

Program Details

Automated Structure Elucidation of Several Kinds of Aliphatic and Alicyclic Compounds. S. I. Sasaki, H. Abe, T. Ouki, M. Sakamoto, and S. Ochiai, *Anal. Chem.*, **40** (1968), 2220.

Heuristic DENDRAL: A Program for Generating Explanatory Hypotheses in Organic Chemistry. B. G. Buchanan, G. L. Sutherland, and E. A. Feigenbaum, in *Machine Intelligence*, Vol. 4, B. Meltzer and D. Michie, (eds.), Edinburgh University Press, Edinburgh, 1969, p. 209.

Symbolic Logic Methods for Spectrochemical Investigations. L. A. Gribov and M. E. Elyashberg, *J. Molec. Struct.*, **5** (1970), 179.

Solution of Spectral Problems by Methods of Symbolic Logic. L. A. Gribov, M. E. Elyashberg, and L. A. Moscovkina, *J. Molec. Struct.*, **9** (1971), 357.

Automated Chemical Structure Analysis of Organic Compounds. An Attempt to Structure Determination by the Use of NMR. S. Sasaki, Y. Kudo, S. Ochiai, and H. Abe, *Mikrochimica Acta* **1971**, 726.

An Application of Artificial Intelligence to the Interpretation of Mass Spectra B. G. Buchanan, A. M. Duffield, and A. V. Robertson, in *Mass Spectrometry: Techniques and Applications.*, G. W. A. Milne, (ed.), Wiley-Interscience, New York, 1971.

Automation of Spectrochemical Investigations. L. A. Gribov, V. A. Dementyev, M. E. Elyashberg, and E. Z. Yakapov, *J. Molec. Struct.*, **22** (1974), 161.

Automated Chemical Structure Analysis of Organic Molecules with a Molecular Structure Generator and Pattern Recognition Techniques. H. Abe and P. C. Jurs, *Anal. Chem.*, **47** (1975), 1829.

Structural Interpretation of Spectra. N. A. B. Gray, *Anal. Chem.*, **47** (1975), 2426.

Applications of Artificial Intelligence to Chemical Inference. XVII. An Approach to Computer-Assisted Elucidation of Molecular Structure. R. E. Carhart, D. H. Smith, H. Brown, and C. Djerassi, *J. Am. Chem. Soc.*, **97** (1975), 5755.

Applications of Artificial Intelligence for Chemical Inference. XX. Intelligent Use of Constraints in Computer Assisted Structure Elucidation. R. E. Carhart and D. H. Smith, *Comput. Chem.*, **1** (1976), 79.

A Structure Isomers Enumeration and Display System. Y. Kudo, Y. Hirota, S. Aoki, Y, Takada, T. Taji, I. Fuioka, K. Higashino, H. Fujishina, and S. Sasaki, *J. Chem. Inf. Comput. Sci.*, **16** (1976), 50.

Computer System for Structure Recognition of Polyatomic Molecules by IR, NMR, UV, and MS Methods. L. A. Gribov, M. E. Elyashberg, and V. V. Serov, *Anal. Chim. Acta*, **95** (1977), 75.

CHEMICS-F: A Computer Program System for Structure Elucidation of Organic Compounds. S. Sasaki, H. Abe, Y. Hirota, Y. Ishida, Y, Kudo, S. Ochiai, K. Saito, and T. Yamasaki, *J. Chem. Inf. Comput. Sci.*, **18** (1978), 211.

A New Approach to the Determination of Molecular Spatial Structure Based on the Use of Spectra and Computers. L. A. Gribov, M. E. Elyashberg, and M. M. Raikhshtat, *J. Molec. Struct.*, **53** (1979), 81.

A Computer Program System—New Chimics—for Structure Elucidation of Organic Compounds by Spectral and Other Structural Information. S. Sasaki, I. Fujiwara, H. Abe, and T. Yamasaki, *Anal. Chim. Acta*, **122** (1980), 87.

Chemics-UBE, a Modified System of Chemics. T. Oshima, Y. Ishida, K. Saito, and S. Sasaki, *Anal. Chim. Acta*, **122** (1980), 95.

Applications of Artificial Intelligence Systems in Molecular Spectrosccopy. L. A. Gribov, *Anal. Chim. Acta*, **122** (1980), 249.

Computer-aided Structural Analysis of Organic Compounds by an Artificial Intelligence System. B. Debska, J. Duliban, B. Guzowska-swider, and Z. Hippe, *Anal. Chim. Acta*, **133** (1981), 303.

GENOA: A Computer Program for Structure Elucidation Utilizing Overlapping and Alternative Substructures. R. E. Carhart, D. H. Smith, N. A. B. Gray, J. G. Nourse, and C. Djerassi, *J. Org. Chem.*, **46** (1981), 1708.

Computer-Aided Structure Elucidation Methods. H. Abe, T. Yamasaki, I. Fujiwara, and S. Sasaki, *Anal. Chim. Acta*, **133** (1981), 499.

CASE, a Computer Model of the Structure Elucidation Process. C. A. Shelley and M. E. Munk, *Anal. Chim. Acta*, **133** (1981), 507.

Computer Assisted Structure Elucidation. M. E. Munk, C. A. Shelley, H. B. Woodruff, and M. O. Trulson, *Fresenius Z. Anal. Chem.*, **313** (1982), 473.

Recent Advances in the Structure Elucidation System CHEMICS. H. Abe, I. Fujiwara, T. Nishimura, T. Okayama, T. Kida, and S. Sasaki, *Comput. Enhanced Spectrosc.*, **1** (1983), 55.

A Dialogue Computer Program System for Structure Recognition of Complex Molecules by Spectrosccopic Methods. L. A. Gribov, M. E. Elyashberg, V. N. Koldashov, and I. V. Pletnjov. *Anal. Chim. Acta*, **148** (1983), 159.

Applications of Artificial Intelligence for Chemical Inference. 43. Applications of the Program GENOA and 2-Dimensional NMR Spectrosccopy to Structure Elucidation. M. R. Lindley, J. N. Shoolery, D. H. Smith, and C. Djerassi, *Org. Magn. Reson.*, **21** (1983), 405.

Applications

Applications of Artificial Intelligence for Chemical Inference. II. Interpretation of the Low Resolution Mass Spectra of Ketones. A. M. Duffield, A. V. Robertson, C. Djerassi, B. G. Buchanan, G. L. Sutherland, E. A. Feigenbaum, and J. Lederberg, *J. Am. Chem. Soc.*, **91** (1969), 2977.

Applications of Artificial Intelligence for Chemical Inference. III. Aliphatic Ethers Diagnosed by the Low Resolution Mass Spectra and Nuclear Magnetic Resonance Data. G. Schroll, A. M. Duffield, C. Djerassi, B. G. Buchanan, G. L. Sutherland E. A. Feigenbaum, and J. Lederberg, *J. Am. Chem. Soc.*, **91** (1969), 7440.

Applications of Artificial Intelligence for Chemical Inference. VI. Approach to a General Method of Interpreting Low Resolution Mass Spectra with a Computer. A. Buchs, A. B. Delfino, A. M. Duffield, C. Djerassi, B. G. Buchanan, E. A. Feigenbaum, and J. Lederberg. *Helv. Chim. Acta*, **53** (1970), 1394.

Applications of Artificial Intelligence for Chemical Inference. IV. Saturated Amines Diagnosed by Their Low Resolution Mass Spectra and Nuclear Magnetic Resonance Spectra. A. Buchs, A. M. Duffield, G. Schroll, C. Djerassi, A. B. Delfino, B. G. Buchanan, G. L. Sutherland, E. A. Feigenbaum, and J. Lederberg, *J. Am. Chem. Soc.*, **92** (1970), 6831.

Photochemical Rearrangements of some Benzo C9H10 Isomers. R. C. Hahn and R. P. Johnson, *J. Am. Chem. Soc.*, **98** (1976), 2600.

Applications of Artificial Intelligence for Chemical Inference. XXI. Chemical Studies of Marine Invertebrates. XVII. The Computer Assisted Identification of (+)Palustrol in the Marine Organism *Cespitularia* sp. aff. *subviridis*. *Tetrahedron*, **32** (1976), 1807.

Marine Natural Products: dactylol, a New Sesquiterpene Alcohol from a Sea Hare. F. J. Schmitz, K. H. Hollenbeak, and D. J. Vanderah, *Tetrahedron*, **34** (1978), 2719.

Terpenoids. LXXVI. Precapnelladiene, a Possible Biosynthetic Precursor of the Capnellane Skeleton. E. Ayanoglu, T. Gebreyesus, C. M. Beecham, and C. Djerassi, *Tetrahedron*, **35** (1979), 1035.

Marine Natural Products: Dihydroxydeodactol Monoacetate, a Halogenated Sesquiterpene Ether from the Sea Hare *Aplysia dactylomela*. F. J. Schmitz, D. P. Michaud, and K. H. Hollenbeak, *J. Org. Chem.*, **45** (1980), 1525.

STRUCTURE-SUBSTRUCTURE REPRESENTATION

Topological Mapping of Organic Molecules. J. Lederberg, *Proc. Natl. Acad. Sci.* (USA), **53** (1965), 134.

Computer Methods in Analytical Mass Spectrometry. Structure Codes in Processing Mass Spectral Data. L. R. Crawford and J. D. Morrison, *Anal. Chem.*, **41** (1969), 994.

Economical Procedure for Coding Molecular Structures for the Handling of Spectra Collections. J. Franzen, H. Hillig, W. Riepe, and S. Stavrido, *Anal. Chem.*, **45** (1973), 475.

Simulation and Evaluation of Chemical Synthesis. Computer Representation and Manipulation of Stereochemistry. W. T. Wipke and T. M. Dyott, *J. Am. Chem. Soc.*, **96** (1974), 4834.

The Connectivity Stack, a New Format for Representation of Organic Structures. Y. Kudo and S. Sasaki, *J. Chem. Doc.*, **14** (1974), 200.

Computer-Aided Coding and Compact Direct Access Storage of Chemical Structures. W. Bremser, E. Frank, B. Franke, and H. Wagner, *J. Chem. Res.* **1979**, S113/M1401.

BCT Representation of Chemical Structures. T. Nakayama and Y. Fujiwara, *J. Chem. Inf. Comput. Sci.*, **20** (1980), 23.

A Graph Theory Data Base for Storage of Chemical Structures Organized by the Block-Cutpoint Tree Technique. Y. Fujiwara and T. Nakayama, *Anal. Chim. Acta*, **133** (1981), 647.

Simple Stereochemical Structure Code for Organic Chemistry. H. Beierbeck, *J. Chem. Inf. Comput. Sci.*, **22** (1982), 215.

Computer Representation of Generic Chemical Structures by an Extended Block Cutpoint Tree. T. Nakayama and Y. Fujiwara, *J. Chem. Inf. Comput. Sci.*, **23** (1983), 80.

Chemical Inference 1. Formalization of the Language of Organic Chemistry. Generic Structural Formulae. J. E. Gordon and J. C. Brockwell, *J. Chem. Inf. Comput. Sci.*, **23** (1983), 117.

A Concise Connection Table Based on Systematic Nomenclatural Terms. J. D. Rayner, *J. Chem. Inf. Comput. Sci.*, **25** (1985), 108.

SUBSTRUCTURE MATCHING AND FUNCTIONAL GROUP PERCEPTION

Techniques for Perception by a Computer of Synthetically Significant Structural Features in Complex Molecules. E. J. Corey, W. T. Wipke, R. D. Cramer, and W. J. Howe, *J. Am. Chem. Soc.*, **94** (1972), 431.

A Program for Rapid Automatic Functional Group Recognition. A. Esack and M. Bersohn, *J. Chem. Soc. Perkin Trans. I*, **1974**, 2463.

Functional Group Discovery Using the Concept of Central Atoms. M. Bersohn and A. Esack, *Chim. Scrip.*, **9** (1976), 211.

An Interactive Substructure Search System. R. J. Feldmann, G. W. A. Milne, S. R. Heller, A. Fein, J. A. Miller, and B. Koch, *J. Chem. Inf. Comput. Sci.*, **17** (1977), 157.

Graph-Based Fragment Searches in Polycyclic Systems. M. Randic and C. L. Wilkins, *J. Chem. Inf. Comput. Sci.*, **19** (1979), 23.

Molecular Substructure Searching: Computer Graphics and Query Entry Methodology. W. J. Howe and T. R. Hagadone, *J. Chem. Inf. Comput. Sci.*, **22** (1982), 8.

Molecular Substructure Searching: Minicomputer Based Query Execution. T. R. Hagadone and W. J. Howe, *J. Chem. Inf. Comput. Sci.*, **22** (1982), 182.

The CAS Online Search System. 1. General System Design and Selection, Generation and Use of Search Screens. P. G. Dittmar, N. A. Farmer, W. Fisanick, R. C. Haines, and J. Mockus, *J. Chem. Inf. Comput. Sci.*, **23** (1983), 93.

DARC Substructure Search System: A New Approach to Chemical Information. R. Attias, *J. Chem. Inf. Comput. Sci.*, **23** (1983), 102.

Rapid Subgraph Search Using Parallelism. W. T. Wipke and D. Rogers, *J. Chem. Inf. Comput. Sci.*, **24** (1984), 255.

Substructure Searching of Heterocycles by Computer Generation of Potential Aliphatic Precursors. R. L. M. Synge, *J. Chem. Inf. Comput. Sci.*, **25** (1985), 50.

IDENTIFICATION OF COMMON SUBSTRUCTURES

Molecular Structure Comparison Program for the Identification of Maximal Common Substructures. M. M. Cone, R. Venkataraghavan, and F. W. McLafferty, *J. Am. Chem. Soc.*, **99** (1977), 7668.

Computer Assisted Examination of Chemical Compounds for Structural Similarities. T. H. Varkony, Y. Shiloach, and D. H. Smith, *J. Chem. Inf. Comput. Sci.*, **19** (1979), 104.

An Algorithm for Finding the Intersection of Molecular Structures. M. Bersohn, *J. Chem. Soc. Perkin Trans. 1*, **1982**, 631.

Computer-Assisted Examination of Compounds for Common Three Dimensional Substructures. C. W. Crandell and D. H. Smith, *J. Chem. Inf. Comput. Sci.*, **23** (1983), 186.

CANONICALIZATION AND SYMMETRY

Generation of Unique Machine Description for Chemical Structures, a Technique Developed at Chemical Abstracts Service. H. L. Morgan, *J. Chem. Doc.*, **5** (1965), 107.

A Method for Generating Unique Structural Representations of Stereoisomers. A. E. Petraca, M. F. Lynch, and J. E. Rush, *J. Chem. Doc.*, **7** (1967), 154.

A Canonical Connection Table Representation of Molecular Structure. M. Bersohn and A. Esack, *Chim. Scrip.*, **6** (1974), 122.

Stereochemically Unique Naming Algorithm. W. T. Wipke and T. M. Dyott, *J. Am. Chem. Soc.*, **96** (1974), 4834.

On Unique Numbering of Atoms and Unique Codes for Molecular Graphs. M. Randic, *J. Chem. Inf. Comput. Sci.*, **15** (1975), 105.

Status of Notation and Topological Systems and Potential Future Trends. J. E. Rush, *J. Chem. Inf. Comput. Sci.*, **16** (1976), 202.

Computer Perception of Topological Symmetry. C. A. Shelley and M. E. Munk, *J. Chem. Inf. Comput. Sci.*, **17** (1977), 110.

Canonical Numbering and Constitutional Symmetry. C. Jochum and J. Gasteiger, *J. Chem. Inf. Comput. Sci.*, **17** (1977), 113.

On Canonical Numbering of Atoms in a Molecule and Graph Isomorphism. M. Randic, *J. Chem. Inf. Comput. Sci.*, **17** (1977), 171.

A Sum Algorithm for Numbering the Atoms of a Molecule. M. Bersohn, *Comput. Chem.*, **2** (1978), 112.

Constitutional Symmetry and Unique Descriptors of Molecules. W. Schubert and I. Ugi, *J. Am. Chem. Soc.*, **100** (1978), 37.

A Unique Computer Representation for Molecular Structures. C. A. Shelley, M. E. Munk, and R. V. Roman, *Anal. Chim. Acta*, **103** (1978), 245.

Erroneous Claims Concerning the Perception of Topological Symmetry. R. E. Carhart, *J. Chem. Inf. Comput. Sci.*, **18** (1978), 108.

Darstellugn Chemischer strukturen fur die Computergestutzte deduktive losung Chemischer Probleme. W. Schubert and I. Ugi, *Chimia*, **33** (1979), 183.

The Configuration Symmetry Group and its Applications to Stereoisomer Generation, Specification and Enumeration. J. G. Nourse, *J. Am. Chem. Soc.*, **101** (1979), 1210.

On the Misinterpretation of our Algorithm for the Perception of Constitutional Symmetry. C. Jochum and J. Gasteiger, *J. Chem. Inf. Comput. Sci.*, **19** (1979), 43.

Canonical Numbering. T. M. Dyott and W. J. Howe, *J. Chem. Inf. Comput. Sci.*, **19** (1979), 187

An Approach to the Assignment of Canonical Connection Tables and Topological Symmetry Perception. C. A. Shelley and M. E. Munk, *J. Chem. Inf. Comput. Sci.*, **19** (1979), 247.

Algorithms for Unique and Unambiguous Coding and Symmetry Perception of Molecular Structure Diagrams, I Vector Functions for Automorphism Partitioning, II Basic Algorithm for Unique Coding and Computation of Group, III Method of Subregion Analysis for Unique Coding and Symmetry Perception. M. Uchino, *J. Chem. Inf. Comput. Sci.*, **20** (1980), 116.

A Topological Code for Molecular Structures. A Modified Morgan Algorithm. G. Moreau, *Nouv. J. Chem.*, **4** (1980), 17.

Computer Perception of Topological Symmetry via Canonical Numbering of Atoms. M. Randic, G. M. Brissey, and C. L. Wilkins, *J. Chem. Inf. Comput. Sci.*, **21** (1981), 52.

Algorithms for Unique and Unambiguous Coding and Symmetry Perception of Molecular Structure Diagrams. 5. Unique Coding by Method of Orbit Graphs. M. Uchino, *J. Chem. Inf. Comput. Sci.*, **22** (1982), 201.

A New System for the Designation of Chemical Compounds. 1. Theoretical Preliminaries and the Coding of Acyclic Compounds. R. C. Read, *J. Chem. Inf. Comput. Sci.*, **23** (1983), 135.

Unique Numbering and Cataloguing of Molecular Structures. J. B. Hendrickson and A. G. Toczko, *J. Chem. Inf. Comput. Sci.*, **23** (1983), 171.

A New System for the Designation of Chemical Compounds. 2. Coding of Cyclic Compounds. R. C. Read, *J. Chem. Inf. Comput. Sci.*, **25** (1985), 116.

RING FINDING

An Algorithm for Machine Perception of Synthetically Significant Rings in Complex Organic Structures. E. J. Corey and G. A. Petersson, *J. Am. Chem. Soc.*, **94** (1972), 460.

An Algorithm for Finding the Synthetically Important Rings of a Molecule. M. Bersohn, *J. Chem. Soc. Perkin Trans. I*, **1973**, 1239.

Use of Ring Assemblies in Ring Perception Algorithms. W. T. Wipke and T. M. Dyott, *J. Chem. Inf. Comput. Sci.*, **15** (1975), 105.

A Procedure for Rapid Recognition of the Rings of a Molecule. A. Esack, *J. Chem. Soc. Perkin Trans. I*, **1975**, 1120,

An Algorithm for Finding the Smallest Set of Smallest Rings. A. Zamora, *J. Chem. Inf. Comput. Sci.*, **16** (1976), 40.

A FORTRAN-IV Program for Finding the Smallest Set of Smallest Rings. B. Schmidt and J. Fleischauer, *J. Chem. Inf. Comput. Sci.*, **18** (1978), 204.

An Algorithm for the Perception of Synthetically Important Rings. J. Gasteiger and C. Jochum, *J. Chem. Inf. Comput. Sci.*, **19** (1979), 43.

Computer-Assisted Mechanistic Evaluation of Organic Reactions. 2. Perception of Rings, Aromaticity and Tautomers. B. L. Roos-Kozel and W. L. Jorgensen, *J. Chem. Inf. Comput. Sci.*, **21** (1981), 101.

Condensed Structure Identification and Ring Perception. J. B. Hendrickson, D.L. Grier, and A. G. Toczko, *J. Chem. Inf. Comput. Sci.*, **24** (1984), 195.

STRUCTURE GENERATORS

Algorithms and Program Details

Constructive Graph Labelling Using Double Cosets. H. Brown, L. Hjelmeland, and L. M. Masinter, *Discrete Math.*, **7** (1974), 1.

Algorithm for the Construction of Graphs of Organic Molecules. H. Brown and L. M. Masinter, *Discrete Math.*, **8** (1974), 227.

Applications of Artificial Intelligence for Chemical Inference. XII. Exhaustive Generation of Cyclic and Acyclic Isomers. L. M. Masinter, N. S. Sridharan, J. Lederberg, and D. H. Smith, *J. Am. Chem. Soc.*, **96** (1974), 7702.

Applications of Artificial Intelligence for Chemical Inference. XIII. Labelling of Objects Having Symmetry. L. M. Masinter, N. S. Sridharan, R. E. Carhart, and D. H. Smith, *J. Am. Chem. Soc.*, **96** (1974), 7714.

Applications of Artificial Intelligence for Chemical Inference. XVI. Computer Generation of Vertex Graphs and Ring Systems. R. E. Carhart, D. H. Smith, H. Brown, and N. S. Sridharan, *J. Chem. Inf. Comput. Sci.*, **15** (1975), 124.

Mathematical Synthesis and Analysis of Molecular Structures. V. V. Serov, M. E. Elyashberg, and L. A. Gribov, *J. Molec. Struct.*, **31** (1976), 381.

Principle for Exhaustive Enumeration of Unique Structures Consistent with Structural Information. Y. Kudo and S. Sasaki, *J. Chem. Inf. Comput. Sci.*, **16** (1976), 43.

Molecular Structure Elucidation. III. H. Brown, *SIAM J. Appl. Math.*, **32** (1977), 534.

An Approach to Automated Partial Structure Expansion. C. A. Shelley, T. R. Hays, M. E. Munk, and R. V. Roman, *Anal. Chim. Acta*, **103** (1978), 121.

Exhaustive Generation of Stereoisomers for Structure Elucidation. J. G. Nourse, R. E. Carhart, D. H. Smith, and C. Djerassi, *J. Am. Chem. Soc.*, **101** (1979), 1216.

Computer Assisted Elucidation of Molecular Structure with Stereochemistry. J. G. Nourse, D. H. Smith, R. E. Carhart, and C. Djerassi, *J. Am. Chem. Soc.*, **102** (1980), 6289.

A Computer Program for the Enumeration of Substitutional Isomers. H. Dolhaine, *Comput. Chem.*, **5** (1981), 41.

Computer Enumeration and Generation of Trees and Rooted Trees. J. V. Knop, W. R. Muller, Z. Jericevic, and N. Trinajstic, *J. Chem. Inf. Comput. Sci.*, **21** (1981), 91.

Exhaustive Generation of Structural Isomers of a Given Formula—A New Algorithm. S. Y. Zhu and J. P. Zhang, *J. Chem. Inf. Comput. Sci.*, **22** (1982), 34.

Generation of Stereoisomeric Structures Using Topological Information Alone. H. Abe, H. Hayasaka, Y. Miyashita, and S. Sasaki, *J. Chem. Inf. Comput. Sci.*, **24** (1984), 217.

A Computer Program for Generation of Constitutionally Isomeric Structural Formulas. H. Abe, T. Okuyana, I. Fujiwara, and S. Sasaki, *J. Chem. Inf. Comput. Sci.*, **24** (1984), 220.

Applications

Actinobolin. 1. Structure of Actinobolamine. M. E. Munk, C. S. Sodano, R. L. McLean, and T. H. Hasketh, *J. Am. Chem. Soc.*, **89** (1967), 4158.

Alanylactinobicyclone. An Application of Computer Techniques to Structure Elucidation. D. B. Nelson, M. E. Munk, K. B. Gash, and D. L. Herald, *J. Org. Chem.*, **34** (1969), 3800.

Applications of Artificial Intelligence to Chemical Inference. I. The Number of Possible Organic Compounds. Acyclic Structures Containing C, H, O and N. J. Lederberg, G. L. Sutherland, B. G. Buchanan, E. A. Feigenbaum, A. V. Robertson, A. M. Duffield, and C. Djerassi, *J. Am. Chem. Soc.*, **91** (1969), 2973.

Applications of Artificial Intelligence to Chemical Inference. V. An Approach to the Computer Generation of Cyclic Structures. Differentiation Between All the Possible Isomeric Ketones of Composition C6H10O. Y. M. Sheikh, A. Buchs, A. B. Delfino, G. Schroll, A. M. Duffield, C. Djerassi, B. G. Buchanan, G. L. Sutherland, E. A. Feigenbaum, and J. Lederberg, *Org. Mass Spectrom.*, **4** (1970), 493.

Applications of Artificial Intelligence for Chemical Inference. Constructive Graph Labelling Applied to Chemical Problems. Chlorinated Hydrocarbons. D. H. Smith, *Anal. Chem.*, **47** (1975), 1176.

The Scope of Structural Isomerism. D. H. Smith, *J. Chem. Inf. Comput. Sci.*, **15** (1975), 203.

Applications of Artificial Intelligence for Chemical Inference. XIX. Computer Generation of Ion Structures. D. H. Smith, J. P. Konopeksi, and C. Djerassi, *Org. Mass Spectrom.*, **11** (1976), 86.

Structural Isomerism of Mono and Sesquiterpenoid Skeletons. D. H. Smith and R. E. Carhart, *Tetrahedron*, **32** (1976), 2513.

SPECTRUM PREDICTION

Applications of Artificial Intelligence for Chemical Inference. II. Interpretation of the Low Resolution Mass Spectra of Ketones. A. M. Duffield, A. V. Robertson, C. Djerassi, B. G. Buchanan, G. L. Sutherland E. A. Feigenbaum, and J. Lederberg, *J. Am. Chem. Soc.*, **91** (1969), 2977.

Applications of Artificial Intelligence for Chemical Inference. III. Aliphatic Ethers Diagnosed by the Low Resolution Mass Spectra and Nuclear Magnetic Resonance Data. G. Schroll, A. M. Duffield, C. Djerassi, B. G. Buchanan, G. L. Sutherland E. A. Feigenbaum, and J. Lederberg, *J. Am. Chem. Soc.*, **91** (1969), 7440.

Generation of Simulated Mass Spectra of Small Organic Molecules by Computerized Pattern Recognition Techniques. J. Schechter and P. C. Jurs, *Appl. Spectrosc.*, **27** (1973), 30.

Calculation of the Spectral Distribution of Absorption Coefficients in Vibrational Spectra of Complicated Molecules, Polymers and Molecular Crystals. L. A. Gribov, *J. Molec. Struct.*, **22** (1974), 353.

Generation of Mass Spectra Using Pattern Recognition Techniques. G. S. Zander and P. C. Jurs, *Anal. Chem.*, **47** (1975), 1562.

A Minicomputer Program Based on Additivity Rules for the Estimation of 13-C NMR Chemical Shifts. J. T. Clerc and H. Sommerauer, *Anal. Chim. Acta*, **95** (1977), 33.

Uniqueness of Carbon-13 Nuclear Magnetic Resonance Spectra of Acyclic Saturated Hydrocarbons. H. L. Surprenant and C. N. Reilley, *Anal. Chem.*, **49** (1977), 1134.

Calculation of Spectral Absorption Curves for Polyatomic Molecules. L. A. Gribov, V. A. Dementiev, and A. T. Todorovsky, *J. Molec. Struct.*, **50** (1978), 389,

On the Solution of One Classical Problem in Vibrational Spectrosccopy. L. A. Gribov, M. E. Elyashberg, and V. V. Serov, *J. Molec. Struct.*, **50** (1978), 371.

Signal Number Prediction in Carbon-13 Nuclear Magnetic Resonance Spectrometry. C. A. Shelley and M. E. Munk, *Anal. Chem.*, **50** (1978), 1522.

Computerized Mass Spectrum Prediction and Ranking. N. A. B. Gray, R. E. Carhart, A. Lavanchy, D. H. Smith, T. Varkony, B. G. Buchanan, W. C. White, and L. Creary, *Anal. Chem.*, **52** (1980), 1095.

Rule Based Mass Spectrum Prediction and Ranking. Applications to Structure Elucidation of Novel Marine Sterols. A. Lavanchy, T. Varkony, D. H. Smith, N. A. B. Gray, W. C. White, R. E. Carhart, B. G. Buchanan, and C. Djerassi, *Org. Mass Spectrom.*, **15** (1980), 355.

Determination of the Structure of Organic Molecules by Computer Evaluation and Simulation of Infrared and Raman Spectra. B. Schrader, D. Bougeard, and W. Niggemann, in *Computational Methods in Chemistry*, J. Bargon, (ed.), Plenum Press, New York, 1980, p. 37.

Computer-Aided Structure Elucidation of Organic Compounds with the CHEMICS System: Removal of Redundant Candidates by 13-C NMR Prediction. I. Fujiwara, T. Okuyama, T. Yamasaki, H. Abe, and S. Sasaki, *Anal. Chim. Acta*, **133** (1981), 527.

Structure Evaluation Using Predicted 13-C Spectra. C. W. Crandell, N. A. B. Gray, and D. H. Smith, *J. Chem. Inf. Comput. Sci.*, **27** (1982), 48.

A Computerized Approach to the Verification of C-13 NMR Spectral Assignments. M. R. Lindley, N. A. B. Gray, D. H. Smith, and C. Djerassi, *J. Org. Chem.*, **47** (1982), 1027.

Use of 13-C NMR Rules for the Ranking of Chemical Structures. G. Szalontai, Zs. Recsey, and Z. Csapo, *Anal. Chim. Acta*, **140** (1982), 309.

Interactive Computer System for the Simulation of C-13 NMR Spectra. G. W. Small and P. C. Jurs, *Anal. Chem.*, **55** (1983), 1121.

Simulation of C-13 NMR Spectra of Cycloalkanols with Computer Based Structural Descriptors. G. W. Small and P. C. Jurs, *Anal. Chem.*, **55** (1983), 1128.

Computer Assisted Assignment of C-13 NMR Spectra. W. Robien, *Monatsch. Chem.*, **114** (1983), 365.

Data Reduction in the Simulation of Carbon-13 NMR Spectra of Steroids. G. W. Small and P. C. Jurs, *Anal. Chem.*, **56** (1984), 2307.

Automated Selection of Models for the Simulation of Carbon-13 NMR Spectra. G. W. Small, T. R. Stouch, and P. C. Jurs, *Anal. Chem.*, **56** (1984), 2314.

MEASURES OF STRUCTURAL CHARACTER AND ACTIVITY

A Graph Theoretical Basis for Structural Chemistry. M. Randic, *Acta Crystallogr*, **A34** (1978), 275.

Fragment Search in Acyclic Structures. M. Randic, *J. Chem. Inf. Comput. Sci.*, **18** (1978), 101.

Search for All Self-avoiding Paths for Molecular Graphs. M. Randic, G. M. Brissey, R. B. Spence, and C. L. Wilkins, *Comput. Chem.*, **3** (1979), 5.

Graph Theoretical Approaches to Recognition of Structural Similarity in Molecules. M. Randic and C. L. Wilkins, *J. Chem. Inf. Comput. Sci.*, **19** (1979), 31.

Connectivity Parameters as Predictors of Retention in Gas Chromatography T. R. McGregor, *J. Chromatogr. Sci.*, **17** (1979), 314.

Use of Self-avoiding Paths for Characterization of Molecular Graphs with Multiple Bonds. M. Randic, G. M. Brissey, R. B. Spence, and C. L. Wilkins, *Comput. Chem.*, **4** (1980), 27.

Graph Theoretic Characterization and Computer Generation of Certain Carcinogenic Benzenoid Hydrocarbons and Identification of Bay Regions. K. Balasubramanian, J. J. Kaufmann, W. S. Koski, and A. T. Balaban, *J. Comput. Chem.*, **1** (1980), 149.

A Procedure for Characterization of the Rings of a Molecules. M. Randic and C. L. Wilkins, *J. Chem. Inf. Comput. Sci.*, **20** (1980), 36.

Generalization of the Graph Center Concept and Derived Topological Centric Indexes. D. Bonchev, A. T. Balaban, and O. Mekenyan, *J. Chem. Inf. Comput. Sci.*, **20** (1980), 106.

Substructure Retrieval and the Analysis of Structure Activity Relations on the Basis of a Complete Ordered Set of Fragments. J. Friedrich and I. Ugi, *J. Chem. Res.* **1980**, S70,M1301.

Isomer Discrimination by Topological Information Approach. D. Bonchev, Ov. Mekanyan, and N. Trinajstic, *J. Comput. Chem.*, **2** (1981), 127.

A Graph-Theoretic Approach to Quantitative Structure-Activity/Reactivity Studies. C. L. Wilkins, M. Randic, S. M. Schuster, R. S. Markin, S. Steiner, and L. Dorgan, *Anal. Chim. Acta*, **133** (1981), 637.

Structure-Activity Relationship Oriented Languages for Chemical Structure Representation. V. V. Avidon, I. A. Pomerantser, V. E. Golender, and A. B. Rozenbit, *J. Chem. Inf. Comput. Sci.*, **22** (1982), 207.

Computerized Chemical Structure Handling Techniques in Structure Activity Studies and Molecular Property Prediction. D. Bawden, *J. Chem. Inf. Comput. Sci.*, **23** (1983), 26.

Atom Pairs as Molecular Features in Structure Activity Studies: Definitions and Applications. R. E. Carhart, D. H. Smith, and R. Venkataraghavan, *J. Chem. Inf. Comput. Sci.*, **25** (1985), 64.

ABSTRACT MODELS OF CHEMICAL SYNTHESIS

A Systematic Characterization of Structures and Reactions for Use in Organic Synthesis. J. B. Hendrickson, *J. Am. Chem. Soc.*, **93** (1971), 6847.

Representations of Chemical Systems and Interconversion by 'be' Matrices and their Transformation Properties. I. Ugi and P. Gillespie, *Agnewandte Chemie International Edition*, **10** (1971), 914.

Matter Preserving Synthetic Pathways and Semi-empirical Computer Assisted Planning of Syntheses. I. Ugi and P. Gillespie, *Agnewandte Chemie International Edition*, **10** (1971), 915.

An Algebraic Model of Constitutional Chemistry as a Basis for Chemical Computer Programs. J. Dugundji, *Top. Curr. Chem.*, **39** (1973), 19.

An Algebraic Model for the Rearrangements of 2-bicyclo[2.2.1]heptyl Cations. C. K. Johnson and C. J. Collins, *J. Am. Chem. Soc.*, **96** (1974), 2514.

Representation of the Constitutional and Stereochemical Features of Chemical Systems in the Computer Assisted Design of Synthesis. J. Blair, J. Gasteiger, C. Gillespie, P. D. Gillespie, and I. Ugi, *Tetrahedron*, **30** (1974), 1845.

Systematic Synthesis Design. III. The Scope of the Problem. J. B. Hendrickson, *J. Am. Chem. Soc.*, **97** (1975), 5763.

Systematic Synthesis Design. IV. Numerical Codification of Construction Reactions. J. B. Hendrickson, *J. Am. Chem. Soc.*, **97** (1975), 5784.

A Heuristic Solution to the Functional Group Switching Problem in Organic Synthesis. H. W. Whitlock, *J. Am. Chem. Soc.*, **98** (1976), 3225.

The Steric Courses of Chemical Reactions. 3. Computer Generation of Product Distributions, Steric Courses and Permutational Isomers. M. G. Hutchings, J. B. Johnson, W. G. Klemperer, and R. R. Knight, *J. Am. Chem. Soc.*, **99** (1977), 7126.

EROS: A Computer Program for Generating Sequences of Reactions. J. Gasteiger and C. Jochums, *Top. Curr. Chem.*, **74** (1978), 93.

A Systematic Organization of Synthesis Reactions. J. B. Hendrickson, *J. Chem. Inf. Comput. Sci.*, **19** (1979), 129.

Systematic Synthesis Design 8: Generation of Reaction Sequences. J. B. Hendrickson and E. Braun-Keller, *J. Comput. Chem.*, **1** (1980), 323.

A Logic for Synthesis Design. J. B. Hendrickson, E. Braun-Keller, and G. A. Toczko, *Tetrahedron*, **37** (Suppl. 9) (1981), 359.

CONVENTIONAL AND AUTOMATIC SYNTHESIS PLANNING

Reaction Libraries, Details of Special Transforms, Storage of Synthetic Paths, and Structure Lists

Computer Assisted Synthetic Analysis. Methods for Machine Generation of Synthetic Intermediates Involving Multistep Lookahead. E. J. Corey, W. J. Howe, and D. A. Pensak, *J. Am. Chem. Soc.*, **96** (1974), 7724.

A Computer Representation of Synthetic Reactions. M. Bersohn and A. Esack, *Comput. Chem.* **1** (1976), 103.

Computer Assisted Synthetic Analysis. Performance of Long Range Strategies for Stereoselective Olefin Synthesis. E. J. Corey and A. K. Long, *J. Org. Chem.*, **43** (1978), 2208.

Hash Functions for Rapid Storage and Retrieval of Chemical Structures. W. T. Wipke, S. Krishnan, and G. I. Ouchi, *J. Chem. Inf. Comput. Sci.*, **18** (1978), 32.

Steps toward the Automatic Compilation of Synthetic Organic Reactions. M. Bersohn and K. MacKay, *J. Chem. Inf. Comput. Sci.*, **19** (1979), 137.

Computer Assisted Synthetic Analysis. Long Range Search Procedures for Antithetic Simplification of Complex Targets by the Application of Halolactonization Transforms. E. J. Corey, A. K. Long, J. Mulzer, H. W. Orf, A. P. Johnson, and A. P. W. Hewett, *J. Chem. Inf. Comput. Sci.*, **20** (1980), 221.

Computer Assisted Synthetic Analysis. Techniques for Efficient Long Range Retrosynthetic Searches Applied to the Robinson Annulation Process. E. J. Corey, A. P. Johnson, and A. K. Long, *J. Org. Chem.*, **45** (1980), 2051.

Storage and Retrieval of Synthetic Trees, F. Choplin, S. Goundin, and G. Kaufmann, *J. Chem. Inf. Comput. Sci.*, **23** (1983), 26.

Synthesis Strategies

General Methods for the Construction of Complex Molecules, E. J. Corey, *Pure Appl. Chem.*, **14** (1967), 19.

Computer-Assisted Design of Complex Organic Syntheses. E. J. Corey and W. T. Wipke, *Science*, **166** (1969), 178.

Computer-Assisted Analysis of Complex Synthetic Problems. E. J. Corey, *Quart. Rev. Chem. Soc.*, **25** (1971), 455.

Automatic Problem Solving Applied to Synthetic Chemistry. M. Bersohn, *Bull. Chem. Soc. Jap.*, **45** (1972), 1897.

The Discovery of Organic Synthetic Routes by Computer. H. Gelernter, N. S. Sridharan, A. J. Hart, F. W. Fowler, and H. Shue, *Top. Curr. Chem.*, **41** (1973), 113.

General Methods of Synthetic Analysis. Strategic Bond Disconnections for Bridged Polycyclic Structures. E. J. Corey, W. J. Howe, H. W. Orf, D. A. Pensak, and G. Petersson, *J. Am. Chem. Soc.*, **97** (1975), 6116.

Ableitung Organisch-chemischer Reaktionen mit der Simulationsprogramm AHMOS. A. Weise, *Zeit. Chemie*, **15** (1975), 333.

Computer Assisted Synthetic Analysis. Synthetic Strategies Based on Appendages and the Use of Reconnective Transforms. E. J. Corey and W. L. Jorgensen, *J. Am. Chem. Soc.*, **98** (1976), 189.

Computer Assisted Synthetic Analysis. Generation of Synthetic Sequences Involving Sequential Functional Group Interchanges. E. J. Corey and W. L. Jorgensen, *J. Am. Chem. Soc.*, **98** (1976), 203.

Computer Assisted Synthetic Analysis. The Identification and Protection of Interfering Functionality in Machine Generated Synthetic Sequences. E. J. Corey, H. W. Orf, and D. A. Pensak, *J. Am. Chem. Soc.*, **98** (1976), 210.

Simulation and Evaluation of Chemical Synthesis. Congestion: A Conformation Dependent Function of Steric Environment at a Reaction Center. Application with Torsional Terms to Stereoselective Nucleophilic Additions to Ketones. W. T. Wipke and P. Gund, *J. Am. Chem. Soc.*, **98** (1976), 8107.

Empirical Explorations with SYNCHEM. H. L. Gelernter, A. F. Sanders, D. L. Larsen, K. K. Agarwal, R. H. Boivie, G. A. Spritzer, and J. E. Searleman, *Science*, **197** (1977), 1041.

Simulation and Evaluation of Chemical Synthesis—SECS: An Application of Artificial Intelligence Techniques. W. T. Wipke, G. I. Ouchi, and S. Krishnan, *J. Artif. Intel.*, **11** (1978), 173.

Some Early Developments in Programming Synthetic Strategies. M. Bersohn, A. Esack, and J. Luchini, *Comput. Chem.*, **2** (1978), 105.

Synthese Assistee par Ordinateur en Chemie des Composes Organophosphores. F. Choplin, C. Laurenco, R. Marc, G. Kaufmann, and W. T. Wipke, *Nouv. J. Chem.*, **2** (1978), 285.

Computer Assisted Mechanistic Evaluation of Organic Reactions. 1. Overview. T. D. Salatin and W. L. Jorgensen, *J. Org. Chem.*, **45** (1980), 2043.

Self-Adapting Computer Program System for Designing Organic Syntheses. Z. Hippe, *Anal. Chim. Acta*, **133** (1981), 677.

Artificial Intelligence in Organic Synthesis. SST: Starting Material Selection Strategies. An Application of Superstructure Search. W. T. Wipke and D. Rogers, *J. Chem. Inf. Comput. Sci.*, **24** (1984), 71.

Program Details

Computer Assisted Synthetic Analysis. Facile Man-Machine Communication of Chemical Structure by Interactive Computer Graphics. E. J. Corey, W. T. Wipke, R. D. Cramer, and W. J. Howe, *J. Am. Chem. Soc.*, **94** (1972), 421.

Techniques for Perception by a Computer of Synthetically Significant Structural Features in Complex Molecules. E. J. Corey, W. T. Wipke, R. D. Cramer, and W. J. Howe, *J. Am. Chem. Soc.*, **94** (1972), 431.

Computer Assisted Synthetic Analysis for Complex Molecules. Methods and Procedures for Machine Generation of Synthetic Intermediates. E. J. Corey, R. D. Cramer, and W. J. Howe, *J. Am. Chem. Soc.*, **94** (1972), 440.

Simulation and Evaluation of Chemical Synthesis. Computer Representation and Manipulation of Stereochemistry. W. T. Wipke and T. M. Dyott, *J. Am. Chem. Soc.*, **96** (1974), 4834.

Computer Manipulation of Central Chirality. A. Esack and M. Bersohn, *J. Chem. Soc. Perkin Trans. I*, **1975**, 1124.

Computer Representation of the Stereochemistry of Organic Molecules. H. W. Davis, Birkhauser, Basel, 1976.

Application of Chemical Transforms in SYNCHEM2. A Computer Program for Organic Synthesis Route Discovery. K. K. Agarwal, D. L. Larsen, and H. L Gelernter, *Comput. Chem.*, **2** (1978), 75.

Computer Design of Synthesis in Phosphorous Chemistry: Automatic Treatment of Stereochemistry. F. Choplin, R. Marc, G. Kaufmann, and W. T. Wipke, *J. Chem. Inf. Comput. Sci.*, **18** (1978), 110.

Computer Assisted Simulation of Chemical Reaction Sequences: Applications to Problems of Structure Elucidation. T. H. Varkony, R. E. Carhart, D. H. Smith, and C. Djerassi, *J. Chem. Inf. Comput. Sci.*, **18** (1978), 168.

Ordinateur et Synthese Organique: Approche Analyticque. Exemple de L'azaadamantane. R. Barone, A. Boche, M. Chanon, and J. Metzger, *Comput. Chem.*, **3** (1979), 83.

The Computer Derivation of Stereochemical Relationships from the Chirality of Ring Atoms. M. Bersohn, *J. Chem. Soc. Perkin Trans. I*, **1979**, **1975**.

Strategie en Synthese Chimique Assistee par Ordinateur. P. Bonnet, J. C. Derniame, M. H. Zimmer, F. Choplin, and G. Kaufmann, *RAIRO Informatique/Comput. Sci.*, **13** (1979), 287.

Computer Assisted Synthetic Analysis. A Rapid Computer Method for the Semi-quantitative Assignment of Conformation of Six Membered Rings. 1. Derivation of a Preliminary Conformation Description of the Six Membered Ring. E. J. Corey and N. F. Feier, *J. Org. Chem.*, **45** (1980), 757.

Computer Assisted Synthetic Analysis. A Rapid Computer Method for the Semi-quantitative Assignment of Conformation of Six Membered Rings. 2. Assessment of Conformational Energies. E. J. Corey and N. F. Feier, *J. Org. Chem.*, **45** (1980), 765.

Computer-Assisted Mechanistic Evaluation of Organic Reactions. 2. Perception of Rings, Aromaticity and Tautomers. B. L. Roos-Kozel and W. L. Jorgensen, *J. Chem. Inf. Comput. Sci.*, **21** (1981), 101.

Microcomputers and Organic Synthesis. R. Barone, M Chanon, P. Cadicot, and J. M. Cense, *B. S. Chem. Belg.*, **91** (1982), 333.

Applications

Computer Assisted Graph Theoretic Analysis of Complex Mechanistic Problems in Polycyclic Hydrocarbons. Mechanism of Diamantane Formation from various Pentacyclotetradecanes. T. M. Gund, P. R. Schleyer, P. H. Gund, and W. T. Wipke, *J. Am. Chem. Soc.*, **97** (1975), 743.

Ordinateru et Synthese Organique. Application d'un Programme non Interactiv a la Synthese du thiazole. R. Barone, M. Chanon, and J. Metzger, *Chimia*, **32** (1978), 216.

Computer Assisted Structure Manipulation. Studies in the Biosynthesis of Natural Products. T. H. Varkony, D. H. Smith, and C. Djerassi, *Tetrahedron*, **34** (1978), 841.

Synthese Organique Assistee par Ordinateur: Application aux indazoles. R. Barone, P. Camps, and J. Elguero, *Anales de Quimica*, **75** (1979), 736.

Automatic Strategy in Computer Design of Synthesis. An Example in Organophosphorous Chemistry. M. H. Zimmer, F. Choplin, P. Bonnet, and G. Kaufmann, *J. Chem. Inf. Comput. Sci.*, **19** (1979), 235.

Computer Assisted Synthetic Analysis at Merck. P. Gund, E. J. J. Grabowski, D. R. Hoff, G. M. Smith, J. D. Andose, J. B. Rhodes, and W. T. Wipke, *J. Chem. Inf. Comput. Sci.*, **20** (1980), 88.

Educational Applications

Computer-Assisted Instruction in Organic Synthesis. H. W. Orf, *J. Chem. Ed.*, **52** (1975), 464.

Computer-Assisted Teaching of Organic Synthesis. R. D. Stolow and L. J. Joncas, *J. Chem. Ed.*, **57** (1980), 868.

BOOKS

Computer Handling of Chemical Structure Information, M. F. Lynch, J. M. Harrison, W. G. Town, and J. E. Ash, American Elsevier, New York, 1971.

Chemical Applications of Pattern Recognition, P. C. Jurs and T. L. Isenhour, Wiley, New York, 1975.

Chemical Applications of Graph Theory, A. T. Balaban, (ed.), Academic Press, New York, 1976.

Computers in Mass Spectrometry, J. R. Chapman, Academic Press, London, 1978.

Computer Assisted Studies of Chemical Structure and Biological Function, A. J. Stuper, W. E. Brugger, and P. C. Jurs, Wiley, New York, 1979.

Applications of Artificial Intelligence for Organic Chemistry, R. K. Lindsay, B. G. Buchanan, E. A. Feigenbaum, and J. Lederberg, McGraw-Hill, New York, 1980.

SYMPOSIA

Computer Representation and Manipulation of Chemical Information, W. T. Wipke, S. R. Heller, R. J. Feldmann, and E. Hyde, (eds.), Wiley, New York, 1974.

Algorithms for Chemical Computations, R. E. Christoffersen, (ed.), ACS Symposium Series 46, American Chemical Society, Washington, DC, 1977.

Computer-Assisted Structure Elucidation, D. H. Smith, (ed.), ACS Symposium Series 54, American Chemical Society, Washington, DC, 1977.

Computer-Assisted Organic Synthesis, W. T. Wipke and W. J. Howe, (eds.), ACS Symposium Series 61, American Chemical Society, Washington, DC, 1977.

High Performance Mass Spectrometry: Chemical Applications, M. L. Gross, (ed.), ACS Symposium Series 70, American Chemical Society, Washington, DC, 1978.

Retrieval of Medicinal Chemical Information, W. J Howe, M. M. Milne, and A. F. Pennell, (eds.), ACS Symposium Series 84, American Chemical Society, Washington, DC, 1978.

Computer Assisted Drug Design, E. C. Olson and R. E. Christoffersen, (eds.), ACS Symposium Series 112, American Chemical Society, Washington, DC, 1979.

Proceeding of the International Converence on Chemometrics in Analytical Chemistry, 1982, *Anal. Chim. Acta*, **150** (1983).

REVIEWS

Some Chemical Applications of Machine Intelligence. T. L. Isenhour and P. C. Jurs, *Anal. Chem.*, **43** (1971), 20A.

Isomer Enumeration Methods. D. H. Rouvray, *Chem. Rev.*, **3** (1973), 355.

The Coming of the Computer Age to Organic Chemistry. Recent Approaches to Systematic Synthesis Analysis. A. J. Thakkar, *Top. Curr. Chem.*, **39** (1973), 3.

Applications of Pattern Recognition to Chemistry. T. L. Isenhour, B. R. Kowalski, and P. C. Jurs, *CRC Crit. Rev. Anal. Chem.*, **4** (1974), 1.

Computers and Organic Synthesis. M. Bersohn and A. Esack, *Chem. Rev.*, **76** (1976), 269.

Interpretation Automatique des Spectres de Masse de Composes Organiques. M. Bachiri and G. Mouvier, *Org. Mass Spectrom.*, **11** (1976), 1272.

Computerunterstutzte SpektrenInterpretation fur Strukturaufklarung Organischer Verbindungen. J. T. Clerc, *Chimia*, **31** (1977), 353.

Use of the Computer for the Solution of Structural Problems in Organic Chemistry by the Methods of Molecular Spectrosccopy. V. A. Koptyug, V. S. Bochkarev, B. G. Derendyaev, S. A. Nekoroshev, V. N. Piottukh-Peletskii, M. I. Podgornaya, and G. P. Ul'yanov. *J. Struct. Chem.*, **18** (1977), 355.

The Deductive Solution of Chemical Problems by Computer Programs on the Basis of a Mathematical Model of Chemistry. I. Ugi, J. Brandt, J. Friedrich, J. Gasteiger, C. Jochum, P. Lemmen, and W. Schubert, *Pure Appl. Chem.*, **50** (1978), 1303.

Computer-Aided Identification of Organic Molecules by Their Molecular Spectra. L. A. Gribov and M. E. Elyashberg, *CRC Crit. Rev. Anal. Chem.*, **8** (1979), 111.

A Novel Role of Computers in the Natural Products Field. C. Djerassi, D. H. Smith, and T. H. Varkony, *Naturwissenschaften*, **66** (1979), 9.

Pattern Recognition in Analytical Chemistry. K. Varmuza, *Anal. Chim. Acta*, **122** (1980), 227.

Computer Retrieval of Spectral Data. Z. Hippe and R. Hippe, *Appl. Spectrosc. Rev.*, **16** (1980), 135.

Survey of Computer Aided Methods for Mass Spectral Interpretation. D. P. Martinsen, *Appl. Spectrosc.*, **35** (1981), 255.

Computer Assisted Synthesis Design: Present State and Future Perspectives. J. Gasteiger, *Chem. Ind. M.*, **64** (1982), 714.

INDEX